EXCURSION FLORA
OF THE
BRITISH ISLES

EXCURSION FLORA
OF THE
BRITISH ISLES

BY

A. R. CLAPHAM
University of Sheffield

T. G. TUTIN
University of Leicester

AND

†E. F. WARBURG
University of Oxford

SECOND EDITION

CAMBRIDGE
AT THE UNIVERSITY PRESS
1968

Published by the Syndics of the Cambridge University Press
Bentley House, P.O. Box 92, 200 Euston Road, London, N.W. 1
American Branch: 32 East 57th Street, New York 22, N.Y. 10022

Library of Congress Catalogue Card Number: 68–10329

Standard Book Number: 521 04656 4

First edition 1959
Reprinted 1965
Second edition 1968

Printed in Great Britain
at the University Printing House, Cambridge
(Brooke Crutchley, University Printer)

CONTENTS

CONTENTS

PREFACE TO THE SECOND EDITION

In preparing a second edition of this Excursion Flora we have had to take account of some recent additions to the British species-list and some notable extensions to the recorded range of many native plants as well as of a large body of taxonomic and nomenclatural investigations completed since the appearance of the first edition. We have endeavoured to include some reference to all new species and to the more important modifications of known range. We have, for example, added *Minuartia recurva*, *Salix hibernica* and *Trifolium occidentale* and we have taken note of the rediscovery of *Erica ciliata* in western Ireland, of the interesting find in the Isle of Man of the orchid *Neotinea intacta*, previously known only in Ireland, and of the increase in localities for *Phyllodoce caerulea* from one mountain in Scotland to two.

Taxonomic and nomenclatural studies are always in progress and make some changes inevitable if the Flora is to be kept up-to-date. Since the publication of the first edition, however, the Flora Europaea project has occasioned an exceptionally large and important volume of relevant research and has led to changes in knowledge and outlook that are bound to have implications for national floras within the area covered. The first volume of Flora Europaea includes, for instance, a modern treatment of European ferns so much in advance of anything previously available that we have decided to adopt it in this second edition of the Excursion Flora. We have made some other taxonomic and a few nomenclatural changes but we have refrained from incorporating all the innovations that will be found in Flora Europaea, even though most of them are likely to be generally adopted in the future. Our chief reason for this conservatism is our belief that many users of the Excursion Flora refer to the larger British Flora for resolving difficulties or for more detailed accounts of critical groups. This being so it seemed to us wrong to allow the Excursion Flora to diverge very far from the second edition of the larger Flora. The authors of plant names have been given in the form used in Flora Europaea, so that further information about them can be readily obtained by reference to Appendix I in that work.

We wish to record our gratitude to the many friends who have drawn our attention to errors and omissions and have helped us to make keys and descriptions more useful.

The initials of only two of us appear at the foot of this Preface. The untimely death in July 1966, of Dr. E. F. Warburg, our collaborator and close friend for so many years, has been a very sad blow to us personally and to all concerned with taxonomic studies of the British flora.

A. R. C.
T. G. T.

September 1967

PREFACE TO THE FIRST EDITION

We have for some time been aware of the need for a shortened and therefore more readily portable and less expensive British Flora which would nevertheless retain many features of our *Flora of the British Isles*. In preparing this *Excursion Flora* we have had particularly in mind the requirements of the upper forms of schools and of university students taking botany as a degree subject. We have therefore provided descriptions of all species that are generally common in lowland districts of the British Isles and also of some others likely to be encountered in the neighbourhood of field centres and field stations. These descriptions are shorter than those in the larger *Flora* chiefly through the omission of certain categories of information provided there, including pollination mechanism, life-form, chromosome number and extra-British distribution. We have been careful not to shorten descriptions until they are mere lists of *differentiae* because we believe that confirming an identification from a key by reference to a fairly full description is an important part of taxonomic practice, and also that the drawing up of plant descriptions is a valuable exercise which we hope to encourage by setting this example.

Space has been saved chiefly by restricting the number of species described in detail and by omitting all text-figures except those illustrating the glossary of botanical terms. To avoid the serious disadvantages of incompleteness, and to increase the general usefulness of the work as an 'excursion flora', we have included in the keys all native species (apart from those of such critical genera as *Alchemilla*, *Sorbus* and *Euphrasia*, and of *Rubus* and *Hieracium*), many naturalized and a few casual species. Care has been taken to distinguish clearly in the keys between species of which full descriptions are provided later and those which are merely named in the key. For two genera only, *Alchemilla* and *Hieracium*, are keys provided which are incomplete in the sense that they serve only to distinguish between commonly encountered lowland species or aggregates. In each instance there is a clear warning that this is so.

Comparison of this *Excursion Flora* with the First Edition of the larger *Flora* will reveal three further kinds of difference. In the first place certain species are included which cannot be found in the larger *Flora*. These are for the most part native plants which had not yet been detected at the time of writing the larger *Flora*, and they include *Diapensia lapponica*, *Artemisia norvegica* and *Koenigia islandica*. Second, it will be seen that some taxonomic changes have been introduced, such as the separation of *Dactylorchis* from *Orchis*, *Tripleurospermum* from *Matricaria* and *Chamaemelum* from *Anthemis*, *Galium sterneri* from *G. pumilum* and *Callitriche platycarpa* from *C. stagnalis*; and, on the other hand, the merging of *Tanacetum* with *Chrysanthemum* and part of *Claytonia* with *Montia*. Third, there will be found purely nomenclatural changes, such as the substitution of *Cerastium atrovirens* Bab. for *C. tetrandrum* Curt., and several others. We intend to make these same changes in the Second Edition of the larger *Flora*. It is inevitable, even though regrettable, that continued research should necessitate periodic modifications of this kind.

The reduction in length to little more than one-third of that of the larger *Flora* has involved the omission of much that is of great interest to the serious student of the British flora. The Synopses of Classification, general and special, have had to be dropped, no mention is made of the rarer naturalized aliens and few casuals are included. More important is the scanty reference to intraspecific variation, the restriction to keys of all uncommon species and the very elementary treatment of 'difficult' genera such as *Euphrasia*, *Rhinanthus*, *Rubus* and *Hieracium*. It is hoped that many of those who are introduced to the study of British plants through this *Excursion Flora* will be encouraged to widen and deepen their knowledge by later resort to the *Flora of the British Isles*.

In conclusion we have again to express our thanks to the many botanists who have given us expert advice. To the list of names in the larger *Flora* we should like to add P. W. Ball (*Salicornia*), C. Cook (*Sparganium*), K. M. Goodway (*Galium*), N. M. Pritchard (*Gentianella*), M. C. F. Proctor (*Helianthemum*) and D. P. Young (*Oxalis*). We owe a very special debt of gratitude to Mr J. E. Dandy, Keeper of Botany in the British Museum (Natural History), who has very generously allowed us to consult proofs of the forthcoming *British Plant List*, embodying the fruits of his invaluable nomenclatural and taxonomic researches, and has always been willing to discuss problems with us.

A. R. C.
T. G. T.
E. F. W.

January 1958

SEQUENCE OF ORDERS AND FAMILIES

PTERIDOPHYTA

LYCOPSIDA

LYCOPODIALES   1. Lycopodiaceae
SELAGINELLALES   2. Selaginellaceae
ISOETALES   3. Isoetaceae

SPHENOPSIDA

EQUISETALES   4. Equisetaceae

FILICOPSIDA

OPHIOGLOSSALES   5. Ophioglossaceae
FILICALES   6. Osmundaceae   12. Thelypteridaceae
    7. Adiantaceae   13. Aspleniaceae
    8. Cryptogrammaceae   14. Athyriaceae
    9. Gymnogrammaceae   15. Aspidiaceae
    10. Hypolepidaceae   16. Blechnaceae
    11. Hymenophyllaceae   17. Polypodiaceae
MARSILEALES   18. Marsileaceae
SALVINIALES   19. Azollaceae

GYMNOSPERMAE

CONIFEROPSIDA

CONIFERAE   20. Pinaceae   22 Taxaceae
    21. Cupressaceae

ANGIOSPERMAE

DICOTYLEDONES

ARCHICHLAMYDEAE

RANALES   23. Ranunculaceae   26. Nymphaeaceae
    24. Paeoniaceae   27. Ceratophyllaceae
    25. Berberidaceae
RHOEADALES   28. Papaveraceae   30. Cruciferae
    29. Fumariaceae   31. Resedaceae
VIOLALES   32. Violaceae
POLYGALALES   33. Polygalaceae
CISTIFLORAE   34. Hypericaceae   35. Cistaceae
TAMARICALES   36. Tamaricaceae   37. Frankeniaceae

ARTIFICIAL KEY TO FAMILIES

(For abbreviations see p. xxxvii)

1   Plant reproducing by spores; fls 0; always herbs.   **2**
Plant reproducing by seeds; fls with stamens or carpels or both;
    often woody.   **29**

2   Stems jointed; lvs not green, forming a sheath at the nodes.
                                4. EQUISETACEAE
Stems not jointed; lvs green, not connate into a sheath.   **3**

3   Plants free-floating on water, much-branched; lvs small imbricate.
                                19. AZOLLACEAE
Plants rooted to the ground, terrestrial or aquatic.   **4**

4   Lvs not differentiated into lamina and petiole.   **5**
Lvs with distinct lamina and petiole.   **8**

5   Lvs forming a basal rosette.   **3. ISOETACEAE**
Lvs not forming a basal rosette.   **6**

6   Lvs filiform, with circinate vernation.   **18. MARSILEACEAE**
Lvs lanceolate to ovate, vernation not circinate.   **7**

7   Stem robust; plant homosporous; lvs not ligulate.
                                1. LYCOPODIACEAE
Stem slender; plant heterosporous; lvs ligulate.
                                2. SELAGINELLACEAE

8   Fertile lvs, or fertile parts of lvs, differing markedly from the
    sterile lvs or parts of lvs.   **9**
Fertile lvs not markedly different from the sterile parts.   **13**

9   Lf looking like a stem with a fertile upper portion and a sterile
    lower portion, both of which may be simple or pinnate.
                              5. OPHIOGLOSSACEAE
Lvs crowded at the end of a stout stock, the inner fertile some-
    times with a few pairs of sterile pinnae at base, the outer
    sterile.   **10**

10   Lvs 1-pinnate; pinnae entire.   **16. BLECHNACEAE**
Lvs 2- to 4-pinnate.   **11**

11   Fertile lvs with 2–3 pairs of sterile pinnae at base; growing in
    damp, ± peaty places.   **6. OSMUNDACEAE**
Fertile lvs without sterile pinnae at base.   **12**

12   Sori on or near the margin of the lf; growing on rocks, screes,
    or stone walls.   **8. CRYPTOGRAMMACEAE**

Sori on lower surface of lf. (Growing on shady, damp banks. Channel Is; very local.
Anogramma leptophylla (L). Link) 9. GYMNOGRAMMACEAE

13	Lvs not more than 1 cell thick (except for midrib), translucent. 11. HYMENOPHYLLACEAE	
	Lvs thicker, not translucent.	14
14	Lvs entire, or pinnatifid, or palmately lobed, or dichotomously forked 1–3 times.	15
	Lvs pinnately divided.	17
15	Lvs not pinnatifid. 13. ASPLENIACEAE	
	Lvs pinnatifid.	16
16	Lvs covered with scales beneath. 13. ASPLENIACEAE	
	Lvs not covered with scales beneath. 17. POLYPODIACEAE	
17	Sori covered by the inflexed margin of the lf.	18
	Sori not covered by the inflexed margin of the lf.	19
18	Rhizome long, subterranean; pinnae not fan-shaped. Common. 10. HYPOLEPIDACEAE	
	Rhizome short, erect; pinnae fan-shaped. (Damp crevices of cliffs and rocks, chiefly by the sea; very local and rare. Maidenhair fern. *Adiantum capillus-veneris* L.) 7. ADIANTACEAE	
19	Indusium absent.	20
	Indusium present.	23
20	Pinnae entire. 17. POLYPODIACEAE	
	Pinnae divided.	21
21	Lvs forming a crown. 14. ATHYRIACEAE	
	Lvs solitary.	22
22	Lf divided into 3 nearly equal portions. 15. ASPIDIACEAE	
	Lf pinnately divided. 12. THELYPTERIDACEAE	
23	Indusium a ring of hair-like scales surrounding the base of the sorus. Small mountain plants; rare. 4. ATHYRIACEAE	
	Indusium not as above.	24
24	Indusium hood-like, attached at lower side of sorus. 14. ATHYRIACEAE	
	Indusium not hood-like.	25
25	Indusium peltate. 15. ASPIDIACEAE	
	Indusium not peltate.	26
26	Sori orbicular.	27
	Sori ovate or linear.	28
27	Sori marginal; indusium lying along vein. 12. THELYPTERIDACEAE	
	Sori not marginal; indusium lying across vein. 15. ASPIDIACEAE	

28 Sori ovate; lower margin of indusium bent in the middle.
14. ATHYRIACEAE
   Sori linear or ovate; lower margin of indusium straight.
13. ASPLENIACEAE

29 Ovules naked, either on the upper surface of scales arranged
   in cones or solitary and terminal on a short scaly axillary
   shoot; pollen-sacs two or more on the lower surface of a flat
   sporophyll, or several pendent from the apex of a peltate sporophyll, the male sporophylls always in cones; monoecious or
   dioecious trees or shrubs with small needle-like or scale-like
   (but green) lvs; perianth 0.   CONIFERAE   30
   Ovules completely enclosed in a carpel; pollen-sacs 4 (or occasionally fewer) surrounding and adnate to a connective at the apex
   of a usually slender filament,   ANGIOSPERMAE   32

30 Lvs opposite or whorled; short shoots 0.   21. CUPRESSACEAE
   Lvs alternate or in clusters on short lateral shoots.   31

31 Ovules on the surface of scales arranged in cones; pollen-sacs
   two on the lower surface of a flat sporophyll; trunk usually
   single.   20. PINACEAE
   Ovules solitary and terminal on short axillary shoots; pollen-sacs
   several on a peltate sporophyll; trunks usually several.
22. TAXACEAE

32 Herbs without chlorophyll, the lvs reduced to scales.   256 (J)
   Green plants (if lfless at flowering time either trees or shrubs,
   or else herbs with only the fls showing above ground).   33

33 Plant free-floating on or below surface of water, not rooted in mud.   34
   Land-plants or aquatics rooted in mud.   36

34 Plant consisting of a discoid thallus (1–15 mm. diam.), with or
   without roots from the lower surface; propagation mainly
   vegetative, so that several plants are often found joined together.   141. LEMNACEAE
   Plants with obvious stems and lvs.   35

35 Plant with small bladders on lvs, or on apparently lfless stems;
   lvs divided into filiform segments. 108. LENTIBULARIACEAE
   Plant without bladders; lvs sessile, in a rosette, or long-petiolate
   and orbicular.   123. HYDROCHARITACEAE

36 Small herb with lvs linear and all basal; fls solitary, unisexual,
   axillary, the male on long stalks, the female sessile (*Littorella*).
112. PLANTAGINACEAE
   Not as above.   37

37 Perianth of two (rarely more) distinct whorls, differing markedly
   from each other in shape, size or colour.   38
   Perianth 0, or of 1 whorl, or of 2 or more similar whorls, or
   segments numerous and spirally arranged.   42

38 Petals free (very rarely cohering at apex, free at base),   *39*
    Petals united at least at the base.   *41*

39 Ovary superior.   *40*
    Ovary inferior or partly so.   *99* (C)

40 Carpels and styles free, or carpels slightly united at the extreme
      base.   *44* (A)
    Carpels or styles or both obviously united, or ovary of one
      carpel.   *51* (B)

41 Ovary superior.   *116* (D)
    Ovary inferior.   *147* (E)

42 Perianth corolla-like, at least the inner segments usually brightly-
      coloured or white.   *161* (F)
    Perianth green and calyx-like, or scarious, or 0.   *43*

43 Trees or shrubs.   *184* (G)
    Herbs.   *203* (H)

GROUP A

Petals free, ovary superior, carpels and styles free or nearly so.

44 Sepals and petals 3.   *45*
    Sepals or petals more than 3.   *46*

45 Aquatic plants; fls conspicuous; at least the upper lvs broad, flat,
      stalked; carpels ±numerous.   121. ALISMATACEAE
    Small land plants of mossy appearance; fls axillary, incon-
      spicuous; lvs small, oblong, rather fleshy, sessile; carpels 3
      (*Crassula*).   58. CRASSULACEAE

46 Stamens numerous.   *47*
    Stamens twice as many as petals or fewer.   *49*

47 Herbs; stipules 0; fls hypogynous.   *48*
    Herbs with stipules, or else shrubs; fls perigynous (sometimes
      only slightly so).   57. ROSACEAE

48 Fls c. 10 cm. diam. (Perennial herb with biternate lvs, usually red
      petals and usually 5 white-downy carpels. Naturalized on Steep
      Holm, Severn Estuary. Peony. *Paeonia mascula* (L.) Miller)
                                            24. PAEONIACEAE
    Fls much smaller.   23. RANUNCULACEAE

49 Lvs ternate, not fleshy; alpine plant (*Sibbaldia*).
                                            57. ROSACEAE
    Lvs simple.   *50*

50 Lvs ±succulent; carpels in 1 whorl.   58. CRASSULACEAE
    Lvs not succulent; carpels spirally arranged on a slender elongated
      receptacle (*Myosurus*).   23. RANUNCULACEAE

GROUP B

Petals free, ovary superior, carpels or styles or both united, or ovary of one carpel.

51 Fls actinomorphic. 52
Fls zygomorphic. 91

52 Stamens more than twice as many as petals (always more than 6), or stamens and petals both numerous. 53
Stamens at most twice as many as petals (never more than 12); or petals 2, stamens 6. 61

53 Aquatic plants with large cordate floating lvs and floating fls; petals more than 10.   26. NYMPHAEACEAE
Plant not aquatic. 54

54 Stamens all united below into a tube; fls pink or purple; lvs usually palmately lobed.   45. MALVACEAE
Stamens free or in bundles; lvs never palmately lobed. 55

55 Lvs very succulent, 3-angled; fls c. 5 cm. diam., with numerous narrow magenta or yellow petals. (Trailing perennial herb introduced and established on sea-side cliffs and banks in S.W. England. 'Mesembryanthemum'. *Carpobrotus edulis* (L.) N.E. Br.)   41. AIZOACEAE
Lvs not succulent; petals 5 or fewer. 56

56 Ovary surrounded by a cup-shaped perigynous zone; ovule 1.   57. ROSACEAE
No cup-shaped perigynous zone; ovules 2 or more. 57

57 Carpel 1; lvs 2-ternate, the lower lflets stalked.   23. RANUNCULACEAE
Carpels 2 or more; lvs not as above. 58

58 Trees; infl. with a conspicuous bract partly adnate to the infl.-stalk.   44. TILIACEAE
Herbs or low shrubs; bracts, if present, not adnate to the infl.-stalk. 59

59 Styles free; stamens united into bundles below.   34. HYPERICACEAE
Style 1 or 0, stigma simple; stamens free. 60

60 Sepals 2; petals 4; lvs toothed to pinnate.   26. PAPAVERACEAE
Sepals 5 (3 large, 2 small); petals 5; lvs entire.   35. CISTACEAE

61 Trees or shrubs. 62
Herbs. 69

62 Fls on the middle of lf-like cladodes; true lvs scale-like, colourless (*Ruscus*).   132. LILIACEAE
Fls not on cladodes; lvs green. 63

63 Per. segs in 2 or more whorls of 3; stamens 3 or 6. 64
Per. segs not in whorls of 3; stamens not 3 or 6. 65

64   Per. segs in more than 2 whorls; stamens 6; lvs broad.
                                25. BERBERIDACEAE
     Per. segs in 2 whorls; stamens 3; lvs linear. 93. EMPETRACEAE

65   Lvs small and scale-like; fls numerous in dense spikes.
                                36. TAMARICACEAE
     Lvs not scale-like, not particularly small.   66

66   Lvs opposite.   67
     Lvs alternate.   68

67   Lvs palmately lobed.   50. ACERACEAE
     Lvs simple, not lobed.   53. CELASTRACEAE

68   Plant with rusty tomentum; fls cream; stamens more than 5
     (*Ledum*).   90. ERICACEAE
     Plant not tomentose; fls greenish; stamens 4–5.
                                55. RHAMNACEAE

69   Sepals 2, petals 5.   40. PORTULACACEAE
     Sepals more than 2; sepals and petals equal in number.   70

70   Lvs modified into pitchers, 10–15 cm.; stigma very large, um-
     brella-like. (Perennial, with all lvs basal. Introduced and
     naturalized on bogs in C. Ireland. Pitcher Plant. *Sarracenia
     purpurea* L.)   64. SARRACENIACEAE
     Lvs not modified into pitchers.   71

71   Sepals and petals normally 6; fls perigynous with a long tubular
     or bell-shaped receptacle.   65. LYTHRACEAE
     Sepals and petals normally fewer than 6; fls hypogynous, or if
     perigynous then with receptacle flat to cup-shaped.   72

72   Lvs opposite or whorled.   73
     Lvs alternate or all basal.   80

73   Lvs compound or lobed.   47. GERANIACEAE
     Lvs entire.   74

74   Lvs in a single whorl of usually 4 on the stem; fl. solitary,
     terminal.   134. TRILLIACEAE
     Lvs opposite or in numerous whorls.   75

75   Stipules present.   76
     Stipules 0.   77

76   Stipules scarious; land plants.   39. CARYOPHYLLACEAE
     Stipules not scarious; submerged aquatic plants.
                                38. ELATINACEAE

77   Sepals free or united at the base; petals always white.   78
     Sepals united to above the middle; petals white, pink or purple. 79

78   Ovary 1-celled with free-central placentation; stamens usually
     twice as many as petals, if as many or fewer then lvs narrowly
     linear or plant ± hairy or sepals scarious-margined.
                                39. CARYOPHYLLACEAE

Ovary 4–5-celled with axile placentation; fertile stamens as many as petals; lvs obovate to oval; plant glabrous; sepals not scarious.   **46. LINACEAE**

79 Style long, simple (but stigmas free); placentation parietal; fls 5 mm. diam., pink; stamens usually 6. (Procumbent red-tinged perennial with heath-like lvs. Very local on sandy salt-marshes in S. and E. England. Sea Heath. *Frankenia laevis* L.)
    37. FRANKENIACEAE
    Styles free; placentation free-central. **39. CARYOPHYLLACEAE**

80 Lvs ternate with obcordate or cuneiform and emarginate lflets.
    48. OXALIDACEAE
    Lvs not ternate.   *81*

81 Sepals and petals 2–3; fls greenish or reddish, in many-fld terminal panicles.   **80. POLYGONACEAE**
    Sepals and petals 4–5.   *82*

82 Both floral whorls green and sepal-like (calyx and epicalyx); fls small, with conspicuous hollow receptacle; lvs palmate or palmately lobed (*Alchemilla* and *Aphanes*).   **57. ROSACEAE**
    Petals ± brightly coloured, never sepal-like.   *83*

83 Sepals and petals 4; stamens 6, rarely 4.   **30. CRUCIFERAE**
    Sepals and petals 5; stamens 5 or 10.   *84*

84 Lvs covered with conspicuous red insectivorous glandular hairs.
    63. DROSERACEAE
    Lvs not conspicuously glandular.   *85*

85 Style 1, stigma simple or shallowly lobed; anthers opening by pores.   **91. PYROLACEAE**
    Styles, or at least the stigmas, more than 1, free; anthers opening by slits.   *86*

86 Stigmas 5; petals blue, pink or purple, rarely white.   *87*
    Stigmas 2–4; petals white or yellow.   *89*

87 Lvs lobed or pinnate.   **47. GERANIACEAE**
    Lvs entire.   *88*

88 Calyx-funnel shaped or obconic, scarious; lvs all ± basal; fls in heads or panicles.   **95. PLUMBAGINACEAE**
    Sepals free, not scarious or scarious only at the margins; stem lfy; fls in loose cymes.   **46. LINACEAE**

89 Fls with conspicuous glandular-fimbriate staminodes; lvs ovate, cordate, entire.   **60. PARNASSIACEAE**
    Staminodes 0; lvs not as above.   *90*

90 Stamens 5; procumbent plant; lvs entire, linear-lanceolate; stipules scarious; fls very small (*Corrigiola*).
    39. CARYOPHYLLACEAE
    Stamens 10; fls conspicuous; other characters not as above.
    59. SAXIFRAGACEAE

91 Fls saccate or spurred at base.   *92*
    Fls not saccate or spurred.   *94*

92 Lvs much divided; corolla (apparently) laterally compressed;
    stamens 2, each with 3 branches bearing anthers, not connivent.
                                                29. FUMARIACEAE
    Lvs simple; corolla not compressed; stamens 5, connivent round
    the style.   *93*

93 Sepals 5, ± equal, not spurred; petals 5 one spurred; stipules
    present; fls solitary, axillary; stem not succulent.
                                                32. VIOLACEAE
    Sepals 3, very unequal, one spurred; petals 3, not spurred; stipules
    0; fls in few-fld infls.; stem ± succulent.
                                                49. BALSAMINACEAE

94 Stamens 8 or more all, or all but 1, united into a long tube; fls
    very zygomorphic, the petals ± erect.   *95*
    Stamens free; fls less zygomorphic, petals spreading.   *96*

95 Fl. with upper sepal; anthers opening by pores; stigma tufted.
                                                33. POLYGALACEAE
    Fl. with upper petal; anthers opening by slits; stigma not tufted.
                                                56. LEGUMINOSAE

96 Trees; lvs palmate.   51. HIPPOCASTANACEAE
    Herbs; lvs not palmate.   *97*

97 Fls in cymes (often umbel-like); ovary 5-lobed with long beak.
                                                47. GERANIACEAE
    Fls in racemes; ovary not lobed or 2-lobed, rarely beaked.   *98*

98 Petals fimbriate or lobed; stamens more than 6.
                                                31. RESEDACEAE
    Petals entire or emarginate; stamens 6.   30. CRUCIFERAE

GROUP C

Petals free, ovary inferior or partly so.

99   Petals numerous.   *100*
     Petals 5 or fewer.   *101*

100 Aquatic plants with floating fls and lvs. 26. NYMPHAEACEAE
     Land plants with very succulent lvs.   41. AIZOACEAE

101 Petals and sepals 3.   *102*
     Petals and sepals 2, 4 or 5.   *105*

102 Fls zygomorphic.   138. ORCHIDACEAE
     Fls actinomorphic.   *103*

103 Both whorls of per. segs petaloid.   *104*
     Outer or both whorls of per. segs sepaloid.
                                                123. HYDROCHARITACEAE

104 Stamens 6.   135. AMARYLLIDACEAE
      Stamens 3.   136. IRIDACEAE

105 Stamens numerous.   57. ROSACEAE
      Stamens 10 or fewer.   *106*

106 Submerged aquatic with lvs pinnately divided into filiform
      segments; fls monoecious or polygamous, in terminal spikes
      projecting above water-surface.   69. HALORAGACEAE
      Land plants, or, if aquatic, then fls hermaphrodite and in umbels.   *107*

107 Trees or shrubs.   *108*
      Herbs.   *112*

108 Woody climber; fls in subglobose umbels, green.
                                                        76. ARALIACEAE
      Not climbing; fls not in umbels.   *109*

109 Lvs palmately lobed; petals shorter than sepals.
                                                  62. GROSSULARIACEAE
      Lvs simple, not lobed.   *110*

110 Both perianth-whorls petaloid; receptacle ('calyx-tube') long
      and tubular (*Fuchsia*).   68. ONAGRACEAE
      Outer perianth-whorl sepaloid.   *111*

111 Calyx-teeth very small; fls in corymbs; carpels 2, each with one
      ovule.   74. CORNACEAE
      Calyx-teeth large; fls not in corymbs; ovules numerous in each
      carpel. (Evergreen shrub with glandular twigs, toothed lvs and
      bright red fls 15 mm. diam. Introduced hedge-plant in S.W.
      England and Ireland and often self-sown. *Escallonia
      macrantha* Hooker)   61. ESCALLONIACEAE

112 Both perianth-whorls green and sepaloid (calyx and epicalyx),
      or with an epicalyx as well as sepals and petals, or with a crown
      of long spines on the receptacle below the calyx; carpels 1 or
      2, free from the receptacle and thus not truly inferior.
                                                        57. ROSACEAE
      Inner perianth-whorl always petaloid, no epicalyx or crown of
      spines; ovary truly inferior.   *113*

113 Petals 5; styles normally 2, rarely 3.   *114*
      Petals 4 or 2; style simple.   *115*

114 Fls in heads or umbels; stamens 5; ovules 1 in each carpel.
                                                     76. UMBELLIFERAE
      Fls not in heads or umbels; stamens 10; ovules numerous.
                                                     59. SAXIFRAGACEAE

115 Fls deep purple, in umbels subtended by 4 conspicuous white
      petaloid involucral bracts.   76. CORNACEAE
      Fls not in umbels; no petaloid involucral bracts.
                                                     68. ONAGRACEAE

GROUP D

Petals united, ovary superior.

116 Stamens more than 10; outer per. segs longer than inner (*Consolida*). 23. RANUNCULACEAE
Stamens 10 or fewer. 117

117 Stamens united into a tube, or 9 united, 1 free. 118
Stamens all free. 119

118 Lvs simple; fl. with upper sepal; stamens 8.
33. POLYGALACEAE
Lvs 3-foliolate; fl. with upper petal; stamens 10.
56. LEGUMINOSAE

119 Stamens twice as many as corolla-lobes (i.e. 8–10). 120
Stamens as many as or fewer than corolla-lobes (i.e. 5 or fewer). 121

120 Shrubs or trees; lvs not peltate; carpels united.
90. ERICACEAE
Succulent herb; lvs peltate; carpels free (*Umbilicus*).
58. CRASSULACEAE

121 Sepals 2; fls actinomorphic. 122
Sepals more than 2 or fls zygomorphic (sometimes 2 conspicuous
sepal-like bracts occur outside the calyx). 123

122 Petals 2; fls in heads; lvs linear, terete. (Perennial ± aquatic
scapigerous herb with translucent lvs 5–10 cm. Inner Hebrides
and W. Ireland. Pipewort. *Eriocaulon aquaticum* (Hill) Druce
(*E. septangulare* With.) 131. ERIOCAULACEAE
Petals 5; fls not in heads; lvs flat. 40. PORTULACACEAE

123 Ovary deeply 4-lobed with 1 ovule in each lobe. 124
Ovary not 4-lobed. 125

124 Lvs spirally arranged. 103. BORAGINACEAE
Lvs opposite. 111. LABIATAE

125 Trees or erect shrubs. 126
Herbs or creeping or cushion-like undershrubs. 129

126 Lvs opposite. 127
Lvs alternate. 128

127 Stamens 2. 98. OLEACEAE
Stamens 4. 97. BUDDLEJACEAE

128 Lvs usually spiny; fls actinomorphic; anthers opening by slits.
52. AQUIFOLIACEAE
Lvs never spiny; fls zygomorphic; anthers opening by pores.
90. ERICACEAE

129 Stamens opposite the corolla-lobes. 130
Stamens alternating with the corolla-lobes. 131

130 Style 1; stigma 1.   96. PRIMULACEAE
     Styles or stigmas more than 1.   95. PLUMBAGINACEAE

131 Lvs opposite.   *132*
     Lvs alternate or all basal.   *136*

132 Carpels 2, free; style expanded into a ring below the stigma;
     trailing evergreen plants.   99. APOCYNACEAE
     Carpels united; style not expanded into a ring below the stigma.   *133*

133 Cushion-like or creeping undershrubs (high mountains).   *134*
     Herbs.   *135*

134 Creeping; lvs elliptical or oblong; fls pink (*Loiseleuria*).
                            90. ERICACEAE
     Cushion-like; lvs spathulate; fls white (Very rare Scottish alpine.
     Diapensia lapponica L.).   94. DIAPENSIACEAE

135 Land plants; lvs sessile.   100. GENTIANACEAE
     Aquatic plants with floating lvs on long petioles (*Nymphoides*).
                           101. MENYANTHACEAE

136 Calyx- and corolla-lobes 4; stamens 4 or 2.   *137*
     Calyx- and corolla-lobes and stamens 5.   *142*

137 Stamens 2; lvs and bracts not spine-toothed.   *138*
     Stamens 4.   *139*

138 Ovary 1-celled; corolla spurred; carnivorous bog or aquatic
     plants with lvs all basal or else divided into filiform segments.
                      108. LENTIBULARIACEAE
     Ovary 2-celled; corolla not spurred; lvs not as above.
                      107. SCROPHULARIACEAE

139 Lvs all basal.   112. PLANTAGINACEAE
     Lvs not all basal.   *140*

140 Bracts spine-toothed; corolla 1-lipped. (Lvs 20–60 cm., pinnatifid;
     robust perennial. Introduced and established in Cornwall and
     Scilly. *Acanthus mollis* L.)   109. ACANTHACEAE
     Bracts not spine-toothed; corolla weakly zygomorphic or
     2-lipped.   *141*

141 Ovules numerous.   106. SCROPHULARIACEAE
     Ovules 4.   110. VERBENACEAE

142 Ovary 3-celled; stigmas 3, or if only 1 then 3-lobed.   *143*
     Ovary 2-celled; stigmas 2 or 1, not 3-lobed.   *144*

143 Erect herb; lvs pinnate.   102. POLEMONIACEAE
     Cushion-like; lvs spathulate; fls white (*Diapensia*: see *134* above).
                           94. DIAPENSIACEAE

144 Ovules 4 or fewer; twining or prostrate herbs; lvs cordate or
     hastate at base; corolla shallowly lobed.
                      104. CONVOLVULACEAE

Ovules numerous; ± erect herbs or woody climbers; corolla-lobes
conspicuous.   145

145 Aquatic or bog plants; lvs orbicular or ternate; corolla fringed.
                              101. MENYANTHACEAE
Land plants; lvs neither orbicular nor all ternate (but some may
be ternate in a woody climber) corolla not fringed.   146

146 Fls numerous, in terminal spikes or racemes (sometimes aggregated into panicles); corolla-tube very short; stamens spreading
(*Verbascum*).   106. SCROPHULARIACEAE
Fls solitary or in cymes (sometimes scorpioid); corolla-tube long,
or, if short, then anthers connivent.   105. SOLANACEAE

GROUP E

Petals united, ovary inferior

147 Stamens 8–10, or 4–5 with filaments divided to the base.   148
Stamens 5 or fewer, filaments not divided.   149

148 Herb; fls in heads, green; lvs ternate.   117. ADOXACEAE
Low shrubs or prostrate creeping undershrubs; fls pink or
white, not in heads; lvs simple.   90. ERICACEAE

149 Fls in heads surrounded by an involucre; herbs (rarely slightly
woody).   150
Fls not in heads, or if in heads then with 2 bracts only and plant
a woody climber.   153

150 Anthers coherent into a tube round the style.   151
Anthers free.   152

151 Ovules numerous; calyx-lobes conspicuous, green; fls blue
(*Jasione*).   113. CAMPANULACEAE
Ovule 1; calyx represented by hairs or scales; fls rarely blue.
                                   120. COMPOSITAE

152 Ovules numerous; corolla-lobes long and narrow, longer than tube.
                                   113. CAMPANULACEAE
Ovule 1; corolla-lobes much shorter than tube.
                                   119. DIPSACACEAE

153 Lvs in whorls or 4 or more; fls actinomorphic; petals 4.
                                   115. RUBIACEAE
Lvs not in whorls; fls zygomorphic, or if actinomorphic then
petals 5.   154

154 Fls zygomorphic.   155
Fls actinomorphic.   157

155 Fls in corymbs.   118. VALERIANACEAE
Fls in terminal racemes or spikes.   156

156 Anthers coherent into a tube round the style; pollen powdery.
114. LOBELIACEAE
    Anthers 2, free; pollen cohering in pollinia.
138. ORCHIDACEAE

157 Herb, climbing by tendrils. 77. CUCURBITACEAE
    Herbs, shrubs or woody climbers; tendrils 0.   158

158 Lvs opposite.   159
    Lvs spirally arranged.   160

159 Stamens 4 or 5; usually shrubs or woody climbers; if herbs either
    prostrate and creeping or with lf-like stipules.
116. CAPRIFOLIACEAE
    Stamens 1–3; herbs, ± erect and without lf-like stipules.
118. VALERIANACEAE

160 Stamens opposite corolla-lobes; stigmas capitate; fls white
    (Samolus). 96. PRIMULACEAE
    Stamens alternating with corolla-lobes; stigmas 2–5; fls normally
    blue or purple. 113. CAMPANULACEAE

GROUP F

Perianth entirely petaloid or in several series, the inner petaloid.

161 Stamens numerous.   162
    Stamens 12 or fewer, or fls female.   165

162 Aquatic plants with floating lvs and fls. 26. NYMPHAEACEAE
    Terrestrial plants.   163

163 Succulent prostrate plant with 3-angled lvs.
41. AIZOACEAE (see 55 above, p. xix)
    Lvs not 3-angled.   164

164 Carpels free, rarely united and then per. segs numerous.
23. RANUNCULACEAE
    Carpels united; petals usually 4; sepals 2, falling as fl. opens.
28. PAPAVERACEAE

165 Fls crimson, in ovoid heads without an involucre; lvs pinnate
    (Sanguisorba). 57. ROSACEAE
    Fls not in heads, or if so then with an involucre.   166

166 Ovary superior.   167
    Ovary inferior or fls male.   175

167 Perianth strongly zygomorphic, spurred or saccate at base;
    stamens 2, each with 3 anther-bearing branches; lvs much
    divided (sepals 2, but bract-like and soon falling).
29. FUMARIACEAE
    Perianth actinomorphic or slightly zygomorphic, and then
    neither spurred nor saccate.   168

168 Shrubs.   *169*
     Herbs.   *172*

169 Fls borne on the surface of lf-like cladodes; true lvs small and
     scale-like (*Ruscus*).   132. LILIACEAE
     Fls not on cladodes.   *170*

170 Per. segs 4, continued below into a coloured receptacular tube.
               66. THYMELAEACEAE
     Per. segs 6 or more, free.   *171*

171 Low heath-like shrubs with inconspicuous axillary fls (if per. segs 8,
     pink-purple, in 2 differing whorls, see *Calluna* in Ericaceae,
     p. 280).   93. EMPETRACEAE
     Tall shrubs with yellow fls in racemes or panicles.
               25. BERBERIDACEAE

172 Per. segs 5   *173*
     Per. segs 6, rarely 4.   *174*

173 Stigma 1, capitate; stipules 0 (*Glaux*).   96. PRIMULACEAE
     Stigmas 2–3; stipules sheathing, scarious.   80. POLYGONACEAE

174 Stamens 8(–9); ovules scattered over whole inner surface of
     carpels; aquatic plant.   122. BUTOMACEAE
     Stamens 6, rarely 4; ovules on axile placentae; plants not aquatic.
               132. LILIACEAE

175 Trees or shrubs; calyx present but very small and rim-like or with
     minute teeth. See *147* (Group E, p. xxvi).
     Herbs.   *176*

176 Lvs in whorls of 4 or more.   115. RUBIACEAE
     Lvs not in whorls.   *177*

177 Fls in heads surrounded by a common involucre.   *178*
     Fls not in heads though sometimes shortly stalked in compact
     umbels.   *179*

178 Stamens free; fls hermaphrodite.   119. DIPSACACEAE
     Anthers cohering in a tube round the style, or fls unisexual.
               120. COMPOSITAE

179 Per. segs 3, or perianth with a long tube swollen below and a uni-
     lateral entire limb; lvs ± orbicular, cordate, entire.
               78. ARISTOLOCHIACEAE
     Per. segs 5 or 6; lvs not as above.   *180*

180 Per. segs 5; fls small; ovules 1 or 2.   *181*
     Per. segs 6; fls large, ovules numerous.   *183*

181 Fls in simple cymes; lvs spirally arranged, narrowly linear, small.
               73. SANTALACEAE
     Fls in umbels or superposed whorls, or if in cymes then lvs
     opposite.   *182*

182 Stamens 5; per. segs free; fls in umbels or superposed whorls; lvs
    spirally arranged.   76. UMBELLIFERAE
    Stamens 1–3; per. segs united; fls in cymes or panicles; lvs
    opposite.   118. VALERIANACEAE

183 Stamens 6.   135. AMARYLLIDACEAE
    Stamens 3.   136. IRIDACEAE

GROUP G

Trees or shrubs; perianth sepaloid or 0.

184 Parasitic on the branches of trees; lvs opposite, obovate or
    oblong, thick, leathery; stems green.
          72. LORANTHACEAE
    Not as above.   185

185 Root-climber; fls in umbels.   75. ARALIACEAE
    Not climbing; fls not in umbels.   186

186 Fls borne on the surface of flattened evergreen lf-like cladodes;
    true lvs colourless, scale-like (*Ruscus*).   132. LILIACEAE
    Fls not on cladodes; lvs green.   187

187 Lvs opposite or subopposite.   188
    Lvs spirally arranged or in 2 ranks (alternate).   192

188 Lvs evergreen, thick, leathery, entire; styles 3.
          54. BUXACEAE
    Lvs deciduous; styles 4, 2 or 1.   189

189 Fls in catkins.   89. SALICACEAE
    Fls not in catkins.   190

190 Lvs pinnate; perianth 0; stamens 2 (*Fraxinus*).
          98. OLEACEAE
    Lvs simple; perianth present; stamens 4 or more.   191

191 Lvs palmately lobed; plant not thorny.   50. ACERACEAE
    Lvs simple, not lobed; plant thorny.   55. RHAMNACEAE

192 Lvs evergreen, small (less than 10×2 mm.), dense, oblong or
    linear, entire; shrubs to 1 m. or less.   193
    Lvs relatively large (longer or broader), not particularly dense,
    usually deciduous and if evergreen then 30 mm., or more.   194

193 Procumbent; stamens 3; stigmas 6–9; lvs leathery; moors, etc.
          93. EMPETRACEAE
    Erect; stamens 5; stigmas 2; lvs fleshy; maritime (*Suaeda*).
          43. CHENOPODIACEAE

194 Lvs pinnate (present at flowering time).   84. JUGLANDACEAE
    Lvs simple (sometimes 0 at flowering time).   195

195 Fls, at least in the male, in catkins or in tassel-like heads on
    long pendulous stalks.   *196*
    Fls not in catkins or stalked heads.   *201*

196 Fls dioecious; perianth 0; fls always solitary in the axil of each
    bract.   *197*
    Fls monoecious, though usually in separate infls; perianth present
    at least in the fls of one or other sex.   *199*

197 Scales of catkins fimbriate or lobed at the tip; fls of both sexes with
    a cup-like disk; ovules numerous (*Populus*).   89. SALICACEAE
    Scales of catkins entire; disk 0.   *198*

198 Ovules numerous; lvs without resin glands, not aromatic when
    crushed; fls of both sexes without bracteoles but with nectaries at
    the base, placed above or below the fl.; stamens with long
    filaments (*Salix*).   89. SALICACEAE
    Ovule 1; lvs dotted with resin glands, strongly aromatic when
    crushed; male fl. without nectaries or bracteoles, female fl.
    with 2 lateral bracteoles; filaments short.   85. MYRICACEAE

199 Fls of both sexes with perianth; styles 3 or more; fr. large and
    nut-like, partly or completely enclosed in a hard cup or shell.
                                   88. FAGACEAE
    Perianth present in one sex only; styles 2; fr. small, or large and
    nut-like; cup if present papery or lf-like.   *200*

200 Male fls 3 to each bract; perianth present; fr. small, in the axils
    of the accrescent bracts which persist till maturity and form
    cone-like structures.   86. BETULACEAE
    Male fls solitary in the axil of each bract; perianth 0; fr. not borne
    in cones, surrounded by a papery or lf-like cup formed from the
    bracts.   87. CORYLACEAE

201 Lvs and twigs densely covered with silvery or brown peltate scales;
    fls very small, dioecious; male with 2 free per. segs; female with
    tubular perianth having 2 small lobes at its apex.
                         67. ELAEAGNACEAE
    Plant without peltate scales; fls hermaphrodite; per. segs 4 or more.   *202*

202 Deciduous trees; fls in sessile clusters, appearing before the lvs;
    perianth ± bell-shaped, the stamens inserted at its base;
    styles 2.   83. ULMACEAE
    Evergreen shrub; fls in short-stalked racemes; perianth continued
    downwards into a long, cylindrical tube, the stamens inserted
    high on the tube; style 1.   66. THYMELAEACEAE

GROUP H

Herbs, perianth sepaloid or 0.

203 Perianth 0 or represented by scales or bristles, minute in fl. but
sometimes elongating in fr.; the fls in the axils of specialized
chaffy bracts which are usually arranged along the rhachis of
spikelets, sometimes themselves aggregated into compound infls;
lvs always ± linear and grass-like, sheathing below.   204
    Perianth present, or if minute or absent then fls not arranged in
spikelets nor the bracts chaffy; lvs various.   205

204 Fls with bract above and below; lvs ± jointed at the junction with
the sheath, commonly with a prominent projecting ligule;
sheaths usually open; stems terete or flattened, usually with
hollow internodes.   144. GRAMINEAE
    Fls with a bract below only; lvs not jointed at the junction with
the sheath; ligule, if present, not projecting, sheaths usually
closed; stem often 3-angled; internodes nearly always solid.
                                                      143. CYPERACEAE

205 Aquatic plants; lvs submerged or floating; infl. sometimes rising
above the surface of the water.   206
    Land plants, or if aquatic then with stiffly erect stems and with lvs
as well as fls rising above the surface of the water.   221

206 Lvs divided into numerous filiform segments.   207
    Lvs entire or toothed.   208

207 Lvs pinnately divided; fls in a terminal spike (bracts sometimes
lf-like).   69. HALORAGACEAE
    Lvs dichotomously divided; fls solitary, axillary.
                                                27. CERATOPHYLLACEAE

208 Fls in a spike surrounded by a petaloid spathe (*Calla*).
                                                     140. ARACEAE
    Without petaloid bracts or spathe.   209

209 Fls sessile or nearly so, arranged in heads.   210
    Fls in spikes or in the axils of the lvs.   212

210 Heads with many small fls, solitary at the ends of a lfless stalk.
                                131. ERIOCAULACEAE (see *122* above, p. xxiv)
    Heads few-fld and terminal, or lateral on lfy stems.   211

211 Fls unisexual, the male heads above, the female heads below.
                                                  141. SPARGANIACEAE
    Fls hermaphrodite.   134. JUNCACEAE

212 Fls in spikes.   213
    Fls axillary, solitary or in few-fld clusters.   215

213 Fls monoecious, arranged on one side of a flattened spadix;
perianth 0; marine.   126. ZOSTERACEAE

Fls hermaphrodite, arranged all round or on two sides of a terete
rhachis; fresh or brackish water but not truly marine.   *214*

214 Per. segs 4; carpels remaining sessile; usually freshwater.
                              128. POTAMOGETONACEAE
Perianth 0; fruiting carpels on long stalks; brackish pools and
ditches.   128. RUPPIACEAE

215 Female fls with very long filiform perianth-tube, resembling a
pedicel and raising them to the surface of the water.
                              123. HYDROCHARITACEAE
Tube and pedicel short or 0.   *216*

216 Carpels 2–6; free, lvs narrowly linear, quite entire, not whorled.
                              129. ZANNICHELLIACEAE
Carpels united or 1 only; lvs broader, or if narrowly linear then
finely toothed or whorled.   *217*

217 Perianth with 4–6 segments; stamens 4 or more.   *218*
Perianth 0, or entire, or with 2 segments; stamen 1.   *219*

218 Per. segs 4; ovary inferior; lvs ovate (*Ludwigia*).
                              68. ONAGRACEAE
Per. segs 6; ovary superior; lvs obovate (*Peplis*).
                              65. LYTHRACEAE

219 Lvs in whorls of 8 or more; fls hermaphrodite; style 1.
                              70. HIPPURIDACEAE
Lvs opposite or in whorls of 3; fls unisexual; styles 2–3.   *220*

220 Lvs narrowly linear with sheathing base, finely (or minutely) spiny-
toothed, apex acute; ovary terete, not lobed.
                              130. NAJADACEAE
Lvs (at least the upper) usually spathulate; if all linear, then entire
and with an emarginate apex; base not sheathing; ovary
flattened, 4-lobed.   71. CALLITRICHACEAE

221 Twining plants; fls unisexual.   *222*
Not climbing or, if climbing, fls hermaphrodite.   *223*

222 Lvs opposite, palmately lobed; per. segs 5.   82. CANNABACEAE
Lvs spirally arranged, cordate, entire; per. segs 6.
                              137. DIOSCOREACEAE

223 Lvs linear, ± grass-, rush- or iris-like; plants of wet places.   *224*
Lvs not linear or, if so, small and not at all grass-like.   *229*

224 Fls monoecious, the male and female in separate infls or in parts
of the same infl.   *225*
Fls hermaphrodite.   *226*

225 Fls in globose heads, the male and female in separate heads.
                              141. SPARGANIACEAE
Fls in dense cylindrical spikes, male above and female below.
                              142. TYPHACEAE

226 Fls in dense spikes borne laterally on a flattened lf-like stem
    (*Acorus*).   139. ARACEAE
    Infl. not as above.   227

227 Carpels united only at extreme base; fls in racemes. (A perennial
    bog plant with linear ligulate sheathing lvs, each with a con-
    spicuous pore at the tip, and yellowish-green fls c. 4 mm. diam.
    Very rare in Scotland and Ireland. *Scheuchzeria palustris* L.)
                                        124. SCHEUCHZERIACEAE
    Carpels ± completely united.   228

228 Fls in spikes; perianth herbaceous.   125. JUNCAGINACEAE
    Fls not in spikes or racemes; perianth scarious.
                                             134. JUNCACEAE

229 Lvs compound.   230
    Lvs simple or 0.   233

230 Fls in heads.   231
    Fls not in heads.   232

231 Lvs simply pinnate; style 1 (rarely 2), stamens 4 or numerous.
                                             57. ROSACEAE
    Lvs ternate (sometimes 2 or 3 times); styles 3–5; stamens ap-
    parently 8–10 (4 or 5 with filaments divided to base).
                                            117. ADOXACEAE

232 Stamens numerous; no epicalyx.   23. RANUNCULACEAE
    Stamens 4 or 5 (rarely 10); epicalyx present. 57. ROSACEAE

233 Infl. umbellate, consisting of several male fls (each of 1 stamen)
    and one female fl. (appearing as a stalked ovary) all surrounded
    by 4 or 5 crescent-shaped or roundish glands; juice milky
    (*Euphorbia*).   79. EUPHORBIACEAE
    Infl. not as above; juice not milky.   234

234 Infl. a dense spike with female fls below and male fls above; lvs
    hastate (*Arum*).   139. ARACEAE
    Infl. not as above; lvs not hastate.   235

235 Lvs 0; stems green and succulent, jointed; perianth flush with the
    stem; salt-marsh plants (*Salicornia*).
                                         43. CHENOPODIACEAE
    Lvs obvious, green; stems not succulent.   236

236 Lvs spirally arranged or all basal (rarely the lower opposite).   237
    Lvs all opposite or whorled.   247

237 Stamens 12 or more.   238
    Stamens 8 or fewer.   239

238 Per. segs 5, with a whorl of honey-lvs within; lvs palmately lobed
    (*Helleborus*).   23. RANUNCULACEAE
    Per. segs 3, without honey-lvs; lvs reniform, entire (*Asarum*).
                                         78. ARISTOLOCHIACEAE

239 Stipules ± scarious, united into a sheath. 80. POLYGONACEAE
    Stipules free or 0.   *240*

240 Lvs large and rhubarb-like, all basal; fls in dense, many-fld
    spikes from the base, much shorter than the lvs (*Gunnera*).
        69. HALORAGACEAE
    Lvs not rhubarb-like; fls not in basal spikes.   *241*

241 Stamens twice as many as per. segs; lvs reniform, cordate
    (*Chrysosplenium*).   59. SAXIFRAGACEAE
    Stamens as many as per. segs or fewer; lvs neither reniform nor
    cordate.   *242*

242 Stipules lf-like; perianth of 4 segments with an epicalyx of 4
    segments outside; lvs palmately lobed (*Aphanes* and *Alchemilla*).
        57. ROSACEAE
    Stipules very small or 0; perianth without epicalyx.   *243*

243 Ovary inferior.   73. SANTALACEAE
    Ovary superior.   *244*

244 Fls in simple ebracteate racemes (*Lepidium*). 30. CRUCIFERAE
    Fls not in simple, ebracteate racemes.   *245*

245 Styles 2 or more, free or united below; stigmas simple; fls
    mostly 5-merous.   *246*
    Style 1; stigma feathery, tufted; fls 4-merous (*Parietaria*).
        81. URTICACEAE

246 Perianth herbaceous.   43. CHENOPODIACEAE
    Perianth scarious.   42. AMARANTHACEAE

247 Lvs toothed or lobed.   *248*
    Lvs entire.   *251*

248 Fls hermaphrodite; stems creeping or decumbent.   *249*
    Fls unisexual; aerial stems erect.   *250*

249 Ovary inferior, not lobed; styles 2; fls in dichotomous cymes
    (*Chrysosplenium*).   59. SAXIFRAGACEAE
    Ovary superior, 5-lobed, prolonged into a long beak bearing
    5 stigmas; fls solitary or very few on long axillary peduncles
    (*Erodium*).   47. GERANIACEAE

250 Plant with stinging hairs; per. segs 4 or 2; stamens 4; style 1;
    stigmas feathery (*Urtica*).   81. URTICACEAE
    Plant without stinging hairs; per. segs 3; stamens 9 or more;
    styles 2, simple (*Mercurialis*).   79. EUPHORBIACEAE

251 Perianth 0 or obscurely 2-lobed or of 2–3 segments.   *252*
    Perianth of 4 or more segments.   *254*

252 Per. segs 3; stamens 3 (*Koenigia*).   80. POLYGONACEAE
    Perianth 0 or of fewer than 3 segments; stamen 1 (plants
    ± aquatic).   *253*

253 Lvs whorled; fls hermaphrodite; style 1.
                                        70. HIPPURIDACEAE
    Lvs opposite; fls monoecious; styles 2.
                                        71. CALLITRICHACEAE
254 Ovary inferior; style 1; per. segs 4 (*Ludwigia*).
                                        68. ONAGRACEAE
    Ovary superior.   255
255 Per. segs 6 or 12, inserted on a bell-shaped receptacle; style 1;
    plant ± aquatic; lvs obovate (*Peplis*).   65. LYTHRACEAE
    Per. segs 4 or 5, usually free (if on a bell-shaped receptacle, then
    lvs linear); styles 2 or more, free; land-plants.
                                        39. CARYOPHYLLACEAE

GROUP J

Herbs without chlorophyll; lvs scale-like.

256 Fls zygomorphic.   257
    Fls actinomorphic.   258
257 Per. segs free.   138. ORCHIDACEAE
    Per. segs united into a tubular corolla. 107. OROBANCHACEAE
258 Erect saprophyte.   92. MONOTROPACEAE
    Twining parasites (*Cuscuta*).   104. CONVOLVULACEAE

SIGNS AND ABBREVIATIONS

agg.	aggregate, incl. 2 or more spp. which resemble each other closely.
C.	central.
c.	about (circa).
f.	forma, *filius*.
fl.	flower, flowering time; plural fls.
-fid	-flowered.
fr.	fruit, fruiting.
incl.	including.
infl.	inflorescence, inflorescences.
lf	leaf; plural lvs.
lfless	leafless.
lflet	leaflet.
lfy	leafy.
per. seg.	perianth segment.
p.p.	*pro parte*.
sp.	species; plural spp.
ssp.	subspecies; plural sspp.
var.	variety.
0	absent.
×	Preceding the name of a genus or sp. indicates a hybrid.
±	more or less.
*	Preceding the name of a sp. or genus indicates that it is certainly introduced.

Measurements without qualification (e.g. lvs 4–7 cm.) refer to lengths; lvs 4–7 × 1–2 cm. means lvs 4–7 cm. long and 1–2 cm. wide. Measurements or numbers enclosed in brackets (e.g. lvs 4–7(–10) cm.) are exceptional ones outside the normal range.

Note on Keys

Species of which a full description is provided are in heavy type, the trivial name being preceded by a serial number. Where there is no full description the name of the species is in italics and unnumbered, the trivial name being preceded by the initial letter of the generic name and followed by the authority for the name. But where, in keys to genera, certain species are keyed out individually, their names are printed in italics, preceded by the serial number of the genus with authorities omitted.

3. ISOETACEAE

Herbs, often aquatic, with short stout stems concave above and 2–3-lobed below. Lvs ligulate, crowded in a dense rosette, awl-shaped or thread-like, usually terete and hollow with frequent septa; base broadened, sheathing. Sporangia ±embedded in the leaf-base below the ligule and of 2 kinds; first lvs of the season bear megasporangia with many large spores, later lvs microsporangia with very numerous small spores.

1. Isoetes L.

The only genus in the northern hemisphere.

1   Plant aquatic, never completely dormant and lfless; stem without persistent lf-bases; lvs 4–20 cm., 2–3 mm. wide.   2
    Plant terrestrial, dormant and lfless in summer; stem ±covered by persistent lf-bases; lvs up to 3 cm., 1 mm. wide. In peaty and sandy places wet only in winter: Lizard district (Cornwall); Channel Is.   *I. histrix* Bory

2   Lvs 8–20 or more cm., very stiff; megaspores with short blunt tubercles.   **1. lacustris**
    Lvs 4–12 cm., rather limp; megaspores with long sharp spines. Lakes and tarns, usually on peaty substrata; local in S.W. England, Wales, Scotland and Ireland.
     I. setacea Lam. (*I. echinospora* Durieu)

1. I. lacustris L.   Quill-wort·

Submerged aquatic. Stem without persistent lf-bases. Lvs 8–20(–45) cm., 2–3 mm. wide, awl-shaped, subterete, stiff, dark green, with 4 longitudinal septate internal tubes. Megaspores usually yellowish, covered with short blunt tubercles. Spores ripe 5–7. In lakes and tarns of base-poor water, chiefly on stones or sand with little or no silt or peat. Locally abundant in mountain districts of Wales, Lake District, Scotland north to Shetland, and Ireland, and in a few scattered places elsewhere.

SPHENOPSIDA

4. EQUISETACEAE

Herbs with creeping rhizomes. Stems hollow, grooved, simple or branched, all alike and green or fertile ones simple and not green. Lvs in whorls, small, joined into a sheath in lower part, rarely green.

Sporangia all alike, on the under side of peltate sporangiophores which are borne in whorls and form a compact terminal cone.

1. EQUISETUM L.

The only genus.

1   All stems present fertile and not green.   2
     Green sterile stems present (fertile ones present or not, like or
          unlike the sterile).   3

2   Sheaths numerous, with 20–30 teeth; cone 4–8 cm.   **7. telmateia**
     Sheaths 4–6, each with 6–12 teeth; cone 1–4 cm.   **6. arvense**

3   Sterile stems with whorls of branches.   4
     Sterile stems simple or with few branches.   10

4   Sterile stem c. 10 mm. diam.; grooves 20–40, very fine; branches
          very numerous.   **7. telmateia**
     Sterile stem rarely more than 7 mm. diam.; grooves less than 20
          (if stouter or grooves more than 20 then branches few and
          ceasing several nodes below the top).   5

5   Branches usually again branched; sheaths of main stems with 3–6
          broad subacute teeth, fewer than the grooves.   **5. sylvaticum**
     Branches usually simple; sheaths of main stem with awl-shaped
          teeth, as many as the grooves.   6

6   Stem with 10–30 very fine shallow grooves; teeth not ribbed; sterile
          and fertile stems alike; central hollow at least $\frac{4}{5}$ diam. of stem.
                                                              3. fluviatile
     Stem with 4–20 deep grooves; teeth ribbed; central hollow less than
          $\frac{2}{3}$ diam. of stem.   7

7   Fertile and sterile stems alike; branches hollow.   8
     Fertile and sterile stems normally differing; branches solid.   9

8   Stem with 4–8 grooves; central hollow less than $\frac{1}{2}$ diam. of stem;
          cone blunt (common).   **4. palustre**
     Stem with 8–20 grooves; central hollow more than $\frac{1}{2}$ diam. of stem;
          cone pointed. Lincolnshire; very rare.   *E. ramosissimum* Desf.

9   Branches mostly 4-angled; fertile stems dying after fr.   **6. arvense**
     Branches mostly 3-angled; fertile stems continuing to grow and
          branching after fr. Local in the north.   *E. pratense* Ehrh.

10 Sheaths finally with a blackish band at top and bottom; teeth soon
          falling; cone pointed.   **1. hyemale**
     Sheaths without a blackish band at bottom (though sometimes
          wholly black); teeth persistent; cone pointed or not.   11

11 Stems smooth or nearly so, dying in autumn; cone blunt.   12
     Stems very rough, persisting through the winter; cone pointed.
                                                              2. variegatum

12 Stems with 4–8 deep grooves; teeth 1-ribbed.   **4. palustre**
    Stems with 10–30 fine grooves; teeth not ribbed.   **3. fluviatile**

1. E. hyemale L.   Dutch Rush.

Stems 30–100 cm., erect, 4–6 mm. diam., simple, rough, persisting through the winter; grooves 10–30; sheaths about as long as broad, whitish with black bands at top and bottom. Spores ripe 7–8. Shady banks, etc., local in the north, very rare in the south.

E. × moorei Newm., with stems dying in autumn and sheaths longer than broad, occurs in Surrey and in Ireland. **E. × trachyodon** A.Br., with persistent stems, wholly black sheaths and persistent teeth, is very local in Scotland and Ireland. Both probably have *E. hyemale* as one parent.

2. E. variegatum Weber & Mohr   Variegated Horsetail.

Stems 10–60 cm., usually decumbent, up to 3 mm. diam., simple or sparsely branched, persisting through the winter; grooves 4–10; sheaths green with a black band at top. Spores ripe 7–8. Dunes, wet ground on mountains, etc., local in the north, rare elsewhere.

3. E. fluviatile L.   Water Horsetail.

Stems 50–140 cm., ± erect, 2–12 mm., diam., all green, simple or with irregular whorls of branches in the middle, smooth; grooves 10–30, very fine; sheaths fitting close to stem; teeth small, not ribbed; central hollow at least $\frac{4}{5}$ diam. of stem. Branches simple, hollow. Spores ripe 6–7. In shallow water, less often in marshes, common and widespread.

E. × litorale Kühlew. (*E. arvense × fluviatile*), with loose sheaths and central hollow $\frac{1}{2}$–$\frac{2}{3}$ diam. of stem, occurs locally.

4. E. palustre L.   Marsh Horsetail.

Stems 10–60 cm., erect or decumbent, 1–3 mm. diam., all green, usually irregularly branched, slightly rough; grooves 4–8, deep; sheaths loose; teeth 1-ribbed; central hollow small. Branches simple, hollow. Damp places, common and widespread.

5. E. sylvaticum L.   Wood Horsetail.

Fertile and sterile stems produced at the same time. Fertile stems at first simple and without chlorophyll, becoming green and branching after spore-dispersal. Sterile stems 10–80 cm., erect, c. 1–4 mm. diam., nearly or quite smooth; grooves 10–18; teeth united into 3–6 broad, several-ribbed lobes. Branches solid, usually branched. Sheaths of fertile stems greenish below, brown above, with 3–6 broad brown teeth. Spores ripe

4–5. Shady places on acid soils, common in the north, very local in the south.

6. E. arvense L.   Common Horsetail.

Fertile stems produced before the sterile ones, not green, simple, dying when spores are shed. Sterile stems 20–80 cm., erect or decumbent, c 3–5 mm. diam., green, slightly rough; grooves 6–19, deep; sheaths 3–8 mm., teeth 1-ribbed. Branches numerous, solid, usually simple. Fertile stems with 4–6 sheaths. Cone 1–4 cm. Spores ripe 4. Fields, roadsides, etc., common and widespread.

7. E. telmateia Ehrh.   Great Horsetail.

Differs from *E. arvense* in being much larger, with sterile stems 1–2m., c. 10–12 mm. diam., whitish; grooves 20–40, fine; sheaths 15–40 mm.; teeth 2-ribbed. Fertile stems with numerous sheaths. Cone 4–8 cm. Spores ripe 4. Damp shady places, etc., rather local, becoming rarer in the north.

FILICOPSIDA

In the following descriptions of members of the Filicopsida or ferns the degree of dissection of the lf (or 'frond') always refers to the lower part of the lf or pinna; the upper part is almost always less divided. 'Pinnate' is used throughout though 'pinnatisect' would be more correct. Lengths of pinnae, etc., refer to the best developed ones and lf-lengths (unless otherwise stated) to fertile lvs. It is unwise to attempt the identification of young or sterile plants until considerable experience has been gained. There is considerable variation in many ferns and no attempt is made to cover the extremes in the following descriptions.

5. OPHIOGLOSSACEAE

Herbs with short rhizomes; scales 0. Lvs 1 or more, not circinate in bud. Fertile lvs with a sterile blade and spore-bearing spike or panicle which, in British spp., appears as if terminal on a stem. Sporangia all alike, borne in 2 rows on the margin of the spike or of the panicle-branches.

Sterile blade pinnate; fertile portion a panicle.   1. BOTRYCHIUM
Sterile blade simple, entire; fertile portion a spike.
                                                          2. OPHIOGLOSSUM

6. OSMUNDACEAE

1. BOTRYCHIUM Sw.

Sterile blade pinnate or pinnately lobed. Sporangia subsessile, on the branches of a panicle.

1. B. lunaria (L.) Swartz   Moonwort.

Lvs (2–)5–15(–30) cm., solitary (–2), with a brown sheath at base; blade pinnate, oblong in outline; pinnae fan-shaped, entire or somewhat crenate, without midrib; panicle 1–3 times branched, overtopping blade. Spores ripe 6–8. Dry grassland and rock ledges; widespread but local.

2. OPHIOGLOSSUM L.

Sterile blade simple, net-veined; Sporangia sunken, in a simple spike or spikes.

1 Lvs usually single, rarely 2; blade more than 3 × 2 cm.   **1. vulgatum**
    Lvs often 2–3 together; blade usually less than 3 × 2 cm.   **2**

2 Blade lanceolate to ovate, acute, 5–20(–30) mm. wide, net-veined, with free vein-endings inside the network; 6–15 sporangia on each side of the fertile spike. Sandy ground and short turf near the sea in the north and west; very local. *O. azoricum* C. Presl (*O. vulgatum* ssp. *ambiguum* (Cosson & Germ.) E. F. Warburg)

   Blade lanceolate to linear-lanceolate, blunt, 2–7 mm. wide, net-veined, with no free vein-endings within the network; 5–10 sporangia on each side of the fertile spike; spores ripe 1–2. Channel and Scilly Is.   *O. lusitanicum* L.

1. O. vulgatum L.   Adder's Tongue.

Lf solitary, rarely 2; blade 3–15 × 2–6 cm., fleshy, yellow-green, sheathing the stalk of the fertile spike at base, net-veined, with free vein-endings inside the network; spike usually 2–5 cm., overtopping the blade, its apex sterile, acute; sporangia 12–40 on each side of the spike. Spores ripe 5–8. Widely distributed in damp grassland, fens and scrub throughout the British Is.

6. OSMUNDACEAE

Rhizome stout, not scaly. Lvs pinnately divided, with glandular hairs at the enlarged base of the petiole; veins not anastomosing. Sporangia on the margins or surface of the lf, all alike and developing simultaneously; indusium 0.

1. Osmunda L.

Sporangia marginal on reduced pinnules with no green blade; fertile pinnules at the top or in the middle of the lf, or the whole lf fertile. Outer lvs sterile.

1. O. regalis L.   Royal Fern.

Lvs 30–300 cm., tufted, 2-pinnate, ± lanceolate in outline; fertile lvs with c. 2–3 pairs of sterile pinnae and 5–14 fertile pairs, markedly decreasing in size upwards. Sterile pinnae with narrowly winged rhachis. Pinnules 2–6·5 cm., oblong, ± truncate at base and often with a rounded lobe on the lower side; margins irregularly crenulate-serrate or occasionally slightly lobed. Fertile pinnules up to 3 cm., densely covered with brown sporangia. Spores ripe 6–8. In damp ± peaty places, local but widely distributed.

7. ADIANTACEAE

Rhizomes with opaque scales. Petiole dark and shining, with 2 vascular strands at the base, uniting distally. Sori borne on the down-turned lf-margins; spores tetrahedral.

1. Adiantum L.

Rhizomes with narrow brown scales. Lvs all alike, usually with black glossy petiole and broad ± fan-shaped segments.

For *Adiantum capillus-veneris* L. see p. xvi.

8. CRYPTOGRAMMACEAE

Rhizomes with opaque scales. Petioles with a single cylindrical vascular strand. Sori submarginal on the vein-ends, at first distinct but forming a continuous band when mature; spores tetrahedral.

1. Cryptogramma R.Br.

Rhizome stout, scaly. Sterile and fertile lvs dissimilar, 2–4 pinnate. Sori covered by the reflexed continuous lf-margin.

1. C. crispa (L.) Hooker   Parsley Fern.

Lvs 15–20 cm., densely tufted, the outer sterile, inner fertile and with longer petioles. Sterile lvs 3-pinnatisect, triangular-ovate; pinnae 3–7 on each side; segments obovate, blunt, cuneate at base, lobed or toothed; petiole about twice as long as the blade, with a few scales at base. Fertile lvs 3–4-pinnate, segments oblong-linear. Sori becoming confluent when mature. Spores ripe 6–8. Screes, stone-walls, etc., on acid rocks usually in mountain districts, local.

9. GYMNOGRAMMACEAE

Rhizomes with opaque scales. Petioles with 2 vascular strands. Sporangia along veins, not in well-defined sori; indusium 0; spores tetrahedral.

1. ANOGRAMMA Link

Annual fern with a short stock bearing few scales. Sterile and fertile lvs somewhat different, thin, 2–3-pinnate; lf-margin flat.

For *Anogramma leptophylla* (L.) Link see p. xvi.

10. HYPOLEPIDACEAE

Rhizomes covered with hairs. Petioles with several vascular strands which fuse into a single U-shaped strand. Sori marginal, covered by the down-curved lf-margin but also with an inner indusial flap; spores tetrahedral.

1. PTERIDIUM Scop.

Rhizome far-creeping, hairy but not scaly, with short ascending branches. Sori continuous round the margin of the lf-segment.

1. P. aquilinum (L.) Kuhn   Bracken.

Lvs (15–)30–180(–400) cm., (2–)3-pinnate, dying in autumn; petiole up to 2 m. and c. 1 cm. diam., erect, dark and hairy at base; blade bent towards the horizontal, ± deltate in outline, scaly and hairy when young; segments 5–15 mm., oblong, sessile, broad-based, entire or lobed at base, rather thick. Spores ripe 7–8. Woods, heaths, moorlands, etc., mainly on light acid soils and often dominant in such places. Common throughout the British Is.

11. HYMENOPHYLLACEAE

Rhizome creeping, wiry. Lvs thin and translucent, their ultimate segments 1-nerved. Sori marginal at the vein-endings. Indusium ± cuplike, 2-lipped or 2-valved, surrounding the base or the whole of the sorus. Sporangia all alike, developing in regular succession from apex to base of the sorus.

Lvs mostly over 10 cm.; rhizome over 1 mm. diam.; pinnae 1–2-pinnatisect; indusium not 2-valved; receptacle projecting as a long bristle. Very rare and local in Ireland; nearly extinct in Gt Britain. Killarney Fern.   *Trichomanes speciosum* Willd.
Lvs rarely as much as 10 cm.; rhizome not more than 1 mm. diam.; pinnae irregularly dichotomously divided; indusium of 2 valves; receptacle not projecting.   1. HYMENOPHYLLUM

1. HYMENOPHYLLUM Sm.

Indusium of 2 valves; receptacle not projecting.

Valves of indusium orbicular, toothed; lf ± flat.   **1. tunbrigense**
Valves of indusium ovate, entire; pinnae bent back from the rhachis.
                                                          2. wilsonii

1. H. tunbrigense (L.) Sm.   Tunbridge Filmy Fern.

Habit moss-like. Lvs 2·5–8(–12) cm., persistent for some years; blade ± oblong in outline; pinnae divided into oblong, remotely serrulate segments, with the veins ceasing slightly below the apex; rhachis winged; petiole ⅓–½ of the whole lf-length, wiry. Sori at the tips of the segments, mostly near the rhachis. Indusium flattened. Spores ripe 6–7. Rocks, tree trunks, etc., in humid places. Locally abundant in the west, north to W. Inverness and Mull, absent from much of the east and from the Midlands.

2. H. wilsonii Hooker   Wilson's Filmy Fern.

Like *H. tunbrigense* but pinnae bent back from rhachis, with fewer and more unilateral segments; veins reaching apex of segments; indusium not flattened. Spores ripe 6–7. In similar places to *H. tunbrigense* but commoner and extending to Shetland but absent from S.E. England.

12. THELYPTERIDACEAE

Rhizomes with hairs or scales. Petioles with 3–7 vascular strands; rhachis uninterruptedly grooved above (cf. Aspidiaceae). Sori submarginal; spores lens-shaped.

1. THELYPTERIS Schmidel

Rhizomes creeping or ascending. Lvs pinnate with pinnatifid pinnae, hairy at least beneath, veins free. Sori circular or oblong, small, close to the margin of the lf-segment; indusium reniform or 0.

1 Lvs bent at junction of rhachis and petiole; lowest pair of pinnae directed forward and downwards; remaining pinnae broad at base, hairy on both surfaces.   **3. phegopteris**
   Lvs not bent at junction of rhachis and petiole; lowest pair of pinnae not deflexed; pinnae with a narrow attachment to rhachis, subglabrous or with a few hairs beneath.   *2*

2 Lvs in a crown at the summit of the short stout ascending rhizome, with numerous sessile glands beneath and lemon-scented when crushed.   **1. limbosperma**
   Lvs solitary from a far-creeping slender rhizome, neither glandular beneath nor lemon-scented when crushed.   **2. palustris**

1. T. limbosperma (All.) H.P. Fuchs (*T. oreopteris* (Ehrh.) Slosson) Mountain Fern.

Lvs 30–100 cm., dying in autumn, lanceolate or oblanceolate, firm, yellowish-green; upper side of rhachis yellowish-green; pinnae deeply pinnatifid, scarcely stalked, longest about the middle of lf, decreasing downwards, the lower spaced out, short (c. 1 cm.); segments oblong, ± obtuse, sinuate-crenate or subentire; lower surface covered with yellowish glands. Sori small and close to margin of segment, which is flat; indusium irregularly toothed, falling early, or 0. Spores ripe 7–8. Woods, banks, etc., especially in mountain districts, not on limestone; common in the west and north, very local elsewhere.

2. T. palustris Schott   Marsh Fern.

Lvs 15–120 cm., sterile often smaller and with broader segments than the fertile, dying in autumn, oblong-lanceolate, light green; pinnae deeply pinnatifid, shortly stalked, longest about the middle of lf, decreasing only slightly downwards; segments oblong or triangular-lanceolate, sinuate, rarely with small lobes at base. Sori close to the strongly reflexed part of the segment-margin, small and ± confluent;

indusium small, irregularly toothed, often falling early. Spores ripe 7–8. Marshes, fens and fen-woods; locally abundant in England and Ireland, rare in Scotland.

3. T. phegopteris (L.) Slosson   Beech Fern.

Rhizome long, creeping. Lvs 10–50 cm., solitary, dying in autumn, 1-pinnate, triangular-ovate; blade almost at right angles to the petiole; pinnae deeply pinnatifid, hairy on both surfaces, the lowest pair directed forward and downwards and with a narrow attachment to the rhachis, the remainder decreasing rapidly in length and having broad bases decurrent on the rhachis. Sori small, close to the margins of the segments; indusium 0. Spores ripe 6–8. Damp woods and shady rocks, not on limestone; rather common in the north, rare in the south.

13. ASPLENIACEAE

Rhizome with firm scales showing under the hand-lens a heavy network of dark cell-walls. Petioles with 2 vascular strands which often fuse to form a single X-shaped strand. Sori borne along one or both sides of veins on the underside of the lf; spores lens-shaped.

1 Lvs densely covered beneath with brown overlapping scales which
     conceal the lf-surface; small plant with deeply pinnatifid lvs.
                                                   2. CETERACH
     Lvs not or sparsely scaly beneath.   2
2 Lvs entire or slightly lobed.   3. PHYLLITIS
     Lvs pinnately lobed or divided or forked irregularly.
                                                   1. ASPLENIUM

1. ASPLENIUM L.

Lvs in British spp. with free veins. Sori elliptical to linear, borne directly on a lateral vein, solitary, not paired; indusium a flap shaped like the sorus, usually opening towards the middle of the lf-segment.

1 Lvs irregularly dichotomous, forked into very narrow linear
     segments. Crevices of hard rocks or walls, not on limestone;
     rare. Forked Spleenwort.   *A. septentrionale* (L.) Hoffm.
     Lvs 1–3-pinnate, segments not linear.   2
2 Lvs simply pinnate; pinnae toothed or rarely lobed.   3
     Lvs 2–3-pinnate.   6
3 Pinnae roundish, ovate or oblong, with a distinct midrib.   4
     Pinnae narrowly wedge-shaped, without a distinct midrib. With the
     parent spp.; rare. *A. septentrionale × trichomanes*

4 Rhachis with a green wing; pinnae 1 cm. or more.   **2. marinum**
   Rhachis not winged or with a narrow brownish wing; pinnae less
   than 1 cm.   *5*

5 Rhachis black.   **3. trichomanes**
   Rhachis green.   **4. viride**

6 Basal pair of pinnae longest.   *7*
   Basal pair of pinnae shorter than the rest. Rocks and walls, usually
   near the sea. Lanceolate Spleenwort.
                          A. billotii F. W. Schultz (*A. obovatum* auct.)

7 Blade with petiole 3–12(–15) cm., dark dull green except for the
   blackish base of the petiole; lf-segments with no distinct midrib.
                                                          5. ruta-muraria
   Blade with petiole 10–50 cm., petiole dark purplish-brown; lf-segments with ± distinct midrib.   *8*

8 Blades of lvs and pinnae triangular in outline, straight-sided; segments ovate to lanceolate with acute teeth.
                                                        1. adiantum-nigrum
   Blades of lvs and pinnae concave-sided above with a long tapering
   point; segments narrowly lanceolate with long acuminate teeth.
   S.W. England and Ireland, local. Narrow Black Spleenwort.
                                                          A. onopteris L.

1. A. adiantum-nigrum L.   Black Spleenwort.

Lvs 10–50 cm., tufted, persistent, tough, triangular in outline, naked,
bright-green; pinnae triangular, stalked, straight, decreasing in size
upwards; pinnules lobed or pinnate; segments variable in shape, narrowed at base, serrate, teeth acute. Sori 1–2 mm., narrow, nearer the
indistinct midrib than the margin of the segment but occupying the
greater part of the segment when old. Spores ripe 6–10. Rocky woods,
shady walls and hedge-banks, common except in the drier parts.

2. A. marinum L.   Sea Spleenwort.

Lvs 6–30(–100) cm., tufted, persistent, tough, lanceolate in outline,
naked, dark green; pinnae oblong, obtuse, markedly asymmetric at
base; apex of lf often with a long lobed or pinnatifid point. Sori 3–5
mm., narrow, about midway between midrib and margin of pinna.
Spores ripe 6–9. Rock crevices near the sea, S. and W. coasts from the
Isle of Wight to Shetland and E. coast south to N. Yorks.

3. A. trichomanes L.   Maidenhair Spleenwort.

Lvs 4–20(–40) cm., tufted, persistent, linear, deep green; pinnae oval or
oblong, obtuse, somewhat asymmetric at base, crenate or ± toothed

near the top, finally falling from the blackish rhachis, ± the same size for some distance in the middle of the lf. Sori 1–2 mm., narrowly oblong, about midway between midrib and margin. Spores ripe 5–10. Walls and rock crevices, widely distributed.

4. A. viride Hudson   Green Spleenwort.

Differs from *A trichomanes* in having the rhachis green, not winged; pinnae roundish, more deeply toothed all round, not falling from the rhachis; sori nearer the midrib than the margin. Spores ripe 6–9. Crevices of basic rocks, local and mainly in the north and west.

5. A. ruta-muraria L.   Wall-Rue.

Lvs 3–12(–15) cm., tufted, persistent, tough, dark dull green except for the blackish petiole-base, triangular-ovate or triangular-lanceolate in outline; pinnae stalked, the largest at the base; pinnules rarely more than 5 and often 3 on each pinna, usually undivided, variable in shape but apex obtuse and base cuneate, crenate or dentate above the middle. Sori finally confluent. Spores ripe 6–10. Walls and rocks, widely distributed but local in the drier parts.

2. CETERACH DC.

Like *Asplenium* but lvs densely scaly beneath; indusium 0 or rudimentary.

1. C. officinarum DC.   Rusty-back Fern.

Lvs 3–8 cm., tufted, persistent, tough; pinnae ovate or oblong, rounded at apex, somewhat widened at the broad base, entire or crenate. Spores ripe 4–10. Mortared walls and crevices of limestone rocks, very local except in S. and W. England and Wales.

3. PHYLLITIS Hill

Rhizome short, scales many, firm. Lvs entire or slightly lobed. Sori borne in pairs direct on the vein, each pair appearing like a single sorus.

1. P. scolopendrium (L.) Newman   Hart's-tongue Fern.
Lvs 10–60 cm., tufted, persistent, narrow-oblong, cordate at base, tapering to the usually blunt apex, with scattered scales when young; veins dichotomous, free. Sori linear, usually occupying more than half the width of the lf. Spores ripe 7–8. Rocky woods, hedge-banks, etc., widely distributed but uncommon in the drier parts and in N. Scotland.

14. ATHYRIACEAE

Rhizomes with thin-walled opaque scales. Petioles with 2 vascular strands which unite distally. Sori on the underside of the lf, on a side-branch from the vein; indusium variable in form or 0; spores lens-shaped.

1 Indusium rudimentary and falling early or 0; Scottish mountains.
              1. ATHYRIUM
 Indusium distinct.   *2*

2 Sori circular; indusium of hair-like scales surrounding the base of the sorus; small and very rare mountain ferns. **3. WOODSIA**
 Sori circular or oblong; indusium not of hair-like scales. *3*

3 Sori oblong; indusium flap-like, oblong (the lower reniform or hooked), attached laterally.   **1. ATHYRIUM**
 Sori circular; indusium acuminate, attached on the inner side of the sorus and covering it at first like a hood but later becoming shrivelled.   **2. CYSTOPTERIS**

1. ATHYRIUM Roth

Rhizome short, with many large soft acuminate scales. Lvs in a crown, 1–3-pinnate, broadly lanceolate, with free veins. Sori oblong or circular; indusium either rudimentary to absent or flap-like and opening towards the middle of the pinnule.

 Sori oblong; indusium conspicuous at maturity; spores minutely papillose, not winged.   **1. filix-femina**
 Sori circular; indusium 0 at maturity; spores reticulate, winged. Screes and rocky outcrops on Scottish mountains. Mountain Lady-fern. *A. distentifolium* Opiz (*A. alpestre* (Hoppe) Rylands)

1. A. filix-femina (L.) Roth   Lady-fern.

Lvs 20–100(–150) cm., crowded, dying in autumn, usually spreading, 2(–3)-pinnate; pinnae linear-lanceolate, acute, the longest about the middle of the lf, rhachis winged; pinnules oblong, sessile, truncate at the narrow base, lobed or pinnatifid, lobes often toothed; rhachis of lf green or purplish, sometimes with scattered scales or hairs; petiole scaly, at least below. Sori in a row down either side of the pinnule; the indusium of the lowest sorus in each row reniform or ± strongly hooked, the uppermost oblong or linear, nearly straight, all persistent and covering the sori until near maturity. Spores ripe 7–8. Shady places on acid soils, widespread and common, but local in E. England.

2. CYSTOPTERIS Bernh.

Differs from *Athyrium* mainly in the vaulted hood-like indusium which becomes reflexed and exposes the sporangia as they ripen.

1   Rhizome long, creeping, lvs solitary; basal pinnae longest. Mountains; rare. Mountain Bladder-fern.   *C. montana* (Lam.) Desv.
  Rhizome short, lvs tufted; basal pinnae not the longest.   *2*
2   Pinnules overlapping; spores rugose. Kincardine; very rare.
                                                    C. dickieana R. Sim
  Most pinnules not overlapping; spores with narrow acute tubercles.
                                                        1. fragilis

1. C. fragilis (L.) Bernh.   Brittle Bladder-fern.

Lvs 5–35(–45) cm., dying in autumn, 2(–3)-pinnate, lanceolate in outline; pinnae ovate to lanceolate, longest about the middle of the lf; pinnules variable in shape and toothing, cuneate at the narrow base; petiole fragile, dark brown and scaly at base, paler and nearly naked above. Sori in 2 rows on either side of the midrib of the pinnule; indusium pale, inflated and glistening when young. Spores ripe 7–8. Rocks, walls, etc., common in the north and west, local in the south and east.

3. WOODSIA R.Br.

Two very rare mountain spp. in Britain, distinguished as follows:

Largest pinnae oblong or ovate-oblong, 1½–2 times as long as broad; lf densely covered with long (2–3 mm.) scales beneath.
                                                    W. ilvensis (L.) R.Br.
Largest pinnae triangular-ovate, 1–1½ times as long as broad; lf with sparse short (c. 1 mm.) scales beneath, at least on the rhachis.
  Alpine Woodsia.   *W. alpina* (Bolton) S. F. Gray

15. ASPIDIACEAE

Rhizomes with thin-walled opaque scales. Petioles with 4–7 small vascular strands; rhachis grooved above, the groove interrupted at the attachments of pinnae. Sori usually circular, on or terminating a vein on the underside of the lf; indusium peltate, reniform or 0; spores lens-shaped.

1   Indusium 0; rhizome slender, far-creeping.   3. GYMNOCARPIUM
  Indusium peltate or reniform; rhizome stout.   *2*
2   Indusium peltate.   2. POLYSTICHUM
  Indusium reniform.   1. DRYOPTERIS

1. DRYOPTERIS Adanson

Rhizome short, erect or ascending, with many large and soft scales. Lvs in a crown, 1–4-pinnate, ±scaly but not hairy. Sori large; indusium reniform.

1 Lvs pinnate with deeply pinnatifid pinnae, or 2-pinnate; lobes of
    pinnae, or pinnules, not narrowed at base.   2
    Lvs 2-pinnate with pinnatifid pinnules, or 3–4-pinnate; pinnules
    strongly narrowed at base or stalked.   3

2 Pinna-lobes or pinnules 20–35 on each side, not more than 5 mm.
    broad, teeth not mucronate; fertile lvs similar to sterile.
                                   1. filix-mas group
    Pinna-lobes or pinnules 10–20 on each side, 7 mm. or more broad,
    teeth usually mucronate; fertile lvs longer and more erect than
    sterile. Wet heaths and marshes northwards to Renfrew; very
    local and rare. Crested Buckler-fern.   *D. cristata* (L.) A. Gray

3 Basal pair of pinnules on the lowest pinna very unequal; lvs
    3-pinnate or nearly so.   4
    Basal pair of pinnae on the lowest pinna nearly equal; lvs 2-pinnate
    with pinnatifid pinnules, glandular on both surfaces. Clefts of
    limestone pavement in N. England and N. Wales; very local. Rigid
    Buckler-fern.   *D. villarii* (Bellardi) Woynar

4 Scales on petioles pale with a dark centre; glands on the indusium
    stalked; pinnules flat or convex.   **3. dilatata** group
    Scales on petioles uniformly pale; glands on the indusium sessile or
    0; pinnules flat or concave.   5

5 Lf ±lanceolate in outline, eglandular; petiole pale above; scales
    ovate or ovate-lanceolate; indusium entire, eglandular.
                                       2. carthusiana
    Lf triangular in outline, minutely glandular; petiole uniformly dark
    brown; scales narrow-lanceolate; indusium toothed, margin
    glandular. Shady places; local and chiefly in Ireland and the
    western half of Gt. Britain. Hay-scented Buckler-fern.
                             D. aemula (Aiton) O. Kuntze

1. D. filix-mas group   Male Fern.

Lvs 15–180 cm., tufted, dying in autumn or lasting till spring, pinnate with deeply pinnatifid pinnae or 2-pinnate, the larger ones ovate to lanceolate; pinnae 20–30 on each side, decreasing in size towards top and bottom of lf; pinnules (except the lowest pair of each pinna) attached at the base by their whole width and often somewhat widened, obtuse or nearly truncate, ±toothed or rarely lobed; petiole with ±numerous pale brown or orange-brown scales. Sori c. 0·5–2 mm. diam., in a row down each side of the pinnule; indusium entire. Spores ripe 7–8. Woods, hedge-banks, rocks and scrub; common and widely distributed.

Includes three species between which hybrids sometimes occur. They may be distinguished as follows:

1 Petiole and rhachis densely covered with orange-brown scales; pinnules toothed mainly at apex; pinnae blackish above at junction with rhachis. Common, particularly in the western half of the country, mainly on acid soils.   **D. borreri** Newm.
 Petiole and rhachis rather sparsely covered with pale brown scales; pinnules toothed all round or more strongly on the margins than at the apex; pinnae not blackish at junction with rhachis.   *2*

2 Lvs usually over 50 cm.; rhizome generally unbranched; pinnules with sharp teeth; sori usually 4–5 on each side of the midrib of the large pinnules. Common.   **D. filix-mas** (L.) Schott
 Lvs usually 30–50 cm.; rhizome generally branched; pinnules with blunt teeth; sori rarely more than 3 on any pinnule, often only 1. In mountain districts, particularly on screes, local.
                                       D. abbreviata (DC.) Newm.

2. **D. carthusiana** (Vill.) H. P. Fuchs (*D. spinulosa* Watt)
                                       Narrow Buckler-fern.

Lvs 30–150 cm., dying in autumn, the base of the petiole decaying first, 2-pinnate with deeply pinnatifid pinnules or 3-pinnate, light or yellow-green, lanceolate or ovate-lanceolate, with a few scales on the rhachis; pinnae triangular-ovate or triangular-lanceolate; pinnules of the lowest pinna deeply pinnatifid or pinnate, narrowed at base; segments with incurved teeth; petiole dark brown at base, with uniformly pale brown scales. Sori c. 0·5–1 mm.; indusium entire or sinuate, eglandular. Spores ripe 7–9. Wet woods, marshes and wet heaths; common in suitable habitats.

3. **D. dilatata** group

Lvs 30–150 cm., dying in autumn, 3-pinnate, the ultimate pinnules narrow-based, pinnately lobed, with mucronate or awned teeth curving towards the apex of the pinnule; petiole-scales dark-centred. Sori 0·5–1 mm. diam.; indusium fringed with stalked glands and usually irregularly toothed.

Includes two species which may be distinguished thus:

1 Lowest basally-directed pinnule of the basal pinna usually less than half as long as the pinna; lf dark green; spores dark brown with blunt tubercles. Woods, hedge-banks, moorlands, mountains, etc.; common and widespread. Broad Buckler-fern.

D. dilatata (Hoffm.) A. Gray

2 Lowest basally-directed pinnule of the basal pinna at least half as long as the pinna; lf pale green; spores pale brown with acute tubercles. Recently recognized on Ben Lawers and probably on other mountains in Scotland and N. England.

D. assimilis S. Walker

2. POLYSTICHUM Roth

Like Dryopteris but lvs usually leathery, pinnules asymmetrical, teeth ending in a bristle-like point; indusium peltate, ± circular.

1 Lvs 2-pinnate (sometimes only the lowest pinnae pinnate).   2
Lvs 1-pinnate; basic rocks on mountains, very local. Holly Fern.

P. lonchitis (L.) Roth

2 Lvs soft; most pinnules of lower pinnae shortly stalked, the margins meeting at an obtuse angle at base.   **1. setiferum**
Lvs stiff and somewhat leathery; most pinnules of lower pinnae sessile, though markedly narrowed at base, the, margins meeting at an acute angle.   **2. aculeatum**

1. P. setiferum (Forskål) Woynar   Soft Shield-fern.

Lvs 30–150 cm., ± persistent, usually arching or drooping, lanceolate, scaly on the rhachis; pinnae up to 40 on either side, lanceolate, fully pinnate, longest about the middle of the lf; pinnules ovate-oblong, sharply toothed with a rounded lobe on the upper side near the base, the majority short-stalked with the blade not reaching the rhachis. Sori c. 0·5–1 mm., in a row on either side of the midrib of the pinnule and of its basal lobe; vein on which the sorus is borne not or only shortly continued beyond it. Spores ripe 7–8. Woods, hedge-banks, etc.; common in the west and south, local in the north and east.

2. P. aculeatum (L.) Roth   Hard Shield-fern.

Like *P. setiferum* but lvs stiff, rather leathery, somewhat glossy; pinnae sometimes pinnately lobed with only a single free pinnule, sometimes fully pinnate; most pinnules sessile though markedly narrowed at base. Vein on which the sorus is borne continuing well beyond the sorus. Spores ripe 7–8. Woods, hedge-banks; rather common in the wetter districts, local in the drier ones.

3. GYMNOCARPIUM Newman

Rhizomes slender, creeping, with pale broadly ovate fringed scales on the younger part only. Lvs not in a crown, solitary, erect; blade shorter than the petiole, triangular in outline, 2–3-pinnate; veins free. Lowest pair of pinnae distinctly jointed to the rhachis. Sori circular or somewhat elongated, submarginal, sometimes confluent; indusium 0.

Lvs glabrous, eglandular, bright green.   **1. dryopteris**
Lvs with short glandular hairs, dull green, rolled up when young to form
    a single ball; each of the lowest pair of pinnae distinctly smaller than
    the rest of the blade. On limestone rocks and screes; local. Limestone
    Polypody.
            G. robertianum (Hoffm.) Newman (*Thelypteris robertiana*
                                                (Hoffm.) Slosson)

1. G. dryopteris (L.) Newman (*Thelypteris dryopteris* (L.) Slosson)
                                                          Oak Fern.

Lvs 10–40 cm., rolled up when young to form 3 small balls; petiole erect, slender, brittle, 1½–3 times as long as the horizontal blade which is bright yellowish-green and glabrous, 3-pinnate below but pinnatifid above, with each of the lowest pair of pinnae often as large as the rest of the blade. Sori circular or the lowest somewhat elongated. Spores ripe 7–8. Damp woods, shaded rocks and stream-banks; frequent in the north, rare in the south.

16. BLECHNACEAE

Rhizomes with firm scales showing under the hand-lens a network of thickened cell-walls. Petioles with 2 vascular strands. Sori free or confluent; indusium flap-like; spores lens-shaped.

1. BLECHNUM L.

Rhizome short, almost erect, densely scaly. Lvs pinnate, fertile and sterile dissimilar; veins of sterile lf free. Sori joined into a continuous line parallel to the pinna-margin; indusium linear, opening towards the midrib of the pinna.

1. B. spicant (L.) Roth   Hard-fern.

Outer lvs sterile and spreading, inner fertile and suberect. Sterile lvs up to 50 cm., narrowly lanceolate; pinnae numerous, tough, entire or nearly so, longest about the middle of the blade, narrow-oblong, somewhat curved, widening at base, the lowest short and often broader than long; petiole dark brown, scaly at base; rhachis green. Fertile lvs up to 75 cm.; pinnae very narrow; sporangia when ripe appearing to cover the whole under-surface of the pinnae. Spores ripe 6–8. Woods, heaths, mountain grassland, etc., on acid soils. Generally distributed but uncommon in lowland districts.

17. POLYPODIACEAE

Rhizomes with thin-walled scales. Lvs 2-ranked on the upper side of the rhizome and jointed to it; petioles with 1–3 main vascular strands. Sori on the lower surface of the blade; indusium 0; spores lens-shaped.

1. POLYPODIUM L.

Rhizomes creeping. Lvs pinnatifid or pinnate, without scales; pinnae entire or toothed. Sori circular or elliptical.

P. vulgare group   Polypody.

Rhizome creeping on or below the surface, densely scaly. Lvs solitary, persistent, almost or quite pinnate; petiole ± erect; blade 5–45 cm., ovate-lanceolate to narrowly oblong in outline. Sori in 1 row on either side of the pinna-midrib, about halfway between midrib and margin, large, often bright yellow. Rocks, walls or on the ground; often epiphytic on trees; common throughout the British Is but especially so in wetter districts.

It has been shown recently that this is a complex of three species and their hybrids. The three species may be distinguished thus:

1   Sori circular when young; lf-blade oblong, with pinnae equally long for a considerable distance round the middle of the blade, or narrowly lanceolate; pinnae subentire; longest pinnae usually 15–35 mm.; basal pair of pinnae not projecting forward; annulus of sporangia with 10–15 thick-walled cells. Common throughout the British Is.   **1. P. vulgare** L.
   Sori oval when young; lf-blade triangular-ovate to ovate-lanceolate; pinnae subentire to deeply toothed; longest pinnae usually 40–70 mm.; basal pair of pinnae projecting forward; annulus of sporangia with 4–11 thick-walled cells.   2

2   Rhizome-scales 3–6 mm.; lf-blade ovate to ovate-lanceolate; pinnae ± acute at apex, subentire to deeply toothed; annulus with 6–11 thick-walled cells. Throughout the British Is.
                                          2. P. interjectum Shivas

Rhizome-scales 5–11 mm.; lf-blade triangular-ovate to ovate; pinnae narrow, acute, usually toothed; annulus with 4–7 thick-walled cells. Very local on limestone in S.W. England.

3. P. australe Fée

Branched paraphyses accompany the sporangia constantly in *P. australe*, rarely in *P. interjectum*. The three species differ in chromosome number.

18. MARSILEACEAE

Perennial herbs with creeping hairy rhizome, ± aquatic. Lvs alternate, in 2 rows, circinate when young, subulate and entire or with 2 or 4 palmately arranged lflets. Sporangia in globose, hard, hairy sporocarps at base of petiole. Sporangia of 2 kinds, both occurring in the same sorus.

1. PILULARIA L.

Lvs subulate. Sporocarp solitary, divided into 2 or 4 compartments each containing one sorus.

1. P. globulifera L.   Pillwort.

Rhizome up to 50 cm., slender, often with short branches. Lvs 3–15 cm. Sporocarps c. 3 mm. diam., shortly stalked, brown at maturity. Sori 4. Spores ripe 6–9. Edges of ponds, etc., on acid soils, often submerged; local, and absent from many counties.

19. AZOLLACEAE

Small free-floating plants with branched stems bearing roots and lvs. Lvs 2-ranked, imbricate, 2-lobed, the upper lobe floating, the lower thin, submerged and bearing the sori in pairs. Sori enclosed in an indusium from the base and comprising either numerous microsporangia or a single megasporangium.

1. AZOLLA Lam.

The only genus.

*1. A. filiculoides Lam.

Plant 1–5 cm. diam., often growing in large masses, bluish-green, usually becoming red in autumn. Upper lf-lobes c. 1 mm., ovate, obtuse, with hyaline margins; covered with unicellular hairs making the surface unwettable. Sori of a pair either both with megasporangia or one of each kind. Introduced and naturalized in ditches, etc., in many places, mainly in S. England.

GYMNOSPERMAE

CONIFEROPSIDA

20. PINACEAE

Trees, rarely shrubs; branches usually whorled. Lvs spirally arranged, linear, entire or very finely toothed, usually evergreen. Male and female fls in separate cones formed of numerous spirally arranged scales. Cones usually large and woody in fr.; seeds winged.

1   Lvs all solitary on the main stems; short shoots 0.   2
    Green lvs all or mainly on the short lateral shoots.   5

2   Lvs without persistent woody bases, the scars not or scarcely pro-
    jecting after the lvs have fallen.   3
    Lvs with persistent woody bases which project as pegs after the lvs
    have fallen.   4

3   Cones erect, their scales falling with the seeds; lf-scars quite flat; lvs
    rigid; buds ovoid, blunt. Several spp. planted in parks and occa-
    sionally for forestry. Silver Firs.   *Abies* Miller
    Cones pendulous, their scales persistent; lf-scars slightly projecting;
    lvs rather soft; buds fusiform, acute. Rather frequently planted.
    Douglas Fir   *Pseudotsuga menziesii* (Mirb.) Franco

4   Main branches all in whorls; lvs without a stalk above the persistent
    base, not 2-ranked on the upper side of the shoot; cones usually
    large.   1. PICEA
    Main branches not all in whorls; lvs with a distinct stalk above the
    persistent base, appearing 2-ranked on the upper side of the
    shoot; cones small. Planted. Western Hemlock.
    Tsuga heterophylla (Rafin.) Sargent

5   Short shoots with numerous deciduous lvs, similar to the lvs on the
    long shoots.   2. LARIX
    Short shoots with 2(–5) evergreen lvs; lvs on long shoots brownish
    and scale-like.   3. PINUS

1. PICEA A. Dietr.

Bark scaly. Twigs covered with lf-cushions which are separated by grooves and have peg-like projections on which the lvs are inserted and which remain after they fall. Lvs 4-angled or flat, not in 2 lateral rows when seen from above. Introduced. The two following are the most commonly planted:

Lvs 4-sided with inconspicuous lines on each side; buds not resinous.
  Norway Spruce.   *P. abies* (L.) Karsten
Lvs flat, dark green above, with conspicuous glaucous lines beneath;
  buds resinous. Sitka Spruce.   *P. sitchensis* (Bong.) Carrière

2. Larix Miller

Deciduous. Bark scaly. Long shoots with spirally arranged lvs; short
lateral shoots growing very slowly and producing numerous lvs at the
apex each year. Lvs flat. Cones erect, often brightly coloured in fl.
Introduced. The two following are commonly planted:

Twigs yellowish, not pruinose; lvs bright green, without white bands
  beneath; ripe cone scales erect. European Larch.   *L. decidua* Miller
Twigs reddish, pruinose; lvs glaucous, with 2 conspicuous white bands
  beneath; ripe cone scales curved back near the tip. Japanese Larch.
    L. kaempferi (Lamb.) Carrière (*L. leptolepis* (Sieber & Zucc.) Endl.)

3. Pinus L.

Bark furrowed or scaly. Long shoots with spirally arranged brownish
scale lvs; short lateral shoots bearing a definite number (usually 2, 3 or
5) green, needle-like lvs surrounded at the base by brown scale lvs, not
growing further and finally falling entire. Male cones lateral, clustered
at the base of a year's growth; female cones terminal.

1   Lvs glaucous, less than 10 cm.; bark of upper part of trunk bright
      reddish-brown.   **1. sylvestris**
    Lvs dark green, mostly more than 10 cm.; bark not bright reddish-
      brown.   2
2   Buds resinous, with appressed scales; lvs not more than 2 mm.
      broad; cone 5–8 cm. Commonly planted. Corsican Pine.
                                                        P. nigra Arnold
    Buds not resinous, the tips of the scales strongly recurved; lvs more
      than 2 mm. broad; cone 9–18 cm. Locally naturalized near the
      sea in S. England. Maritime Pine.   *P. pinaster* Aiton

1. P. sylvestris L.   Scots Pine.
Old trees 30(–50) m., usually with a flat crown. Buds resinous, upper
scales free but not reflexed at tips. Lvs 3–8(–10) cm. × 1–2 mm., in pairs
on the short shoots, glaucous, stiff, finely toothed. Cones 3–7 cm.,
ovoid-conic, ± symmetrical. Fl. 5–6.

Ssp. **scotica** (Schott) E. F. Warburg, with lvs and cones c. 4 cm., the tree remaining pyramidal for a long time and finally developing a round crown, is common in the Highlands. Trees of varied origin, usually with larger needles and cones, are widely planted elsewhere.

21. CUPRESSACEAE

Evergreen trees or shrubs. Lvs opposite or in whorls of 3 or 4 (rarely spiral on sterile shoots), small, scale-like, appressed, on young plants (and in some spp. throughout their life) needle-like. Fls monoecious or sometimes dioecious, in small cones formed of opposite or whorled scales. Scales of female cones woody or rarely fleshy when ripe.

1. JUNIPERUS L.

Trees or shrubs. Bark thin, usually shed in long strips. Lvs in whorls of 3, needle-like or subulate in our sp. but commonly scale-like in mature plants of most spp. Female fls of 3–8 scales which become fleshy and coalesce forming a berry-like fr.

1. J. communis L.   Juniper

Shrub with very varied habit, rarely a small tree to 10 m. Bark reddish-brown. Lvs 5–19 mm., narrow-subulate, sessile, jointed at base, tip spiny, spreading or ascending, concave with a broad white band above, keeled and green beneath. Dioecious. Male cones c. 8 mm., solitary, cylindric. Female cones c. 2 mm. in fl., solitary. Fr. c. 5–6 mm., ripening in the 2nd or 3rd year, ± globose, blue-black, pruinose. Fl. 5–6. Fr. 9–10.

Ssp. **communis**. Erect or spreading. Lvs 8–19 × c. 1 mm., spreading almost at right angles, gradually tapering to a long prickly point. Fr. globose. The lowland form on chalk, limestone and slate; widespread but local.

Ssp. **nana** Syme. Prostrate. Lvs 4–10 × c. 1·5 mm., ascending or loosely appressed, rather suddenly narrowed to a scarcely prickly point. Fr. longer than broad. Rocks and moors on mountains, local.

22. TAXACEAE

Evergreen trees or shrubs. Lvs narrow, spirally inserted. Fls usually dioecious; male in small cones; female solitary or in pairs in the lf-axils. Fr. wholly or partly surrounded by a fleshy aril when ripe.

1. Taxus L.

Lvs ± spreading in 2 rows. Male cones axillary, head-like, stalked, surrounded by scales at base. Ovule solitary, with scales at base. Seed surrounded by a red fleshy cup-like aril, ripening the first year.

1. T. baccata L.   Yew.

Tree to 20 m. with massive trunk and usually rounded outline. Bark reddish-brown, thin, scaly, Twigs green. Lvs 1–3 cm., shortly stalked, mucronate, dark green above, paler beneath, midrib prominent on both sides, margins recurved. Seeds c. 6 mm., olive-brown, ellipsoid. Fl. 3–4. Fr. 8–9. Woods and scrub, mainly on chalk and limestone where it occasionally forms pure woods; often planted.

ANGIOSPERMAE

DICOTYLEDONES: ARCHICHLAMYDEAE

23. RANUNCULACEAE

Chiefly herbs; some woody climbers. Lvs usually spirally arranged and lacking stipules, often palmately lobed or dissected. Fls usually hermaphrodite and actinomorphic. Perianth undifferentiated or of ± distinct calyx and corolla, the latter of variously shaped free nectar-secreting organs ('nectaries', 'honey-leaves', 'petals'); stamens hypogynous, usually numerous, sometimes few; carpels 1 to numerous, usually free. Fr. various but usually 1 or more follicles or a cluster of achenes.

Key to Genera

1 Lvs opposite; woody climbers.   6. CLEMATIS
  Lvs spirally arranged or in 2 ranks (rarely ± opposite on creeping
    stems); not climbers.   2

2 Carpel 1.   3
  Carpels 2 or more, free or united below.   4

3 Fr. a berry; fls small, whitish, not spurred.   4. ACTAEA
  Fr. a follicle; fls blue, spurred. Casuals. Larkspurs.
                Consolida (*Delphinium*) spp.

4 Each carpel ripening to a 1-seeded achene.   5
  Each carpel ripening to a many-seeded follicle.   10

5 Petals equalling or exceeding the sepals.   6
  Petals shorter than the sepals or 0; per. segs often all petaloid.   8

6 Sepals 3; lvs simple.   7. *Ranunculus ficaria*
  Sepals 5 or more.   7

7 Petals red with no basal nectary; lvs bi- or tri-pinnate with linear
    segments. Introduced and naturalized as a cornfield weed in a
    few southern English counties. Pheasant's Eye.
                Adonis annua L.
  Petals yellow or white, each with a nectary near its base; lvs simple
    or palmately lobed or divided.   7. RANUNCULUS

8 Fl.-stems with a whorl of 3 lvs below the fl.; per. segs petaloid,
    conspicuous, not spurred.   5. ANEMONE
  Fl.-stem not with a whorl of 3 lvs; per. segs inconspicuous.   9

9   Fl. solitary, terminal on a lfless stem; sepals greenish, spurred; petals very narrow; achenes in a long slender spike.

                                                      8. Myosorus

     Fls not solitary, on lfy fl.-stems; petals 0; achenes in a ± globular cluster.   10. Thalictrum

10   Fls strongly zygomorphic with a large helmet-shaped hood. A tall perennial herb, 50–100 cm., with palmately divided lvs whose 3–5 lobes are wedge-shaped at base and deeply divided into narrow segments; fls bluish, with 2 nectar-secreting spurs included under the hood. Local on shady stream-banks in S.W. England. Monkshood. *Aconitum napellus* L. (*A. anglicum* Stapf)

     Fls actinomorphic.   *11*

11   Each of the 5 petals spurred.   9. Aquilegia

     Petals, if present, not spurred.   *12*

12   Fls with a single series of yellow per. segs.   1. Caltha

     Fls with both sepals and petals or 'honey-lvs'.   *13*

13   Petals represented by tubular nectaries.   *14*

     Petals narrow, not tubular; sepals yellow, numerous, forming a globular fl.; fr. a spherical cluster of many small follicles.

                                                 2. Trollius

14   Fl. solitary, with a green involucre of deeply-cut lvs; sepals and nectaries yellow. Naturalized. Winter Aconite.

                          Eranthis hyemalis (L.) Salisb.

     Fls not solitary and without an involucre; nectaries greenish.

                                           3. Helleborus

1. Caltha L.

Fls actinomorphic. Per. segs petaloid; petals 0; carpels few. Fr. a group of follicles.

1. C. palustris L.   Kingcup, Marsh Marigold.

Perennial herb with stout creeping rhizome and prostrate to erect glabrous aerial stems. Lvs chiefly basal, orbicular, reniform or triangular, cordate at the base, long-stalked; upper lvs subsessile; all ± crenate or toothed. Fls 16–50 mm. diam., bright golden yellow. Carpels 5–13, nectar-secreting. Fl. 3–7. Very variable in size and habit. In marshes, fens, ditches and wet woods. Common throughout British Is except Channel Is.

2. Trollius L.

Fls actinomorphic. Sepals yellow, petaloid; petals small, narrow; carpels numerous. Fr. a group of small many-seeded follicles.

1. T. europaeus L.   Globe Flower.

Perennial herb with a short erect woody stock and an erect glabrous leafy shoot, 10–60 cm. Basal lvs stalked, pentagonal in outline, palmately 3–5-lobed with the lobes deeply cut and toothed; stem-lvs ± sessile; all paler beneath. Fls 2·5–3 cm. diam., terminal, usually solitary, almost globose. Sepals about 10, pale or greenish yellow, strongly incurved; petals yellow, hidden by the sepals, nectar-secreting. Follicles c. 12 mm., transversely wrinkled. Fl. 6–8. Locally common in wet pastures in mountain districts northwards from S. Wales and Derby and in N.W. Ireland.

3. HELLEBORUS L.

Bear's-foot,
Hellebore.

Fls actinomorphic. Sepals green or petaloid, persistent; petals tubular, green, shorter than the sepals; carpels few, sessile, free or slightly joined below. Fr. a group of many-seeded follicles.

Basal lvs 0; uppermost stem-lvs (bracts) simple, entire; fls numerous; perianth almost globular.   **1. foetidus**
Basal lvs usually 2; uppermost stem-lvs digitately lobed, serrate; fls 2–4; perianth spreading, almost flat.   **2. viridis**

1. H. foetidus L.   Stinking Hellebore.

Perennial foetid herb with a stout ascending stock and a robust over-wintering lfy stem 20–80 cm., glabrous below, glandular above. No basal lvs. Lower stem-lvs evergreen, pedate with 3–9 narrow acute toothed segments, long-stalked, sheathing-based; middle stem-lvs with enlarged sheaths and reduced blades, transitional to the uppermost (bracts) which are entire or with a small vestigial blade. Fls 1–3 cm diam., numerous, drooping, campanulate or globular, in a branched cyme. Sepals yellowish-green usually bordered with reddish-purple. Nectaries 5–10. Carpels usually 3. Fl. 3–4. Local in woods and scrub on calcareous soil, or scree, probably native northwards to Lancashire and Northants, naturalized elsewhere.

2. H. viridis L. ssp. **occidentalis** (Reuter) Schiffner   Green Hellebore.

Perennial herb with a short ascending stock and an erect stem 20–40 cm., glabrous or sparsely hairy above, not overwintering. Basal lvs usually 2, arising with the fls and dying before winter, long-stalked, digitate-pedate with 7–11 narrowly elliptical acute toothed segments; stem-lvs smaller, sessile; all glabrous beneath. Fls 3–5 cm. diam., 2–4, half-drooping. Sepals spreading, pale green. Nectaries 9–12. Carpels 3. Fl. 3–4. Local in moist calcareous woods; probably native northwards to Westmorland and N. Yorks., naturalized elsewhere.

4. ACTAEA L.

Fls actinomorphic, small, in short racemes. Sepals petaloid, deciduous; petals small or 0; carpel 1. Fr. a berry with numerous seeds.

1. A. spicata L.   Baneberry, Herb Christopher.
Perennial foetid herb with stout ascending rhizome and erect glabrous stem, 30–65 cm. Basal lvs large, long-stalked, biternate or bipinnate, the segments ovate, acute, often 3-lobed, finely toothed; stem-lvs much smaller. Infl. 25–30 cm., terminal, dense-fld, elongating in fr. Sepals usually 4, whitish; petals 4–6 or 0, white, short, not nectar-secreting; stamens white, club-shaped. Berry c. 1 cm. diam., at first green, then blackish. Fl. 5–6. Local in ashwoods and on limestone pavement in Yorks., Lancashire and Westmorland.

5. ANEMONE L.

Stem-lvs, in a whorl beneath the usually solitary terminal actinomorphic fl. Per. segs 4–20, petaloid; outermost stamens sometime sterile and nectar-secreting; carpels numerous. Fr. a cluster of 1-seeded achenes.

Fl. bell-shaped, violet; achenes plumed.   **2. pulsatilla**
Fl. not bell-shaped, white or pinkish; achenes not plumed.
                                                          1. nemorosa

1. A. nemorosa L.   Wood Anemone.
Perennial herb with slender brown rhizome and erect glabrous or sparsely hairy flowering stems, 6–30 cm. Basal lvs 1–2, borne on the rhizome, long-stalked, palmately 3-lobed, the lobes cut into cuneate coarsely-toothed acute segments; stem-lvs stalked, like the basal lvs, borne about ⅔ up from the base of the flowering stem. Fl. 2–4 cm. diam., solitary. Per. segs usually 6 or 7, white, pinkish or reddish-purple. Achenes in a globular cluster, downy. Fl. 3–5. Abundant, chiefly in deciduous woodland, throughout the British Is.

2. A. pulsatilla L. (*Pulsatilla vulgaris* Miller)   Pasque Flower.
Perennial herb with erect stock and erect flowering stem, 10–30 cm. Basal lvs in a rosette, long-stalked, bipinnate, the lflets pinnately cut

into linear segments; stem-lvs sessile, ± erect, deeply divided into long linear segments; all hairy, the basal woolly, stalked. Fl. solitary, bell-shaped. Per. segs 6, violet, paler and silky on the outside; outer stami-nodes nectar-secreting. Achenes silky, the styles persisting as long silky plumes. Fl. 4–5. Very local on dry calcareous grassy slopes from Gloucester and Essex to N. Lincs.

6. CLEMATIS L.

Herbs, shrubs or woody climbers with opposite lvs which often end in tendrils or have a winding stalk and rhachis. Fls actinomorphic. Per. segs usually 4, valvate, ± petaloid; outer stamens usually sterile; carpels numerous. Fr. a cluster of achenes with persistent long plumose styles.

1. C. vitalba L.   Traveller's Joy, Old Man's Beard.
Perennial woody climber with stems up to 30 m. Lvs pinnate, usually with 5 rather distant lflets; lflets narrowly ovate, acute, rounded at the base, entire or coarsely toothed, ± glabrous. Fls 2 cm. diam., in termi-nal and axillary panicles, fragrant. Per. segs greenish-white, downy outside. Achenes downy. Fl. 7–8. In hedgerows, thickets and wood-margins chiefly on calcareous soils southwards from Denbigh, Stafford and S. Yorks. Naturalized elsewhere.

7. RANUNCULUS L.

Fls actinomorphic. Sepals 3–5; petals 5 or more, rarely 0, usually yellow or white, each with a nectar-secreting depression near its base; carpels numerous. Fr. a head of achenes.

1	Aquatic plants with white fls; achenes strongly wrinkled trans-versely.
	Terrestrial or marsh plants with yellow fls; achenes not strongly wrinkled transversely.
2	Sepals 3; petals 7–12; lvs simple, cordate, **20. ficaria**
	Sepals 5; petals usually 5; lvs various.
3	Lvs palmately lobed or divided.
	Lvs simple, lanceolate or spathulate, entire or toothed.
4	Achenes smooth or with small tubercles.
	Achenes with spines or hooked hairs.
5	Sepals strongly reflexed during flowering.
	Sepals not strongly reflexed.
6	Stem tuberous (corm-like) at base, ± appressed-hairy above; achenes quite smooth, with hooked beak. **3. bulbosus**

1 ... 15
  ... 2
2 ... 3
3 ... 4
  ... 12
4 ... 5
  ... 11
5 ... 6
  ... 7

   Stem not tuberous, with spreading hairs; achenes usually with a
      few tubercles just within the conspicuous green border; beak
      almost straight.   **5. sardous**

7 Lvs glabrous or nearly so.   8
   Lvs hairy.   9

8 Fls 0·5–1 cm. diam.; achenes in an oblong head; in damp places,
      especially on bare mud.   **10. sceleratus**
   Fls 1·5–2·5 cm. diam. (when petals are present); achenes in a
      spherical cluster; in woods and shady hedge-banks.
                                                        7. auricomus

9 Plant with fleshy root-tubers at the base of a short erect stock; fls
      2·5–3 cm. diam., very glossy; stem-lvs 1(–2), small. Dry places in
      Jersey. Fan-leaved Buttercup.   *R. paludosus* Poiret
   Plant without root-tubers; stem-lvs usually more than 2, the lower
      ones large.   10

10 Plant with long runners; fl-stalk furrowed.   **2. repens**
   Plant without runners; fl-stalk not furrowed.   **1. acris**

11 Plant decumbent or ascending, diffusely branched; fls 3–6 mm.
      diam.; achenes with shortly hooked tubercles on the faces.
                                                       6. parviflorus
   Plant erect with erect branches; fls 4–12 mm. diam.; achenes spiny
      with the largest spines on the margin.   **4. arvensis**

12 Plant 60–90 cm. high; fls 2–3 cm. diam.   **8. lingua**
   Plant not exceeding 60 cm.; fls less than 2 cm. diam.   13

13 Plant erect with broadly ovate ± cordate basal lvs and fls 6(–9) mm.
      diam.; achenes tubercled. A rare plant of marshes in Gloucester.
      Snaketongue Crowfoot.   *R. ophioglossifolius* Vill.
   Plant ascending, decumbent or creeping; basal lvs rarely cordate;
      fls 5–20(–25) mm. diam.; achenes not tubercled.   14

14 Plant with thread-like stem rooting at every node and often arch-
      ing in the internodes; fls solitary, 5–10 mm. diam.; achenes
      1 mm. Very rare plant of lake shores.   *R. reptans* L.
   Plant, if creeping, not rooting at every node; fls 1–several; achenes
      1–2 mm.   **7. flammula**

15 Plant with no finely dissected submerged lvs.   16
   Finely dissected submerged lvs present.   18

16 Lvs deeply 3(–5)-lobed; receptacle hairy; fls 6–12 mm. diam.
      Local in muddy ditches and ponds chiefly in S. England, Wales
      and S. Ireland. Forms of *R. tripartitus* DC. and hybrid-swarms
      of *R. tripartitus* × *omiophyllus* (formerly named *R. lutarius*
      (Revel) Bouvet).
   Lvs shallowly lobed; receptacle glabrous.   17

17   Lf-lobes broadest at their base; fls 3–6 mm. diam.   **11. hederaceus**
       Lf-lobes narrowest at their base; fls 8–12 mm. diam.
                                                            12. omiophyllus

18   Floating lvs less than 1·5 cm. across, all deeply 3(–5)-lobed, the
       lobes cuneate, distant; submerged lvs usually few, with very fine
       collapsing segments; fls 3–7 mm. diam. Shallow ponds in S.W.
       England, Cheshire, Wales and S. Ireland. Three-lobed Water-
       Crowfoot.   *R. tripartitus* DC.
       (For plants having submerged lvs with flattened non-collapsing
       segments and rather larger fls see under 16 above.)
       Floating lvs, if present, 1–3 cm. across; submerged lvs numerous.   *19*

19   Segments of submerged lvs very long (8–30 cm.) and parallel;
       floating lvs 0; receptacle glabrous.   **13. fluitans**
       Segments of submerged lvs rarely exceeding 8 cm.; floating lvs
       present or 0; receptacle hairy.   *20*

20   Submerged lvs circular in outline, 0·5–2 cm. diam., their rigid
       segments all lying in one plane; floating lvs 0.   **14. circinatus**
       Submerged lvs with segments not lying in one plane.   *21*

21   Petals rarely exceeding 5 mm.; nectaries crescent-shaped; floating
       lvs 0.   **15. trichophyllus**
       Petals usually exceeding 5 mm.; nectaries not crescent-shaped;
       floating lvs present or 0.   *22*

22   Achenes completely glabrous even when immature, winged,
       c. 1 mm. long, very numerous; maritime.   **19. baudotii**
       Achenes pubescent at least when immature, not winged, c. 2 mm.
       long, usually fewer than 40.   *23*

23   Petals up to 10 mm.; nectaries circular; floating lvs, when present,
       ±deeply cut into cuneate straight-sided usually dentate seg-
       ments.   **16. aquatilis**
       Petals usually exceeding 10 mm.; nectaries pear-shaped; floating
       lvs, when present, usually 3–5-lobed, the lobes crenate with
       rounded sides.   *24*

24   Submerged lvs longer than the stem-internodes; floating lvs
       usually 0; fr.-stalk up to 15 cm.   **18. pseudofluitans**
       Submerged lvs shorter than the stem-internodes; floating lvs
       usually present; fr.-stalk commonly exceeding 5 cm.
                                                             17. peltatus

1. R. acris L.   Meadow Buttercup.

Perennial with erect or creeping stock and a much-branched hairy stem,
15–100 cm., hollow below; not stoloniferous. Basal lvs long-stalked,
pentagonal in outline, palmately lobed, the terminal lobe sessile, all
lobes further cut into 3-toothed segments; stem-lvs shorter-stalked or

sessile; all hairy. Fls 18–25 mm. diam. in an irregular cyme, their stalks hairy, not furrowed. Sepals not reflexed. Fl. 6–7. Common in meadows and pastures throughout the British Is. Reaching 4000 ft. in Scotland.

2. R. repens L.   Creeping Buttercup, Crowfoot.

Perennial with long stout roots and strong lfy above-ground stolons (runners) which root at the nodes; flowering stems erect, 15–60 cm., hairy. Basal and lower stem-lvs stalked, triangular-ovate in outline, 3-lobed with the terminal lobe long-stalked, the lobes further cut into 3-toothed segments; upper lvs sessile with narrow segments; all hairy. Fls 2–3 cm. diam. their stalks hairy, furrowed. Sepals not reflexed. Fl. 5–8. Common throughout the British Is. in wet meadows, pastures and woods, in dune-slacks, by roadsides and as a weed, especially of heavy soils.

3. R. bulbosus L.   Bulbous Buttercup.

Perennial with an erect or ascending stem, 15–40 cm., which forms a somewhat flattened tuber (corm) at its base; not stoloniferous. Basal and lower stem-lvs stalked, ovate in outline, 3-lobed with the terminal lobe long-stalked, all lobes further cut into toothed segments; upper lvs sessile with narrow segments; all usually hairy. Fls 2–3 cm. diam., their stalks hairy, furrowed. Sepals yellowish, strongly reflexed. Fl. 5–6. Abundant in dry pastures and on grassy slopes in the south but less common in the north though found throughout the British Is except in Shetland.

4. R. arvensis L.   Corn Crowfoot.

Annual with an erect lfy stem, 15–60 cm., ± hairy. Lvs stalked, the lowest simple, broadly spathulate to obovate, toothed near the tip; the rest 3-lobed and further cut into narrow segments; uppermost with few linear segments. Fls 4–12 mm. diam. Sepals pale, spreading. Petals bright lemon-yellow. Achenes 6–8 mm., conspicuously bordered, spiny. Fl. 6–7. Native or introduced. Formerly common as a cornfield weed in the south but less so towards the north. Rare in Ireland.

5. R. sardous Crantz

Annual with an erect lfy ± hairy stem, 10–45 cm., with the habit of *R. bulbosus* but with no stem-tuber. Basal lvs stalked, shining, ± deeply 3-lobed, or ternate with stalked 3-lobed further cut or toothed lflets, the middle lflet longest stalked; uppermost stem-lvs sessile with few narrow segments; all hairy. Fls many, 1·2–2·5 cm. diam., their stalks hairy, furrowed. Sepals reflexed. Petals 5 or more, pale yellow. Achenes 3–4 mm., their brownish faces usually with a ring of tubercles close to the conspicuous green border. Fl. 6–10. A local and often casual weed of damp places. From Argyll and Angus southwards.

6. R. parviflorus L.   Small-flowered Buttercup.

Annual with numerous spreading, ascending or decumbent, hairy stems, 10–40 cm. Basal lvs roundish-cordate with 3–5 cuneate toothed lobes; upper lvs with fewer and narrower lobes, uppermost oblong; all softly hairy, yellowish-green. Fls 3–6 mm diam., their hairy furrowed stalks opposite the lvs or in the forks of branches. Sepals reflexed. Petals 5 or fewer, pale yellow. Achenes few, 2·5–3 mm., narrowly bordered, with hooked tubercles all over the reddish-brown faces. Fl. 5–6. A local lowland plant of dry grassy banks, etc. Throughout England and Wales; rare in Ireland.

7. R. auricomus L.   Goldilocks.

Perennial with a short stock and numerous erect sparsely hairy stems, 10–40 cm. Lvs variable in form, chiefly basal; the lowest long-stalked, roundish or reniform, crenate or coarsely toothed but hardly lobed; the rest ± deeply 3-lobed, the lobes crenate or further cut; stem-lvs few, ± sessile, deeply divided into very narrow segments. Fls few, 1·5–2·5 cm. diam. when petals are present. Sepals hairy. Petals 5 or fewer or 0; golden yellow. Achenes 3·5–4 mm., downy, with a hooked beak. Apomictic. Fl. 4–5. Very variable. A local woodland herb, occasionally on rocks. Throughout most of Great Britain; Ireland.

8. R. lingua L.   Great Spearwort.

Perennial stoloniferous herb with stout stems, 50–120 cm., creeping below then erect, hollow, ± glabrous. Basal lvs up to 20 × 8 cm., produced in autumn and often submerged, ovate-cordate, blunt, disappearing before flowering; stem-lvs up to 25 × 2·5 cm., distichous, short-stalked or sessile, half-clasping, oblong-lanceolate, acute, entire or remotely denticulate. Fls 2–5 cm. diam., bright yellow. Achenes 2·5 mm., glabrous, minutely pitted. Fl. 6–9. A local plant of marshes and fens throughout Great Britain; Ireland.

9. R. flammula L.   Lesser Spearwort.

Perennial with stems 8–50 cm., creeping or ascending, rooting at irregular intervals or only at the base, glabrous. Basal lvs stalked, broadly ovate to elliptical, usually rounded below; stem-lvs short-stalked to sessile, half-clasping, lanceolate to linear-lanceolate, acute, entire or distantly toothed. Fls 7–18 mm. diam., pale yellow. Achenes 1·5–2·0 mm., glabrous, minutely pitted. Fl. 6–8. Variable. Common in wet places throughout the British Is.

10. R. sceleratus L.   Celery-leaved Crowfoot.

Annual or overwintering, with a stout erect stem, 20–60 cm., hollow, ± glabrous, furrowed. Lower lvs long-stalked, reniform or pentagonal in outline, palmately 3-lobed, the lateral lobes often again lobed, crenate, glabrous, shining; stem-lvs short-stalked to sessile with fewer and narrower segments, slightly hairy below. Fls 5–10 mm. diam., numerous. Sepals reflexed. Petals pale yellow, hardly exceeding the sepals. Achenes 70–100, 1 mm., glabrous, each face with a faintly wrinkled central area; the head of achenes oblong-ovoid. Fl. 5–9. Common on damp mud at the edge of ponds, etc., throughout most of the British Is.

11. R. hederaceus L.   Ivy-leaved Water-Crowfoot.

Annual or perennial with a branched stem, 10–40 cm., creeping in mud or with the upper part floating. Lvs 1–3 cm. wide, usually opposite, stalked, reniform or roundish-cordate, often with dark basal markings, shallowly 3–5 lobed, the lobes broadest at their base, roundish to triangular, blunt, entire. No dissected lvs. Fls 3–6 mm. diam. Petals whitish, scarcely exceeding the sepals. Achenes numerous, 1–1·5 mm., usually glabrous. Fl. 6–9. Rather local on mud and in shallow water throughout the British Is.

12. R. omiophyllus Ten. (*R. lenormandii* F. W. Schultz)
                                                Lenormand's Water-Crowfoot.

Annual or perennial with a branched stem usually 5–25 cm., creeping in mud or the upper part floating. Lvs 8–30 mm. wide, often opposite, stalked, roundish-reniform, never with dark basal markings, 3-lobed, the lobes shallow, rounded, narrowest at their base; the lateral lobes slightly 2-lobed, all crenate. No dissected lvs. Fls 8–12 mm. diam. Petals white, about twice as long as the sepals. Achenes numerous, 1–1·5 mm., usually glabrous. Fl. 6–8. Locally common in non-calcareous streams and muddy places from Argyll and Stirling southwards; Ireland.

13. R. fluitans Lam.   Long-leaved Water-Crowfoot.

Large robust perennial herb with creeping rhizome and submerged lfy stems up to 6 m. long. Submerged lvs 8–30 cm., greenish-black, with rather few very long firm slender sub-parallel occasionally forking segments; floating lvs very rarely present. Fls 2–3 cm. diam. Petals 5–10, white; stamens numerous, shorter than the head of carpels. Receptacle glabrous. Achenes inflated, glabrous. Fl. 6–8. In rapidly flowing streams and rivers throughout Great Britain from the Clyde southwards; Antrim.

14. R. circinatus Sibth.   Stiff-leaved Water-Crowfoot.

Slender perennial herb with erect submerged stems. Submerged lvs 0·5–2 cm. diam., sessile, circular in outline, the short rigid segments all lying in one plane; no floating lvs. Fls 8–18 mm. diam. Petals 5, white. Stamens exceeding the head of carpels. Fr. stalks much exceeding the lvs, tapering upwards. Achenes ± glabrous. Receptacle hairy. Fl. 6–8. Locally common in ditches, canals, slow streams, ponds and lakes with a high mineral content. From Aberdeen southwards; Ireland.

15. R. trichophyllus Chaix   Short-leaved Water-Crowfoot.

Perennial with short (usually 2–3 cm.) sessile or shortly stalked submerged lvs whose segments do not lie in one plane; no floating lvs. Fls 5–10(–15) mm. diam. Petals white, evanescent. Fr. stalks not or hardly exceeding the lvs, strongly recurved from the base. Achenes usually hairy. Receptacle hairy. Fl. 5–6. Variable. In ponds, ditches and slow streams throughout the British Is.

16. R. aquatilis L.

Annual or perennial with branched submerged stems and finely dissected submerged lvs, 3–6(–8) cm., shorter than the stem-internodes, their segments not lying in one plane; floating lvs ± deeply cut into cuneate usually straight-sided segments, dentate distally; floating lvs sometimes 0. Petals up to 10 mm.; nectaries circular. Fr.-stalk rarely more than 5 cm. and usually shorter than the opposed lf-stalk. Achenes ± distinctly pubescent, at least when immature. Fl. 5–6. In ponds, ditches and streams throughout lowland Britain and Ireland.

17. R. peltatus Schrank   Common Water-Crowfoot.

Annual or perennial, resembling *R. aquatilis* but floating lvs semicircular with a truncate base to ± circular and having 3–7 crenate lobes with rounded sides; floating lvs rarely 0. Petals usually more than 10 mm., often more than 5 in number; nectaries pear-shaped. Fr.-stalk usually more than 5 cm. and exceeding the opposed lf-stalk, sometimes shorter. Achenes ± distinctly pubescent to hispid. Fl. 5–8. Lakes, ponds and slow streams except in the extreme north.

18. R. pseudofluitans (Syme) Baker & Foggitt

Like *R. peltatus* but more robust, with finely dissected submerged lvs longer than the stem-internodes but usually not exceeding 8 cm.; floating lvs usually 0. Fr.-stalk very long (to 15 cm.), exceeding the opposed lf-stalk. In fast-flowing streams in England, S. Scotland and Ireland.

19. R. baudotii Godron   Brackish Water-Crowfoot.

Perennial herb whose stout stems bear submerged lvs with slender spreading rigid segments; small ± reniform deeply 3-lobed floating lvs, with the lobes crenate or further cut, may be present, often with transitional lvs. Fls 12–18 mm. diam. Petals white. Stamens not exceeding the head of carpels. Fr.-stalks much exceeding the lvs, stout, tapering, strongly curved downwards. Achenes 1–1·2 mm., 40–100, glabrous, inflated, on an elongated hairy receptacle. Fl. 5–9. In brackish streams, ditches and ponds in coastal districts throughout the British Is; rarely inland.

20. R. ficaria L.   Lesser Celandine.

Perennial, with numerous fusiform or clavate root-tubers, 10–25 mm. Stems 5–25 cm., branched, ascending. Lvs glabrous, all stalked and with sheathing bases; basal lvs 1–4 cm., in a rosette, long-stalked, cordate, bluntly angled or crenate, rarely toothed; stem-lvs similar but smaller and shorter-stalked. Fls 2–3 cm. diam., solitary, terminal. Sepals 3. Petals 8–12, bright golden-yellow, fading white. Achenes numerous, to 2·5 mm. but often aborting, downy. Fl. 3–5. Variable: some forms have small bulbils in the axils of the lvs. A common plant of woods, meadows, grassy banks and stream-sides throughout the British Is.

8. MYOSURUS L.

Small annual herbs with linear lvs confined to a basal rosette. Fls terminal, solitary, small, hermaphrodite, actinomorphic. Sepals with a small basal spur; petals tubular, nectar-secreting, or 0; stamens few. Fr. of numerous 1-seeded achenes in an elongated spike.

1. M. minimus L.   Mousetail.

Glabrous, with entire lvs and numerous erect lfless flowering stems, 5–12·5 cm. Fls very small, pale greenish-yellow. Sepals 5 or more, their spurs appressed to the stem. Petals 5, greenish, about equalling the sepals. Achenes 1–1·5 mm., on a threadlike receptacle, 2·5–7 cm. Fl. 3–5. Very local in damp arable fields throughout lowland England.

9. AQUILEGIA L.

Perennial herbs with erect woody stocks and spirally arranged bi- or tri-ternately compound lvs. Fls hermaphrodite, actinomorphic, with all parts in 5-merous whorls. Sepals petaloid; petals each with a nectar-secreting spur; stamens numerous, the innermost staminodal; carpels 5 or 10, sessile, free. Fr. a group of many-seeded follicles.

Secondary lflets stalked; spur hooked.   **1. vulgaris**
Secondary lflets sessile; spur almost straight. A blue-flowered plant, 20–25 cm. Introduced and naturalized in one locality in Angus. Pyrenean Columbine.   **A. pyrenaica* DC.

1. A. vulgaris L.   Columbine.

Flowering stems erect, lfy, glabrous or softly hairy, 40–100 cm. Basal lvs long-stalked, biternate, the lflets with crenate lobes; uppermost stem-lvs (bracts) ± sessile and narrowly 3-lobed; all somewhat glaucous. Fls in a cyme, 3–5 cm. diam., drooping, usually blue, sometimes white or reddish. Petal-spur sharply curved and knobbed at its tip. Follicles 5(–10), 15–25 mm., erect, with black shining seeds. Fl. 5–6. Local in woods and wet places on calcarous soil or fen peat throughout the British Is.

10. THALICTRUM L.

Meadow Rue.

Perennial herbs usually with repeatedly pinnate stipulate lvs. Fls numerous, small, hermaphrodite, actinomorphic. No nectaries. Per. segs 4–5 petaloid, readily falling; stamens with long filaments. Fr. a group of sessile or stalked 1-seeded achenes.

1 Fls in a simple raceme; plant usually not exceeding 15 cm.
                                                         2. alpinum
   Fls panicled; plant usually exceeding 15 cm.   2
2 Fls in dense clusters; stamens erect; achenes 1·5–2·5 mm.
                                                         1. flavum
   Fls not densely clustered; stamens drooping; achenes 3–6 mm.
                                                         3. minus

1. T. flavum L.   Common Meadow Rue.

Rhizomatous, with erect robust furrowed stem, 50–100 cm. Lower stem-lvs stalked, upper sessile, all bi- or tri-pinnate with stipule-like struc-tures at each branching; lflets longer than broad, 3–4-lobed distally, pale beneath. Per. segs 4, whitish. Stamens yellow, their anthers not apiculate. Achenes sessile, 6-ribbed. Fl. 7–8. Variable. Common in marshy meadows and fens throughout Great Britain, northwards to Inverness; Ireland.

2. T. alpinum L.   Alpine Meadow Rue.

Rhizomatous, with erect slender wiry stem, usually 8–15 cm. Lvs chiefly basal, stalked, biternate; lflets roundish, shallowly lobed or crenate, whitish beneath. Per. segs 4, pale purplish. Stamens with long slender pendulous pale violet filaments and yellow apiculate anthers. Fr. stalks recurved. Achenes 2–3, 3–3·5 mm., stalked. Fl. 6–7. Rocky slopes and ledges on mountains but reaching sea-level in the north-west. Great Britain northwards from Westmorland, Yorks and Durham; N. Wales; Hebrides, Orkney, Shetland.

3. T. minus L.   Lesser Meadow Rue.

Rhizomatous, tufted or far-creeping, with erect rigid often flexuous lfy stems, 15–150 cm., often glandular above. Lvs 3 or 5 times pinnate; lflets usually about as long as broad, 3–7 lobed or toothed, glabrous or glandular. Fls in a loose spreading panicle. Per. segs 4, yellowish- or purplish-green. Stamens long and pendulous, with apiculate anthers. Achenes 3–15, 3–6 mm., sessile, 8–10 ribbed. Fl. 6–8. A very variable group. Small forms on dry limestone slopes, rocks, ledges, scree and shingle, chalk quarries and banks and sand-dunes; larger forms by streams and lakes, often in shade. Local throughout the British Is.

24. PAEONIACEAE (p. xviii)

25. BERBERIDACEAE

Herbs or shrubs with simple or compound, spirally arranged, usually exstipulate lvs. Fls regular, hermaphrodite, usually 3-, rarely 2-merous. Per. segs free, in 2–7 whorls. Stamens usually in 2 whorls opposite the per. segs; anthers usually opening by valves. Ovary of 1 carpel with 1 to many ovules; style short or 0. Fr. a berry or capsule; seeds endospermic.

Lvs simple; stem spiny.   1. BERBERIS
Lvs pinnate with spiny-toothed lflets; stem unarmed; fls in clustered
  terminal racemes, yellow; fr. a blue-black berry. A commonly planted
  shrub, native of western N. America, locally naturalized. Oregon
  Grape.   *Mahonia aquifolium* (Pursh) Nutt.

1. Berberis L.

Evergreen or deciduous shrubs with yellow wood. Shoots of 2 kinds:
long shoots with the lvs represented by 3-partite spines and short
axillary shoots bearing clusters of simple lvs. Fls 3-merous. Per. segs
yellow or orange, usually in 5 whorls, those of the outermost very small,
those of the 2 inner usually smaller than the intermediate and each
bearing 2 nectaries near the base. Stamens springing inwards when
touched near the base. Ovules 1–few. Fr. a berry.

1. B. vulgaris L.   Barberry.

Shrub 1–2·5 m.; twigs grooved, yellowish, spines 1–2 cm., usually 3-
partite. Lvs 2–4 cm., obovate, usually blunt, finely spine-toothed,
reticulate; stalk to 1 cm. Infl. a pendulous raceme, 4–6 cm. Fls 6–8
mm. diam., yellow. Fr. 8–12 mm., oblong, red. Fl. 5–6. Probably native.
Local in hedges and scrub throughout Great Britain; introduced in
Ireland.

26. NYMPHAEACEAE

Water-lilies with stout creeping rhizomes and floating lvs. Fls solitary,
terminal, floating; hermaphrodite, actinomorphic. Perianth of 3–6
green sepals and numerous petals. Stamens numerous. Ovary syncarpous, superior or ±inferior, many-celled, with numerous ovules
inserted all over the inner walls of the cells. Fr. a spongy capsule.

Petals white, the outermost longer than the sepals; lateral veins of lvs
  anastomosing.   1. Nymphaea
Petals yellow, much shorter than the sepals; lateral veins of lvs forking
  but not anastomosing.   2. Nuphar

1. Nymphaea L.

Lvs stipulate, almost circular, with a deep basal sinus. Sepals 4; petals
inserted on the side of the ovary, the innermost grading into the stamens;
stamens inserted near top of ovary; summit of ovary concave with a
central boss from which the stigmatic surfaces radiate and with marginal
stylar processes. Fr. ripening and splitting under water.

1. N. alba L.   White Water-lily.

Lvs all floating, 10–30 cm. diam. Fls 5–20 cm. diam. Sepals white inside.
Petals about 20, white. Fr. 16–40 mm. diam., obovoid or ±globose.
Fl. 7–8. In lakes and ponds throughout the British Is. Plants from N.W.
Scotland and W. Ireland often have very small lvs and fls.

2. NUPHAR Sm.

Lvs not stipulate, broadly elliptical or oblong with a deep basal sinus. Sepals 5–6; petals yellow, shorter than the sepals, hypogynous like the stamens; ovary superior, its summit convex with a central depression and many radiating sessile stigmas. Fr. flask-shaped, ripening above water and splitting irregularly into parts each comprising the contents of a single ovary-cell.

1   Fls 4–6 cm. diam.; stigma-rays 10–25; margin of ovary-top entire.
                                                      1. lutea
    Fls 1·5–4 cm. diam.; stigma-rays 7–14; margin of ovary-top wavy to
    deeply lobed.   *2*

2   Ovary-top 6–8·5 mm. diam., star-shaped. A local plant of lakes in
        C. and N. Scotland, Shropshire and Merioneth. Least Yellow
        Water-lily.   *N. pumila* (Timm) DC.
    Ovary-top 7·5–11 mm. diam., circular with a wavy margin. A rare
        plant of lakes in C. and S. Scotland and Northumberland. Hybrid
        Yellow Water-lily.   *N. × intermedia* Ledeb.

1. N. lutea (L.) Sm.   Yellow Water-lily, Brandy-bottle.

Foating lvs 12–40 × 9–30 cm., submerged lvs broadly ovate-cordate, thin and translucent. Fls 4–6 cm. diam., rising out of the water, smelling of alcohol. Sepals yellowish-green outside, bright yellow within. Petals one-third as long as the sepals. Fr. 3·5–6 cm. Fl. 6–8. In lakes, ponds and streams throughout the British Is.; rare in N. Scotland.

27. CERATOPHYLLACEAE

Perennial submerged aquatic herbs with whorled lvs divided into linear or filiform segments. Fls unisexual, sessile, usually borne singly just above a lf-whorl, male and female fls at different nodes. Perianth of 6–12 greenish segments, toothed above, joined at the base. Male fls with 10–20 stamens; anthers subsessile, 2-celled, with prolonged connective. Female fls with a sessile unilocular ovary with 1 ovule. Fr. a nut.

1. CERATOPHYLLUM L.
                                                            Hornwort.

Lvs once or twice forked; fr. with 2 spines at base.   **1. demersum**
Lvs 3–4 times forked; fr. without basal spines.   **2. submersum**

1. C. demersum L.

Stems 20–100 cm., slender. Lvs rigid, dark green, 1–2 cm., segments linear, denticulate, those of winter lvs shorter and broader than of summer lvs. Fr. c. 4 mm., not exceeding the persistent style. Fl. 7–9. Local in ponds, ditches and streams throughout England and Ireland; rare in Wales and Scotland.

2. C. submersum L.

Like *C. demersum* but softer and brighter green. Fr. longer than the persistent style. Fl. 7–9. In ponds, ditches and streams, chiefly in coastal counties of S. and E. England; less common than *C. demersum*.

28. PAPAVERACEAE

Usually herbs with spirally arranged often deeply lobed or divided lvs and a milky or coloured latex; stipules 0. Fls hermaphrodite, actinomorphic, usually hypogynous. Sepals 2, falling early; petals 2+2, crumpled in bud, fugacious; stamens usually numerous; ovary superior, syncarpous, 1-celled (rarely 2-celled). Fr. usually a capsule opening by valves or pores; ovules many, on 2 or more parietal placentae.

1 Stigmas sessile on a disk at the top of the ovary; fr. a capsule opening by pores just beneath the disk; latex usually milky; fls red or
   whitish, rarely dull orange-brown.   **1. Papaver**
   Stigmas not sessile on an expanded disk; fr. a long pod-like capsule
   opening by 2 valves; latex yellow or orange; fls yellow or orange.   *2*

2 Fls not more than 2·5 cm. diam.; fr. 3–5 cm., opening by 2 valves
   from below.   **4. Chelidonium**
   Fls more than 2·5 cm. diam.; fr. opening from above.   *3*

3 Lvs green; fr. 2·5–3 cm., opening only at the top; not maritime.
                                                      2. Meconopsis
   Lvs glaucous; fr. 15–30 cm., opening almost to the base; maritime.
                                                      3. Glaucium

1. Papaver L.

Poppies.

Annual to perennial herbs, usually with white but sometimes with yellowish or orange latex. Fls solitary, showy, often bright red. Stigmas sessile over the placentae. Fr. a capsule opening by pores just beneath the expanded stigmatic disk.

1 Annuals; rosette-lvs usually deeply pinnatifid to once or more times
   pinnatisect, or, if less deeply lobed, then ovate-oblong and
   glaucous; petals scarlet, pale lilac or white.   *2*

Perennial; rosette-lvs lanceolate, ± coarsely toothed or at most
shallowly pinnatifid, stiffly hairy, not glaucous; petals 3–4 cm.,
dull orange or brownish; capsules up to 2·5 cm., club-shaped,
narrowing from just beneath the disk. Introduced and established in gardens and waste places in S. England.

P. atlanticum (Ball) Cosson

2   Lvs toothed or slightly lobed, clasping the stem at their base,
glaucous; fls usually white or pale lilac.   **5. somniferum**
Lvs once or twice pinnatifid or pinnatisect, green; fls usually bright
red.   3

3   Capsule glabrous   4
Capsule with stiff hairs or bristles.   6

4   Capsule almost globose, about as long as wide.   **1. rhoeas**
Capsule obovoid-oblong, at least twice as long as wide.   5

5   Latex white; roots white; anthers ± violet-tinged, usually falling
short of the stigmatic disc; petal-bases usually overlapping.

2. dubium

Latex turning yellow on exposure; roots yellow; anthers yellow,
equalling the stigmatic disc; petal-bases usually not overlapping.
An infrequent weed of roadsides, field-margins, quarries, etc.,
chiefly on calcareous soils in S. and E. England. Babington's
Poppy.   *P. lecoqii* Lamotte

6   Capsule almost globose; bristles numerous, spreading.

3. hybridum

Capsule narrowly obovoid-oblong; bristles few, erect.

4. argemone

1. P. rhoeas L.   Field Poppy.

Usually annual, with erect or ascending stems, 20–60 cm., glabrous or
with stiff spreading hairs. Lvs stiffly hairy; basal stalked, once or twice
pinnately cut with narrow bristle-pointed segments; upper sessile,
usually 3-lobed, the central lobe long, lanceolate or elliptical, all lobes
further lobed, cut or toothed. Fls 7–10 cm. diam., scarlet, rarely pink or
white, their stalks with spreading hairs. Capsule not more than twice as
long as wide, glabrous; stigma-rays 8–12. Fl. 6–8. A weed of arable
fields and waste places throughout the British Is. but rare and local in
N. Scotland and W. Ireland.

2. P. dubium L.   Long-headed Poppy.

Annual, with erect stems, 20–60 cm., with stiff hairs spreading below,
appressed above. Lvs as in *P. rhoeas* but with shorter, broader and more
abruptly acute segments and with a smaller terminal segment. Fls 3–
7 cm., diam. pale scarlet, their stalks with appressed hairs. Capsule

obovoid-oblong, narrowing from near the top and more than twice as long as wide, glabrous; stigma-rays 4–12, not reaching the margin of the disk. Fl. 6–7. A weed of arable land and waste places throughout the British Is.; commoner than *P. rhoeas* in the north.

3. P. hybridum L.   Round Prickly-headed Poppy

Annual, with erect stems 15–50 cm., stiffly hairy. Lvs 2–3 times pin-nately lobed, the ultimate segments acuminate, bristle-pointed; basal lvs stalked, upper sessile. Fls 2–5 cm. diam., crimson, their stalks with appressed hairs. Capsule almost globose, densely covered with stiff yellow bristles; stigma-disk convex with 4–8 rays. Fl. 6–7. A local weed of arable land and waste places in lowland Great Britain northwards to S. Scotland; Ireland.

4. P. argemone L.   Long Prickly-headed Poppy.

Annual, with erect or ascending stems, 15–45 cm., stiffly hairy with appressed hairs. Lvs stiffly hairy, twice pinnately lobed, the ultimate segments suddenly acuminate, bristle-pointed; basal lvs stalked, upper sessile. Fls 2–6 cm. diam., scarlet, their stalks with appressed hairs. Capsule narrowly obovoid-oblong, strongly ribbed, with a few erect bristles, especially above; stigma-disk convex with 4–6 rays. Fl. 6–7. A weed of light soils throughout Great Britain and Ireland, common in the south, rare in the north.

*5. P. somniferum L.   Opium Poppy.

Annual, with erect stems, 30–100 cm., glabrous or with a few spreading bristles. Lvs undulate, ovate-oblong, ± pinnately lobed, the lobes shallow, irregular, coarsely toothed; lowest lvs short-stalked, upper sessile and clasping; all glaucous and ± glabrous. Fls to 18 cm. diam., white or pale lilac. Capsule globular or ovoid, large but very variable in size. Fl. 7–8. Introduced. A relic of cultivation in some parts of England, and a garden escape.

2. MECONOPSIS Vig.

Annual or perennial herbs with yellow latex. Petals usually 4, rarely more; stamens numerous; capsule subglobose or cylindrical with a distinct style and 4–6 stigmas, opening by valves which usually reach only a short distance below the top; seeds numerous.

1. M. cambrica (L.) Vig.   Welsh Poppy.

Perennial with branched tufted stock covered with old lf-bases, and erect ± glabrous lfy stems 30–60 cm. Basal lvs long-stalked, pinnately divided, with pinnatifid ovate acute segments; upper short-stalked. Fls 5–7·5 cm. diam., single in axils of upper lvs. Petals yellow. Anthers yellow. Capsule 2·5–3 cm., ovoid to ellipsoid, 4–6-ribbed and splitting into 4–6 valves for about ¼ its length. Damp shady rocky places in S.W. England and Wales and scattered throughout Ireland.

3. GLAUCIUM Miller

Annual to perennial glaucous herbs with yellow latex. Lvs pinnately lobed or cut. Fls axillary, large. Petals 4, yellow or red; stamens numerous; stigma sessile, 2-lobed. Capsule long, slender, 2-celled, opening almost to the base by 2 valves and leaving the seeds embedded in the septum.

Stem and capsule ±glabrous; fls yellow, 6–9 cm. diam.   **1. flavum**
Stem and capsule stiffly hairy; fls orange to scarlet, 3–5 cm. diam.
      Annual, 25–30 cm. Introduced. A casual.
                              G. corniculatum (L.) J. H. Rudolph

1. G. flavum Crantz   Yellow Horned-poppy.

Biennial or perennial with deep stout tap-root and erect glaucous glabrous stem, 30–90 cm. Basal lvs roughly hairy, stalked, pinnately lobed or divided, the lobes pointing various ways and further lobed or coarsely toothed; upper lvs sessile, half-clasping, rough; all glaucous. Fls 6–9 cm. diam., short-stalked. Petals yellow. Capsule 15–30 cm., glabrous but rough. Fl. 6–9. Chiefly on maritime shingle all round the coast of Great Britain northwards to Argyll and Kincardine; Ireland northwards to Galway and Down; Channel Is.

4. CHELIDONIUM L.

Perennial herbs with short branching stocks and erect lfy stems. Latex bright orange. Lvs pinnately cut or lobed. Petals 4; stamens numerous; style very short with 2 spreading stigma-lobes. Capsule slender, 1-celled, with no septum, opening from below by 2 valves which separate from the placentae; seeds with a fleshy crested appendage.

1. C. majus L.   Greater Celandine.

Stems 30–90 cm., brittle, slightly glaucous, sparsely hairy. Lvs almost pinnate with 5–7 ovate or oblong lflets, terminal often 3-lobed, laterals usually with a stipule-like lobe on lower side; all crenate-toothed, ± glabrous, somewhat glaucous beneath. Fls 2–2·5 cm. diam., terminal. Sepals greenish-yellow; petals bright yellow; filaments thickened above.

Capsule 3–5 cm.; seeds black with white appendage. Fl. 5–8. Probably native. Frequent on banks, hedgerows, walls, etc., chiefly near houses. Northwards to Inverness; throughout Ireland: Channel Is.

29. FUMARIACEAE

Herbs, usually with brittle stems, sometimes climbing; juice not milky or coloured. Lvs spirally arranged, usually much divided. Fls commonly in racemes or spikes, zygomorphic, hermaphrodite. Sepals 2, small, caducous. Petals 4, in two dissimilar whorls; those of the outer whorl connivent, one or both spurred or saccate; of the inner narrower and often cohering. Stamens 2, tripartite, the central branch bearing a 2-celled anther, the lateral branches each bearing a 1-celled anther. Ovary 1-celled with 2 parietal placentae, each bearing 1–many ovules. Fr. a capsule or nutlet; seeds endospermic.

Fr. a many-seeded capsule; petals without dark tips.　　1. CORYDALIS
Fr. a 1-seeded nutlet; inner petals with dark purple tips.　2. FUMARIA

1. CORYDALIS Medic.

Glabrous ± glaucous herbs. Lvs variously divided. Fls zygomorphic, in racemes; only the upper petal spurred. Ovules ± numerous. Fr. a 2-valved capsule.

Lvs ending in a branched tendril; fls 5–6 mm., cream.　　**1. claviculata**
Lvs without tendrils; fls 12–18 mm., yellow. Commonly naturalized on
old walls. Yellow Fumitory.　　　　　　　　　　*C. lutea* (L.) DC.

1. C. claviculata (L.) DC.　　　　　　White Climbing Fumitory.

Annual climbing much-branched herb 20–80 cm. Lvs pinnate, ending in a branched tendril; lflets distant, long-stalked, digitately divided into 3–5 segments each 5–12 mm., elliptic, mucronate, entire. Infl. c. 6-fld, lf-opposed. Fls 5–6 mm.; petals cream, the upper with a very short blunt spur. Fr. c. 6 mm., with 2–3 seeds. Fl. 6–9. Woods and shady rocks on acid soils over most of Great Britain, rather local; very local in Ireland.

2. FUMARIA L.

Glabrous ± glaucous annual herbs, often climbing by the lf-stalks. Lvs irregularly 2–4-pinnatisect, none basal. Fls zygomorphic, in lf-opposed racemes; only the upper petal spurred. Ovule 1, or 1 on each placenta. Fr. a nutlet with 2 apical pits when dry.

The spp. are most readily distinguished in healthy specimens growing in the open. Features of importance for species-identification are the dark-coloured lateral wings of the upper petal, which may be turned upwards so as to hide the green keel or be ± spreading; the shape of the lower petal, whether spathulate or ± parallel-sided, and the direction of its margins, whether ± erect or spreading laterally to form a rim; and the presence or absence of a distinct narrow fleshy neck between the fr. and the expanded top of its stalk.

1   Fls 9 mm. or more; lower petal not spathulate; lf-lobes broadish.   2
      Fls 5–8(–9) mm.; lower petal distinctly spathulate; lf-lobes lanceo-
      late to linear.   7

2   Lower petal with broad spreading margins; fr. c. 3 mm.; corolla
      12–14 mm., white at first, becoming pink, with wings and tips
      blackish-red. Arable land and waste places in Cornwall.
                                                F. occidentalis Pugsley
      Lower petal with narrow margins; fr. 2–2½(–2⅔) mm.   3

3   Fr.-stalks rigidly recurved; infl. not longer than its stalk; fls
      numerous (c. 20 or more); fr. with a distinct fleshy neck when
      fresh.   4
      Fr.-stalks rarely recurved and then usually wavy; infl. longer than its
      stalk or fls few (c. 12); fr. with an indistinct neck.   5

4   Fls white; upper petal laterally compressed, its wings not concealing
      the keel.   **1. capreolata**
      Fls purple; upper petal not laterally compressed, its wings con-
      cealing the keel; corolla 10–13 mm. A robust plant of cultivated
      and waste ground, widely scattered over the British Is but very
      local.   *F. purpurea* Pugsley

5   Lower petal with spreading margins; fls numerous (c. 20).   6
      Lower petal with erect margins; fls usually few (c. 12), fr. smooth
      when dry.   **3. muralis**

6   Fr. wrinkled when dry; fls 9–11 mm.; upper petal laterally compres-
      sed, often not dark-tipped; sepals serrate.   **2. bastardii**
      Fr. smooth even when dry; fls 11–13 mm.; upper petal not laterally
      compressed, always dark-tipped; sepals subentire. Very rare; in
      cultivated ground in S.W. England and Channel Is.
                                                F. martinii Clavaud

7   Sepals at least 2 × 1 mm.; fls at least 6 mm.   8
      Sepals not more than 1½ × ¾ mm.; fls 5–6 mm.   9

8   Bracts longer than fr.-stalks; fr. rounded at apex; lf-segments
      channelled.   **4. densiflora**
      Bracts shorter than fr.-stalks; fr. truncate or emarginate at apex; lf-
      segments flat.   **5. officinalis**

9   Lf-segments flat; bracts shorter than fr.-stalks; fls pink; racemes
      lax, short-stalked. Arable land, usually on chalk in S. England;
      very local.   *F. vaillantii* Loisel.

Lf-segments channelled, very narrow; bracts about equalling fr.-stalks; fls white or pink-flushed; racemes dense, c. 20-fld, subsessile. Arable land, usually on chalk, in S. and E. England, rare elsewhere. *F. parviflora* Lam.

1. F. capreolata L. ssp. **babingtonii** (Pugsley) P. D. Sell

Ramping Fumitory.

Robust, climbing, often to 1 m. Lf-segments with flat oblong or wedge-shaped lobes. Infl. rather dense, many (c. 20)-fld, usually shorter than its stalk. Sepals 4–6 × 2½–3 mm., ± oval, toothed at base, entire above. Corolla 10–12(–14) mm., creamy white, with wings and tips blackish-red; upper petal strongly compressed laterally, its wings not concealing the keel; lower petal with erect narrow margins, not spathulate. Fr. c. 2½ × 2 mm., truncate, with a distinct fleshy neck, ± smooth when dry; its stalk rigidly arcuate-recurved. Fl. 5–9. Cultivated and waste ground and hedge-banks, scattered over the British Is but local and commoner in the west than in the east.

2. F. bastardii Boreau

Rather robust, not or scarcely climbing. Lf-segments with flat oblong lobes. Infl. rather lax, usually 15–25-fld, longer than its stalk. Sepals c. 3 × 1½ mm., oval ± serrate nearly all round. Corolla 9–11(–12) mm., pink; tips of lateral petals blackish-red, wings blackish-red or more frequently pink; upper petal laterally compressed; lower petal with narrow spreading margins, not spathulate. Fr. c. 2(–2½) mm., suborbicular, with an indistinct fleshy neck, wrinkled into broad apical shallow pits when dry, its stalk straight and ascending. Fl. 4–10. Cultivated and waste ground. Widespread and locally common in W. England, Wales and Ireland; scattered and rare elsewhere.

3. F. muralis Sonder

Variable in habit. Lf-segments with flat broadish lobes. Infl. rather lax, usually 10–15-fld. Sepals 3–5 × 1½–3 mm., ovate, usually toothed towards the base, ± entire above. Corolla 9–12 mm., pink, tips and wings blackish-red; upper petal dorsally compressed, its wings concealing the keel; lower petal with narrow erect margins, not spathulate. Fr. 2–2½ mm., with an indistinct fleshy neck, ± smooth when dry, its stalk slender, usually straight and ascending or spreading. Fl. 5–10. Cultivated and waste ground, hedge-banks and old walls over the whole of Gt Britain and common in many places in the west; local in Ireland. Variable.

4. F. densiflora DC. (*F. micrantha* Lag.)

Rather robust, suberect. Lf-segments with channelled linear to linear-oblong lobes. Infl. very dense, c. 20–25-fld, much longer than its stalk. Bracts longer than the fr.-stalks. Sepals $2\frac{1}{2}$–$3\frac{1}{2}$ × 2–3 mm., orbicular or broadly ovate, subentire or toothed at base. Corolla 6–7 mm., pink, tips and wings blackish-red; upper petal somewhat laterally compressed, wings ascending; lower petal spathulate. Fr. 2–$2\frac{1}{2}$ mm., subglobose, ± blunt, wrinkled when dry. Fl. 6–10. Arable land on dry soils in E. England and E. Scotland, local and rare elsewhere.

5. F. officinalis L.   Common Fumitory.

Robust. Lf-segments with flat lanceolate or linear-oblong lobes. Infl. dense, 10–40-fld, longer than its stalk. Bracts shorter than the fr.-stalks. Sepals 2–$3\frac{1}{2}$ × 1–$1\frac{1}{2}$ mm., ± ovate, irregularly toothed at base, acuminate. Corolla 7–8(–9) mm., pink, tips and wings blackish-red; upper petal dorsally compressed, wings concealing keel; lower petal spathulate. Fr. 2–$2\frac{1}{2}$ mm., usually considerably broader than long, truncate or retuse, wrinkled when dry. Fl. 5–10. Cultivated ground especially on lighter soils; common throughout the British Is.

30. CRUCIFERAE

Annual to perennial herbs, rarely woody, with spirally arranged lvs and racemose infls; stipules 0; bracts usually 0. Fls usually hermaphrodite and actinomorphic. Sepals 4, in 2 decussate pairs, the inner pair often with saccate nectar-collecting bases; petals 4, free, placed diagonally and so alternating with the sepals; stamens hypogynous, usually 6, an outer transverse pair with short filaments and 2 inner pairs with long filaments, one at the back and one at the front of the fl., sometimes 4 or fewer; ovary superior, syncarpous with 1–many ovules on each of 2 parietal placentae, usually 2-celled by the meeting of outgrowths from the placentae; style single or 0; stigma capitate, discoid or ± 2-lobed. Fr. usually a specialized capsule (called a *siliqua*, or, if less than 3 times as long as wide, a *silicula*) opening from below by 2 valves which leave the seeds attached to a framework consisting of the placentae with adjoining wall tissue (*replum*) and the septum; sometimes 1 or more seeds develop also in an indehiscent beak at the base of the style, and sometimes the whole fr. is indehiscent either because the valves do not open or because seeds are restricted to the beak. Seeds usually in 1 or 2 rows in each cell, non-endospermic.

1   A woodland plant with pinnate basal lvs, ternate or simple stem-lvs bearing brownish-violet axillary bulbils, and purple, pink or rarely white fls; rhizome white and scaly. Very local, usually in calcareous woods. Coral-wort.

Cardamine bulbifera (L.) Crantz (*Dentaria bulbifera* L.)

  Not as above.   *2*

2   A robust plant of gardens, waysides and waste places with a very stout tap-root, large oblong waved lvs up to 60 cm., white fls in a large panicle and obovoid fr. which do not ripen in this country.

21. ARMORACIA

  Not as above.   *3*

3   A small submerged aquatic plant with awl-shaped lvs, minute white fls and ovoid scarcely flattened siliculae. Local in lakes and pools in Wales, N. England, Scotland and W. Ireland. Awlwort.   *Subularia aquatica* L.

  Not as above.   *4*

4   Infls borne opposite the pinnately cut lvs (as well as terminal and at forkings of the stem); small prostrate plants with whitish fls.

10. CORONOPUS

  Infls not lf-opposed.   *5*

5   Fr. 8–20 mm., pendulous, flattened, winged; tall plants with glaucous clasping stem-lvs and yellow fls. Probably introduced and locally naturalized. Woad.   **Isatis tinctoria* L.

  Fr. not pendulous, flattened and winged.   *6*

6   Fr. at least 3 times as long as wide.   *7*

  Fr. less than 3 times as long as wide.   *33*

7   Stigma deeply 2-lobed, not capitate or discoid.   *8*

  Stigma capitate or discoid, at most shallowly 2-lobed.   *11*

8   Stigma-lobes spreading; fls yellow or red, without deep violet veins.

29. CHEIRANTHUS

  Stigma-lobes erect, facing one another, not spreading or spreading only at their tips; fls purplish or white, or, if pale yellow, then with deep violet veins.   *9*

9   Stigma-lobes thickened or horned at the back; maritime plants.

26. MATTHIOLA

  Stigma-lobes not thickened at the back.   *10*

10   Biennial to perennial, with simple toothed lvs; petals violet or white, not veined with deep violet; fr. not beaked.   27. HESPERIS

  Annual or overwintering, with lyrate-pinnatifid lvs; petals pale yellow or white, with deep violet veins; fr. erect, short-stalked, with a broad flat sabre-shaped beak. Introduced and a frequent casual.   **Eruca sativa* Miller

11   Fr. strongly flattened.   *12*

  Fr. quadrangular, cylindrical or slightly flattened.   *14*

<table>
<tr><td>12</td><td colspan="2">Valves of fr. veinless, rolling suddenly into spirals on dehiscence;
lvs pinnate; fls white, greenish or mauve.   22. CARDAMINE</td></tr>
<tr><td></td><td>Valves of fr. with a ±conspicuous central vein, not rolling into spirals on dehiscence; lvs simple, sometimes pinnatifid but not pinnate; fls white or very pale yellow.</td><td>13</td></tr>
</table>

13   Valves of fr. with a strong central vein and prominent seeds; basal lvs distinctly long-stalked, often lyrate-pinnatifid; a rare alpine plant with white fls. Northern Rockcress.

                            Cardaminopsis petraea (L.) Hiitonen

         Valves of fr. with a weak central vein; seeds not prominent; basal lvs not pinnatifid, narrowed gradually into a short stalk.

                                            24. ARABIS

| *14* | Fls white or lilac. | 15 |
| | Fls yellow, apricot or cream. | 19 |

| *15* | Lvs simple. | 16 |
| | At least the basal lvs pinnate or pinnatifid. | 17 |

16   Plant with both simple and branched hairs; basal lvs small, elliptical or spathulate.   32. ARABIDOPSIS

         Plant glabrous or with simple hairs; basal lvs reniform or cordate, long-stalked; plant smelling of garlic when crushed.

                                      30. ALLIARIA

<table>
<tr><td>17</td><td>Fls white or lilac, dark-veined, or white at first but later turning lilac; weeds of arable or waste land, with pinnatifid lvs.</td><td>18</td></tr>
<tr><td></td><td>Fls white; plants of wet places, with pinnate lvs.   25. RORIPPA</td><td></td></tr>
</table>

18   Fr. either distinctly constricted between the seeds or inflated and with a spongy wall; not flattened; indehiscent.   5. RAPHANUS

         Fr. a siliqua with somewhat flattened valves and seeds in 2 rows in each cell; dehiscent.   4. DIPLOTAXIS

| *19* | Plant with branched or stellate hairs. | 20 |
| | Plant glabrous or with simple hairs only. | 21 |

| *20* | Lvs deeply and finely pinnatisect.   34. DESCURAINIA | |
| | Lvs simple, entire or toothed.   28. ERYSIMUM | |

<table>
<tr><td>21</td><td>Fr. evidently with 2 valves (by which it will eventually open upwards from the bottom).</td><td>22</td></tr>
<tr><td></td><td>Fr. not valved (and not opening from below upwards), cylindrical, constricted between the seeds and ultimately breaking into 1-seeded joints.   5. RAPHANUS</td><td></td></tr>
</table>

<table>
<tr><td>22</td><td>Valves of fr. veinless or with an indistinct central vein which vanishes above.   25. RORIPPA</td><td></td></tr>
<tr><td></td><td>Valves of fr. with a distinct central vein.</td><td>23</td></tr>
</table>

<table>
<tr><td>23</td><td>Valves of fr. 1-veined, or with lateral veins very much weaker than the central so that the valves appear 1-veined.</td><td>24</td></tr>
<tr><td></td><td>Valves of fr. with lateral veins only slightly weaker than the central vein and so appearing 3–7-veined, at least when young.</td><td>30</td></tr>
</table>

24   Seeds in 2 rows in each cell of the fr.   *25*
       Seeds in 1 row in each cell of the fr.   *26*

25   Upper stem-lvs sagittate at base, clasping the stem; fls cream or very pale yellow; fr. 3–6 cm., erect, appressed to the stem. A tall glaucous plant, 30–100 cm. Local on dry banks, rocks, roadsides, etc. Tower Mustard.

                     Arabis glabra (L.) Bernh. (*Turritis glabra* L.)
       Upper stem-lvs not clasping the stem; fls bright yellow; fr. not appressed to the stem.   **4. DIPLOTAXIS**

26   Fr. with convex rounded valves.   **1. BRASSICA**
       Central veins of valves prominent so that the valves are keeled and the fr. ± quadrangular in section.   *27*

27   Stem and lvs ± glabrous.   *28*
       Stem hairy, at least below.   *29*

28   Lvs all glaucous, entire; fls cream-coloured; fr. 6–10 cm. Introduced. A frequent casual, established locally. Hare's Ear Cabbage.   **Conringia orientalis* (L.) Dumort.
       Lvs green, the basal lyrate-pinnatifid; fls yellow.   **23. BARBAREA**

29   Stem-lvs deeply pinnatifid; fr. 2–4 cm., curving upwards but not stiffly erect nor appressed to the stem. Introduced. A frequent casual.   **Erucastrum gallicum* (Willd.) O. E. Schulz
       Stem-lvs shallowly lobed or ± entire; fr. held stiffly erect, appressed to the stem when ripe.   **1. *Brassica nigra***

30   Fr. with a distinct beak, often containing 1 or more seeds, between the top end of the valves and the stigma.   *31*
       Fr. not beaked and never with seeds beyond the ends of the valves, which open almost to the stigma.   **31. SISYMBRIUM**

31   Fr. short-stalked, erect, appressed to the stem; beak short, swollen; seeds ovoid. Tall annual, 30–100 cm., stiffly hairy below and lower lvs grey with dense hairs. Introduced; naturalized in Channel Is and casual in S. England. Hoary Mustard.

                     **Hirschfeldia incana* (L.) Lagrèze-Fossat
       Fr. not appressed to the stem; beak flat or conical, not swollen; seeds spherical.   *32*

32   Sepals spreading horizontally; weeds of arable land and waste places.   **3. SINAPIS**
       Sepals erect; maritime plants, rarely casuals in waste places.
                     2. RHYNCHOSINAPIS

33   Fls yellow.   *34*
       Fls not yellow, or petals 0.   *41*

34   Fr. of 2 distinct segments, the lower resembling a short stalk, the upper ovoid or almost spherical; weeds of arable land or waste places.   **7. RAPISTRUM**
       Fr. not of 2 distinct segments.   *35*

35 Fr. almost spherical, tiny (1·5–3 mm. diam.); casuals.   *36*
      Fr. not spherical and tiny.   *37*

36 Lvs grey-green with stellate hairs, ±entire. Introduced; a casual
          weed of arable land.   *Neslia paniculata* (L.) Desv.
      Lvs glabrous or with simple hairs, irregularly toothed.
                                              25. *Rorippa austriaca*

37 Fr. distinctly flattened.   *38*
      Fr. not or slightly flattened.   *39*

38 Tufted plant with lvs confined to basal rosettes; fl-stem lfless; fr.
          elliptical.   19. *Draba aizoides*
      Fl-stems lfy; fr. 3–4 mm., almost circular, notched above. Annual
          herb with entire narrow lvs, grey with stellate down. Introduced;
          an occasional weed of grassy fields and arable land. Small
          Alison.   *Alyssum alyssoides* (L.) L.

39 Lvs entire or nearly so; fr. pear-shaped.   33. CAMELINA
      At least the basal lvs pinnatifid or distinctly toothed.   *40*

40 Fr. irregularly ovoid, with either crested wings or warty promi-
          nences.   17. BUNIAS
      Fr. neither winged nor warty.   25. RORIPPA

41 Large glaucous cabbage-like maritime plant with much-branched
          infl. of greenish-white fls and fr. having 2 segments, the lower
          slender, stalk-like, the upper almost spherical (and indehiscent).
                                                          6. CRAMBE
      Not as above.   *42*

42 Annual herb of sandy or shingly sea-shores with succulent lvs, at
          least the lower pinnatifid; fr. of 2 segments, the lower small, top-
          shaped, the upper larger, mitre-shaped.   8. CAKILE
      Not as above.   *43*

43 Petals bifid almost to half-way, white.   *44*
      Petals entire, or notched but not deeply bifid, or 0.   *45*

44 Stem lfless, 2–20 cm.   20. EROPHILA
      Stem lfy, 20–60 cm., lvs lanceolate, grey with stellate hairs; fr.
          7–10 mm., broadly elliptical, stellate-hairy. Introduced.
          Naturalized in a few places, casual elsewhere.
                                              Berteroa incana (L.) DC.

45 Two adjacent petals of each fl. conspicuously larger than the other
          two.   *46*
      Petals of equal size, or 0.   *47*

46 Lvs scattered up the stem; smaller petals twice as long as sepals;
          fr. 4–5 mm., distinctly winged.   12. IBERIS
      Lvs almost or quite confined to a basal rosette; smaller petals
          scarcely longer than sepals; fr. 3–4 mm., very narrowly winged.
                                                        14. TEESDALIA

47   Fr. rounded, not or slightly flattened.   16. COCHLEARIA
      Fr. distinctly flattened.   48

48   Cells of fr. 1-seeded.   49
      Cells of fr. each with 2 or more seeds.   51

49   Lvs 2–4 cm., all simple, narrow, entire, not clasping the stem; fr.
      2·5 mm., obovate, latiseptate, i.e. with the septum across the
      widest diam. Introduced. Naturalized in scattered coastal
      localities. Sweet Alison.   *Lobularia maritima* (L.) Desv.
      At least some lvs pinnatifid or toothed or clasping the stem; fr.
      angustiseptate, i.e. with the septum across the narrowest diam.   50

50   Infl. a corymbose panicle of small white fls; fr. broadly cordate or
      deltate, tapering above into the persistent style, not winged,
      indehiscent.   11. CARDARIA
      Infl. not corymbose; fr. not or hardly cordate below, often winged,
      dehiscent.   9. LEPIDIUM

51   Fr. 2–9 cm., with thin-walled translucent valves and silvery
      septum; petals purple (rarely white); lvs cordate. 18. LUNARIA
      Fr. usually less than 2 cm., its valves not translucent.   52

52   Fr. winged.   13. THLASPI
      Fr. not winged.   53

53   Fr. triangular-obcordate.   15. CAPSELLA
      Fr. oval or elliptical.   54

54   Lvs deeply pinnatifid to pinnate; fr. 2–4 mm., angustiseptate,
      i.e. with the septum across the narrowest diam., elliptical to
      oblong-obovate, on spreading stalks. A small overwintering
      annual, 2–14 cm., very local on limestone rocks and sand-dunes.
      Rock Hutchinsia.   *Hornungia petraea* (L.) Reichenb.
      Lvs simple, entire or toothed; fr. latiseptate, i.e. with the septum
      across the widest diam.   19. DRABA

1. BRASSICA L.

Annual or biennial, rarely perennial, herbs, glabrous or with simple hairs. Sepals ± erect; petals clawed, usually yellow; stamens 6. Fr. a narrow beaked siliqua with convex valves each with 1 dominant central vein; seeds almost spherical, in 1 row in each cell.

1   Upper stem-lvs stalked; fr.-stalks short, erect; fr. held close to the
      stem.   **3. nigra**
      Upper stem-lvs rounded or deeply cordate at base, often broadened
      and ±clasping; fr.-stalks spreading or ascending; fr. not held
      close to the stem.   2

2   All lvs glabrous; all stamens erect. A glaucous plant, woody below
     and covered with old lf-scars. Local on maritime cliffs in S.
     England and Wales. Wild Cabbage.   *B. oleracea* L.
     Lowest lvs always somewhat bristly; filaments of outer stamens
     curved outwards at base.   3

3   All lvs glaucous; buds slightly overtopping the open fls; petals pale
     yellow or buff.   **1. napus**
     Lowest lvs grass-green; open fls overtopping the buds; petals bright
     yellow.   **2. rapa**

1. B. napus L.   Rape, Cole; Swedish Turnip, Swede.

Annual or biennial with strong and often tuberous tap-root and erect
fl-stems up to 1 m. Lowest lvs stalked, lyrate, sparsely bristly; others
sessile, oblong with a broadened somewhat clasping base. Petals almost
twice as long as sepals. Fl. 5–8. Probably escaped from cultivation. The
biennial non-tuberizing variety (Rape, Cole, Coleseed) is often found as
an arable or wayside weed and on the banks of streams and ditches.
Swedish Turnip or Swede, with tuberized stem-base and tap-root, may
also escape.

2. B. rapa L.   Turnip, Navew.

Annual or biennial, with erect flowering stems to 1 m. Lowest lvs
stalked, lyrate, bristly; others sessile, oblong, almost completely clasping the stem with the broadened cordate base. Petals about half as long
again as the sepals. Fl. 5–8. Perhaps escaped from cultivation. Ssp.
sylvestris (L.) Janchen, with a non-tuberizing tap-root, is common
throughout the British Is, annual forms as weeds of arable land and
biennial forms on stream-banks. Ssp. **rapa** is the cultivated turnip.

3. B. nigra (L.) Koch   Black Mustard.

Annual, with narrow, erect flowering shoot up to 1 m., bristly below and
glaucous above, with numerous ascending branches. Lvs all stalked;
the lowest lyrate-pinnatifid with a large terminal lobe, grass-green,
bristly; other lvs narrow, sinuate or entire, glabrous, glaucous. Sepals
half-spreading, half as long as the bright yellow petals. Fr. erect and
appressed to the stem, ± quadrangular, with strongly keeled glabrous
valves and a slender seedless beak. Fl. 6–9. Probably native. On cliffs by
the sea in S.W. England, on stream-banks throughout England and
Wales, and common as an escape in waysides and waste places; in
Scotland only in the south; in Ireland mainly in the south and east.

2. RHYNCHOSINAPIS Hayek

Perennials with long tap-root and stems becoming woody below, glabrous or with simple hairs. Sepals erect; petals yellow; stamens 6. Fr. a narrow siliqua with distinctly 3-veined valves and a flattened sword-shaped beak. Seeds spherical, in 1 row in each cell.

1   Plant glabrous, except for a few hairs on the sepals; stem ascending with spreading branches from the basal rosette; stem-lvs few. Locally common on the west coast of Great Britain from N. Devon and Glamorgan to Kintyre and in the Isle of Man and Clyde Is. Isle of Man Cabbage.   *R. monensis* (L.) Dandy

    Stem hairy, at least below; lvs hairy, at least beneath; stem erect, branched above; stem-lvs several.   *2*

2   Stem with scattered hairs; lvs hairy beneath, rarely also above; sepals equalling or exceeding the fl.-stalks; siliqua 1·5–2 mm. wide. Introduced. A casual in S. England and S. Wales; naturalized in Channel Is. Tall Wallflower Cabbage.
                           **R. cheiranthos* (Vill.) Dandy

    Stem densely hairy; lvs hairy on both sides; sepals shorter than fl.-stalks; siliqua 3–4 mm. wide. Lundy Island. Lundy Cabbage.
                          R. wrightii (O. E. Schulz) Dandy

3. SINAPIS L.

Usually annual herbs with simple hairs. Sepals spreading; petals bright yellow, clawed; stamens 6. Fr. a siliqua with distinctly 3–7-veined valves and a long beak. Seeds spherical, in 1 row in each cell.

Upper lvs lanceolate, toothed; beak of fr. conical, straight, rather more than half as long as the valves.   **1. arvensis**
All lvs pinnately lobed or cut; beak of fr. strongly flattened, sabre-shaped, equalling or exceeding the valves.   **2. alba**

1. S. arvensis L.   Charlock, Wild Mustard.

Annual. Stem erect, 30–80 cm., usually stiffly hairy at least below. Lvs roughly hairy, the lower stalked, lyrate, the upper sessile, ± simple, lanceolate, coarsely toothed. Seeds dark brown. Fl. 5–7. A weed of arable land, especially in spring-sown crops on calcareous and heavy soils, throughout the British Is.

*2. S. alba L.   White Mustard.

Annual. Stem erect, 30–80 cm., with stiff downwardly directed hairs, rarely glabrous. Lvs stiffly hairy, all stalked and lyrate-pinnatifid or pinnate with the terminal lobe largest. Valves of fr. stiffly hairy, strongly 3-veined. Seeds yellow or pale brown. Fl. 6–8. Introduced. Naturalized as a weed of arable and waste land, especially on calcareous soils, throughout Great Britain and Ireland.

4. DIPLOTAXIS DC.

Annual to perennial herbs with pinnatifid lvs. Sepals somewhat spreading; petals clawed; stamens 6. Fr. a long slender short-beaked siliqua with flattened 1-veined valves. Seeds numerous, ovoid, in 2 rows in each cell.

1 Fls white or pale lilac; beak of fr. conical; annual; lvs with whitish
    horny-tipped teeth. Introduced. Naturalized as a weed of arable
    and waste land in S. England; casual elsewhere. White Wall
    Rocket.   *D. erucoides (L.) DC.
    Fls yellow; beak of fr. ±cylindrical.   2

2 Stem glabrous at base; lvs glaucous, the lower pinnatifid with long
    narrow lobes; fr. stipitate, i.e. distinctly stalked above the
    insertion-scars of the sepals.   2. tenuifolia
    Stem sparsely hairy towards base; lvs yellowish-green with short
    triangular teeth; fr. not stipitate.   1. muralis

*1. D. muralis (L.) DC.   Wall Rocket, Stinkweed.

Annual or perennial, with slender tap-root and erect or ascending stems, 15–60 cm., branched at the base, with a few stiff hairs especially towards the base. Lvs up to 10 cm., elliptic-spathulate, toothed or with triangular lobes up to twice as long as broad, yellowish-green, foetid when crushed; lvs at first almost confined to a basal rosette but these dying and replaced by stem-lvs in perennial plants. Fls c. 10 mm. diam., lemon yellow; petals 5–8 mm., twice as long as sepals. Fr. 30–40 × 2–4 mm., not stalked above the insertion-scars of the sepals; stigma hardly broader than style. Fr.-stalk much shorter than fr. Seeds yellow-brown. Fl. 6–9. Introduced. Naturalized especially in S. England on limestone rocks and walls and as a weed of arable and waste land; by railways in Ireland.

2. D. tenuifolia (L.) DC.   Perennial Wall Rocket.

Perennial, with long stout tap-root and erect stem, 30–80 cm., branched above, entirely glabrous and often glaucous. Rosette-lvs 0; lower stem-lvs deeply and narrowly pinnatifid with linear lobes never less than 3 times as long as broad; upper less deeply divided to almost entire; all glabrous, glaucous and foetid when crushed. Fls lemon yellow; petals 8–15 mm., twice as long as sepals. Fr. 25–35 × 2 mm., with stalk 1–

3 mm. between the insertion-scars of the sepals and the base of the valves; stigma much broader than style. Fr.-stalk usually almost as long as fr. Seeds as in *D. muralis*. Fl. 5–9. Doubtfully native on old walls and in waste places in S. England and a casual further north and in Ireland.

5. RAPHANUS L.

Annual to perennial herbs, somewhat glaucous above, stiffly hairy below, with lyrate-pinnatifid lvs. Sepals usually erect; petals long-clawed; stamens 6. Fr. elongated, jointed, the lowest joint resembling a short stalk, the upper part of the fr. either cylindrical and indehiscent, or constricted between the seeds and then often breaking into 1-seeded joints at maturity; fr.-apex narrowing into a seedless beak.

1   Tuberous tap-root; fr. inflated, not markedly constricted between the seeds; fls white or lilac with dark veins. Cultivated and sometimes escaping. Radish.   **R. sativus* L.
     Tap-root not tuberous; fr. markedly constricted between seeds.   2

2   Fr. with long but not very deep constrictions between the 3–8 seeds, breaking readily into 1-seeded joints; beak 5 times as long as the top joint.   **1. raphanistrum**
     Fr. with short deep constrictions between the 1–5 seeds, not breaking readily into 1-seeded joints; beak not more than twice as long as the top joint.   **2. maritimus**

1. R. raphanistrum L.   Wild Radish, White Charlock, Runch.

Annual, with slender whitish tap-root and erect stem, 20–60 cm., rough with spreading or reflexed bristles, especially below. Lower lvs with large terminal lobe and 1–4 pairs of much smaller distant laterals; upper lvs smaller; all grass-green and ± bristly. Petals twice as long as the sepals, yellow, lilac or white, usually dark-veined. Fr. 3–6 mm., weakly ribbed. Fl. 5–9. Probably native. A common and troublesome weed, especially of acid soils, throughout the British Is.

2. R. maritimus Sm.   Sea Radish.

Biennial or perennial, with stout tap-root and erect stem, 20–80 cm., bristly. Lower lvs with large terminal lobe and usually 4–8 pairs of contiguous lateral lobes, often alternating in size; upper lvs smaller; all dark green. Petals usually yellow, not strongly veined. Fr. 5–8 mm. diam., strongly ribbed. Fl. 6–8. On drift-lines and cliffs of sandy and rocky shores from Argyll and Durham southwards; Hebrides; Ireland; Channel Is.

6. CRAMBE L.

Herbs with swollen tap-roots, large lvs and white fls in large much-branched infls. Sepals spreading; petals short-clawed; stamens 6. Fr. indehiscent, 2-jointed, the lower joint slender, stalk-like, seedless; the upper spherical or ovoid, 1-seeded.

1. C. maritima L.   Sea-Kale.

Perennial, stoloniferous, with branched fleshy root-stock and erect fl.-stems, 40–60 cm., branching from the base, glabrous. Lower lvs up to 30 cm., ovate, long-stalked, glabrous, glaucous, ± pinnately lobed, with irregularly toothed and wavy margins; upper lvs narrow, ± entire. Fls 10–16 mm. diam., petals white with green claws. Fr. 12–14 × 8 mm. ascending to make an angle with the spreading stalk. Fl. 6–8. On coastal sands, shingle, rocks and cliffs from Fife and Islay southwards; Ireland.

7. RAPISTRUM Crantz

Annual to perennial bristly herbs with pinnately lobed or cut lvs. Sepals half-spreading; petals short-clawed; stamens 6. Fr. of 2 joints: the lower ± slender, with 0–2 seeds, separated by a constriction from the larger upper joint which is ovoid to spherical, usually 1-seeded, falling when ripe.

1   Fr. with upper joint narrowing gradually into a broadly conical
      beak 0·5–1 mm.; perennial; petals bright yellow. Introduced.
      Naturalized locally as a weed of arable land; casual elsewhere.
                                                    R. perenne (L.) All.
    Upper joint of fr. surmounted by the slender persistent style,
      1–5 mm.; annual.   2

2   Fr.-stalk 1–3 times as long as the lower joint; upper joint strongly
      wrinkled and ribbed, narrowing suddenly into the slender style;
      petals lemon-yellow with darker veins. Introduced. Naturalized
      locally as a weed of arable and waste land and frequent as a
      casual.   *R. rugosum* (L.) All.
    Fr.-stalk 2–4 times as long as the lower joint; upper joint hardly
      wrinkled, slightly ribbed, narrowing gradually into the slender
      style. Introduced. Casual.   *R. hispanicum* (L.) Crantz

8. CAKILE Miller

Annual herbs with glabrous succulent lvs. Sepals erect; petals clawed; stamens 6. Fr. indehiscent, of 2 unequal 1-seeded joints: the upper larger, mitre-shaped, the lower smaller, top-shaped. At maturity the upper joint breaks off, the lower remaining attached.

1. C. maritima Scop.   Sea Rocket.

Very long slender tap-root and prostrate or ascending branched stems, 15–45 cm. Lower lvs narrowed into a stalk base, ± deeply pinnatifid, the lobes oblong, distant, entire or distantly toothed; upper lvs sessile, less lobed or entire. Petals purple, lilac or white, twice as long as the sepals. Fr. 10–25 mm. overall, on short thick stalks; upper joint up to twice as long as the lower, with two broadly triangular basal teeth fitting over the convex top of the lower joint which is sometimes seedless and stalk-like. Fl. 6–8. On drift-lines of sandy and shingly shores all round Great Britain, Ireland and the Channel Is.

9. LEPIDIUM L.

Annual to perennial herbs often with simple hairs. Fls small, whitish, in dense infls. Petals sometimes 0; stamens 2, 4 or 6. Fr. an angusti-septate silicula with strongly keeled or winged valves and usually 1 seed in each cell.

1   Fr. at least equalling its stalk, broadly winged above, the wing
    involving the lower part of the style.   *2*
    Fr. shorter than its stalk, not winged or, if so, then with the style free
    from the narrow wing.   *3*
2   Annual or biennial; fr. densely covered with small scale-like vesicles;
    style usually not projecting beyond the apical notch of the wing.
                              1. campestre
    Perennial; vesicles on fr. few or 0; style projecting beyond the apical
    notch of the wing.   **2. heterophyllum**
3   Annual or biennial, not exceeding 30 cm.; basal lvs deeply pinnate-
    cut; petals usually 0; fr. with very short style at base of apical
    notch.   **3. ruderale**
    Perennial, 50–150 cm.; lvs simple, ovate to lanceolate; petals white,
    longer than sepals; fr. with the short style projecting beyond the
    top of the shallow apical notch. Salt-marshes and wet sand; local.
    Dittander.   *L. latifolium* L.

1. L. campestre (L.) R.Br.   Pepperwort.

Annual or biennial, with a single erect stem, 20–60 cm., usually grey-green with dense short spreading hairs; branches curving upwards. Basal lvs entire or lyrate, falling early; lower stem-lvs narrowed into a short stalk; middle and upper stem-lvs narrowly triangular, sessile, with long narrow pointed basal lobes; all softly hairy and with small distant marginal teeth. Fls 2–2·5 mm. diam. Petals white, little longer than the sepals. Stamens 6; anthers yellow. Fr. 5 × 4 mm.; style not or slightly projecting beyond the apical notch. Fl. 5–8. In dry pastures, on walls and banks, by waysides and in arable and waste land throughout Great Britain from Moray and Lanark southwards; rare in Ireland.

2. L. heterophyllum Bentham (*L. smithii* Hooker)

Smith's Pepperwort.

Perennial, with stout woody root-stock and many ascending stems, 15–45 cm., grey-green with short spreading hairs; branches curving upwards. Basal lvs oblanceolate or elliptical, falling early; lower stem-lvs narrowed into a short stalk; middle and upper stem-lvs narrowly triangular, sessile, with long narrow pointed basal lobes. Fls 3–3·5 mm. diam. Petals white, half as long again as the sepals. Stamens 6; anthers violet. Fr.-stalks hairy; fr. 5 × 4 mm.; style projecting beyond the apical notch. Fl. 5–8. In arable fields, on dry banks and waysides, throughout Great Britain from Moray and Dunbarton southwards; Ireland.

3. L. ruderale L.　　Narrow-leaved Pepperwort.

Annual or biennial, foetid, with a single erect or ascending stem, 10–30 cm., almost glabrous or with sparse short spreading hairs. Basal lvs long-stalked, deeply pinnately divided, the narrow segments further divided or lobed; lower stem-lvs pinnate with narrow entire segments; middle and upper stem-lvs sessile, narrowly oblong, entire. Petals usually 0. Stamens 2 or 4. Fr. 2–2·5 × 1·5–2 mm., deeply notched above; style very short, at the base of the notch. Fl. 5–7. In waste places and by waysides, generally near the sea, throughout England but especially in E. Anglia; rare in Scotland.

10. CORONOPUS Haller

Annual to perennial herbs often with deeply pinnatifid lvs. Infl. often opposite the lvs. Sepals half-spreading; petals whitish, often small or 0; stamens 2, 4 or 6. Fr. indehiscent or splitting into 1-seeded halves; valves hemispherical, ridged or pitted; septum narrow.

Fr. not notched above, longer than its stalk.　　**1. squamatus**
Fr. with an apical notch, shorter than its stalk.　　**2. didymus**

1. C. squamatus (Forskål) Ascherson (*Senebiera coronopus* (L.) Poiret)

Swine-cress, Wart-cress

Annual or biennial with prostrate branched stems, 5–30 cm. Lvs stalked, deeply pinnate-cut, the segments of the lower lvs obovate to oblanceolate, pinnately lobed especially on the upper side; those of the upper lvs narrower, ± entire. Infls terminal and lf-opposed. Fls 2·5 mm. diam. Petals white, longer than the sepals. Stamens usually 6. Fr. 2·5–3 × 4 mm., emarginate below but narrowed abruptly above into the short pointed style; valves deeply reticulate-pitted or strongly and irregularly ridged, indehiscent. Fl. 6–9. Waste ground and especially trampled places; common in S. England, rarer northwards; throughout Ireland but rare in the centre.

*2. C. didymus (L.) Sm. Lesser Swine-cress.

Annual or biennial, foetid, with prostrate or ascending stems, 15–30 cm. Basal and lower lvs stalked, very deeply pinnately cut, the segments oblanceolate, pinnatifid, with slender acute lobes especially on the upper side; upper lvs sessile with narrower and ± entire segments. Infls terminal and lf-opposed or at forkings of the stem. Fls 1–1·5 mm. diam. Petals white, shorter than the sepals, or more usually 0. Fertile stamens usually 2, rarely 4. Fr. 1·5 × 2·5 mm., emarginate above and below; valves reticulate-pitted; style 0; valves separating into 1-seeded nutlets. Fl. 7–9. Introduced. A weed of cultivated and waste ground especially in S. England and S. Ireland, extending northwards to Ross.

11. CARDARIA Desv.

Like *Lepidium* but with an indehiscent cordate fr.

*1. C. draba (L.) Desv. (*Lepidium draba* L.) Hoary Pepperwort.

A perennial herb with a deep tap-root and fine spreading roots which bear adventitious stem-buds. From the branched woody stock arise several erect densely lfy stems, 30–90 cm., glabrous or with short simple hairs. Basal lvs stalked, obovate, ± sinuate-toothed, falling early; middle and upper stem-lvs ovate-oblong, enlarging below into auricles that clasp the stem; all ± sinuate-toothed and ± hoary with short simple hairs. Fls 5–6 mm. diam., in a dense corymbose panicle. Petals white, twice as long as the sepals. Stamens 6. Fr. 4 × 4 mm., ± emarginate below but tapering above into the persistent style; valves rounded, unwinged, somewhat keeled above, reticulate when dry, indehiscent but often separating into 1-seeded nutlets; fr. asymmetric if only 1 seed matures. Fl. 5–6. Introduced. Spreading rapidly as a weed of arable land and now throughout England and Wales to Yorkshire and Cumberland; Kintyre; local in Ireland.

12. Iberis L.

Annual to perennial herbs with narrow lvs and simple hairs. Infl.
± corymbose. Petals white or pinkish, the 2 towards the outside of the
infl. much larger than the other 2; stamens 6. Fr. an angustiseptate
silicula, the valves keeled and usually winged above. Seeds 1 in each
cell, flat, large, often winged.

1. I. amara L.   Wild Candytuft.

Annual with ± erect lfy stem, 10–30 cm., ± hairy below. Lvs scattered,
the lower spathulate narrowing into a stalk-like base; the upper
oblanceolate-cuneate, sessile; all distantly pinnatifid or toothed, sometimes entire, ciliate but otherwise almost glabrous. Fls 7–8 mm. diam.,
white or mauve, the infl. elongating in fr. Siliculae 4–5 mm., suborbicular, the wings broadening upwards and leaving a deep triangular
apical notch; style about equalling the notch. Seeds slightly winged.
Fl. 7–8. Locally common on dry calcareous hillsides and in cornfields
southwards from Gloucester and S. Lincs.; naturalized further north.

13. Thlaspi L.

Annual to perennial glabrous herbs with simple lvs, the stem-lvs usually
± clasping. Petals short-clawed; stamens 6. Fr. an angustiseptate silicula with the valves keeled and usually winged. Seeds 1–8 in each cell.

1 Fr. almost circular, 12–22 mm.   **1. arvense**
     Fr. obovate or obcordate, less than 8 mm.   2

2 Annuals; style less than 0·5 mm., included within the apical notch of
     the fr.   *3*
     Biennial to perennial, with short branched woody stock; style
     variable in length but usually more than 0·5 mm. and equalling or
     exceeding the apical notch of the fr. A local alpine or subalpine
     plant of basic rocks; often common on spoil from old lead-
     workings. Alpine Pennycress.   *T. alpestre* L.

3 Stem 20–60 cm., hairy below, grooved; stem-lvs with acute auricles;
     fls white; fr. narrowly obovate with narrow wing and shallow
     apical notch. A garlic-scented weed of arable land. Introduced
     and locally naturalized in S. England.   **T. alliaceum* L.
     Stem 5–25 cm., glabrous, terete; stem-lvs glaucous, with blunt
     auricles; fls white; fr. broadly obovate with broad wing and deep
     apical notch. A rare plant of limestone spoil in W. England;
     casual elsewhere. Perfoliate Penny-cress.   *T. perfoliatum* L.

1. T. arvense L.   Field Penny-cress.

Annual, with erect lfy stem, 10–60 cm., foetid when crushed. Basal lvs oblanceolate to obovate, narrowed into a stalk-like base; stem-lvs oblong or lanceolate, sessile, with sagittate clasping base; all glabrous, entire or distantly sinuate-toothed. Infl. greatly lengthening in fr. Petals white, twice as long as the sepals. Fr. on upwardly curving stalks; wing broad, especially above; style very short, at base of the deep narrow apical notch; seeds 5–8 in each cell. Fl. 5–7. Probably native. A weed of arable land and waste places throughout most of the British Is.

14. TEESDALIA R.Br.

Annual herbs, glabrous or with simple hairs. Lvs usually pinnatifid, all or most in a basal rosette. Petals white, sometimes unequal; stamens 4 or 6, their filaments broadly white-winged below. Fr. an angustiseptate roundish-cordate silicula, narrowly winged above, concave on the upper and convex on the lower face; style very short. Seeds 2 in each cell.

1. T. nudicaulis (L.) R.Br.   Shepherd's Cress.

Stem erect, 6–45 cm., glabrous or shortly pubescent, often with ascending basal branches. Basal lvs 2–5 cm., stalked, narrowly lyrate-pinnatifid with few short blunt lateral segments and a broader often 3-lobed terminal segment; stem-lvs 1–3, on laterals only. Fls 2 mm. diam. Inner petals slightly longer than the sepals, outer twice as long. Stamens 6. Fr. 3–4 mm., somewhat broader above the middle, emarginate above. Fl. 4–6. Locally common on sand and gravel throughout Great Britain northwards to Ross; rare and local in Ireland.

15. CAPSELLA Medicus

Annual or biennial herbs with entire or pinnatifid basal lvs and clasping stem-lvs. Petals usually white; stamens 6. Fr. an angustiseptate silicula, usually obcordate, with keeled valves. Seeds several in each cell.

1. C. bursa-pastoris (L.) Medicus   Shepherd's Purse.

Stem 3–40 cm., glabrous or with simple and branched hairs. Basal lvs in a rosette, oblanceolate in outline, narrowed into a stalk, varying from very deeply pinnatifid to quite entire; stem-lvs with acute basal auricles. Fls 2·5 mm diam., white. Petals up to twice as long as the sepals. Fr. 6–9 mm., triangular-obcordate. Fl. 1–12. Common on cultivated land, waysides and waste places everywhere in the British Is.

16. Cochlearia L.

Scurvy-grass.

Annual to perennial herbs with usually simple lvs, glabrous or with simple hairs. Petals short-clawed; stamens 6. Fr. a swollen or angusti-septate silicula, its valves strongly convex with a strong midrib. Seeds in 2 rows in each cell.

1 Basal lvs wedge-shaped at base; petals 5–7 mm.; fr. much flattened laterally, its septum at least 3 times as long as broad.

3. anglica

Basal lvs not wedge-shaped at base; petals usually not exceeding 5 mm.; fr. little flattened laterally, its septum usually not more than twice as long as broad.   2

2 Fls 3–5 mm. diam.; seeds 1–1·5 mm.; upper lvs usually stalked.

2. danica

Fls more than 5 mm. diam.; seeds more than 1·5 mm.; upper lvs usually sessile.   3

3 Plants variable in size, often large and straggling, with cordate basal lvs; fls usually white.   **1. officinalis** group

Dwarf compact plants with reniform or truncate-based lvs, not cordate; fls often pale mauve. Local on northern coasts, includ-ing N. and W. Ireland. Scottish Scurvy-grass.   *C. scotica* Druce

1. C. officinalis group   Common Scurvy-grass.

Biennial to perennial, with long stout tap-root and one or more ascend-ing glabrous shoots, 5–50 cm. Basal lvs in a loose rosette, long-stalked, orbicular- to reniform-cordate, ±entire; stem-lvs oblong or ovate, coarsely and distantly toothed or sinuate, only the lowest short-stalked, the rest sessile and clasping the stem with broadly cordate base; all glabrous. Fls 8–10 mm. diam. Fr. 3–7×2–6 mm., ovoid-elliptical to nearly spherical, usually narrowed above into the short persistent style; valves with prominent midrib. Seeds usually more than 1·5 mm., 2–4(–6) per cell.

Two species may be recognized:

C. officinalis L.: lvs orbicular-cordate, markedly thick and fleshy; fr. usually subglobose, rounded below, but occasionally ovoid-elliptical, narrowed below as well as above. Fl. 5–8. Widely distributed on drier salt and brackish marshes and on cliffs and banks near the sea throughout the British Is.

C. pyrenaica DC. (*C. alpina* (Bab.) H. C. Watson): lvs usually reniform, rather firm but not markedly thick and fleshy; fr. usually ovoid-elliptical, narrowed at both ends, but occasionally subglobose, rounded below. Fl. 6–8. Local by streams, in flushes and on wet rock-ledges in alpine and subalpine regions of the British Is. Very variable in size and compactness.

Some inland populations do not fall clearly into either species.

2. C. danica L.   Danish Scurvy-grass.

An overwintering annual with ascending shoots, 10–20 cm. or sometimes much shorter. Basal lvs long-stalked, roundish to triangular, c. 1 cm. wide, cordate at the base, fleshy; stem-lvs mostly stalked, entire or distantly toothed, not clasping, the lowest 3–7-lobed like small ivy lvs. Fls 4–5 mm. diam., whitish or mauve. Fr. $3–5 \times 2 \cdot 5–4$ mm., ovoid to ellipsoid, $\pm$ narrowed at both ends, finely net-veined when ripe. Seeds c. 1 mm. Fl. 1–6. Locally common on sandy and rocky shores and on walls and banks by the sea all round the British Is.

3. C. anglica L.   Long-leaved Scurvy-grass.

Biennial to perennial with slender tap-root and stiffish $\pm$ erect shoots, 8–35 cm. Basal lvs in a rosette, ovate to obovate, $\pm$ cuneate at the base, never cordate, entire or with a few distant teeth, fleshy; stem-lvs ovate or elliptical, toothed or entire, mostly sessile, the upper lvs clasping the stem. Fls 10–14 mm. diam., white or pale mauve. Fr. 8–15 mm., much compressed laterally and strongly furrowed at the narrow septum which is 3–5 times as long as wide. Seeds 2–2·5 mm., 5–6 per cell. Fl. 4–7. Locally common on muddy shores and in estuaries all round the British Is except Orkney and Shetland. Hybrids with *C. officinalis* and *C. danica* have been reported.

17. BUNIAS L.

Annual to perennial herbs with branched and simple hairs and some stout glandular hairs. Lvs $\pm$ pinnatifid, hairy. Infl. much elongating in fr. Sepals half-spreading; petals yellow; stamens 6. Fr. irregularly ovoid, warty or with wing-like crests, 1–4-celled with 1–4 seeds, indehiscent.

Annual to biennial, 30–60 cm.; fr. with 4 irregularly crested wings and a
　　long slender style. Introduced. A frequent casual, sometimes
　　established.   *B. erucago* L.
Biennial to perennial, 25–100 cm.; fr. irregularly ovoid, warty, with a
　　short broad asymmetrically placed style. Introduced. Casual,
　　sometimes established.   *B. orientalis* L.

18. LUNARIA L.

Annual to perennial herbs with simple hairs and large cordate lvs.
Petals usually purple; stamens 6. Fr. a large latiseptate silicula, with
quite flat thin-walled translucent net-veined valves and a shining white
septum. Seeds large, strongly flattened, in 2 rows in each cell.

L. annua L. (Honesty), a usually biennial herb, 30–100 cm., is much
  grown in gardens for its decorative broadly elliptical to almost
  circular fr., 30–45 × 20–25 mm., and often escapes.

19. DRABA L.

Annual to perennial often densely tufted plants with stellate-hairy or
rarely glabrous shoots and rosettes of simple lvs. Fls small, white or
yellow; petals entire or slightly notched; stamens 6 or 4. Fr. a latiseptate
silicula, its valves ± flat with the central vein conspicuous only in the
lower half. Seeds numerous, in 2 rows in each cell.

1 Lvs glabrous apart from marginal bristles; fls yellow. A rare tufted
     plant of limestone rocks and slopes in S. Wales. Yellow Whitlow
     Grass.   *D. aizoides* L.
     Lvs stellate-hairy; fls white.   2

2 Fl.-stems lfless or with 1–2 small lvs. A rare alpine near the tops of a
     few Scottish mountains. Rock Whitlow Grass.
                          D. norvegica Gunn. (*D. rupestris* R.Br.)
     Fl.-stem lfy.   3

3 Stem-lvs lanceolate or narrowly ovate, not or hardly clasping; fr.
     twisted.   **1. incana**
     Stem-lvs broadly ovate, clasping; fr. straight.   **2. muralis**

1. D. incana L.   Hoary Whitlow Grass.

Biennial to perennial, with short prostrate occasionally branched stock
bearing the remains of dead rosettes, and erect shoots, 7–50 cm. Rosette-
lvs oblong-lanceolate, narrowing to a stalk, entire or distantly toothed,
densely stellate-hairy, ciliate; stem-lvs ± erect, sessile, narrowly elliptical
or ovate, rounded at the base, entire to coarsely toothed, densely stellate-
hairy and ciliate. Infl. dense, elongating later. Fls 3–5 mm. diam., white.
Fr. 7–9 × 2–2·5 mm., elliptical to lanceolate, twisted, glabrous or stellate-
hairy, held erect. Fl. 6–7. Local on screes and rock-ledges, especially of
limestone, and on shelly sand-dunes from Caernarvon, Stafford and
Derby northwards; N.W. Ireland, to Antrim and Galway; Kerry.

2. D. muralis L.   Wall Whitlow Grass.

Annual to biennial, with an ascending lfy stem, 8–30 cm., densely stellate-hairy below, sparsely above. Rosette-lvs obovate or oblanceolate, narrowed into a stalk, entire or toothed; stem-lvs sessile, broadly ovate with a broad rounded clasping base and sharply toothed margins; all lvs rough with stellate and simple hairs. Fls 2·5–3 mm. diam., white. Fr. 3–6 × 1·5–2 mm., elliptical-oblong, straight, spreading on almost horizontal stalks; glabrous. Fl. 4–5. Local on limestone rocks and walls in W. and N. England from Somerset to Westmorland; a garden weed elsewhere.

20. EROPHILA DC.

Small annual or overwintering herbs with lvs confined to a basal rosette. Petals white, deeply bifid; stamens 6. Fr. a latiseptate silicula, the valves with a thin midrib vanishing above the middle. Seeds numerous, in 2 rows in each cell. Differs from *Draba* chiefly in the deeply bifid petals.

1. E. verna (L.) Chevall.   Spring Whitlow Grass.

Lvs 4–15 mm., lanceolate to spathulate, ± acute, narrowed below into a stalk, with very varying numbers of stellate and simple hairs. Flowering stems 2–20 cm., hairy below. Fls 3–6 mm. diam. Fr. 3–9 × 1·5–3 mm., obovoid to ellipsoid, ± compressed. Fl. 3–6. Very variable. Common and widely distributed on rocks, walls, sand-dunes and other dry places throughout Great Britain and Ireland.

21. ARMORACIA Gilib.

Tall glabrous perennial herb with stout cylindrical sharp-tasting roots and panicled fls. Petals white; stamens 6. Fr. an almost spherical silicula with indistinctly net-veined valves; seeds in 2 rows in each cell.

*1. A. rusticana P. Gaertner, B. Meyer & Scherb.   Horse-Radish.

Root-stock thick, branched above and continuous downwards with the very long fleshy cylindrical roots, which produce adventitious buds. Stems to 125 cm., erect, lfy. Basal lvs 30–50 cm., ovate to ovate-oblong, long-stalked, crenate-serrate; stem-lvs short-stalked to sessile, elliptical or oblong-lanceolate, lower often pinnatifid, upper coarsely toothed to entire. Infl. a much-branched corymbose panicle. Fls 8–9 mm. diam. Fr. (rarely ripening) 4–6 mm., on slender ascending stalks; seeds 8–12 in each cell. Fl 5–6. Introduced for the condiment prepared from the roots and widely naturalized in fields, roadsides and waste places and on stream-banks throughout Great Britain northwards to Moray; very local in Ireland.

22. Cardamine L.

Herbs with pinnate or trifoliate lvs. Petals white or purple; stamens 4–6. Fr. a siliqua with flattened inconspicuously veined valves which open suddenly, coiling spirally from the base and flinging the seeds to some distance; seeds in 1 row in each cell.

1	Petals broad, spreading, about 3 times as long as the sepals.	2
	Petals narrow, ± erect, little longer than the sepals or 0.	3
2	Petals usually white; anthers purple.	**2. amara**
	Petals purplish, rarely white; anthers yellow.	**1. pratensis**
3	Lvs with slender pointed basal lobes which clasp the stem; fls 6 mm. diam., with narrow whitish petals scarcely longer than the sepals, often 0; anthers greenish; fr. 18–30 mm. Local in woods and on moist limestone. Narrow-leaved Bitter-cress. *C. impatiens* L.	
	Lvs not with basal stem-clasping lobes.	4
4	Annual; stamens 4.	**4. hirsuta**
	Biennial or perennial; stamens 6.	**3. flexuosa**

For *C. bulbifera* (L.) Crantz (*Dentaria bulbifera* L.) see p. 51.

1. C. pratensis L.   Cuckoo Flower, Lady's Smock.

Perennial, with short prostrate or ascending stock and erect or ascending terete ± glabrous stems, 30–60 cm. Basal lvs in a rosette, long-stalked, pinnate, the terminal lflet reniform and much larger than the ovate or roundish laterals, all with distant teeth; stem-lvs short-stalked with narrowly lanceolate lflets, all ± entire or the terminal 3-toothed; all lvs sparsely hairy. Vegetative reproduction in damp places by forma-tion of small plantlets in axils of basal lvs. Infl. of 7–20 fls, corymbose at first. Fls 12–18 mm. diam. Petals lilac, rarely white, c. three times as long as the violet-tipped sepals; anthers yellow. Fr. 25–40 × 1–1·5 mm. Fl. 4–6. Common throughout the British Is. in damp meadows and pastures and by streams.

2. C. amara L.   Large Bitter-cress.

Perennial, with horizontal creeping stock from which arise stems, 10–60 cm., prostrate at the base then erect or ascending, angled, usually glabrous. Basal lvs not in a rosette, stalked, pinnate, the 5–9 lflets ovate or orbicular, ± cordate, short-stalked, the terminal largest; upper lvs very short-stalked with 5–11 ovate to lanceolate lflets; all lflets angular to crenate in outline. Fls c. 12 mm. diam. Petals white, rarely purple, twice as long as the sepals; anthers violet. Fr. 20–40 × 1–2 mm. Fl. 4–6. Locally abundant in springs, flushes, fens and on streamsides, chiefly on peat and often in alder woods, throughout Great Britain from Aberdeen southwards; N.E. Ireland.

3. C. flexuosa With.   Wood Bitter-cress.

Usually perennial, with short branched ascending stock and erect or ascending very lfy flexuose stem, 10–50 cm., hairy especially below. Basal lvs few in a loose rosette, stalked, pinnate, with c. 5 pairs of rounded or reniform lflets and a larger terminal lflet; stem-lvs short-stalked to sessile with 5 or more pairs of lflets becoming narrower up the stem; all lflets gland-toothed or ± lobed and sparsely ciliate. Fls 3–4 mm. diam. Petals white, twice as long as the sepals. Stamens usually 6. Young frs usually not overtopping the unopened fls. Fr. 12–25 × 1 mm. on slender upwardly curved stalks. Fl. 4–9. Common in moist shady places, by streams, etc., throughout the British Is.

4. C. hirsuta L.   Hairy Bitter-cress

Usually annual, with erect ± glabrous stem, 7–30 cm. Basal lvs numerous in a compact rosette, stalked, pinnate, with 3–7 pairs of obovate or orbicular lflets and a larger reniform terminal lflet; stem-lvs few, almost sessile, with smaller and narrower lflets; all lflets lobed or angled and sparsely hairy above and on the margins. Fls 3–4 mm. diam. Petals white, twice as long as the sepals, sometimes 0. Stamens usually 4, sometimes 6. Young frs usually overtopping the unopened fls. Fr. 18–25 × 1 mm., on slender upwardly curved stalks. Fl 4–8. Common throughout the British Is on bare ground, rocks, screes, walls, etc.

The two preceding species are not always clearly distinguishable.

23. BARBAREA R.Br.

Biennial or perennial ± glabrous herbs with yellowish tap-roots. Basal lvs lyrate-pinnatisect; stem-lvs ± sessile, clasping. Fls bright yellow; stamens 6. Fr. a bluntly 4-angled siliqua whose valves have a strong midrib and a lateral network of veins. Seeds in 1 row in each cell.

1   Upper stem-lvs simple, toothed or shallowly lobed.   2
     Upper stem-lvs pinnately divided or lobed.   3
2   Terminal segment of basal lvs rounded, shorter than the rest of the
     lf; fl.-buds glabrous; petals about twice as long as the sepals.
                                           1. vulgaris

Terminal segment of basal lvs ovate, equalling the rest of the lf; fl.-
buds hairy; petals about half as long again as the sepals. A rare
plant of stream-banks and waste places. Small-flowered Yellow
Rocket.   *B. stricta* Andrz.

3   Basal lvs with 3–5 pairs of lateral segments; petals up to twice as
long as sepals; fr. 1–3 cm., straight. Introduced. Local in cultivated fields, by streams and on waysides. Intermediate Yellow
Rocket.   **B. intermedia* Boreau
Basal lvs with 6–10 pairs of lateral segments; petals about three times
as long as sepals; fr. 3–6 cm., upwardly curved. Introduced.
Widespread but local in waste and cultivated ground. Early-
flowering Yellow Rocket.   **B. verna* (Miller) Ascherson

1. B. vulgaris R.Br.   Winter Cress, Yellow Rocket.

Biennial or perennial with erect stem, 30–90 cm., stalked rosette-lvs with
a rounded sometimes cordate terminal segment, shorter than the rest of
the lf, and 5–9 lateral segments, the uppermost pair of laterals at least
equalling the width of the terminal. Lower stem-lvs lyrate-pinnatisect
with a few small lateral segments; upper stem-lvs simple, ovate,
± irregularly toothed; all lvs deep green, shining, glabrous. Fl. 7–9 mm.
diam. Petals twice as long as the sepals. Fr. 15–25 × 1·5–2 mm., ± erect
or upwardly curved. Fl. 5–8. Hedges, stream-banks and other damp
places throughout Great Britain and Ireland, but less common in the
north.

24. ARABIS L.

Annual to perennial herbs usually with simple and branched hairs.
Stem-lvs numerous, sessile. Stamens 6. Fr. a strongly flattened unbeaked siliqua, its valves with a slender indistinct midrib; seeds in 1 row
in each cell (2 in *A. glabra* (L.) Bernh.: see p. 53).

1   Fls yellowish; fr. 8–12 cm., all twisted to one side and curving
downwards. Introduced. Established on old walls in Cambridge.
Tower Rock-cress.   **A. turrita* L.
Fls white; fr. ± erect.   2

2   Mat-forming perennials with fls more than 6 mm. diam.   3
Not mat-forming; fls not more than 6 mm. diam.   4

3   Lvs grey-green or whitish, with 2–3 marginal teeth on each side; fls
c. 15 mm. diam. Introduced. Much grown on walls and rock-
gardens and naturalized locally. Arabis.   **A. caucasica* Willd.
Lvs green, with 3–6 marginal teeth on each side; fls 6–10 mm. diam.
Known only from mountains in Skye. Alpine Rock-cress.
                                                            A. alpina L.

4 Lvs hairy only on their margins; stem glabrous; seeds unwinged.
Dunes and other sandy places in W. Ireland. Fringed Rock-cress.
A. brownii Jordan
Lvs hairy on their surfaces; stem roughly hairy, at least below; seeds winged. 5

5 Fls 3–4 mm. diam., white, in a many-fld raceme; seeds roundish, 1·2–1·5 mm. **1. hirsuta**
Fls 5–6 mm. diam., cream-coloured, in a 3–6-fld raceme; seeds ovoid, 1·6 mm. Only in a few places on limestone near Bristol. Bristol Rock-cress. *A. Scabra* All. (*A. stricta* Hudson)

1. A. hirsuta (L.) Scop. Hairy Rock-cress.

Biennial or perennial with long slender tap-root and 1 or more erect stems, 10–60 cm., rough with simple and some stellate hairs. Basal lvs in a rosette, obovate, narrowing gradually into a stalk-like base; stem-lvs erect, ovate to narrowly oblong, truncate or half-clasping at the base; all lvs rough with simple and stellate hairs, and entire or with distant marginal teeth. Fls 3–4 mm. diam. Petals white, twice as long as the sepals. Fr. 15–50 × 1·2–1·5 mm., erect, glabrous. Seeds winged. Fl. 6–8. On chalk and limestone slopes, on limestone rocks and walls, on dry banks and sometimes on sand-dunes throughout Great Britain and Ireland.

25. RORIPPA Scop.

Annual to perennial herbs, glabrous or with simple hairs. Petals white or yellow; stamens 6; ovary with many ovules in each cell. Fr. variable in shape but always with convex valves whose midrib is very slender or indistinct or vanishes above the middle; seeds in 1 or 2 rows in each cell.

1 Fls white. 2
Fls yellow. 4

2 Fr. a well-formed siliqua with many good seeds. 3
Fr. dwarfed and misshapen with few or no good seeds.
3. microphylla × nasturtium-aquaticum

3 Fr. 13–18 mm.; seeds distinctly 2-rowed with c. 25 depressions in each face of the seed-coat. **1. nasturtium-aquaticum**
Fr. 16–22 mm.; seeds ± in 1 row with c. 100 depressions in each face. **2. microphylla**

4 Fr. 9–18 mm. **4. sylvestris**
Fr. usually less than 9 mm. 5

5 Fr. 1·5–3 mm., spherical, with persistent style almost equalling rest of fr. Introduced. Naturalized locally, casual elsewhere. Austrian Yellow-cress. ***R. austriaca** (Crantz) Besser

Fr. 3–9 mm., ovoid or oblong, with style not more than half as long
as rest of fr.   6

6   Fr. 3–6 mm., ovoid, on stalks at least twice as long; petals about
twice as long as sepals.   **6. amphibia**
Fr. 4–9 mm., oblong, about equalling their stalks; petals not or
slightly longer than sepals.   **5. islandica**

1. R. nasturtium-aquaticum (L.) Hayek (*Nasturtium officinale* R.Br.)
Water-cress.

Usually perennial, with hollow angular glabrous stems, commonly
10–60 cm., procumbent and rooting below, then ascending or floating.
Lvs lyrate-pinnate, lowest stalked with 1–3, upper sessile, auricled, with
5–9 or more lflets; terminal lflet roundish or broadly cordate, laterals
narrower, all entire or sinuate-toothed. Lvs and stems remain green in
autumn. Fls 4–6 mm. diam. Petals white, almost twice as long as sepals.
Fr. a siliqua, 13–18 mm., on horizontal or slightly deflexed stalks,
8–12 mm., the valves beaded, with slender midrib; seeds in 2 distinct
rows in each cell, with c. 25 polygonal depressions in each face of the
seed-coat. Fl. 5–10. Streams, ditches, flushes, etc., with moving water.
Common throughout lowland Great Britain and Ireland. Cultivated as
green or summer cress.

2. R. microphylla (Boenn.) Hyl.   One-rowed Water-cress.

Very like *R. nasturtium-aquaticum* but with stems and lvs turning purple-
brown in autumn. Fr. usually 16–22 mm., on stalks 11–15 mm., both
fr. and stalks more slender than in *R. n.-a.* Seeds ± in 1 row in each cell,
with c. 100 polygonal depressions in each face of the seed-coat. Fl. 5–10,
about 2 weeks later than *R. n.-a.* Common throughout Great Britain;
Ireland.

3. R. microphylla × nasturtium-aquaticum   Brown Water-cress.

Like *R. microphylla* in turning purple-brown in autumn, but distinguishable from both parents by the dwarfed and deformed fr. with an average
of less than 1 good seed each. Common, especially in N. England,
Scotland and Ireland. Cultivated as brown or winter cress.

4. R. sylvestris (L.) Besser   Creeping Yellow-cress.

Perennial, with horizontal or ascending stock, spreading roots and erect
or ascending angular shoots, 20–50 cm. Lower lvs stalked, pinnate or
deeply pinnatifid with narrow toothed or lobed segments; upper lvs
sessile, usually pinnatifid with narrow segments, sometimes almost

entire. Infl. corymbose with a zig-zag axis. Fls 5 mm. Petals yellow, twice as long as the sepals. Fr. 9–18 × 1·5–2 mm., ascending on horizontal or somewhat deflexed stalks, 5–12 mm. Fl. 6–8. Frequent on moist ground by streams and where water stands in winter, occasionally a garden weed. Throughout Great Britain northwards to Argyll and Angus; scattered throughout Ireland.

5. R. islandica (Oeder) Borbás   Marsh Yellow-cress.

Annual or biennial with erect hollow angular stem, 8–60 cm., glabrous or very sparsely hairy below. Lower lvs stalked, deeply lyrate-pinnatifid with narrow lateral and broader terminal segments, all ± toothed or lobed; upper lvs short-stalked or sessile. Fls c. 3 mm. diam., in lax corymbs. Petals pale yellow, not or slightly exceeding the sepals. Fr. 4–9 × 1·5–3 mm., oblong, turgid, curved, on horizontal or slightly deflexed stalks 4–10 mm. long. Fl. 6–9. In moist places, especially where water stands only in winter, throughout most of the British Isles.

6. R. amphibia (L.) Besser   Great Yellow-cress.

Perennial, usually glabrous, stoloniferous herb with ± erect stout furrowed hollow stem, 40–120 cm. Lower lvs short-stalked, elliptical to broadly oblanceolate, entire, sinuate-toothed or pinnatifid; upper lvs sessile, narrower; all bright or yellowish green. Fls c. 6 mm. diam., in lax corymbs. Petals bright yellow, about twice as long as the spreading sepals. Fr. 3–6 × 1–3 mm., ovoid, straight, on horizontal or deflexed stalks 6–18 mm. long. Fl. 6–8. Locally common by ponds, ditches and streams northwards to Westmorland and Berwick and throughout Ireland.

26. MATTHIOLA R.Br.

Herbs or dwarf shrubs with entire, sinuate or pinnatifid lvs; stem and lvs grey with branched hairs. Fls large; sepals erect; petals long-clawed; stamens 6; ovary with numerous ovules; stigma of 2 erect lobes each with a swelling or horn-like process on its outer side. Fr. a siliqua with hairy 1-veined valves; seeds flattened, in 1 row in each cell.

Stem woody below; all lvs entire, narrow; fr. without glandular hairs.
  An erect robust plant, to 80 cm., with fragrant purple, red or white fls, 2·5–5 cm. diam. On sea-cliffs in very few places in S. England and S. Wales. Stock.   *M. incana* (L.) R.Br.
Stem not woody; lower lvs sinuate or pinnatifid; fr. glandular. Diffuse biennial, 20–60 cm., with fragrant pale purple fls, 2–2·5 cm. diam. On sea-cliffs and dunes in scattered places in S. England, S. Wales and Ireland. Sea Stock.   *M. sinuata* (L.) R.Br.

27. HESPERIS L.

Tall biennial to perennial herbs with toothed or pinnatifid lvs; stem and lvs with simple, branched and glandular hairs. Petals large, long-clawed; stamens 6; ovary with many ovules; style short; stigma deeply 2-lobed, the lobes ± erect, facing each other, not appendaged at the back. Fr. a narrow cylindrical or somewhat 4-angled siliqua, its valves beaded, with distinct midrib and ± distinct lateral veins; seeds many, in 1 row in each cell.

*1. H. matronalis L.   Dame's Violet.

Biennial or perennial, with tap-root and erect lfy stems, 40–90 cm., with short simple and stellate hairs. Lvs oblong-ovate to lanceolate, narrowing up the stem, all short-stalked, finely toothed and roughly hairy. Fls c. 18 mm. diam., violet or white. Fr. 9 cm., curving upwards on spreading stalks. Fl. 5–7. Introduced. A garden-escape, occasionally naturalized in meadows, hedgerows, plantations, etc.

28. ERYSIMUM L.

Annual to perennial herbs with very lfy shoots covered with branched appressed hairs. Lvs usually narrow. Stamens 6, with nectaries just outside the inner stamens as well as encircling the bases of the outer stamens; stigma slightly 2-lobed. Fr. a 4-angled siliqua with strongly 1-veined hairy valves and seeds in 1 row in each cell.

*1. E. cheiranthoides L.   Treacle Mustard.

Annual, with erect stems, 15–90 cm., with scattered short appressed 2–3-branched hairs. Rosette-lvs dying before the fls open, oblong-lanceolate, acute, narrowed into a short stalk, hairy, irregularly sinuate-toothed; upper lvs narrower, sessile, ± toothed. Fls 6 mm. diam. Petals yellow, twice as long as the sepals. Fr. 12–25 × 1 mm., 4-angled, slightly curved on slender ascending stalks, their valves conspicuously 1-veined, stellate-hairy. Fl. 6–8. Probably introduced, Locally common as a weed of cultivated ground and waste places in S. England, rarer in the north; Ireland.

29. CHEIRANTHUS L.

Perennial herbaceous or suffruticose plants with stems and narrow entire lvs covered with appressed branched hairs. Stamens 6, with nectaries round the bases only of the 2 outer; stigma often with 2 deep spreading lobes. Fr. a ± flattened siliqua, the valves with 1 conspicuous vein; seeds in 1–2 rows in each cell.

Close to *Erysimum* and best distinguished by the absence of median nectaries.

*1. C. cheiri L. Wallflower.

A perennial herb with ± erect stem, 20–60 cm., woody below, angled, covered with forked appressed hairs. Rosette-lvs short-stalked; stem-lvs subsessile, crowded; all oblong-lanceolate, ± entire, with forked hairs especially beneath. Fls 2·5 cm. diam. Petals bright orange-yellow, at least twice as long as the sepals. Fr. 2·5–7 cm. × 2–4 mm., ± erect on ascending stalks. Fl. 4–6. Introduced and well established on walls, etc., throughout lowland Great Britain and Ireland.

30. ALLIARIA Scop.

Annual to perennial herbs with simple hairs. Lvs reniform or cordate. Fls white. Stamens 6. Fr. a linear 4-angled unbeaked siliqua with 3-veined valves.

1. A. petiolata (Bieb.) Cavara & Grande

Hedge Garlic, Garlic Mustard, Jack-by-the-hedge.

Biennial or perennial smelling strongly of garlic, with tap-root and erect stem, 20–120 cm., sparsely hairy below, glabrous and pruinose above. Rosette-lvs long-stalked, reniform, sinuate or distantly toothed; stem-lvs short-stalked, triangular-ovate, cordate, deeply and irregularly sinuate-toothed; all lvs thin, pale green and smelling of garlic when crushed. Fls 6 mm. diam. Petals white, twice as long as the sepals. Fr. 35–60 × 2 mm., curving at the base so as to stand almost erect on short thick spreading stalks; valves glabrous, with prominent midrib and faint lateral veins. Fl. 4–6. Common in hedgerows and wood-margins, shady gardens, wall-bases, etc., and locally frequent in beechwoods on chalk; Great Britain northwards to Ross; Ireland.

31. SISYMBRIUM L.

Annual to perennial herbs with simple hairs and entire to pinnatifid lvs. Petals yellow; stamens 6; ovary with numerous ovules in each cell. Fr. a long slender unbeaked siliqua whose convex valves have a strong midrib and usually 2 weaker lateral veins; seeds not winged.

1   Lvs 3–8 × 1–3 cm., simple, elliptical, ±entire; fls 4–6 mm. diam. in
    large infls; fr. 3–5 cm. Introduced. Established in waste places in
    various parts of Great Britain.   **S. strictissimum* L.
    Lvs ±deeply lobed or divided.   *2*

2   Fr. straight, stiffly erect and appressed to the infl.-axis.
                                                  1. officinale
    Fr. not appressed to the infl.-axis.   *3*

3   Uppermost lvs sessile, pinnately divided into linear or filiform
    segments.   **3. altissimum**
    Uppermost lvs stalked or narrowed into a stalk-like base, lanceolate
    or hastate or, if pinnatifid, then not with linear or filiform
    segments.   *4*

4   Fr. 5–10 cm., scarcely narrowed at the lower end, straight, not
    beaded.   **2. orientale**
    Fr. 3–5 cm., narrowed at both ends, usually curving upwards,
    beaded. Doubtfully native. On roadsides, walls and waste places
    in various parts of the British Is, uncommon. London Rocket.
                                                  S. irio L.

1. S. officinale (L.) Scop.   Hedge Mustard.

Annual or overwintering herb with stiffly erect stem, 30–90 cm., usually
bristly with downwardly-directed hairs. Basal lvs 5–8 cm., in a rosette,
deeply pinnatifid with round terminal lobe and 4–5 smaller laterals on
each side, all toothed; stem-lvs with long hastate terminal and 1–3 small
oblong lateral lobes. Infl. corymbose at first, greatly lengthening in fr.
Fls 3 mm. diam. Petals pale yellow, half as long again as sepals. Fr.
10–15 × 1 mm., short-stalked, held stiffly erect and close to the stem;
valves hairy or glabrous, 3-veined. Fl. 6–7. Hedgebanks, roadsides,
waste places and arable land throughout the British Is except Shetland.

*2. S. orientale L.   Eastern Rocket.

Annual or overwintering herb with erect stem, 25–90 cm., grey with
downwardly directed hairs. Basal lvs in a rosette, dying before flowering,
long-stalked, with large terminal lobe and c. 4 pairs of broadly triangular
lateral lobes; stem-lvs with hastate terminal lobe, upper hastate or
simply lanceolate, entire; all hairy, grey-green, ±stalked. Fls c. 7 mm.
diam. Petals pale yellow, twice as long as sepals. Fr. 4–10 cm. ×
1–2 mm., becoming glabrous, not narrowed at lower end, ±straight and
obliquely erect on short thick stalks, valves 3-veined, not beaded. Fl.
6–8. Introduced. Established on waste ground and bombed sites
especially in S. England, and often confused with the far less frequent
S. irio (see key).

*3. S. altissimum L. Tall Rocket.

Annual or overwintering herb with erect stem, 20–100 cm., ± hairy below but usually glabrous and pruinose above. Basal lvs dying before flowering, roughly hairy, stalked, pinnatifid with 6–8 pairs of narrowly triangular distant-toothed lobes; stem-lvs deeply pinnatisect, uppermost sessile, glabrous, with linear or filiform entire segments. Fls c. 11 mm. diam. Petals pale yellow, about twice as long as spreading sepals. Fr. 5–10 cm. × 1–1·5 mm., ± glabrous, straight, obliquely erect. Fl. 6–8. Introduced. Established in waste places in many parts of Great Britain.

32. Arabidopsis Heynh.

Slender annual to perennial herbs with both simple and branched hairs. Petals white, lilac or yellow; stamens 6; ovary with many ovules; style short. Fr. a slender siliqua, with convex prominently 1-veined valves; seeds in 1 row in each cell.

1. A. thaliana (L.) Heynh. Thale Cress.

Usually annual, with erect stems, 5–50 cm., roughly hairy below with mostly simple hairs. Basal lvs in a rosette, elliptical or spathulate, distantly toothed, stalked, grey-green with simple and branched hairs; stem-lvs narrow, sessile, often with branched hairs beneath. Fls 3 mm. diam. Petals white, twice as long as sepals. Fr. 10–18 × 0·8 mm., on slender spreading stalks. Fl. 4–5 and sometimes 9–10. Fairly common on walls and banks, in hedgerows and waste places and on dry soils throughout the British Is.

33. Camelina Crantz

Annual or overwintering herbs with erect stems and simple hastate sessile clasping stem-lvs; stems and lvs ± hairy with simple and branched hairs. Petals yellow or whitish; stamens 6; ovary with several ovules. Fr. an obovoid or pear-shaped latiseptate silicula with long style, its valves convex, strongly keeled and often winged, with a strong midrib vanishing above; seeds several, in 2 rows in each cell.

*1. C. sativa (L.) Crantz Gold-of-Pleasure.

Stems 30–80 cm., erect, glabrous or hairy. Basal lvs oblong-lanceolate, narrowed to the sessile base, entire or sinuate-toothed, rarely pinnatifid; middle and upper lvs narrow, with short acute basal auricles. Fls 3 mm. diam. Petals yellow, half as long again as sepals. Fr. 6–9 mm., obovoid, rounded above, narrowly winged, on ascending stalks 10–25 mm. Fl. 6–7. Introduced. An occasional weed of corn, flax and lucerne fields throughout Great Britain northwards to Stirling and in Ireland.

34. DESCURAINIA Webb & Berth.

Usually annual herbs with finely pinnatisect lvs and stellate and often glandular as well as simple hairs. Fls small; petals yellowish: stamens 6; ovary with many ovules. Fr. a short siliqua whose somewhat convex valves have a strong midrib and a lateral network of smaller veins.

1. D. sophia (L.) Prantl   Flixweed.

Annual or overwintering herb with erect stem, 30–80 cm., usually grey below with stellate hairs. Lvs grey-green, stellate-hairy, 2 or 3 times pinnatisect with linear segments; uppermost often almost simply pinnate. Fls 3 mm. diam. Petals pale yellow, about as long as sepals. Fr. 15–25 × 1 mm., ± cylindrical, almost erect on very slender ascending stalks. Fl. 6–8. Roadsides and waste places throughout the British Is except the extreme north but nowhere common.

31. RESEDACEAE

Annual to perennial herbs, rarely woody, with spirally arranged simple or pinnately divided lvs and fls in racemes or spikes. Fls zygomorphic, usually hermaphrodite. Sepals 4–7; petals usually 4–7, free, entire or lacinate, those at the upper side of the fl. largest; stamens 3–40, hypogynous or perigynous, inserted on the nectar-secreting disk, the largest and most crowded at the lower side of the fl.; carpels 2–6, superior, free or united below into a 1-celled ovary which often remains open at the top; ovules numerous, on 2–6 parietal placentae. Fr. usually a capsule open at the top; seeds numerous with curved embryo and no endosperm.

1. RESEDA L.

Herbs with lvs having glandular stipules. Petals 4–7, those at the top of the fl. larger and more deeply and repeatedly lobed than those on the lower side. Stamens 7–40, crowded to the front of the disk, where it is narrowest. Carpels 3–6, united below, open above between the stigmabearing lobes. Fr. a 1-celled capsule, opening more widely by the spreading of the apical lobes, but not splitting.

1 Ovary and capsule usually with 4 apical lobes; sepals 5; petals 5, white. A glabrous herb with pinnatifid lvs. Introduced. Casual and occasionally establishing itself. Upright Mignonette.

R. alba L.

Ovary and capsule with 3 apical lobes. 2

2 Petals whitish; capsules drooping. Introduced. Naturalized in Surrey, casual elsewhere. *R. phyteuma* L.

Petals yellow or green; capsules erect. 3

3 Lvs simple, entire; sepals and petals usually 4; capsule 5–6 mm., subglobular. **1. luteola**

Lvs pinnately lobed or divided; sepals and petals usually 6; capsule 12–18 mm., oblong. **2. lutea**

1. R. luteola L. Dyer's Rocket, Weld.

Biennial glabrous herb with deep tap-root, producing only a rosette of lvs in the first season. Fl.-stem 50–150 cm., stiffly erect, ribbed, hollow. Rosette-lvs commonly 5–8 cm., narrowly oblanceolate, sessile; stem-lvs narrowly oblong; all with entire waved margins. Fls 4–5 mm. diam., in long slender spike-like racemes, yellowish-green. Petals usually 4, those at top and sides of the fl. with 3 or more lobes, the bottom petal linear, entire. Fl. 6–8. Common on disturbed ground, arable land, walls, etc., especially on calcareous substrata, throughout most of Great Britain and Ireland but local in the north.

2. R. lutea L. Wild Mignonette.

Biennial to perennial glabrous herb with deep tap-root, woody stock and erect or ascending diffusely branched stems, 30–75 cm., ribbed, solid, rough with whitish tubercles. Basal lvs commonly 2·5–8 cm., in a rosette, withering early; stem-lvs numerous; all pinnatifid and with somewhat waved margins. Fls 6 mm. diam., greenish-yellow, in short compact conical racemes. Sepals unequal; petals usually 6, the top pair 3-fid, the lateral 2–3-fid, the bottom linear, entire. Fl. 6–8. Waste places, disturbed and arable land, especially on calcareous substrata. Great Britain northwards to Lancs. and Durham, probably introduced further north and in Ireland.

32. VIOLACEAE

Trees, shrubs or herbs. Lvs usually stipulate, simple. Fls hermaphrodite. Stamens hypogynous, equalling in number and alternate with the petals. Ovary superior, 1-celled with usually 3 parietal placentae. Fr. a capsule or berry; seeds endospermic.

1. Viola L.

Herbs, rarely undershrubs. Lvs spirally arranged, stipulate (stalked in British spp.). Fls solitary, 5-merous. Sepals prolonged with appendages below their insertion. Corolla zygomorphic, the lower petal spurred. Two lower stamens spurred; connective broad. Ovary with 3 placentae; ovules numerous; style thickened above, straight or curved. Fr. a capsule with 3 elastic valves.

1	Stipules not lf-like, entire to fringed but not lobed; lateral petals spreading ±horizontally; style hooked or obliquely truncate.
	Stipules lf-like, pinnately or palmately lobed; lateral petals directed towards the top of the fl.; style expanded above into a globose head.
2	Style hooked at apex.
	Style straight, obliquely truncate at apex; lvs orbicular-reniform; plant with long slender creeping rhizome. **6. palustris**
3	Lvs and fls all basal; sepals blunt; lf-stalks and capsules pubescent.
	Plant with aerial stem (sometimes very short in small forms); sepals acute; lf-stalks and capsules usually glabrous.
4	Plant with long stolons; fls sweet-scented, dark violet or white; hairs on lf-stalks deflexed. **1. odorata**
	Plant without stolons; fls scentless, usually blue-violet; hairs on lf-stalks spreading. **2. hirta**
5	Lf-stalks and capsules pubescent; lvs all or mostly blunt, ±reniform, truncate at base or shallowly cordate. Only on sugar-limestone in upper Teesdale. Teesdale Violet.
	V. rupestris F. W. Schmidt
	Lf-stalks and capsules glabrous; lvs all or mostly acute or acuminate.
6	Main axis ending in a rosette of lvs, not growing into a fl.-stem; lvs ovate-orbicular; teeth of stipules usually thread-like, wavy, spreading.
	Main axis without basal rosette, growing out into a fl.-stem; lvs ovate to lanceolate; teeth of stipules narrowly triangular, ±straight, ascending.
7	Appendages of sepals $\frac{1}{4}$–$\frac{1}{3}$ as long as sepals, enlarging in fr.; corolla blue-violet, spur paler, stout, furrowed or notched at the end. **3. riviniana**
	Appendages of sepals small, inconspicuous in fr.; corolla lilac, spur darker, slender, not furrowed or notched at the end. **4. reichenbachiana**
8	Corolla deep or bright blue; lvs usually ovate; stipules rarely more than one-third as long as lf-stalk. **5. canina**

Corolla pale or nearly white; lvs lanceolate; stipules of middle and
upper lvs usually half or more as long as the lf-stalks.   9

9 Lvs rounded or broad-cuneate at base, widest at about ⅓ from
apex, rather thick; upper stipules ovate-lanceolate equalling or
exceeding the lf-stalks; fl. pale greyish-violet, spur yellowish or
greenish, thick, blunt. Local on heaths from Yorks southwards,
commonest in S. W. England; S. and W. Ireland. Pale Heath
Violet.   *V. lactea* Sm.
Lvs truncate or cordate at base, widest very near base. Fens.   10

10 Lvs triangular-lanceolate, thin; corolla 10–15 mm., bluish-white or
white, spur greenish, short, ±conical, hardly longer than the
sepal-appendages; plant creeping below the ground. A very
local plant of fens, chiefly of fens in E. England and damp grassy
hollows (turloughs, etc.), on limestone in Ireland. Fen Violet.
                          V. persicifolia Schreber (*V. stagnina* Kit.)
Lvs ovate to ovate-lanceolate, often ±triangular, thick; corolla
blue; spur usually yellowish, about twice as long as sepal-
appendages; plant not creeping.   **5. canina**

11 Petals yellow, violet, or particoloured, longer than the sepals.   12
Petals cream, occasionally violet-tinged, not longer than the sepals. 13

12 Perennial, with long creeping rhizome; fls 2–3·5 cm. vertically;
spur 2–3 times as long as sepal-appendages; upland grassland and rock-ledges.   **6. lutea**
Annual or perennial, without long creeping rhizome; fls 1·5–2·5 cm.
vertically; spur rather longer than sepal appendages (or twice as
long in maritime forms).   **7. tricolor**

13 Corolla 13 mm. or more vertically, flat; a common arable weed.
                                                          8. arvensis
Corolla c. 5 mm., somewhat concave; plant small, grey-pubescent;
lower stipules lf-like. Sandy places near the sea in Scilly Is and
Channel Is.
       V. kitaibeliana Roemer & Schultes (*V. nana* (DC.) Corbière)

In the true violets (spp. 1–6) the stipules are not lf-like, the lateral
petals spread horizontally and the style is hooked or obliquely truncate
at the apex. Cleistogamous fls are produced in summer.

1. V. odorata L.   Sweet Violet.

Perennial herb with short thick rhizome and long procumbent stolons
rooting at the ends; no leafy aerial stem. Lvs 1·5–6 cm., all basal,
ovate-orbicular or broadly ovate, deeply cordate at base, blunt (or
summer ones acute), crenate-serrate, dark green, sparingly hairy; their
stalks long, usually with short deflexed hairs; stipules ovate or ovate-

lanceolate, fimbriate. Fl-stalks with bracteoles about or above the middle. Sepals with spreading appendages. Corolla c. 1·5 cm., deep violet or white, rarely purple, pink or apricot, sweet-scented. Capsule globose, pubescent, lying on the ground. Fl. (9–)2–4. Hedge-banks, scrub, plantations, railway cuttings, etc., usually on calcareous soils. Rather common from Dunbarton and Angus southwards and all over Ireland; Jersey.

2. V. hirta L.   Hairy Violet.

Like *V. odorata* but no stolons; lvs narrower, paler green, their stalks with longer denser spreading hairs and lanceolate stipules; bracteoles usually below the middle of the fl.-stalk; fls scentless, paler and bluer; spur often upwardly curved or hooked. Fl. 4–5. Calcareous pastures, scrub and open woods from Kincardine and Kirkcudbright southwards; rare in Ireland.

3. V. riviniana Reichenb.   Common Violet.

Perennial herb with distinct aerial lfy stems 2–20 cm. in fl. to 40 cm. in fr., glabrous or somewhat pubescent; not creeping but sometimes forming root-buds; with central non-flowering lf-rosette, fls on axillary branches. Lvs ovate-orbicular (slightly longer than broad), shortly acuminate or blunt, deeply cordate at base, crenate; stipules lanceolate, much shorter than the glabrous lf-stalk, fimbriate, mostly with long spreading thread-like wavy fimbriae. Sepal-appendages ¼ to ⅓ as long as the sepals, enlarging in fr. Corolla 14–22 mm., usually blue-violet, the petals broad, overlapping; spur paler, usually whitish or pale violet, rarely yellowish, thick, scarcely tapering, furrowed or notched at the tip. Capsule 6–13 mm., acute, glabrous, held erect. Fl. 4–6. Woods, hedge-banks, pastures and mountain rocks, to 3350 ft. Common throughout the British Is.

4. V. reichenbachiana Jordan   Pale Wood Violet.

Like *V. riviniana* but sepal-appendages very short, obsolete in fr.; corolla lilac (paler and somewhat redder than in *V. riviniana*) with the narrower petals not overlapping, and spur slender, laterally compressed, tapering, not furrowed or notched, dark violet. Fl. 3–5 (earlier than *V. riviniana*). Woods, hedge-banks, etc., usually on calcareous soils and rarely in the open. Rather common in S., C. and E. England, more local in N. England, Wales and Ireland.

5. V. canina L. Heath Violet.

Perennial herb with distinct aerial lfy stems 2–30 cm. (–40 cm. in fr.),
± glabrous; stems decumbent to erect from a short creeping rhizome;
no central non-flowering rosette. Lvs ovate to ovate-lanceolate, often
± triangular in outline and usually distinctly longer than broad, blunt,
truncate or widely and shallowly cordate at base, crenate, rather thick;
stipules rarely more than ⅓ as long as the glabrous lf-stalk, ± lanceolate,
subentire or with usually ascending, straighter, stouter, shorter and
fewer teeth than in *V. riviniana*. Sepal-appendages rather large. Corolla
7–18(–22) mm., blue; spur usually yellowish (rarely greenish-white),
thick, straight, ± cylindric, blunt and sometimes furrowed at the tip.
Capsule c. 8–9 mm., blunt (often apiculate), glabrous, held erect. Fl.
4–6 (later than *V. riviniana*). Heaths, dry grassland, dunes and fens
throughout the British Is, but local.

6. V. palustris L. Marsh Violet.

Perennial herb with long slender creeping rhizome on which all the lvs
are borne: no aerial stem. Lvs 1–4 cm., orbicular-reniform, cordate at
base, obscurely crenate; stipules ovate, with small glandular teeth.
Sepals blunt. Corolla 10–15 mm., lilac (rarely white) with darker veins;
spur blunt, longer than the sepal-appendages. Style straight, obliquely
truncate at tip. Capsule glabrous, held erect. Fl. 4–7. Bogs, fens,
marshes and wet heaths, to 4000 ft. Common almost throughout the
British Is.

In the pansies (spp. 7–9) the stipules are lf-like, pinnately or palmately
lobed or divided, the lateral petals are directed upwards, and the
straight style is expanded at its tip into a subglobose head with a
hollow at one side.

7. V. lutea Hudson Mountain Pansy.

Perennial herb 7–15 cm., with slender creeping rhizome and slender
wavy aerial stems. Lvs rapidly narrowing upwards, the lowest ovate,
blunt, the upper 1–2 cm., oblong-lanceolate, subacute, cuneate at base,
± crenate, sparsely hairy at least on margins and veins beneath; stipules
palmately divided, the middle lobe 2 mm. wide, entire. Sepals acute,
their appendages toothed. Corolla 2–3·5 cm. vertically, flat, bright
yellow, blue-violet, red-violet, or with these colours variously combined,
always yellow at the base of the lowest petal; spur 2–3 times as long
as the sepal-appendages. Fl. 6–8. Grassland and rock-ledges in hilly
districts, especially on base-rich but not strongly calcareous soils, to
3500 ft. Absent from lowland England but widespread and locally
common elsewhere in Great Britain; very local in Ireland.

8. V. tricolor L.   Wild Pansy.

Annual or perennial herb 3–45 cm., glabrous or somewhat pubescent.
Lvs narrowing upwards; lowest oval or ovate, blunt; upper very variable
in shape from ovate to narrowly elliptic, ± crenate; stipules palmately or
pinnately lobed, the middle lobe 2–5 mm. wide, variable, but rarely
quite entire. Sepals acute, their appendages variable. Corolla 15–25 mm.
vertically, longer than sepals, flat, sulphur-yellow, blue-violet, pink, or
of combinations of these colours; spur rather longer than the sepal-
appendages (or up to c. twice as long in the perennial small-fld maritime ssp. *curtisii*). Stigma with a projection below the opening. Fl.
4–9. Annual forms (ssp. **tricolor**) on cultivated and waste ground, mainly
on acid or neutral soils, throughout the British Is; large perennial forms
(15–45 cm.) in short grassland or bare ground mainly in hilly districts,
local (ssp. **subalpina** Gaudin), and smaller perennial forms (3–15 cm.)
on dunes and grassy places near the sea in the north and west and in a
few inland localities (ssp. **curtisii** (E. Forster) Syme).

9. V. arvensis Murray   Field Pansy.

Like annual forms of *V. tricolor* but corolla 8–14 mm. vertically,
equalling or rather shorter than the sepals, cream, sometimes with a
violet tinge on the upper petals, the spur sometimes deep violet; spur
about equalling sepal-appendages; stigma without a projection below
the opening. Fl. 4–10. Cultivated and waste ground, mainly on basic
and neutral soils. Common throughout the British Is.

33. POLYGALACEAE

Herbs or small shrubs. Fls in bracteate racemes, superficially resembling
those of Leguminosae. Inner sepals petaloid, much larger than outer.
Petals 3, outer united with lower and adnate to staminal tube. Stamens
8. Capsule 2-celled, flattened, obcordate and narrowly winged.

1. POLYGALA L.

1	Lower lvs smaller than upper, not forming a rosette.	2
	Lower lvs larger than upper, forming a rosette.	3
2	All lvs alternate.	**1. vulgaris**
	At least the lower lvs opposite.	**2. serpyllifolia**
3	Fls 6–7 mm.; rosette usually with a lfless portion of stem below;	
	stem-lvs usually less than 1 cm. (local).	**3. calcarea**
	Fls not more than 5 mm.; rosette at base of stem; stem-lvs usually	

more than 1 cm. Bitter Milkwort. A rare plant of chalk grassland in S.E. England and limestone pastures in N. England.

P. amara L.

1. P. vulgaris L. Common Milkwort.

Perennial 10–30 cm. with wiry stems and alternate evergreen lvs. Lower
lvs 5–10 mm., narrowly obovate; upper up to c. 35 mm., lanceolate or
linear-lanceolate, all ± acute. Infl. usually many-fld. Fls 6–8 mm., blue,
pink or whitish. Outer sepals c. 3 mm., green with coloured borders;
inner c. 6 mm., ovate, with much-branched anastomosing lateral veins.
Grassland, heaths, dunes, etc., widely distributed.

2. P. serpyllifolia Hose Thyme-leaved Milkwort.

Like 1 but usually smaller and more slender. Lower lvs opposite. Fls 3–
8, commonly bright blue or slaty-blue. Inner sepals 4·5–5·5 mm. Heaths
and grassy places, usually on lime-free soils. Widely distributed and
usually the commonest sp.

3. P. calcarea F. W. Schultz Chalk Milkwort.

Like 1 but the lower lvs 5–20 mm., obovate, obtuse, and crowded into
an irregular rosette from which the unbranched flowering stems arise;
upper lvs smaller than lower, lanceolate, ± acute. Fls 6–7 mm., intense
blue or sometimes bluish-white. Inner sepals c. 5 mm., with little-
branched and not or scarcely anastomosing lateral veins. Calcareous
grassland in S. and E. England north to Rutland, locally common.

34. HYPERICACEAE

Herbs, shrubs or trees with resinous juice. Lvs opposite, simple, usually
entire. Fls showy, actinomorphic, solitary or in branched cymes.
Sepals usually 5. Petals usually 5. Stamens numerous. Ovary superior,
1-, 3- or 5-celled. Fr. a capsule, rarely fleshy; seeds numerous.

1. HYPERICUM L.

Lvs sessile or nearly so. Fls yellow. Petals generally very oblique.
Ovary 1-celled or 3- or 5-celled; styles 3 or 5.

1	Plant ± shrubby; petals deciduous; stamens in 5 bundles united at base only.	2
	Stems not or slightly woody and then at base only; petals persistent; stamens in 3 bundles.	5

2   Rhizome 0; stems freely branched; styles 3.   3
    Rhizome long; stems simple or nearly so; styles 5. Rose of
        Sharon. Naturalized in shrubberies, etc.   *H. calycinum L.

3   Stems quadrangular; sepals deciduous; stamens longer than petals.
        Stinking St John's Wort. Locally naturalized.   *H. hircinum L.
    Stems 2-edged or nearly terete; sepals persistent; stamens about
        equalling petals.   4

4   Stems with 2 raised lines; sepals as long as petals; styles shorter
        than stamens; fr. fleshy.   1. androsaemum
    Stems slightly 2-edged; sepals distinctly shorter than petals; styles
        longer than stamens; fr. dry. Tall St John's Wort. Locally
        naturalized.   *H. inodorum Miller (H. elatum Aiton)

5   Plant hairy, at least on underside of lvs.   6
    Plant glabrous.   8

6   Lvs glabrous above, finely hairy beneath; fls crowded together.
                                                        8. montanum
    Lvs hairy on both sides; fls not crowded together.   7

7   Stems stiff, erect, ± branched; fls numerous; stamens free except at
        base (dry places).   7. hirsutum
    Stems soft, creeping and rooting below; fls few; stamens united in
        3 bundles for ⅓ way up (wet places).   9. elodes

8   Stems quadrangular.   9
    Stems terete or with 2 raised lines.   12

9   Petals shorter or little longer than sepals.   10
    Petals 2–3 times as long as sepals.   11

10 Stems up to 20 cm., very slender; petals deep golden-yellow.
        Canadian St John's Wort. W. Ireland; very local.
                                                        H. canadense L.
    Stems usually 30–70 cm., stout; petals pale yellow.   4. tetrapterum

11 Lvs suddenly narrowed at base, not clasping the stem, with few or 0
        glandular dots; sepals blunt.   3. maculatum
    Lvs half clasping the stem, with abundant glandular dots; sepals
        acute. Wavy-leaved St John's Wort. Cornwall, Devon and
        Pembroke; local.   H. undulatum Schousboe

12 Petals little longer than sepals; stems slender, prostrate or
        ascending.   5. humifusum
    Petals at least twice as long as sepals; stems usually stout and erect.   13

13 Sepals broadly ovate, rounded at apex; lvs of main stems truncate
        to cordate at base.   6. pulchrum
    Sepals lanceolate, acute; lvs of main stems narrowed at base.   14

14 Sepals entire, not fringed with black stalked glands.
                                                        2. perforatum

Sepals fringed with black stalked glands. Toadflax-leaved St John's Wort. S.W. England and Wales, very rare.

H. linarifolium Vahl

1. H. androsaemum L.   Tutsan.

40–100 cm., glabrous, half-evergreen shrub. Lvs 5–10 cm., ovate, obtuse, with minute translucent glands. Fls c. 2 cm. diam.; sepals ovate, obtuse, very unequal. Stamens about as long as petals. Fr. red, turning black. Fl. 6–8. Damp woods and hedges, local.

2. H. perforatum L.   Common St John's Wort.

30–90 cm., rhizomatous. Lvs 1–2 cm., elliptic, oblong or sometimes very narrow, obtuse, with abundant translucent glands. Fls c. 2 cm. diam.; sepals lanceolate, much shorter than petals, glandular. Fl. 6–9. Open woods, hedge-banks and grassland, specially common on calcareous soils.

3. H. maculatum Crantz   Imperforate St John's Wort.

20–60 cm., rhizomatous. Lvs 1–2 cm., elliptic, obtuse, abruptly narrowed at base, not clasping stem, with few or no glands. Fls c. 2 cm. diam.; petals golden yellow; sepals ovate, glandular, ⅓–¼ as long as petals. Fl. 6–8. Damp places, wood margins, hedge-banks, local but widely scattered.

4. H. tetrapterum Fries   Square-stemmed St John's Wort.

30–70 cm., with slender stolons. Lvs 1–2 cm., ovate, obtuse, base half clasping the stem, with small translucent glands. Fls c. 1 cm. diam.; petals pale yellow; sepals ⅔ as long as petals. Fl. 6–9. Damp meadows, marshes, by rivers, etc., widely distributed.

5. H. humifusum L.   Trailing St John's Wort.

5–20 cm., stems very slender, rather woody at base, with 2 raised lines. Lvs 0·5–1(–1·5) cm., elliptic, oblong or obovate, ± glandular. Fls c. 1 cm. diam.; sepals unequal, oblong to lanceolate, obtuse or acute, glandular, entire or toothed, teeth sometimes glandular at tip. Fl. 6–9 Heaths, dry moors and open woods, on non-calcareous soils, locally common.

6. H. pulchrum L.   Slender St John's Wort.

30–60 cm., stems terete, often reddish. Lvs 0·5–1 cm., broadly ovate-cordate, obtuse, base half clasping the stem, with translucent glands.

Fls c. 1·5 cm. diam.; petals red-tinged with a row of black glands near margin; sepals obtuse, $\frac{1}{3}$–$\frac{1}{4}$ as long as petals, margins with nearly sessile black glands. Fl. 6–8. Dry woods and rough grassy places on non-calcareous soils, locally common.

7. H. hirsutum L.   Hairy St John's Wort.

40–100 cm., stems terete, hairy. Lvs 2–5 cm., ovate, obtuse, gland-dotted but without marginal black glands. Bracteoles with short-stalked black glands. Fls c. 1·5 cm. diam.; petals pale yellow, sparsely glandular at tips; sepals oblong-lanceolate, subacute, $\frac{1}{2}$ as long as petals, margins with short-stalked black glands. Fl. 7–8. Locally common in woods and long grass, chiefly on basic soils, widely scattered through Britain.

8. H. montanum L.   Mountain St John's Wort.

40–80 cm., slightly hairy. Stems terete, wiry, simple or nearly so, upper internodes longer than lvs. Lower lvs 3–5 cm., ovate to elliptic, obtuse, base half-clasping the stem, with marginal black glands beneath. Bracts and bracteoles with glandular teeth. Fls 1–1·5 cm. diam., fragrant, crowded; petals pale yellow, nearly eglandular; sepals lanceolate, acute, $\frac{1}{2}$–$\frac{1}{3}$ as long as petals, strongly glandular-toothed. Fl. 6–8. Woods, scrub, hedge-banks, etc., on calcareous or gravelly soils, local.

9. H. elodes L.   Marsh St John's Wort.

10–30 cm., covered with short dense velvety hairs. Lvs suborbicular or oblong, cordate, $\frac{1}{2}$-clasping the stem at base. Fls c. 1·5 cm. diam.; sepals elliptic, obtuse or acute, with fine red or purplish glandular teeth. Fl. 6–9. Scattered through the British Is in bogs, by ponds, etc., on acid soils, commoner in the west and lost through drainage in some places.

35. CISTACEAE

Shrubs or herbs. Lvs usually opposite, stipulate or not, simple, mostly with stellate hairs. Fls hermaphrodite, regular. Sepals 5 or 3. Petals usually 5, often falling within a day. Stamens numerous, hypogynous. Ovary superior, 1-celled or septate only at the base, with 3 or 5 (rarely 10) parietal placentae; ovules few to numerous; style simple or 0; stigmas 3–5. Fr. a loculicidal capsule. Seeds endospermic.

Perennial undershrubs; style very slender.   1. HELIANTHEMUM
Annual herb; style 0.   2. TUBERARIA

1. HELIANTHEMUM Miller

Infl. cymose. Sepals 5, the two outer usually smaller than the three inner. Petals 5. Stamens all fertile. Placentae 3; stigmas large, capitate. Capsule 3-valved.

1 Stipules 0; style markedly bent in the middle; corolla 1–1·5 cm. diam., yellow. Rocky limestone pastures; very local in N. England, Wales and W. Ireland. Hoary Rockrose.
                                                      H. canum (L.) Baumg.
   Stipules present; style nearly straight; corolla c. 2 cm. diam., yellow or white.   2

2 Lvs green above; corolla yellow.   **1. nummularium**
   Lvs grey-tomentose above, oblong-linear, their margins strongly revolute; corolla white. Very local in S.W. England. White Rockrose.   *H. apenninum* (L.) Miller

1. H. nummularium (L.) Miller (*H. chamaecistus* Miller)
                                                      Common Rockrose.

Undershrub 5–30 cm., with procumbent or ascending, often rooting branches from a thick woody stock and vertical taproot. Lvs 0·5–2 cm., oblong or oval, green and glabrous to somewhat pubescent above, densely white-tomentose beneath, bluntish, short-stalked; margins entire, not or slightly revolute; stipulate. Fls 1–12 in lax one-sided cymes, axis and stalks tomentose. Inner sepals ovate, outer subulate, shorter. Corolla c. 2 cm. diam. or smaller, usually bright yellow. Capsule ovoid, tomentose. Fl. 6–9. Basic grassland and scrub, to 2100 ft. Common over most of Great Britain but absent from Cornwall, Isle of Man, N.W. Scotland, Orkney and Shetland and only in one locality in Ireland.

2. TUBERARIA (Dunal) Spach

Like *Helianthemum* but annual herbs with basal lvs in a *Plantago*-like rosette; style 0 or very short.

T. guttata (L.) Fourr., 6–30 cm., has pale yellow fls 8–12 mm. diam., often with a red spot at the base of the petals. Spotted Rockrose. Dry places near the sea in N. Wales, W. Ireland and Channel Is.

36. TAMARICACEAE

Trees or shrubs. Lvs small, scale-like or needle-like, spirally arranged, exstipulate. Fls regular, hermaphrodite, with a disk. Petals imbricate, free. Stamens as many as petals and alternate with them, or twice as many. Ovary superior, 1-celled with 2–5 basal or parietal placentae, each with 2–many ovules; styles free or united below, or stigmas sessile. Fr. a capsule; seeds with long hairs, endospermic or not.

1. TAMARIX L.

Tamarisk.

Deciduous, the smaller twigs falling with the lvs. Fls in long catkin-like racemes. Stamens free or nearly so. Styles short and thick. Endosperm 0.

Tamarix gallica L. (incl. *T. anglica* Webb) is a feathery shrub, 1–3 m., with the lvs on larger twigs c. 2 mm., acute or acuminate, those on ultimate twigs much smaller and densely imbricate, ovate-lanceolate, sessile, green or glaucous. Infl. terminal on the current season's growth, of cylindrical dense-fld spike-like racemes, 1–3 cm., arranged in a panicle. Fls c. 3 mm., pink or white; sepals, petals and stamens 5, styles 3. Capsule ovoid-trigonous. Fl. 7–9. Introduced, often planted near the sea and naturalized in many places in S. and E. England from Cornwall to Suffolk. Variable.

37. FRANKENIACEAE (p. xxi)

38. ELATINACEAE

Small herbs of wet places. Lvs opposite or in whorls, simple. Fls actinomorphic, small, solitary or in cymes, in the lf-axils. Sepals and petals 3–5, stamens as many or twice as many as petals. Ovary superior, 3–5-celled; fr. a capsule.

1. ELATINE L.

Waterwort.

Small submerged herbs. Petals and sepals 3–4. The two British spp. are both rare.

Fl. stalked, 3-merous; seeds 8–12 in each cell.
                                        E. hexandra (Lapierre) DC.
Fl. sessile or nearly so, 4-merous; seeds 4 in each cell. *E. hydropiper* L.

39. CARYOPHYLLACEAE

Annual to perennial, sometimes shrubby below. Lvs usually in opposite decussate pairs, rarely whorled, sometimes spirally arranged; always simple, entire, and commonly without stipules but sometimes with scarious stipules. Fls commonly in forking cymes, usually hermaphrodite and actinomorphic. Sepals 4–5; petals sometimes 0, usually 4–5, free; stamens usually twice as many or as many as the sepals, usually hypogynous; ovary superior, syncarpous, 1-celled at least above, with 1 to many ovules on a basal or free-central placenta; styles 2–5. Fr. a capsule, rarely a berry or a 1-seeded nutlet; seeds usually with a curved embryo surrounding a nutritive tissue derived from the nucellus (perisperm).

1   At least the lower lvs in opposite pairs or whorled.   **2**
    Lvs alternate, linear-oblanceolate, glaucous, with whitish scarious
      stipules; fls tiny whitish; fr. 1-seeded, indehiscent. A very local
      plant of pool-banks in Devon and Cornwall. Strapwort.
                                       Corrigiola litoralis L.

2   Stipules 0.   *3*
    Stipules present.   **26**

3   Fls perigynous; petals 0; fr. 1-seeded, indehiscent, enclosed in the
    perigynous tube; lvs awl-shaped.   18. SCLERANTHUS
    Fls hypogynous; fr. a capsule or berry with few to many seeds.   **4**

4   Calyx of joined sepals.   *5*
    Calyx of free sepals.   **12**

5   Styles 2.   **6**
    Styles 3–5 (or fls with stamens only).   **8**

6   Fls few (opening 1 at a time) in a compact head ±enclosed in a
    loose involucre of large shining brownish membranous scales;
    petals pale purplish-red. A rare annual linear-lvd herb of sandy
    or gravelly places in S. England, Glamorgan and Jersey. Proli-
    ferous Pink.
      Petrorhagia prolifera (L.) P. W. Ball & Heywood (*Kohlrauschia*
                             prolifera (L.) Kunth)
    Fls not in a head, or if so involucre 0 or herbaceous.   **7**

7   Lvs ±linear, glaucous; calyx enclosed below in an epicalyx of 1–3
    pairs of usually awned scales.   4. DIANTHUS
    Lvs broadly ovate to elliptic, green; epicalyx 0.   5. SAPONARIA

8   Fr. a black berry, 6–8 mm. diam.; calyx widely and loosely bell-
    shaped; fls c. 18 mm. diam., greenish-white. Introduced and
    naturalized in a few places. Berry Catchfly.
                                  **Cucubalus baccifer* L.
    Fr. a capsule; calyx not widely and loosely bell-shaped.   **9**

9   Calyx-teeth long and narrow, much longer than the petals.
                                  3. AGROSTEMMA
    Calyx-teeth not longer than the petals.   **10**

10   Styles 3; capsule opening by 6 teeth (or fls with stamens only).
                                      1. SILENE
    Styles 5.   **11**

11 Dioecious; capsule opening by 10 teeth.   1. SILENE
    Hermaphrodite; capsule opening by 5 teeth.   2. LYCHNIS

12 Petals present.   13
    Petals 0.   24

13 Petals ± deeply bifid.   14
    Petals entire, emarginate or irregularly toothed.   17

14 Styles 5.   15
    Styles 3, or varying in number (3–6).   16

15 Lvs 2–5 cm., ovate-cordate; petals bifid almost to the base.
                                                            7. MYOSOTON
    Lvs rarely exceeding 2·5 cm., not cordate; petals bifid to less than
        halfway.   6. CERASTIUM

16 Styles 3–6; a decumbent high-alpine plant with small elliptical lvs;
        capsule oblong.   6. *Cerastium cerastoides*
    Styles 3; capsule narrowly to broadly ovoid.   8. STELLARIA

17 Petals irregularly toothed or jagged. A small annual herb, 3–
        20 cm., with small whitish fls in an umbel-like terminal infl.
        Very rare on walls, roofs and sandy soils, now only in Surrey.
        Jagged Chickweed.   *Holosteum umbellatum* L.
    Petals entire or slightly emarginate.   18

18 As many styles as sepals (4–5).   19
    Fewer styles (2–3) than sepals.   20

19 Styles opposite sepals; capsule opening by 8 short teeth; small
        glaucous herb with strap-shaped or narrow lanceolate lvs.
                                                            9. MOENCHIA
    Styles alternating with sepals; capsule splitting to base into 4–5
        valves; small non-glaucous herbs with linear lvs.   10. SAGINA

20 Succulent maritime plants with broad lvs, greenish fls and globose
        capsules.   12. HONKENYA
    Not as above.   21

21 Lvs linear.   22
    Lvs not linear.   23

22 Fls greenish; petals minute or 0; nectaries 10, conspicuous.
        Compact alpine cushion-plant of mountains in Scotland and
        Inner Hebrides. Mossy Cyphel.
                    Minuartia sedoides (L.) Hiern (*Cherleria sedoides* L.)
    Fls with white petals; nectaries not conspicuous.   11. MINUARTIA

23 Lvs ovate, 1–2·5 cm., 3-veined; seeds with a fleshy oily appendage;
        a slender woodland plant.   13. MOEHRINGIA
    Lvs not exceeding 1 cm.; seeds without oily appendage.
                                                            14. ARENARIA

24 Lvs linear.   25
    Lvs not linear.   8. STELLARIA

25 Styles 3. *Minuartia sedoides* (L.) Hiern (see *22* above).
    Styles 4–5.   10. SAGINA
26 Lower lvs apparently in whorls of 4, obovate; styles joined below.
        A small decumbent herb with tiny white fls, 2–3 mm. diam. Rare
        and local on sandy and waste places in S.W. England. Four-
        leaved All-seed.   *Polycarpon tetraphyllum* (L.) L.
    Lower lvs in opposite pairs, not whorls of 4; styles free to the base.   27
27 Fls in terminal cymes; petals about equalling sepals or exceeding
        them.   28
    Fls in dense lateral or axillary clusters; petals minute or 0.   29
28 Styles 5.   15. SPERGULA
    Styles 3.   16. SPERGULARIA
29 Fls white, in axillary clusters or false whorls; sepals hooded and
        with dorsal awns, resembling small follicles; upper lvs in equal
        pairs. Small decumbent annual with 5 petals and 2 short styles.
        Very local in moist sandy places in S. and S.W. England.
                                          Illecebrum verticillatum L.
    Fls greenish, in lateral clusters; sepals not hooded; upper lvs
        commonly in unequal pairs.   17. HERNIARIA

1. SILENE L.

Herbs or dwarf shrubs with opposite lvs; stipules 0. Fls hermaphrodite
or unisexual (sometimes dioecious). Epicalyx 0; sepals joined below;
stamens 10; styles 3 or 5; ovary 1-celled at least above, usually 3–5-
celled at base. Fr. a capsule opening by twice as many teeth as styles.

1 Dioecious; styles 5.   2
    Styles 3 (rarely 5); usually hermaphrodite, rarely dioecious.   3
2 Fls red; capsule-teeth curled back.   **1. dioica**
    Fls white; capsule-teeth erect.   **2. alba**
    (Fls pink.   **dioica × alba**)
3 Fruiting-calyx with 20–30 veins, strongly inflated at least below.   4
    Fruiting-calyx with 10 veins, not strongly inflated.   7
4 Perennial; fruiting-calyx usually glabrous, bladdery, conspicuously
        net-veined; its teeth broadly triangular.   5
    Annual; fruiting-calyx pubescent, swollen below but not bladdery,
        conical, strongly ribbed but not conspicuously net-veined, its
        teeth long-acuminate to subulate. Local in dunes and inland
        sandy pastures and waste places, chiefly by the south and east
        coasts northwards to Angus. Striated Catchfly.   *S. conica* L.
5 Upper lvs linear-lanceolate; fls dull red, rarely white. Introduced
        and established on Plymouth Hoe. Narrow-leaved Bladder
        Campion.   **S. linearis* Sweet
    Upper lvs elliptic-lanceolate to ovate; fls white.   6

6 Plants with prostrate non-flowering shoots; fls 1–4; calyx hardly
   narrowed at its mouth; petals with distinct coronal scales;
   capsule with recurved teeth.   **5. maritima**
   Plants with no non-flowering shoots; fls many; calyx narrowed at
   its mouth; petals with very small coronal scales or bosses; capsule with erect teeth.   **4. vulgaris**

7 Calyx glabrous.   8
   Calyx hairy, downy or sticky.   9

8 Lvs linear; fls solitary, short-stalked, pink or rarely white.
   A cushion-plant of mountains and maritime cliffs in the north.
   Moss Campion.   *S. acaulis* (L.) Jacq.
   Basal lvs spathulate; fls in long narrow panicles, yellowish-green;
   stamens and styles projecting. An erect ± dioecious perennial,
   local on dry sandy soils of the E. Anglian 'breckland' heaths.
   Spanish Catchfly.   *S. otites* (L.) Wibel

9 Perennial plants with non-flowering shoots; infl. a lax panicle of
   opposite 3–7-fld cymes; fls whitish.   10
   Annuals lacking non-flowering shoots; infls various.   11

10 Non-flowering shoots short, rosette-like; fls horizontal or drooping; capsule with a stalk (concealed by the calyx) about ⅓ its
   length.   **7. nutans**
   Non-flowering shoots elongated; fls ± erect; capsule with a stalk
   about its own length. Rare plant of quarry-sides. Italian Catchfly.   *S. italica* (L.) Pers.

11 Infl. a few-fld dichasium; calyx 20–25 mm.; petals rolled inwards
   during day-time, deeply 2-cleft.   **3. noctiflora**
   Infl. long, raceme-like, of many short-stalked fls; calyx to 10 mm.;
   petals not inrolled, entire or emarginate.   **6. gallica**

1. S. dioica (L.) Clairv. (*Melandrium rubrum* (Wiegel) Garcke)
Red Campion.

Biennial to perennial, with a slender creeping stock, numerous decumbent non-flowering shoots and erect flowering shoots, 30–90 cm., softly
hairy and sometimes sticky above. Basal lvs obovate, narrowed into a
long winged stalk; others ovate-oblong, short-stalked to sessile; all
hairy. Dioecious. Fls numerous, 18–25 mm. diam., scentless. Calyx-
tube of male fls cylindrical, faintly 10-veined; of female fls ovoid and
20-veined, becoming rounded in fr.; hairy, slightly sticky and with tri-
angular-acute teeth. Petals bright rose with 2 narrow acute scales at the
base of the blade. Styles 5. Capsule broadly ovoid, opening widely by
10 down-curled teeth. Fl 5–6. Locally abundant throughout the British
Is in woods on base-rich and well-drained soils or rocky slopes, on cliff-
ledges and in hedgerows.

2. S. alba (Miller) E. H. L. Krause (*M. album* (Miller) Garcke)

White Campion.

Usually a short-lived perennial with a thick stock, a few short non-flowering shoots and erect flowering shoots, 30–100 cm., softly hairy, slightly sticky above. Lower lvs and those of barren shoots oblanceolate or elliptical, acuminate, sessile; all hairy. Dioecious. Fls few, 25–30 mm. diam., slightly night-scented. Calyx-tube of male fls cylindrical, 10-veined; of female fls narrowly ovoid, 20-veined; downy, sticky, with narrow blunt teeth. Petals white, with 2-lobed scales at the base of the blade. Styles 5. Capsule ovoid-conical, opening narrowly by 10 sub-erect teeth. Common throughout most of the British Is in hedgerows, waste places and cultivated land.

Hybrids of *S. dioica* and *S. alba*, often with pink fls and intermediate in other characters, are often seen. They are interfertile with either parent.

3. S. noctiflora L. (*M. noctiflorum* (L.) Fries)

Night-flowering Campion.

Annual, with erect stem, 15–60 cm., softly hairy, sticky especially above. Lower lvs obovate to ovate-lanceolate, narrowed to a stalk, others narrower, sessile, acute; all with scattered hairs. Fls few, c. 18 mm. diam., hermaphrodite. Calyx woolly and sticky, swollen in fr., membranous and whitish between the 10 broad green veins. Petals yellowish beneath and rosy above, inrolled by day, spreading and fragrant by night, deeply 2-cleft. Styles 3. Capsule often bursting the calyx-tube, opening by 6 teeth. Fl. 7–9. A rather local weed of lowland arable fields throughout Great Britain northwards to Aberdeen and in Ireland.

4. S. vulgaris (Moench) Garcke

Bladder Campion.

Perennial herb with several erect or ascending usually glabrous shoots, 25–90 cm., all flowering. Lvs elliptic-lanceolate to ovate, the lowest short-stalked, others sessile; all acute, glabrous or ciliate, often glaucous. Infl. many-fld, with scarious bracts. Fls c. 18 mm. diam., drooping, often unisexual. Calyx bladdery, narrowed at the mouth. Petals white, with small inconspicuous scales or bosses at the base of the blade. Styles 3. Capsule enclosed by the calyx, opening by 6 erect teeth. Fl. 6–8. Common on grassy slopes, arable land, roadsides and broken ground through most of Great Britain and Ireland.

5. S. maritima With.   Sea Campion.

Perennial herb with ascending flowering shoots, 8–25 cm., and prostrate non-flowering shoots forming a loose cushion. Lvs elliptic-lanceolate, usually glabrous, glaucous. Infl. of 1–4 fls with herbaceous bracts. Fls 20–25 mm. diam., erect. Calyx bladdery, its mouth hardly narrowed. Petals white, each with 2 distinct scales at the base of the blade. Styles 3 (rarely 5). Capsule with 6 recurved teeth. Fl. 6–8. Locally abundant on sea-side shingle, cliffs and stony ground all round the British Is; rarely on shingle by inland lakes and alpine streams and on cliff-ledges.

6. S. gallica L.   Small-flowered Catchfly.

Annual, with an erect or ascending hairy shoot, 15–45 cm., sticky above. Lower lvs spathulate, stalked, others lanceolate, sessile; all hairy. Infl. a spike-like cyme with the ± erect short-stalked fls alternating and all turned to one side. Calyx sticky, with 10 hairy veins and 5 long teeth. Petals white or pale pink, sometimes red-blotched, with 2 scales at the base of the blade. Styles 3. Widespread but local in fields and waste places on sand or gravel in Great Britain northwards to Inverness and in Ireland. Other varieties in the Channel Is and as casuals elsewhere.

7. S. nutans L.   Nottingham Catchfly.

Perennial herb with short non-flowering shoots and erect flowering shoots, 25–80 cm., downy below and sticky above. Basal lvs spathulate, long-stalked; others narrowly lanceolate, acute, almost sessile; all softly hairy and ciliate. Infl. a lax panicle with opposite 3–7-fld branches turned somewhat to one side. Fls 18 mm. diam., drooping, opening and fragrant at night. Calyx 10-veined with 5 white-edged teeth, sticky. Petals whitish, 2-cleft, with narrow inrolled lobes and with 2 scales at the base of the blade. Styles 3. Capsule ovoid, enclosed in the calyx. Fl. 5–7. Local on dry slopes, rocks, walls and field-borders, especially on chalk and limestone in S. and C. England but reaching N. Wales and Fife.

2. LYCHNIS L.

Annual to perennial herbs with opposite lvs; stipules 0. Fls hermaphrodite, in infls of various kinds. Epicalyx 0. Sepals joined below into a tube; petals red or white, with coronal scales; stamens 10; ovary 1-celled throughout or 5-celled only at the base; styles 5. Capsule opening by 5 teeth.

1   Stems usually less than 15 cm. high; fls in a compact head-like infl.;
     petals deeply 2-cleft. A very rare alpine. Red Alpine Catchfly.
                          L. alpina L. (*Viscaria alpina* (L.) G. Don fil.)

Stems usually more than 30 cm. high; fls not in a compact head. 2

2 Stems very sticky beneath each node; fls in an interrupted spike-like infl.; petals slightly notched. A very rare plant of cliffs, dry rock and rock-debris; chiefly on basic igneous rocks. Red German Catchfly. *L. viscaria* L. (*Viscaria vulgaris* Bernh.)
  Stems not sticky beneath the nodes; fls long-stalked; petals deeply 4-cleft with narrow segments. **1. flos-cuculi**

1. L. flos-cuculi L.   Ragged Robin.

Stock slender, branching, with decumbent non-flowering shoots and erect flowering shoots, 30–70 cm., rough above and slightly hairy. Lower lvs and those of barren shoots oblanceolate, acute, narrowed to a stalk; others narrower, oblong-lanceolate, almost sessile; all glabrous, rough. Fls numerous, 30–40 mm. diam., long-stalked. Calyx-tube reddish, strongly 10-veined, with 5 ovate-acuminate teeth. Petals rose-red or white, deeply 4-cleft with narrow spreading segs and 2 very narrow 2-cleft scales at the base of the blade. Capsule broadly ovoid, enclosed by the calyx, opening by 5 short down-curled teeth. Fl. 5–6. Common in damp meadows, marshes, fens and wet woods throughout the British Is.

3. AGROSTEMMA L.

Annual herbs with tall erect stems and narrow opposite lvs. Fls solitary or few. No epicalyx; sepals joined below into a 10-ribbed tube with 5 narrow spreading teeth much longer than the petals; petals with no basal scales; stamens 10; styles 5; ovary 1-celled to the base. Fr. a capsule opening by 5 teeth.

*1. A. githago L.   Corn Cockle.

Stem 30–100 cm., covered with appressed white hairs. Lvs narrowly lanceolate, acute, with white hairs. Fls 3–5 cm. diam., usually solitary at the ends of main stem and branches, their stalks long, hairy. Calyx-tube leathery, woolly, 10-ribbed, with long spreading narrow lf-like teeth, 3–5 cm. Petals pale reddish-purple, shorter than the sepals. Capsule ovoid; teeth almost erect. Introduced. A cornfield weed throughout most of lowland Great Britain, formerly common but decreasing.

4. DIANTHUS L.

Usually perennial herbs with very narrow and often glaucous opposite lvs. Base of calyx tightly enclosed by an epicalyx of 1–3 pairs of awn-tipped scales; sepals joined below into a cylindrical tube which is neither strongly ribbed nor has scarious seams; calyx-teeth 5; petals usually pink or red, with no scales at the base of the blade; stamens 10; styles 2. Fr. a 1-celled capsule opening by 4 teeth.

1 Fls in a compact head-like cluster surrounded by an involucre of bracts. A rare plant of hedgerows, waysides and dry pastures especially on light sandy soils northwards to Perth and Angus. Deptford Pink.   **D. armeria* L.

   Fls solitary or 2–5 in a lax infl.   *2*

2 Petals deeply cut (up to half the length of the blade) into long narrow segs; very fragrant. Introduced and naturalized on old walls. Common Pink.   **D. plumarius* L.

   Petals entire to toothed but not deeply cut.   *3*

3 Stem downy below; epicalyx scales long-awned; fls c. 18 mm. diam., not fragrant.   **1. deltoides**

   Stem glabrous; epicalyx scales not long-awned; fls more than 20 mm. diam., fragrant.   *4*

4 Stems 10–20 cm.; lvs with rough edges; fls c. 25 mm. diam.; petals hairy at the base. Confined to limestone cliffs at Cheddar Gorge. Cheddar Pink.   *D. gratianopolitanus* Vill.

   Stems 20–50 cm.; lvs with smooth edges; fls 35–40 mm. diam.; petals smooth. Introduced and sometimes naturalized on walls. Clove Pink.   **D. caryophyllus* L.

1. D. deltoides L.   Maiden Pink.

A perennial loosely tufted green or glaucous herb with short creeping non-flowering shoots and ascending flowering shoots, 15–45 cm., rough with short hairs. Lower lvs and those of barren shoots narrowly oblanceolate, blunt; others very narrowly lanceolate, acute; all roughly hairy on the margins and on the underside of the midrib. Fls c. 18 mm. diam., scentless, solitary or 2–3 at the ends of main stem and branches. Epicalyx about ½ as long as sepals, broadly ovate, long-awned. Calyx-teeth lanceolate, acute. Petals rose, with pale spots and a dark basal band, or white, shortly and irregularly toothed. A local lowland plant of dry grassy fields and banks and hilly pastures throughout Great Britain northwards to Inverness.

5. SAPONARIA L.

Annual or perennial herbs with opposite lvs. No epicalyx; sepals joined below into a green tube without scarious seams and with teeth above; petals 5 with scales at the base of the blade; stamens 10; styles 2(–3); ovary 1-celled. Fr. a capsule opening by 4(–6) teeth.

1. S. officinalis L.   Soapwort.

Perennial with a stout creeping rhizome from which arise long stolons and erect or ascending ± glabrous flowering shoots, 30–90 cm. Lvs broadly ovate to elliptical, acute, 3(–5)-veined, ± glabrous. Fls c. 25 mm. diam., in compact flat-topped cymose infls. Calyx-teeth short, triangular. Petals pinkish, entire or nearly so, with 2 small blunt scales at the base of the blade. Capsule opening by 4(–5) somewhat unequal teeth, but often failing to ripen. Fl. 7–9. Native or introduced. Fairly common on hedge-banks and waysides near villages throughout Great Britain, northwards to Aberdeen; perhaps native by streams in S.W. England and N. Wales.

6. CERASTIUM L.

Annual to perennial herbs, usually hairy, with opposite, entire, sessile lvs. Fls with parts in 5s, sometimes in 4s. Sepals free; petals white, ± deeply notched, sometimes 0; stamens usually 10 (8), sometimes 5 or fewer; ovary 1-celled; styles usually 5 (3–6). Fr. an oblong capsule, longer than the sepals and usually curved, opening by twice as many short teeth as styles; seeds numerous.

1 Styles usually 3 (varying 4–6); capsule-teeth usually 6. A rare plant
         of high mountains in the Lake District and Scotland. Starwort
         Mouse-ear Chickweed.   *C. cerastoides* (L.) Britton
      Styles 4–5; capsule-teeth 8 or 10.   2

2 Petals about twice as long as sepals; perennials with prostrate non-
         flowering shoots and fls at least 12 mm. diam.   3
      Petals not more than 1½ times as long as sepals; chiefly annuals; fls
         usually less than 12 mm. diam.   8

3 Lvs linear-lanceolate to narrowly oblong; lower lvs commonly
         with axillary lf-clusters; chiefly lowland plants.   4
      Lvs narrowly elliptical-oblong to nearly circular; lower lvs usually
         lacking axillary lf-clusters; chiefly alpine plants; at sea-level only
         in the extreme north.   5

4 Stem and lvs densely white-felted. A garden escape. Dusty Miller.
                                                          *C. tomentosum L.
      Stem and lvs shortly hairy or almost glabrous.   **1. arvense**

5 Stem and young lvs ± densely covered with long soft white hairs;
         sepals oblong-lanceolate; capsule narrow and curved in its upper
         half; seeds 1–1·4 mm., tubercled. Mountains; local. Alpine
         Mouse-ear Chickweed.   *C. alpinum* L.
      Stem and lvs with few or no long soft white hairs, though com-
         monly with short whitish hairs and sometimes also with short
         glandular hairs; sepals broadly lanceolate to ovate-lanceolate.   6

6   Bracts and bracteoles wholly herbaceous; fls 18–30 mm. diam.;
    capsule broad and nearly straight in its upper half; seeds more
    than 1·4 mm., wrinkled.   7
       At least the upper bracteoles with narrow scarious margins and
    tips; fls 12–25 mm. diam.; capsule narrow and distinctly curved
    in its upper half; seeds not exceeding 1·4 mm., tubercled.
                                       2. vulgatum forms and hybrids

7   Lvs dark and purple-tinged, almost circular, covered with short
    stout glands. Unst (Shetland) only. Edmondston's Mouse-ear
    Chickweed.
       C. arcticum Lange ssp. *edmondstonii* (H. C. Watson) Á. & D.
                 Löve (*C. nigrescens* Edmondston ex H. C. Watson)
       Lvs not dark and purplish nor densely glandular, usually elliptical.
    A local plant of high mountains in Wales and Scotland. Arctic
    Mouse-ear Chickweed.   *C. arcticum* Lange

8   Perennial usually non-glandular herb with prostrate non-flowering
    shoots; sepals with glabrous tips, non-glandular; capsule
    9–12 mm.   **2. fontanum**
       Annual or biennial herbs lacking non-flowering shoots, usually
    glandular (except *brachypetalum*); sepals glabrous or hairy at
    the tip; capsule not exceeding 10 mm., commonly much shorter.   9

9   Fls in compact clusters which remain compact in fr., so that the
    capsule-stalk does not exceed the sepals.   **3. glomeratum**
       Infl. lax, at least in fr.; fr.-stalk exceeding sepals.   10

10 Plant shaggy with spreading hairs, eglandular or nearly so; petals
    about half as long as sepals; stamens 10, their filaments hairy at
    the base. Introduced and very locally established on railway
    banks, etc.   *C. brachypetalum* Pers.
       Plant sticky with dense glandular hairs, not shaggy; petals more
    than half as long as sepals; stamens 4–5, their filaments
    glabrous.   11

11 Bracts entirely herbaceous; fr.-stalks usually erect throughout;
    parts of fls usually in 4s, sometimes in 5s.   **4. diffusum**
       At least the upper bracts with scarious tips or margins; fr.-stalks at
    first recurved or sharply deflexed; parts of fls in 5s.   12

12 Bracts with the upper half scarious; petals about ⅔ as long as
    sepals, slightly notched; fr.-stalk at first sharply deflexed from
    the base, not curved; plant often decumbent.   **5. semidecandrum**
       Upper bracts with narrow scarious margins; petals about equalling
    sepals, about ¼-cleft; fr.-stalk at first curved downwards, not
    sharply deflexed; plant erect, reddish. Local, chiefly on lime-
    stone in S. and C. England and Wales. Curtis' Mouse-ear Chick-
    weed.   *C. pumilum* Curtis

1. C. arvense L. Field Mouse-ear Chickweed.

Perennial ± glandular herb with a creeping stock, long prostrate rooting non-flowering shoots to 30 cm. and ascending flowering shoots, 4–30 cm.; all ± hairy. Lvs narrowly oblong or lanceolate, hardly narrowed to the base, soft and downy. Fls 12–20 mm. diam., in lax infls. Petals white, 2-cleft, about twice as long as the oblong-lanceolate glandular sepals. Stamens 10. Styles 5. Capsule just exceeding the sepals. Fl. 4–8. On dry banks and waysides and in grassland; especially on calcareous or slightly acid sandy soils in E. England; local elsewhere.

2. C. fontanum Baumg. ssp. **triviale** (Link) Jalas (**C. vulgatum** auct.)
Common Mouse-ear Chickweed.

Perennial herb, ± hairy but very rarely glandular, with a slender creeping stock, decumbent non-flowering shoots to 15 cm. and ascending flowering shoots up to 45 cm. Lvs dark green, densely hairy; those on barren shoots oblanceolate, blunt, narrowed to a stalk; on flowering shoots ovate-oblong, subacute, sessile. Petals white, equalling or somewhat exceeding the ovate-lanceolate hairy sepals, rarely almost twice as long, deeply 2-cleft. Stamens 10. Styles 5. Capsule 9–12 mm., distinctly curved, about twice as long as the sepals. Fl. 4–9. Very common in grassland and on shingle, sand-dunes, waysides, waste places and cultivated ground throughout the British Is.

3. C. glomeratum Thuill. Sticky Mouse-ear Chickweed.

An annual usually glandular herb with erect or ascending shoots, 5–45 cm. Basal lvs oblanceolate to obovate, others broadly ovate; all pale yellowish-green, apiculate, covered with long white hairs. Fls in compact clusters, their stalks very short. Bracts herbaceous. Sepals glandular and with long white hairs extending to and projecting beyond the tip. Petals white, about equalling the sepals, 2-cleft for more than half their length. Stamens 10. Styles 5. Capsule 7–10 mm., curved, about twice as long as the sepals. Fl. 4–9. A common weed of arable land and waste places, on walls and banks and on open sand-dunes throughout the British Is.

4. C. diffusum Pers. (**C. atrovirens** Bab.)
Dark-green Mouse-ear Chickweed.

An annual densely glandular and sticky herb with diffusely branched decumbent or ascending shoots, 7·5–30 cm. Basal lvs oblanceolate, others ovate-oblong, acute; all hairy and glandular. Bracts wholly herbaceous. Fls few, 3–6 mm. diam., usually with parts in 4s, sometimes in 5s. Sepals 4–9 mm., glandular but with hairs not projecting beyond the glabrous tip. Petals white, shorter than the sepals, 2-cleft for hardly ¼ of their length, with branching veins. Stamens 4(–5). Styles 4(–5). Capsule 5–7·5 mm., nearly straight, not much longer than the sepals. Fl. 5–7. Locally common in sandy and stony places near the sea all round the British Is; rarely inland.

5. C. semidecandrum L.   Little Mouse-ear Chickweed.

An annual densely glandular and sticky herb with erect or ascending stems, 1–20 cm., branching at the base. Basal lvs narrowly oblanceolate with a long stalk-like base; others ovate-oblong; all with short white hairs. Lowest bracts with a broad scarious margin; upper largely scarious with a small green central portion. Fls 5–7 mm. diam. Sepals 3–5 mm., narrowly lanceolate, glandular but with hairs not projecting beyond the glabrous tip. Petals white, shorter than the sepals, slightly notched, with unbranched veins. Stamens 5. Styles 5. Capsule 4·5–6·5 mm., slightly curved. Fr.-stalks at first sharply deflexed from the base, later erecting. Fl. 4–5. Common in dry open habitats especially on calcareous or sandy soils (incl. dunes) through most of the British Is; less common in Ireland and chiefly coastal.

7. MYOSOTON Moench

A herb with opposite lvs and a lfy infl. Sepals free; petals white, 2-cleft; stamens 10; ovary 1-celled; styles 5, alternating with the sepals. Fr. an ovoid capsule opening by 5 teeth, each 2-cleft.

1. M. aquaticum (L.) Moench   Water Chickweed.

A usually perennial glandular-hairy herb with prostrate overwintering stem-bases from which arise short decumbent non-flowering and decumbent or ascending flowering shoots, 30–100 cm., weak and fragile and often trailing over other plants. Lvs 2–5 cm., thin, ovate-acuminate, with a cordate base and often with a wavy margin; lower short-stalked, upper almost sessile. Infl. with lfy bracts so that the fls appear solitary at forks of the stem. Fls 12–15 mm. diam., with glandular stalks. Sepals blunt with broad membranous margins. Petals up to half as long again as the sepals, 2-cleft almost to the base. Capsule longer than the sepals, drooping. Fl. 7–8. In marshes and fens, streamsides, ditches and damp woods throughout Great Britain northwards to Stirling.

8. STELLARIA L.

Annual to perennial herbs with opposite lvs. Sepals free; petals white, usually very deeply 2-cleft, sometimes 0; stamens 10 or fewer; styles 3; ovary 1-celled. Fr. a rounded capsule opening by 6 valves.

1 Lvs ovate or cordate, at least the lower stalked; stem cylindrical.   *2*
   Lvs narrow, sessile; stem 4-angled.   *3*

2 Petals about twice as long as the sepals; capsule narrowly ovoid, up
     to twice as long as the sepals.   **1. memorum**
   Petals not or little exceeding the sepals, or 0; capsule ovoid, somewhat longer than the sepals.   **2. media**

3 Lvs 5–10(–20) mm. long; fls c. 6 mm. diam., their petals distinctly
     shorter than the sepals.   **6. alsine**
   Lvs more than 1·5 cm. long; fls more than 6 mm. diam.; petals at
   least as long as the sepals.   *4*

4 Bracts wholly herbaceous; fls 20–30 mm. diam.; petals notched to
     less than half-way; capsule globular.   **3. holostea**
   Bracts with scarious margins; petals 2-cleft at least half-way; capsule
   oblong-ovoid.   *5*

5 Bracts ciliate; plant not glaucous; fls 5–12 mm. diam. **5. graminea**
   Bracts not ciliate; plant usually glaucous; fls 12–18 mm. diam.
                                                             4. palustris

1. S. nemorum L.   Wood Stitchwort, Wood Chickweed.

A perennial herb with subterranean stolons up to 15 cm. long, and weak decumbent or ascending pale green lfy shoots, 15–60 cm., cylindrical, sparsely hairy all round or almost glabrous, slightly glandular above. Lower lvs and those of non-flowering shoots long-stalked, ovate, ± cordate; others 2·5–7·5 cm., almost sessile, ovate-acuminate; all thin, glabrous or sparsely hairy and often somewhat ciliate. Fls 13–18 mm. diam., in lax infls, their stalks slender, glandular. Sepals blunt, with narrow membranous margins. Petals white, twice as long as the sepals, deeply 2-cleft. Stamens 10. Capsule narrowly ovoid, opening almost to the base by 6 valves. In damp woods and by streams in the west and north of Great Britain.

2. S. media (L.) Vill.   Chickweed.

An annual herb with diffusely branched decumbent or ascending shoots, 5–40 cm., with a single line of hairs down each internode. Lower lvs 3–20 mm., ovate-acute, long-stalked; others usually larger, ovate or broadly elliptical, acute, almost sessile; all glabrous or ciliate at the base.

Fls numerous. Sepals with a narrow membranous margin, usually glandular. Petals not exceeding the sepals, deeply 2-cleft, sometimes 0. Stamens 3–8. Capsule ovoid-oblong, somewhat longer than the sepals. Fr.-stalks downwardly curved. Fl. 1–12. An abundant weed of cultivated ground and waste places throughout the British Is.

S. pallida (Dumort.) Piré (*S. apetala* auct.) is a smaller paler green plant with lvs usually less than 7 mm. long and mostly short-stalked, fls never opening widely, petals minute or 0, stamens 1–3 and smaller paler seeds.

S. neglecta Weihe is a larger annual to perennial herb with stems 25–90 cm. high, lvs 1–5 cm., stamens 10, and larger darker seeds than in *S. media*.

3. S. holostea L.   Greater Stitchwort.

Perennial, with a slender creeping stock and weak brittle ascending shoots, some short and non-flowering, others 15–60 cm., flowering; all with 4 rough angles, glabrous or hairy above. Lvs 4–8 cm., slightly glaucous, rigid, lanceolate-acuminate, tapering from a wide base to a long fine point, very rough on the margins and the underside of the midrib. Fls 20–30 mm. diam., long-stalked, their bracts lf-like. Petals white, 2-cleft about half-way, longer than the narrowly membranous-margined sepals. Stamens usually 10. Capsule about equalling the sepals. Fl. 4–6. Common, especially in woods and hedgerows, throughout most of the British Is.

4. S. palustris Retz.   Marsh Stitchwort.

Perennial, with a slender creeping stock and glabrous smoothly 4-angled ± erect flowering and non-flowering shoots, 20–60 cm., weak and brittle. Lvs 1·5–5 cm., usually glaucous, sessile, ascending, very narrowly lanceolate, glabrous and with smooth margins. Fls 12–18 mm. diam., their bracts with broad membranous non-ciliate margins and a narrow green central strip. Petals white, 2-cleft almost to the base, up to twice as long as the broadly white-margined sepals. Stamens 10. Capsule about equalling the sepals. Fl. 5–7. A local plant of marshes and base-rich fens in Great Britain northwards to Perth and in Ireland.

5. S. graminea L.   Lesser Stitchwort.

Perennial, with a slender creeping stock and glabrous smoothly 4-angled decumbent or ascending flowering and non-flowering shoots, the former 20–90 cm., slender and brittle, diffusely branched. Lvs 1·5–4 cm., not glaucous, sessile, very narrowly lanceolate or elliptical, acute, smooth at the margins but often ciliate near the base. Fls 5–12 mm. diam., their bracts wholly scarious with ciliate margins. Petals white, 2-cleft more than half-way, equalling or exceeding the sepals. Stamens 10, of which some or all may be sterile. Capsule longer than the sepals. Fl. 5–8. Common in woods, heaths and grassland, especially on light sandy soils, throughout most of the British Is.

6. S. alsine Grimm   Bog Stitchwort.

Perennial, with a slender creeping stock and numerous decumbent and ascending smoothly 4-angled glabrous shoots, 10–40 cm. Lvs 5–10(–20) mm. sessile, elliptical or oblanceolate, acute, slightly ciliate at the base but otherwise glabrous. Fls c. 6 mm. diam., their bracts scarious with a green central strip. Petals white, shorter than the sepals, 2-cleft almost to the base with very widely divergent lobes. Stamens 10. Capsule equalling the sepals. Fl. 5–6. Frequent on streamsides, flushes, wet tracks and woodland rides, etc., throughout the British Is.

9. MOENCHIA Ehrh.

Annual herbs with opposite narrow lvs. Sepals free; petals white, entire; stamens 8 (10) or 4 (5); styles 4 or 5, opposite the sepals; ovary 1-celled. Fr. an oblong straight capsule opening by twice as many short blunt teeth as styles.

1. M. erecta (L.) P. Gaertner, B. Meyer & Scherb.

Upright Chickweed.

Main stem erect, 3–12 cm., usually with a few ascending or decumbent basal branches. Basal lvs 6–20 mm., strap-shaped, shortly stalked; others shorter, narrowly lanceolate, sessile, ascending; all rigid, acute, glabrous and glaucous. Fls c. 8 mm. diam., 1–3, on long stalks. Petals 4, shorter than the broadly white-margined sepals. Stamens usually 4. Styles 4. Capsule a little longer than the sepals, opening by 8 teeth. Fl. 5–6. A local plant of gravelly pastures, maritime cliffs and dunes throughout England and Wales.

10. SAGINA L.

Pearlwort.

Small annual or perennial herbs, often tufted, with slender prostrate or ascending shoots and small awl-shaped exstipulate opposite lvs. Fls almost spherical in bud. Sepals free; petals white, entire, often minute, sometimes 0; stamens as many or twice as many as the sepals; styles 4–5, alternating with the sepals; ovary 1-celled. Fr. a capsule splitting to the base into 4 or 5 valves.

1   Fls with parts usually in 4s; stamens 4; petals minute or 0.   2
    Fls with parts usually in 5s; stamens usually 10, rarely fewer; petals
    ± conspicuous.   4

2   Perennial; branches long, procumbent, rooting, from a central
    rosette of leaves; main stem not flowering.   **3. procumbens**
    Annual; branches ascending or erect, not rooting; main stem
    flowering.   3

3   Lvs blunt or minutely pointed, not awned.   **2. maritima**
    Lvs awned.   **1. apetala**

4   Petals twice as long as the sepals.   **5. nodosa**
    Petals not exceeding the sepals.   5

5   Lvs with awn at least 3/4 as long as the maximum lf-width; sepals
    and fl.-stalks commonly glandular-hairy.   **4. subulata**
    Lvs with awn less than 3/4 as long as the maximum lf-width; sepals
    and fl.-stalks glabrous.   6

6   Plant forming small tufts, 1–3 cm. high; lvs 3–6 mm.; dehisced
    capsule greenish-yellow, dull. A very rare high alpine plant.
                       S. intermedia Fenzl
    Plant forming tufts or mats usually more than 3 cm. high; basal lvs
    1·5–3 cm.; dehisced capsule straw-coloured, shining. Rare plants
    of Scottish mountains and at lower levels in Shetland.   7

7   Lvs of basal rosette up to 2 cm.; stems decumbent then ascending,
    2–7 cm. high; stamens 10; capsule 3·5–4 mm., setting seeds freely.
    Alpine Pearlwort.   *S. saginoides* (L.) Karsten
    Lvs of basal rosette up to 3 cm.; stems prostrate, rooting, 2·5–
    10(–15) cm.; stamens often fewer than 10; capsule 3–3·5 mm.,
    rarely setting good seed. Druce's Pearlwort.
                       S. normaniana Lagerh.

1. S. apetala Ard.

Annual, with a loose central rosette of lvs which soon wither and a ± erect flowering main stem with several decumbent or ascending branches, 3–18 cm. Lvs 3–7 mm., very narrow, flat above, tapering into an awn. Fl.-stalks glabrous or glandular. Sepals ovate, usually hooded at the tip, spreading or appressed to the capsule in fr. Petals minute, often falling early. Capsule about equalling the sepals. Fl. 5–8.
Very variable. Two sspp. may be recognized:

Ssp. **apetala** (*S. ciliata* Fries) has the ± acute sepals appressed or only slightly spreading in fr. Frequent in dry grassland and on heaths, bare ground, etc., throughout the British Is north to Inverness.

Ssp. **erecta** (Hornem.) F. Hermann (*S. apetala* auct.) has the bluntish sepals wide-spreading in fr. and has somewhat smaller seeds than ssp. *apetala.* Common on bare soil, paths, walls, etc., throughout most of Great Britain and Ireland.

2. S. maritima G. Don fil.

Annual, with the main stem flowering, it and the numerous laterals varying from prostrate to almost erect; usually glabrous. Lvs very narrow, fleshy, blunt or very shortly pointed but not awned; usually glabrous, rarely ciliate, shorter than the internodes. Fl.-stalks remaining erect, glabrous. Sepals hooded, blunt, glabrous, often with a purplish margin, half-spreading in ripe fr. Petals minute or 0. Capsule about equalling the sepals. Fl. 5–9. A local maritime plant of duneslacks, rocks and cliffs all round the British Is.; occasionally on Scottish mountains. Very variable.

3. S. procumbens L.　　　　　　　　　　　Procumbent Pearlwort.

Perennial, with central lf-rosette; main stem short and non-flowering; laterals to 20 cm., prostrate and rooting at base, then ascending. Lvs 5–12 mm., linear, shortly awned, glabrous or ciliate. Fl.-stalks glabrous, erect at first, then curving down at tip, finally re-erecting. Sepals hooded, blunt, horizontal in ripe fr. Petals minute or 0. Capsule longer than sepals. Fl. 5–9. Paths, lawns, banks, streamsides, etc.; common.

4. S. subulata (Swartz) C. Presl　　　　　Awl-leaved Pearlwort.

Perennial, mat-forming, with rosettes of linear lvs 5–15 mm., short and non-flowering main stems and many ascending laterals, 2–7·5(–12·5) cm. Lvs long-awned, ciliate. Fl.-stalks 2–4 cm. Sepals hooded, blunt, appressed to ripe fr. Petals about equalling sepals, capsule somewhat longer. Stems, fl.-stalks and sepals commonly glandular. Fl. 6–8. Dry sandy and rocky places; local, more frequent in north.

5. S. nodosa (L.) Fenzl　　　　　　　　　Knotted Pearlwort.

Perennial, tufted, with procumbent or ascending stems, 5–15(–25) cm., from basal rosettes of lvs 5–20 mm.; stem-lvs smaller with axillary lf-clusters giving 'knotted' appearance; all lvs linear, mucronate. Fls 5–10 mm. diam. Sepals concave, blunt, appressed to ripe fr. Petals nearly twice as long as sepals. Fr. 4 mm. Stems, lvs, fl.-stalks and sepals glabrous or glandular. Fl. 7–9. Frequent in damp sandy places.

11. MINUARTIA L.

Sandwort.

Annual to perennial herbs with usually very narrow opposite exstipulate lvs. Fls hermaphrodite or unisexual. Sepals free; petals usually white, entire or nearly so, sometimes rudimentary or 0; stamens usually 10; styles 3(–5); ovary 1-celled. Fr. a capsule opening by 3(–5) teeth.

1   A slender annual herb of walls and dry places.   **2. hybrida**
     Perennial herbs with non-flowering shoots, often forming tufts or
     ± compact cushions.   *2*

2   Fls greenish, usually unisexual; petals minute or 0; nectaries 10,
     conspicuous. Compact alpine cushion-plant of mountains in
     Scotland and Inner Hebrides. Mossy Cyphel.
                 M. sedoides (L.) Hiern (*Cherleria sedoides* L.)
     Fls white, hermaphrodite; petals not minute.   *3*

3   Lvs obscurely 1-veined or apparently veinless; sepals erect in the
     open fl.; pedicels 1·5–5 cm. A rare loosely tufted plant of cal-
     careous flushes in Upper Teesdale, resembling *M. verna*. Bog
     Sandwort.   *M. stricta* (Swartz) Hiern
     Lvs 3-veined; sepals spreading in the open fl.   *4*

4   Petals twice as long as the 5–7-veined glandular sepals. A densely
     tufted perennial, woody below, with fl.-stems to 5 cm.; lvs 4–
     7 mm., down-curved. On non-calcareous rocks in S. Ireland;
     very local.   *M. recurva* (All.) Schinz & Thell.
     Petals not more than half as long again as the 3-veined sepals,
     sometimes shorter.   *5*

5   Lvs 6–15 mm.; petals longer than sepals.   **1. verna**
     Lvs 4–8 mm.; petals usually ⅔ as long as sepals. Small tufted
     perennial, woody below, with ascending fl.-stems 2–6(–9) cm.
     and glandular pedicels, 4–15 mm.; styles 3–4(–5). A very rare
     Scottish alpine.   *M. rubella* (Wahlenb.) Hiern

1. M. verna (L.) Hiern   Vernal Sandwort.

A perennial cushion-forming herb with a branching stock from which tufts of ascending non-flowering and flowering shoots arise, the latter 5–15 cm., glabrous or glandular-hairy. Lvs 6–15 mm., very narrow, acute, strongly 3-veined, often curved. Fls 8–9 mm. diam., on slender, glandular stalks. Petals white, a little longer than the strongly 3-veined sepals. Fl. 5–9. A local plant of dry calcareous rocks, screes and pastures and often abundant on the rocky debris of old lead workings; scattered through the west and north of Great Britain northwards to Banff; Shetland; in Ireland only in Antrim, Derry and Clare.

2. M. hybrida (Vill.) Schischkin (*M. tenuifolia* (L.) Hiern)

Fine-leaved Sandwort.

An annual herb with slender erect or ascending stems, 5–20 cm., glabrous or glandular above. Lvs 5–15 mm., crowded below, very narrow, 3(–5)-veined near the enlarged base, often curved downwards. Fls c. 6 mm. diam., on slender glabrous (rarely glandular) stalks. Petals white, little more than half as long as the 3-veined sepals. Fl. 5–6. In dry rocky or stony places, walls, railway-tracks, sandy fields, etc., southwards from York and Caernarvon, especially in E. England; S. and C. Ireland.

12. Honkenya Ehrh.

A perennial maritime ± dioecious herb with opposite lvs and greenish fls. Sepals free; petals entire; stamens 10 or abortive; styles usually 3; ovary 1-celled, abortive in male fls. Fr. a globular capsule opening by 3 teeth.

1. H. peploides (L.) Ehrh.   Sea Sandwort.

A succulent stoloniferous herb with lfy flowering and non-flowering shoots, decumbent below, then erect, arising terminally from long pale slender stolons creeping in sand or shingle. Flowering stems 5–25 cm., with sessile, very fleshy, rigid, ovate-acute lvs, 6–18 mm., having translucent wavy margins and downwardly pointing tips. Fls 6–10 mm. diam. Petals greenish-white, equalling the blunt fleshy sepals in male fls but shorter in female. Styles 3 or more. Capsule up to 8 mm. diam., longer than the sepals. Fl. 5–8. Common all round the British Is on mobile sand and sandy shingle.

13. Moehringia L.

Annual to perennial usually slender herbs with opposite exstipulate lvs. Sepals free; petals white, ± entire; stamens 8 or 10; styles 2–5; ovary 1-celled. Fr. a rounded capsule usually opening by 4 or 6 downwardly curved teeth. Seeds with an oily appendage, so differing from *Arenaria*.

1. M. trinervia (L.) Clairv.   Three-nerved Sandwort.

Usually annual, with weak slender diffusely branching shoots, 10–40 cm., prostrate or ascending, downy. Petals white, entire, shorter than the 3-veined ciliate sepals. Stamens 10. Styles 3(–4). Capsule almost globular, shorter than the sepals, opening by 6(–8) valves. Seeds blackish, shining, with a small pale fringed appendage. Fl. 5–6. A woodland herb of well-drained mull soils throughout Great Britain and Ireland.

14. ARENARIA L.

Annual to perennial herbs or dwarf shrubs with small usually ovate or lanceolate opposite exstipulate lvs. Sepals free; petals white or pink, entire or nearly so; stamens 10; styles usually 3; ovary 1-celled. Capsule opening by 6 (rarely 8 or 10) teeth. Seeds without oily appendage.

1 Petals shorter than the sepals.   **1. serpyllifolia**
   Petals longer than the sepals.   2
2 Lvs obscurely 1-veined, often slightly ciliate in the lowest third.
      Perennial (ssp. *norvegica*) or annual to biennial (ssp. *anglica*
      Halliday) loosely tufted herbs, 3–7 cm. Very local in N. England,
      Scotland and W. Ireland.   *A. norvegica* Gunnerus
   Lvs distinctly 1-veined, usually ciliate at least in the lower half.
      Low-growing perennial herb with ascending flowering shoots to
      5 cm. Very local on limestone cliffs in W. Ireland.   *A. ciliata* L.

1. A. serpyllifolia L.   Thyme-leaved Sandwort.

Annual to biennial, with slender decumbent or ascending grey-green shoots, 2·5–25 cm., often bushy in habit. Lvs up to 6 mm., ovate-acuminate, almost sessile, roughly hairy, ciliate. Fls 5–8 mm. diam., numerous. Petals white, shorter than the ovate-lanceolate acute sepals. Stamens 10 or fewer. Styles 3. Capsule flask-shaped, with curving sides, opening by 6 short teeth, its stalk almost erect. Seeds brownish-black. Common on bare ground, arable fields, walls, cliffs, etc., throughout the British Is.

A. **leptoclados** (Reichenb.) Guss. is a more slender and diffuse plant with fls only 3–5 mm. diam., lanceolate sepals, a conical straight-sided capsule and pale brown seeds. It is common in similar habitats to those of *A. serpyllifolia*, with which it often grows, but is rare in the north of the British Is.

15. SPERGULA L.

Annual herbs whose very narrow blunt opposite lvs have small deciduous scarious stipules and conspicuous axillary clusters of lvs. Sepals free, petals white, entire; stamens 5–10; styles 5; ovary 1-celled. Fr. an ovoid capsule opening deeply by 5 valves.

1. S. arvensis L.   Spurrey.

Stems 7·5–40 cm., ascending, geniculate, brittle. Lvs 1–3 cm., grass-green to grey-green, very narrow, blunt, fleshy, convex above and channelled beneath. Stem and lvs more or less sticky with glandular

hairs. Fls 4–7 mm. diam., their stalks 1–2·5 cm.; bracts small, scarious.
Petals white, slightly longer than the ovate blunt glandular sepals.
Stamens 10 or fewer. Capsule ovoid-conical, up to twice as long as the
sepals. Seeds blackish, minutely tubercled or with pale club-shaped
papillae. Fl. 6–8. A locally abundant and often troublesome calcifuge
weed of arable land, throughout the British Is northwards to Aberdeen.

16. SPERGULARIA (Pers.) J. & C. Presl

Annual to perennial herbs whose very narrow opposite lvs have pale
scarious stipules. Sepals free; petals white or pink, entire, sometimes 0;
stamens 5–10; ovary 1-celled; styles 3. Capsule opening by 3 valves.

1　Seeds all broadly winged; capsule usually 7–10 mm.　　**3. media**
　　Seeds not winged, or a few winged at the base of the capsule;
　　　capsule not exceeding 7 mm.　　　　　　　　　　　　　2
2　Perennial; densely glandular-hairy throughout; fls 8–10 mm. diam.;
　　on cliffs, rocks and walls.　　　　　　　　　　　**2. rupicola**
　　Annual; glabrous or glandular only above; fls not exceeding 8 mm.
　　　diam., often much smaller.　　　　　　　　　　　　　3
3　Lvs bright or yellowish green, blunt or short-pointed, fleshy; fls
　　6–7 mm. diam.　　　　　　　　　　　　　　　**4. marina**
　　Lvs grey-green, awned, not fleshy; fls 2–5 mm. diam.　　　4
4　Stipules lanceolate-acuminate, silvery; fls 3–5 mm. diam.; fl.-stalks
　　longer than the sepals.　　　　　　　　　　　　**1. rubra**
　　Stipules broadly triangular, dull; fls 2 mm. diam.; fl.-stalks usually
　　　shorter than the sepals; infl.-branches lengthening to resemble
　　　1-sided racemes. Rare and local in dry sandy and rocky coastal
　　　areas.　　　　　　　*S. bocconii* (Scheele) Ascherson & Graebner

1. S. rubra (L.) J. & C. Presl　　　　　　　　　Sand-spurrey.

Annual or biennial with decumbent stems, 5–25 cm., glandular above.
Lvs 4–25 mm., tapering to an awned tip, not fleshy; stipules silvery and
conspicuous, usually torn at the tip. Fls 3–5 mm. diam. Petals rose-
coloured, shorter than the ovate-lanceolate broadly scarious-margined
glandular sepals. Stamens usually fewer than 10. Capsule about
equalling the sepals, its stalk spreading or downwardly curved at first,
then erecting. A common calcifuge plant of open sandy or gravelly
places throughout the British Is but very local in Ireland.

2. S. rupicola Le Jolis　　　　　　　　　　Cliff Sea-spurrey.

Perennial, with a woody stock and many decumbent cylindrical shoots,
5–15 cm., often dark purple, densely glandular-hairy throughout. Lvs

5–15 mm., narrow, fleshy, somewhat flattened, with the horny tip prolonged into a short point, sparsely glandular especially near the base; stipules silvery; lvs of axillary shoots forming clusters at each node. Fls 8–10 mm. diam., their stalks glandular-hairy. Petals deep pink, about equalling the lanceolate blunt white-margined glandular sepals. Stamens 10. Capsule 5–7 mm., slightly longer than the sepals but shorter than its stalk, which bends strongly downwards after flowering but then re-erects. Seeds unwinged, black, minutely tubercled. Fl. 6–9. Local on maritime cliffs, rocks and walls, chiefly in the S. and W. of Great Britain but reaching to Ross and Aberdeen; all round the coast of Ireland and inland by Lough Neagh.

3. S. media (L.) C. Presl (*S. marginata* (DC.) Kit.)

Greater Sea-spurrey.

Perennial, with a stout stock and many decumbent or geniculate-ascending stout flattened shoots, to c. 30 cm., usually glabrous except in the infl. Lvs 1–2·5 cm., fleshy, horny-tipped and often short-pointed, flat above and rounded beneath, usually glabrous; stipules not silvery. Fls 8–12 mm. diam.; petals whitish or pink above, somewhat exceeding the blunt sepals. Stamens 10. Capsule 7–11 mm., longer than the sepals and not shorter than its stalk. Seeds pale yellowish, almost all with a broad scarious border. Fl. 6–9. In muddy and sandy salt-marshes all round the coasts of the British Is.

4. S. marina (L.) Griseb. (*S. salina* J. & C. Presl) Lesser Sea-spurrey.

Annual, with many prostrate or decumbent slender flattened shoots to 20 cm. or more, almost glabrous or glandular above. Lvs 1–2·5 cm., horny-tipped, acute, flat above and rounded beneath, glabrous or glandular; stipules not silvery. Fls 6–8 mm. diam. Petals pink or deep rose with a white base, shorter than the blunt often pink-tinged sepals. Stamens usually 4–8 or fewer. Capsule somewhat exceeding the sepals but distinctly shorter than its stalk. Seeds brownish, all unbordered or a few at the base with a broad scarious border as in *S. media*. Fl. 6–8. In the drier zones of muddy and sandy coastal salt-marshes and of brackish marshes all round the British Is and in inland salt areas of Worcester and Cheshire; rarely an inland casual.

Intermediates between typical *S. media* and *S. marina* sometimes occur and may be hybrids.

17. HERNIARIA L.

Rupture-wort.

Annual or perennial mat-forming herbs with opposite stipulate lvs, the upper lvs often alternate through abortion of one member of a pair. Fls in dense axillary clusters, perigynous; bracteoles scarious. Sepals 5; petals 5, very narrow, shorter than the sepals; stamens 5, opposite the petals; style deeply 2-branched; ovary free at the base of the perigynous cup which closely surrounds its lower half, 1-celled with 1 ovule. Fr. an indehiscent nutlet with 1 shining black seed.

1 Calyx ±glabrous; plants green.   2
   Calyx with spreading hairs; plants grey- or whitish-hairy.   3

2 Stipules often greenish; fl.-clusters confluent on short lateral branches which resemble lfy spikes; fr. acute, much longer than the calyx. Rupture-wort. A rare and local plant in a few dry sandy places scattered through England.   *H. glabra* L.
   Stipules white; fl.-clusters roundish, separate; fr. blunt, little longer than the calyx. A very rare plant of maritime sands and rocks in Cornwall and the Channel Is.   *H. ciliolata* Melderis

3 Lower lvs opposite; sepals ending in a long bristle, so that the calyx is bur-like. Introduced. Naturalized on sandy ground near Christchurch, Hants. Hairy Rupture-wort.   **H. hirsuta* L.
   Most lvs alternate; sepals not long-pointed. Introduced. Established on waste ground at Burton-on-Trent.   **H. cinerea* DC.

18. SCLERANTHUS L.

Annual to perennial herbs with diffusely branched stems and opposite awl-shaped exstipulate lvs. Fls very small, green or whitish, in dense terminal and axillary infls, hermaphrodite, perigynous. Sepals usually 5; petals 0; stamens 1–10; styles 2; ovary 1-celled, its apex hardly reaching the rim of the flask-shaped perigynous zone; ovules 1(–2). Fr. an indehiscent 1(–2)-seeded nutlet enclosed by the hardened perigynous zone and sepals.

Annual or biennial; sepals acute with narrow scarious margins, suberect in fr.   **1. annuus**
Perennial, woody below; sepals blunt with broad white margins incurved in fr. A rare plant of sandy fields in Norfolk and Suffolk and on rocks in Radnor.   *S. perennis* L.

1. S. annuus L.

Knawel.

An annual or biennial herb with decumbent or ascending glabrous or shortly hairy stems, 2·5–25 cm. Lvs 5–15 mm., acute, usually ciliate, with narrow scarious margins. Fls c. 4 mm. diam., almost sessile, solitary in forks of the stem and in terminal and axillary clusters. Sepals triangular, acute, glabrous, narrowly scarious-margined. Stamens 10 or fewer. Fr. enclosed in the 10-furrowed perigynous tube which is surmounted by the almost erect sepals. Fl. 6–8. In dry sandy and gravelly places and in cultivated and waste ground on sandy soil throughout Great Britain; local in Ireland, and mainly in the north.

40. PORTULACACEAE

Annual to perennial herbs, usually glabrous, sometimes fleshy. Lvs opposite in British spp., simple, entire, with or without stipules. Fls hermaphrodite, actinomorphic, hypogynous to half-epigynous. Sepals 2; petals usually 4–6, free or joined below; stamens 4–6, opposite the petals, or more numerous, or only 3; style simple or branched; ovary 1-celled with 1 or more basal ovules. Fr. a capsule opening by valves or transversely.

Distinguished from Caryophyllaceae by the calyx of only 2 sepals.

1 Fls yellow, sessile; sepals soon falling; stamens 6–15; capsule opening transversely. Fleshy herb with prostrate or ascending stems; uppermost lvs crowded beneath fls. Introduced and occasionally established. Purslane.   **Portulaca oleracea* L.

  Fls white or pink, stalked; sepals persistent; stamens 3–5; capsule opening by valves.   2

2 Small, often aquatic or subaquatic herbs with several pairs of stem-lvs.   1. MONTIA

  Herbs with 1 pair of stem-lvs, the members of a pair free or broadly connate.   3

3 At least the later-formed basal lvs with broadly elliptical or ovate blades.   1. MONTIA

  All basal lvs with linear or linear-oblanceolate blades; fls rose-coloured with darker veins. Introduced as a garden plant and sometimes becoming established.   **Claytonia virginica* L.

1. MONTIA L.

Annual to perennial glabrous herbs with stem-lvs in 1 or more opposite pairs; stipules 0. Fls in terminal infls which may appear lateral through overtopping by a branch. Sepals persistent. Petals (3–)5, sometimes unequal, free or united into a short tube. Stamens 3–5, adhering to the base of the petals. Ovary superior; styles 3-cleft. Fr. a 3-valved capsule with usually 1–3 seeds.

1 Stem-lvs many, narrowed into a short stalk-like base; fls 2–3 mm.
    diam.; stamens 3.   **1. fontana**
    Basal lvs long-stalked; stem-lvs 2; fls more than 4 mm. diam.;
    stamens 5.   *2*
2 Stem-lvs broadly connate; fls 5–8 mm. diam.; petals white, entire
    or slightly notched.   **2. perfoliata**
    Stem-lvs sessile but not broadly connate; fls 15–20 mm. diam.;
    petals pink or white, 2-cleft.   **3. sibirica**

1. M. fontana L.   Blinks.

Annual to perennial herbs with branching shoots, 2–50 cm., short and
erect in land-forms, weaker, decumbent and rooting below in aquatic
forms; sometimes floating. Stem-lvs in few to many pairs, 2–20 ×
1·5–6 mm., narrowly spathulate to obovate, narrowed into a stalk-like
base, the members of a pair free or somewhat connate. Fls 2–3 mm.
diam., inconspicuous in terminal cymes which may appear lateral in
vigorous shoots. Petals 5, white, unequal, joined into a short tube cleft
to the base in front. Stamens usually 3, opposite the smaller petals.
Capsule 1·5–2 mm., globose. Seeds 3, dark brown or black, dull or
shining, tubercled or smooth. Fl. 5–10. Streamsides, flushes, wet places
amongst rocks, moist pastures, etc., especially on non-calcareous substrata, throughout the British Is; also in arable fields in S.W. England.
Very variable.

*2. M. perfoliata (Willd.) Howell (*Claytonia perfoliata* Willd.)

Annual glabrous herb with erect or ascending stems, 10–30 cm. Basal
lvs very long-stalked, their blades 1–2·5 cm., elliptical to ovate-rhombic, entire, rather fleshy, faintly veined; stem-lvs 2, opposite, broadly
connate to form a cup beneath the infl. Fls 5–8 mm. diam. Petals 5,
2–3 mm., white, entire or slightly notched, joined below to a very short
complete tube. Stamens 5. Capsule subglobose, shorter than sepals.
Seeds black, shining. Fl. 5–7. Introduced. Cultivated, disturbed and
waste ground, especially on light sandy soils. Scattered throughout
Great Britain and locally abundant northwards to Aberdeen and
Inverness; Channel Is.

*3. M. sibirica (L.) Howell (*Claytonia sibirica* L.; *C. alsinoides* Sims)

Annual to perennial glabrous herb with erect or ascending stems,
15–40 cm. Basal lvs very long-stalked, their blades 1–3 cm., ovate-
acuminate, entire, rather fleshy, distinctly veined; stem-lvs 2, opposite,
sessile, not connate. Fls 15–20 mm. diam. Petals 5, 8–10 mm., white or
pink with darker veins, deeply notched or bifid, joined below into a very

short complete tube. Stamens 5. Capsule about as long as sepals, ovoid. Fl. 4–7. Introduced. Damp woods, shaded streamsides, etc., especially on sandy soil. Scattered through western and northern Great Britain: Inner Hebrides; very rare and casual in Ireland.

41. AIZOACEAE (p. xix)

42. AMARANTHACEAE

Annual, rarely perennial. Lvs opposite or alternate, entire. Perianth of 3–5, dry and scarious, often brightly-coloured segments. Stamens 3–5, opposite the per. segs. Ovary superior, unilocular; fr. often indehiscent.

1. AMARANTHUS L.

Infl. of many small cymes forming dense axillary spikes or a large terminal lfless panicle. Fls mostly unisexual. Fr. 1-seeded. A few spp. are found, rather rarely and usually as casuals. The 2 commonest ones can be distinguished as follows:

Stems hairy; per. segs 5; terminal part of infl. lfless.   *A. retroflexus L.
Stems nearly or quite glabrous; per. segs 3; infl. lfy to top.
*A. albus L.

43. CHENOPODIACEAE

Annual or perennial herbs or shrubs, frequently ± succulent, or mealy with bladder-like hairs. Lvs usually alternate, simple. Fls small and greenish, hermaphrodite or unisexual, usually actinomorphic. Perianth 3–5-lobed, rarely 0 in female fls, persistent and often enlarging in fr. Stamens usually the same number as the per. segs. Ovary superior or ½-inferior; fr. usually indehiscent; seeds 1.

1 Plant apparently lfless but with green succulent stems.
                                                        7. SALICORNIA
  Plant with obvious lvs.   2
2 Lvs flattened, neither subterete nor ½-terete and succulent.   3
  Lvs subterete or ½-terete and succulent.   6
3 Fls mostly hermaphrodite; fr. ± enclosed in 2–5 persistent per. segs.   4
  Fls all unisexual; fr. enclosed in 2 ± vertical appressed bracteoles.   5
4 Often mealy; lower lvs toothed or lobed, or if entire, hastate to cordate at base or else not more than 5 cm.; fr. not joined firmly together in groups of 2–4.   1. CHENOPODIUM

Glabrous; lvs entire, usually cuneate at base, at least some more than 5 cm.; fr. firmly joined together in groups of 2–4 by thickened bases of per. segs.   2. BETA

5 Lvs toothed or if entire not elliptic; bracteoles in fr. not united above the middle; annual herbs.   3. ATRIPLEX
   Lvs entire, elliptic or nearly so; bracteoles in fr. united above the middle; small shrub or annual herb with long-pedicelled fr. (the latter very rare).   4. HALIMIONE

6 Lvs acute or obtuse, not spinescent at tip.   5. SUAEDA
   Lvs spinescent at tip.   6. SALSOLA

1. CHENOPODIUM L.

Herbs or small shrubs, usually mealy. Fls in small cymes arranged in a ± branched infl. Perianth green, of 2–5 ± joined segments. Seed covered with a thin pericarp; testa variously sculptured.

The markings on the testa of the seeds are useful in identification. The pericarp can often be removed by rubbing the seed between finger and thumb but in some spp. it is necessary to boil or scrape with a needle. When the pericarp has been removed the sculpturing can be seen with the low power of a microscope.

1 Perennial; lvs triangular-hastate; stigmas long; testa rugose.
                                                      1. bonus-henricus
   Annual; lvs rarely hastate; stigmas short.   2

2 Larger stem-lvs cordate or truncate at base, coarsely sinuate-toothed; testa with large deep nearly circular pits. Rare.
                                                      C. hybridum L.
   Lvs never cordate, ± cuneate at base.   3

3 Infl. axis and perianth glabrous (rarely ± mealy but then not enclosing the fr.).   4
   Infl. axis and perianth mealy at least when young; fr. usually nearly or quite enclosed in perianth.   9

4 Lvs entire or at most with a single obscure tooth on each side, green on both sides (or purple); stems 4-angled; seeds black; testa with sinuous markings.   **2. polyspermum**
   All lvs, except the uppermost, toothed, very rarely entire, but then seeds red-brown; stems ridged but not 4-angled.   5

5 All fls with 5 per. segs and 5 stamens; seeds black, 1·2–1·5 mm. diam.; testa with an elongate reticulum. Rare.   *C. urbicum* L.
   All fls, except the terminal ones, with 2–4 per. segs and 2–3 stamens; seeds red-brown, 0·75–1·1 mm. diam.   6

6 Lvs mealy-glaucous beneath, green above; testa with ± diamond-shaped markings. Local.   *C. glaucum* L.
   Lvs green on both sides.   7

7   Fr. perianth fleshy, turning scarlet; fls in sessile heads forming a
      spike; testa minutely pitted. Rare alien.
                                            *C. capitatum (L.) Ascherson
    Fr. perianth not fleshy, not turning scarlet; infl. branched; testa
      rough, with obscure ± diamond-shaped markings.   8

8   Per. segs of lateral fls usually free to middle or below, not or feebly
      ridged on back; lvs usually much toothed.   5. rubrum
    Per. segs of lateral fls joined almost to top, forming a sack closely
      investing the fr., ± distinctly ridged or keeled on back (at least
      when young); lvs entire or ± deltate and slightly toothed. Rare,
      in salt marshes.   C. botryodes Sm.

9   Lvs entire or almost entire.   10
    Lvs toothed or lobed.   12

10 Lvs ovate or rhombic; per. segs rounded on back; plant strongly
      grey-mealy, stinking of bad fish; testa with a ± quadrate
      reticulum. Rare.   C. vulvaria L.
    Lvs linear to linear-lanceolate or narrowly oval; per. segs strongly
      keeled on back; plant not stinking.   11

11 Lvs ± densely white-mealy beneath, linear to linear-oblong;
      petioles, even of larger stem-lvs, all short (up to c. 1 cm.); testa
      with a small rounded reticulum. Rare casual.
                                            *C. pratericola Rydb.
    Lvs normally green beneath; petioles of larger stem-lvs usually
      much longer than 1 cm.   15

12[1] Lvs toothed, but not distinctly 3-lobed.   13
    At least some of the lvs distinctly 3-lobed.   16

13 Infl. lfy almost to top; branches short, spreading; seeds dull,
      sharply keeled; testa with numerous small pits. Local.
                                                        C. murale L.
    Infl. usually lfless in upper part; branches long; seeds shiny with-
      out a sharp keel; testa never densely pitted.   14

14 Lvs usually markedly longer than broad, often longer than 3 cm., if
      nearly entire then scarcely mealy. (album agg.)   15
    Larger stem-lvs often as broad as or broader than long, up to 3 cm.,
      much toothed to nearly entire, but some at least ± 3-lobed,
      usually grey-green and very glaucous-mealy when young (as is the
      infl.); stems never red; testa with a very elongate reticulum.
      Local.   *C. opulifolium Schrader ex Koch & Ziz

15 Plant usually deep green (though often ± masked by grey meal);
      stems often reddish; lvs variable, usually ovate-lanceolate,
      toothed or entire; testa with an elongate or quadrate reticulum.
                                                        3. album

[1] The sculpturing of the testa (which appears to be constant) should be used
to check identifications from this point onwards in the key.

Plant usually rather bright bluish-green; stems not red; larger
stem-lvs always ovate-rhombic with sharp forward-directed
teeth; testa with more numerous, closer and deeper furrows than
in *C. album*. Casual.   *C. suecicum* J. Murr

16   Plant usually smelling strongly of bad fish; young shoots densely
greyish-white and mealy; lvs deeply 3-lobed, lobes spreading;
testa with radial lines and oval pits. Casual.
                                                C. hircinum Schrader
Plant not stinking (rarely so in *C. berlandieri*, but then lvs weakly
3-lobed with short lateral lobes).   17

17   At least the larger stem-lvs as broad as or broader than long, lateral
lobes short; plant usually grey-green; stems never red. Introduced. Waste places; local.
                                      C. opulifolium Schrader ex Koch & Ziz
Lvs usually markedly longer than broad.   18

18   Seeds c. 1·15 mm. diam.; mid-lobes of lvs long, ± parallel-sided,
often blunt; testa with elongate pits.   **4. ficifolium**
Seeds 1·2–1·85 mm. diam.; mid-lobes of lvs very rarely parallel-
sided but then testa not pitted.   19

19   Testa without radial lines but with deep honeycomb-like pitting.
Casual.   *C. berlandieri* Moq.
Testa not at all pitted.   15

*1. C. bonus-henricus L.   All-good, Mercury, Good King Henry.
30–50 cm. Lvs up to 10 cm., mealy when young, margins sinuous,
entire. Infl. narrow and tapering, lfless except at base. Seeds 1·8–
2·2 mm., red-brown, not enclosed by perianth. Fl. 5–7. Naturalized in
farm yards, by roads, etc., widespread.

2. C. polyspermum L.   All-seed.
Up to c. 100 cm., glabrous or slightly mealy. Lvs up to 5 cm., ovate or
elliptic, ± blunt, thin. Cymes in lf-axils, with spreading branches. Per.
segs rounded on the back. Seeds 1·1–1·25 mm., not enclosed by perianth.
Fl. 7–10. Cultivated ground, etc., locally abundant but mainly in the
south.

3. C. album L.   Fat Hen.
Up to c. 100 cm., mealy, usually with rather short erect branches. Lvs
usually ovate-lanceolate, toothed, but varying from rhombic to
lanceolate. Seeds (1·25–)1·5–1·85 mm. diam. Fl. 7–10. Waste places
and cultivated land, common and variable.

4. C. ficifolium Sm.   Fig-leaved Goosefoot.

30–90 cm., mealy. Blades of lower lvs up to 8 cm., 3-lobed; lateral lobes short, oblong or triangular, often with 1 tooth on lower margin; middle lobe oblong, coarsely toothed to subentire; upper lvs subentire. Per. segs keeled on back. Seeds c. 1·15 mm. diam. Fl. 7–9. Waste ground and arable land, locally common in the south.

5. C. rubrum L.   Red Goosefoot.

Up to 70 cm., nearly or quite glabrous, often reddish, prostrate to erect. Lvs commonly 2–5 cm., ovate to rhombic, coarsely and irregularly toothed to subentire. Seed 0·7–1·1 mm. diam. Fl. 7–9. Waste places, cultivated ground and near the sea, often abundant in England, usually casual elsewhere.

2. BETA L.   Beet.

Herbs. Fls hermaphrodite, in small cymes arranged in branched spike-like infl. Per. segs 5, becoming thickened especially towards the base as the fr. ripens. Ovary ½-inferior. Fr. 1-seeded, adhering firmly in groups by the swollen perianth-bases.

1. B. vulgaris L.

30–120 cm., glabrous or nearly so. Root often swollen. Stems ± branched and lfy. Lvs varied in shape, often dark green or reddish and shiny. Infl. usually large and ± branched; cymes sessile in the axils of small narrow lvs, usually 2–4-fld; per. segs green.

Ssp. **maritima** (L.) Arcangeli. Stems usually decumbent and lvs up to c. 10 cm., rhombic, thick and leathery; cymes usually of 2–3 fls. Plant usually reddish.

Ssp. **vulgaris.** Sugar beet, beetroot, spinach beet, chard, mangold. Stems ascending or erect and lvs up to c. 20 cm., often ovate and cordate at base; cymes usually of 3–4 fls. Plant usually green. The 2 sspp. are interfertile and many intermediates occur. Fl. 7–9. Near the sea and widely cultivated.

3. ATRIPLEX L.   Orache.

Annual herbs, often mealy. Stem frequently striped white and green or red and green. Infl. like *Chenopodium* but fls unisexual; male fls with perianth of (3–)5 segments; female fls without perianth but enclosed by 2 bracteoles. Bracteoles not joined above the middle, not obdeltate, entire or else with lateral lobes smaller than middle one.

1 Bracteoles joined only at base.   *2*
     Bracteoles joined up to middle.   *4*

2 Lower lvs very narrow or narrowly oblong.   **1. littoralis**
     Lower lvs triangular or rhombic, often hastate.   *3*

3 Lower lvs cuneate at base, narrowing gradually into the petiole.
                                                              2. patula
     Lower lvs truncate or nearly so at base, abruptly contracted into the
        petiole.   **3. hastata**

4 Plant greenish, mealy; bracteoles in fr. not hardened below.
                                                          4. glabriuscula
     Plant silvery, often with red stems; bracteoles in fr. not hardened below.
                                                          5. laciniata

1. A. littoralis L.   Shore Orache.

Up to 1 m., ± mealy. Lvs entire or shallowly toothed, lower short-
stalked, upper sessile. Infl. narrow, lfless except at base. Bracteoles in
fr. triangular-ovate, warty on back. Fl. 7–8. Near the sea, but not on
sand; often forming dense stands.

2. A. patula L.   Iron-root, Common Orache.

30–90 cm., ± mealy, much-branched, prostrate to erect. Lvs entire or
toothed, lower rhombic–lanceolate, with a projecting lobe on each
side above the cuneate base; upper narrow, entire. Petiole 1–10 mm.
Bracteoles in fr. 2–3(–10) mm., broadly rhombic, lateral angles entire
or toothed, smooth or slightly warty on back. Fl. 8–10. Cultivated
ground and waste places, common and variable.

3. A. hastata L.   Hastate Orache.

Like 2 but lvs glabrous or mealy on underside only, the lower triangular-
hastate, truncate or nearly so at base; petiole usually 10 mm. or more;
bracteoles ovate with a subcuneate to subcordate base. Fl. 8–9. In
similar places to *A. patula*, but commoner near the sea.

4. A. glabriuscula Edmondston   Babington's Orache.

Up to 90 cm., mealy, spreading, rather stout. Lvs triangular with a
small spreading lobe on each side towards the base, ± toothed, rather
fleshy. Infl. lfy nearly to top. Bracteoles in fr. rhombic to suborbicular,
usually inflated, united in lower half, smooth or warty on back. Fl. 7–9.
On sandy or gravelly shores at about high-tide mark, widespread.

5. A. laciniata L.   Frosted Orache.

Up to c. 30 cm., silvery, decumbent, rather slender. Lvs rhombic to ovate, sinuate-toothed, rather fleshy. Infl. very lfy. Bracteoles in fr. 6–7 mm., usually broader than long, lateral angles rounded and sometimes toothed, hardened in lower half when ripe. Fl. 8–9. Sandy shores at about high-tide mark, local but widely distributed.

4. HALIMIONE Aellen

Differs from *Atriplex* as follows: Lvs entire, lower elliptic or nearly so; bracteoles obdeltate in fr., united almost to top, 3-lobed, lateral lobes usually larger than middle one.

Woody much-branched perennial; fr. sessile.   **1. portulacoides**
Little-branched annual; fr. long-stalked. Very rare.
                                 H. pedunculata (L.) Aellen

1. H. portulacoides (L.) Aellen   Sea Purslane.

50–150 cm., very mealy, shrubby. Stems spreading, branches ascending, terete below, angled above. Lower lvs opposite, upper alternate, narrower. Bracteoles in fr. 3–5 mm., rather broader than long. Fl. 7–9. Salt-marshes, especially fringing channels and pools, locally abundant.

5. SUAEDA Forskål

Herbs or small shrubs. Lvs alternate, small. Fls unisexual or hermaphrodite, in lf–axils; bracteoles very small. Per. segs 5, ±succulent. Stamens 5. Stigmas 3–5.

Annual herb; lvs acute or subacute and narrowed at base.   **1. maritima**
Small shrub; lvs rounded at tip and base. S. and E. coasts; local.
                       S. vera J. F. Gmelin (*S. fruticosa* auct.)

1. S. maritima (L.) Dumort.   Herbaceous Seablite.

7–30(–60) cm., prostrate to erect, glaucous or purplish. Lvs 3–25 × 1–2(–4) mm., ½-terete. Fls 1–3 together in lf–axils. Seed 1·1–2 mm. diam., biconvex, black, shiny, with fine reticulate sculpturing. Fl. 7–10. Salt-marshes and sea-shores, common and widely distributed.

6. SALSOLA L.

Herbs. Lvs sessile. Fls sessile; per. segs (4–)5, usually developing a transverse wing in fr. Stamens (3–)5. Stigmas 2–3.

1. S. kali L. Saltwort.

Up to 60 cm., ± decumbent. Lvs 1–4 cm., narrowed into a little spine at tip. Fls usually solitary in lf-axils, each with 2 lf-like bracteoles. Per. segs becoming tough and thickened about the middle in fr., the thickening forming a transverse ridge or wing of variable width. Fr. c. 2·5 mm., top-shaped, enclosed in the perianth. Fl. 7–9. Sandy shores, widespread.

7. SALICORNIA L.

Glasswort, Marsh Samphire.

Annual or perennial salt-marsh plants. Lvs opposite, the pairs joined along their margins and with the stem (except at the tips) forming the 'segments'. Infl. of simple, rarely branched, fleshy spikes with cymes of 3 (1) fls in the lf-axils. Perianth fleshy. Stamens 1 or 2, appearing in succession.

All the annual spp. may be unbranched or little branched when growing in closed communities. Well-grown branched specimens should be examined to begin with and badly-grown plants ignored until some familiarity with the genus is gained. The following key refers mainly to well-grown widely-spaced plants. A gathering of 10–20 apparently similar individuals is desirable to allow for variability in size, etc., and care should be taken to avoid plants whose main stem has been injured.

1 Perennial, with creeping woody stems, rooting at intervals, from which arise ± erect ones; plant often orange or claret-coloured; fl. spikes almost truncate. *S. perennis* Miller

Annual, erect or prostrate, never rooting along the stems; plant green or ± crimson tinged, rarely orange; fl. spikes tapering, at least in the upper ¼. *2*

2 Cymes 1-fld (except rarely the basal ones); perianth nearly circular, often tinged with reddish-brown; terminal spikes mostly 5–15 × 1·5–2 mm. with 1–3(–8) segments. S. and E. coasts, along the drift line; local. *S. pusilla* Woods

Cymes 3-fld; perianth cuneate at base, green or magenta; terminal spikes stouter and commonly longer than the above. *3*

3 Anthers 0·2–0·5 mm., often not exserted; fertile segments with convex sides; lateral smaller than central *4*

Anthers (0·5–)0·6–1 mm., always exserted; fertile segments ± cylindrical; lateral fls almost equalling the central. *5*

4 Lower fertile segments of terminal spike 3–5 mm. wide at the narrowest point; upper edge with inconspicuous scarious margin. Common and widespread.

S. europaea L. (incl. *S. obscura* P. W. Ball & Tutin)

Lower fertile segments of terminal spikes 2·5–4 mm. wide at the
narrowest point; upper edge with conspicuous scarious margin.
Common and widespread.   *S. ramosissima* J. Woods

5   Lower fertile segments of the terminal spike 2–3·5 mm. wide at the
narrowest point; plant usually becoming light brownish- or
orange-purple. E. Suffolk to Hants.   *S. nitens* P. W. Ball & Tutin
Lower fertile segments of the terminal spike usually more than
3·5 mm. wide at the narrowest point; plant usually without
purple coloration.

6   Terminal spikes with 6–15(–22) fertile segments, ±cylindrical;
spikes of the primary lateral branches cylindrical; plant dull
green to yellowish-green, often becoming bright yellow in fr. E.
Suffolk to Kent.   *S. fragilis* P. W. Ball & Tutin
Terminal spikes with 12–30 fertile segments, distinctly tapering;
spike of the primary lateral branches tapering; plant dark dull
green, becoming paler or dull yellow in fr. Widespread but local.
                                             S. dolichostachya Moss

44. TILIACEAE

Trees, shrubs or rarely herbs. Lvs spirally arranged or in 2 ranks, rarely
opposite; stipules usually small and soon falling, often functioning as
bud-scales, sometimes 0. Infl. usually cymose. Fls hermaphrodite,
actinomorphic. Sepals 5(–3), free or united, usually with their edges
touching, not overlapping, in bud; petals as many as sepals, free, rarely
0; stamens 10 or more, their filaments free or united only at base; anthers
with 4 pollen-sacs; ovary superior, 2–10-celled, each cell with 1 or more
ovules; style 1, with as many radiating stigma-lobes as ovary-cells, or
stigmas sessile. Fr. usually a capsule, drupe or nut with 1–5 seeds.

1. TILIA L.

Deciduous trees with sympodial growth owing to abortion of terminal
buds. Winter-buds large, blunt. Lvs in 2 ranks, alternate, usually cordate
or truncate at base and slender-stalked; stipules (bud-scales) soon falling. Fls yellowish or whitish, fragrant, in a cymose infl. whose stalk is
adnate for about half-way to a large oblong rather membranous bracteole. Sepals 5, free; petals 5, free; stamens many, free or in bundles
opposite the petals, the filaments often forked; ovary 5-celled, each cell
with 2 ovules; style slender with 5-lobed stigma. Fr. ovoid, indehiscent
and nut-like, 1-celled, with usually 1–3 seeds; infructescence shed as a
whole, the bracteole acting as a wing.

1   Lvs pubescent beneath and often also above; fls 2–5, usually 3; fr.
     strongly 3–5-ribbed.   **1. platyphyllos**
   Lvs glabrous beneath except for tufts in the vein-axils; fls 4–10; fr.
     slightly 3–5-ribbed or ribs 0.   *2*

2   Lvs 6–10 cm., broadly ovate-acuminate, bright green beneath, with
     the tertiary veins prominent; lf-stalks 3–5 cm; infls hanging; fr.
     slightly ribbed.   **3. × vulgaris**
   Lvs 3–6 cm., roundish, abruptly acuminate, often broader than long,
     somewhat glaucous beneath with the tertiary veins not prominent;
     lf-stalks 1·5–3 cm.; infls obliquely erect; fr. not or barely ribbed.
          2. cordata

1.   T. platyphyllos Scop.   Large-leaved Lime.

Large tree to 30 m. with spreading branches and a ± smooth dark bark.
Young twigs pubescent. Lvs 6–12 cm., broadly ovate, abruptly acuminate, obliquely cordate at base, dark dull green and glabrescent above,
pale green and pubescent beneath with simple hairs all over the surface
as well as in whitish tufts in the vein-axils; margin sharp-toothed; stalk
1·5–5 cm., pubescent. Infl. usually 3-fld (2–5), hanging, the wing
5–12 cm., pubescent on the veins below. Fls yellowish-white. Stamens
exceeding petals. Fr. 8–10 mm., apiculate, densely pubescent, strongly
3–5-ribbed, woody when mature. Fl. late June. Probably native. In
woods on good calcareous or base-rich soils and on limestone cliffs, but
perhaps often planted initially. Established and possibly native in the
Wye Valley and neighbourhood, in E. Wales and in S. Yorks, and
naturalized in old plantations northwards to Perth.

2.   T. cordata Miller   Small-leaved Lime.

Large tree to 25 m., differing from *T. platyphyllos* as follows: Young
twigs downy at first, then becoming glabrous. Lvs 3–6 cm., roundish,
dark shining green and glabrous above, somewhat glaucous and
glabrous beneath except for tufts of rusty hairs in the vein-axils; stalk
1·5–3 cm., glabrous. Infl. 3–10-fld, obliquely erect or spreading, the
wing 3·5–8 cm., glabrous below. Stamens about equalling petals. Fr.
c. 6 mm., thin-walled, ribs obscure or 0. Fl. early July. In woods on a
wide range of fertile soils but especially over limestone; commonly on
wooded limestone cliffs. Scattered throughout England and Wales
northwards to the Lake District and Yorks, but planted farther north.

3.   T. × vulgaris Hayne (*T. × europaea* L., p.p.: *T. cordata × platyphyllos*)
                                                Common Lime.

Large tree to 25 m., with arching lower branches and the trunk often
covered with irregular bosses. Young twigs usually glabrous. Lvs

6–10 cm., broadly ovate, obliquely cordate or truncate at base, glabrous above, paler and glabrous or nearly so beneath, except for tufts of whitish hairs in the vein-axils; stalk 3–5 cm., ± glabrous. Infl. 4–10-fld, hanging. Stamens equalling or somewhat exceeding petals. Fr. c. 8 mm., pubescent, slightly ribbed, woody when mature. Fl. early July. Introduced or very doubtfully native. Widely planted over a long period; throughout the British Is except the far north.

45. MALVACEAE

Herbs, shrubs or trees, usually with mucilage canals and stellate pubescence. Lvs spirally arranged, usually palmately veined and commonly palmately lobed; stipules small, usually falling soon. Fls usually hermaphrodite, actinomorphic. Sepals 5, free or united, their edges touching but not overlapping in bud, often with an 'epicalyx' of 3 to several members resembling an outer set of sepals; petals 5, free, convolute, commonly adherent to the base of the staminal tube; stamens numerous, their filaments united below into a staminal tube which divides above into branchlets each bearing a single 1-celled anther-lobe; ovary superior, with 2–many cells each with 1–many ovules; styles and stigmas as many as cells. Fr. a capsule or separating into 1-seeded nutlets. Seeds with little endosperm.

1	Epicalyx of 3 segments.	3
	Epicalyx of 6–9 segments.	2
2	Fls not more than 5 cm. diam., at least some of them distinctly stalked; staminal tube terete.	3. ALTHAEA
	Fls 6–9 cm. diam., all subsessile in spike-like infls; staminal tube. 5-angled. Tall sparsely hairy garden plant to 3 m., with blunt-lobed lvs; often escaping. Hollyhock.	
	Alcea rosea L. (*Althaea rosea* (L.) Cav.)	
3	Epicalyx-segments free to the base.	1. MALVA
	Epicalyx-segments united below.	2. LAVATERA

1. MALVA L.

Herbs with palmately lobed or divided very mucilaginous lvs. Epicalyx of 3 segments free to the base. Sepals 5, united below. Petals emarginate or deeply notched. Fr. of numerous 1-seeded nutlets arranged in a flat whorl round the short conical apex of the receptacle.

1	Stem-lvs deeply divided into slender palmate segments.	
		1. moschata
	Stem-lvs palmately lobed or roundish-crenate.	2

2 Perennial; fls 2·5–4 cm. diam. with rose-purple petals 3–4 times as
     long as sepals.   **2. sylvestris**
     Annual; fls not exceeding 2·5 cm. diam., with petals usually whitish
       or pale lilac and not more than twice (rarely 3 times) as long as
       sepals.   *3*

3 Fls more than 1·5 cm. diam.   **3. neglecta**
     Fls less than 1·5 cm. diam.   *4*

4 Fls 1–1·5 cm. diam. in dense axillary clusters; petals distinctly
     longer than sepals; nutlets meeting laterally in a straight line;
     stems ± erect, to 80 cm. Introduced. Waste places.
                                     *M. verticillata L.
     Fls c. 0·5 cm. diam., with petals barely exceeding sepals; nutlets
     meeting laterally in a wavy line; stems decumbent or ascending.
     Introduced. Waste places, foreshores, etc.; local northwards to
     Aberdeen.   **M. pusilla* Sm.

1. M. moschata L.   Musk Mallow.

Perennial herb with branching stock and several erect stems, 30–80 cm.,
often purple-spotted, sparsely hairy. Basal lvs 5–8 cm. diam., reniform
in outline, long-stalked, with 3 crenate lobes; stem-lvs successively
shorter-stalked and more deeply divided, the 3–7 primary divisions
deeply pinnatifid into ± linear ultimate segments; all subglabrous;
stipules small, lanceolate. Fls 3–6 cm. diam., usually solitary in axils of
upper lvs and in an irregularly racemose terminal cluster. Epicalyx half
as long as calyx. Calyx-lobes erect and enlarging in fr. Petals rose-pink,
rarely white, deeply emarginate, c. 3 times as long as calyx. Nutlets
blackish when ripe, not wrinkled, roughly hairy on the back, back and
sides not separated by an angle. Fl. 7–8. Grassy places, pastures,
hedge-banks, etc.; not uncommon on the more fertile soils throughout
the British Is northwards to Perth and Inner Hebrides.

2. M. sylvestris L.   Common Mallow.

Perennial herb with erect, ascending or decumbent stem, 45–90 cm., with
sparse spreading hairs. Basal lvs 5–10 cm. diam., roundish, with very
shallow crenate lobes, cordate at base, long-stalked; stem-lvs with 5–7
rather deep crenate lobes, sparsely hairy, ciliate. Fls 2·5–4 cm. diam.,
stalked, in an irregular raceme. Epicalyx-segments two-thirds as long as
calyx. Calyx-lobes connivent and not enlarging in fr. Petals rose-purple
with darker stripes, deeply emarginate, 2–4 times as long as calyx.
Nutlets brownish-green when ripe, the netted back separated by a sharp
angle from the transversely wrinkled sides. Fl. 6–9. Roadsides, waste
places, etc.; throughout the British Is and common in the south, but less
so in the north and not in Outer Hebrides, Orkney or Shetland.

3. M. neglecta Wallr.   Dwarf Mallow.

Annual or longer-lived, with stems 15–60 cm., decumbent or with the
central stem ascending, densely clothed in stellate down. Lvs 4–7 cm.
diam., long-stalked, roundish-reniform, deeply cordate, with 5–7 shallow
acutely crenate lobes, ±pubescent; stipules ovate. Fls 1·8–2·5 cm.
diam. in irregular racemes. Epicalyx-segments linear-lanceolate, $\frac{1}{2}$–$\frac{2}{3}$ as
long as calyx. Calyx-lobes slightly connivent with reflexed tips, not
enlarging in fr. Petals whitish with lilac veins, or pale lilac, with bearded
claws, deeply emarginate, about twice (or rather more) times as long as
calyx. Nutlets brownish-green with the pubescent but smooth back and
sides of each nutlet separated by a blunt angle and with a straight line of
contact between neighbours. Fl. 6–9. Waste places, roadsides, drift-lines,
etc.; throughout the British Is except Outer Hebrides and Shetland,
frequent in the south, less so in the north.

2. LAVATERA L.

Herbs or woody plants usually covered with stellate pubescence. Closely
resembling *Malva* but commonly larger and with the 3 epicalyx-segments united at base to form a lobed cup.

Plant 60–300 cm., erect, woody below; epicalyx enlarging in fr.; nutlets
    transversely wrinkled with an acute raised edge between back and
    sides.   **1. arborea**
Plant 50–150 cm., erect or ascending, not woody, roughly hairy; epicalyx
    not enlarging in fr.; nutlets not transversely wrinkled and with a
    rounded edge between back and sides. A rare plant of waysides and
    waste places near the sea in Cornwall, Scilly and Jersey.
                                                                L. cretica L.

1. L. arborea L.   Tree Mallow.

Biennial almost tree-like plant with stout erect stems, 60–300 cm., woody
below, thinly and softly stellate-pubescent above. Lvs to 20 cm. diam.,
roundish-cordate, stalked, softly velvety with stellate hairs, ±folded
like a fan, with 5–7 broadly triangular crenate lobes. Fls 3–4 cm. diam.
in terminal racemes. Epicalyx with blunt broadly ovate lobes, exceeding calyx, much enlarging in fr. Calyx-lobes ovate, acute. Petals 2–3
times as long as calyx, pale rose-purple with broad deep purple veins
joining below. Nutlets yellowish, transversely wrinkled, with an acute
raised edge between back and sides. Fl. 7–9. Rocks and waste ground
near the sea. S. and W. coasts of Great Britain from Sussex to Ayr and
S. and E. coasts of Ireland from Kerry to Antrim; casual elsewhere.

3. ALTHAEA L.

Herbs resembling *Malva* and *Lavatera* but with an epicalyx of 6–9 segments joined below into a cup-like involucre.

> Annual to 60 cm., rough with swollen-based bristly hairs; fls 2·5 cm. diam., long-stalked, the pale rose-purple petals scarcely longer than calyx; calyx erect in fr. Borders of fields and woods in Somerset, Kent and Jersey. Hispid Mallow. *A. hirsuta* L.
> Biennial to perennial; softly pubescent; lvs with shallow acute lobes; fls 2·5–5 cm. diam. **1. officinalis**

1. A. officinalis L. Marsh Mallow.

Perennial herb with thick stock tapering downwards into a tap-root, and erect densely velvety stems, 60–120 cm. Lower lvs 3–8 cm. across, roundish, stalked, upper narrower, ± ovate, all slightly 3–5-lobed, irregularly toothed, folded like a fan, velvety. Fls 2·5–4(–5) cm. diam. in an irregular raceme, short-stalked. Epicalyx-segments narrowly triangular. Calyx-lobes ovate-acuminate, velvety. Petals pale pink, 2–3 times as long as calyx, somewhat emarginate. Nutlets pubescent, a blunt edge separating the smooth rounded back from the sides; calyx curved over fr. Fl. 8–9. Upper margins of salt and brackish marshes and ditch-sides and banks near the sea. Locally common on coasts of Great Britain northwards to Northumberland, Kirkcudbright and Arran; W. half of Ireland, but not native.

46. LINACEAE

Herbs, shrubs or trees. Lvs usually spirally arranged, simple, ± entire. Fls regular, hermaphrodite, usually 5-merous. Petals usually contorted in bud. Stamens usually equalling in number and alternate with petals, often with small staminodes also; filaments ± joined at base. Ovary (2–)3–5-celled, often with each cell nearly halved by a false septum; ovules usually 2 per original cell; styles commonly free. Fr. a capsule or drupe.

Fls 5-merous; petals longer than calyx; sepals entire at apex.
1. LINUM
Fls 4-merous; petals not longer than calyx; sepals toothed or lobed at apex; plant small and inconspicuous. 2. RADIOLA

1. Linum L.

Usually herbs. Lvs sessile, often narrow. Fls cymose, 5-merous. Sepals entire. Petals clawed. Stamens 5, united into a basal tube, with tooth-like staminodes between. Ovary 5-celled; ovules 2 per cell, separated by false septa. Capsule dehiscing by 10 valves; seeds flat.

L. usitatissimum L., the cultivated flax, is like *L. bienne* but larger and always annual.

1 Petals blue, 8 mm. or more; lvs linear, spirally arranged.   2
  Petals white, 6 mm. or less; lvs oblong or obovate, opposite.
                                                       3. catharticum

2 Sepals all acuminate, more than half as long as ripe capsule, inner
    often glandular-ciliate.   3
  Inner sepals very blunt, outer blunt to acuminate, all entire, less
    than half as long as ripe capsule.   **2. perenne**

3 Petals 8–12 mm.; capsule c. 6 mm.   **1. bienne**
  Petals c. 15 mm.; capsule 10 mm. or more. Cultivated Flax.
                                                    *L. usitatissimum L.

1. L. bienne Miller   Pale Flax.

Glabrous and subglaucous annual to perennial 30–60 cm., with several ascending, rather rigid stems from the base. Lvs 1–2·5 cm. on the main stem, spirally arranged, linear, acute, mostly 3-veined, entire. Sepals all ovate-acuminate, the inner with a scarious glandular-ciliate border, the outer entire; more than ½ as long as the ripe capsule. Petals 8–12 mm., pale blue. Stigmas club-shaped. Capsule c. 6 mm., globose-conic. Fl. 5–9. Dry grassland, especially near the sea; commonest in the south-west but reaching Isle of Man, Lancashire and Yorks; S. Ireland.

2. L. perenne L. (*L. anglicum* Miller)   Perennial Flax.

Glabrous glaucous perennial 30–60 cm., with several ascending, rather rigid, often curved stems. Lvs 1–2 cm. on the main stems, spirally arranged, very numerous, linear, acute, 1-veined, entire. Outer sepals oval or elliptic, blunt; inner broader, rounded at apex, with an entire scarious border; all less than ½ as long as the ripe capsule. Petals 15–20 mm., sky blue. Stigmas capitate. Capsule c. 7 mm., globose or oval. Fl. 6–7. Calcareous grassland chiefly in E. England but extending to Westmorland and Kirkudbright.

3. L. catharticum L.   Purging Flax.

Slender glabrous biennial, with erect wiry usually solitary stems 5–25 cm. Lvs 5–12 mm., opposite, distant, oblong or obovate, 1-veined, entire.

Sepals ovate-lanceolate, acuminate. Petals 4–6 mm., white. Capsule c. 3 mm., globose. Fl. 6–9. Grassland, heaths, moors, rock-ledges and dunes, to 2800 ft.; especially on calcareous soils. Common throughout the British Is.

2. RADIOLA Hill

Small annual herb differing from *Linum* in its 4-merous fls, usually 3-lobed or 3-toothed sepals and ovoid seeds.

1. R. linoides Roth   All-seed.

Very small delicate annual 1·5–8 cm., with very slender usually several times forked stems, giving a bushy habit. Lvs 3 mm. or less, opposite, ovate-elliptic or elliptic, ± acute. Fls very numerous in a dichasial cyme. Sepals scarcely 1 mm. Petals white, about equalling sepals. Capsule globose, scarcely 1 mm. Fl. 7–8. Damp bare sandy or peaty ground on grasslands and heaths. Widespread but local over the whole of Great Britain; in Ireland frequent near the west coast, rare elsewhere.

47. GERANIACEAE

Herbs or undershrubs with lobed or compound usually stipulate lvs. Fls regular or slightly zygomorphic, hermaphrodite, 5-merous. Sepals persistent, usually imbricate. Petals usually imbricate, rarely 0. Stamens usually in 2 whorls, sometimes some sterile; filaments usually ± united at the base. Ovary superior, 3–5-celled, usually lobed and with a long beak ending in the free stigmas; ovules axile, usually 2 per cell. Fr. a lobed capsule, the lobes usually 1-seeded, usually opening septicidally from base to apex; seeds ± non-endospermic.

Lvs palmate or palmately lobed; beak of carpel rolling upwards in
    dehiscence and scattering the seeds.   1. GERANIUM
Lvs pinnate or pinnately lobed; beak of carpel twisting spirally in
    dehiscence and remaining attached to seeds.   2. ERODIUM

1. GERANIUM L.

Herbs. Lvs palmate or palmately lobed. Fls solitary or in pairs, regular or nearly so, without spur. Stamens 10, all fertile (except *G. pusillum*). Seeds 1 per carpel; beak of carpel usually rolling upwards at dehiscence and remaining attached by its tip, releasing the seeds; beak sometimes curling away at its tip but never coiling spirally.

1 Perennial; petals usually 10 mm. or more.   2
   Annual; petals less than 10 mm., or, if to 12 mm., then with a claw
      at least as long as the blade.   10

2 Petals c. 15 mm., pink with claw about as long as blade; stamens
      curving downwards. Introduced and naturalized on walls in S.
      Devon.   *G. macrorrhizum L.
   Petal-claw short or 0; stamens not curved downwards.   3

3 Lvs reniform or orbicular; lf-lobes wedge-shaped; petals 7–10 mm.,
      deeply emarginate.   4. pyrenaicum
   Lf-lobes not wedge-shaped; petals 10 mm. or more.   4

4 Fls solitary; petals purplish-crimson, rarely pink or white; lobes of
      lvs narrow throughout their length.   3. sanguineum
   Fls mostly in pairs; petals not purplish-crimson; lobes of lvs
      broader about the middle.   5

5 Petals blackish-purple, apiculate, spreading or slightly reflexed so
      that the fl. is ±flat; sepals mucronate, the point less than ½ mm.
      Introduced and naturalized in many places.   *G. phaeum L.
   Petals not blackish-purple, rounded or emarginate; fl. ±cup-
      shaped; sepals awned, the point c. 1 mm.   6

6 Petals rounded to ±emarginate, veins not darker than rest of
      petal; lvs 5–7-lobed.   7
   Petals emarginate, veins darker than the rest of petal; lvs 3–5-
      lobed.   9

7 Fls pink; rhizome horizontal, creeping; plant almost eglandular;
      stems 30–80 cm., ±erect with ±spreading hairs. Introduced
      and locally naturalized by roadsides.   *G. endressii Gay
   Fls violet or blue; rhizome oblique, not creeping; plant glandular
      above.   8

8 Fls 3–4 cm. diam.; fl.-stalks recurved after fl., re-erecting in ripe
      fr.; lvs dark green, their lowest secondary lobes c. 4–5 times as
      long as the basal entire portion of the lf.   1. pratense
   Fls 2·5–3 cm. diam.; fl.-stalks remaining erect after fl.; lvs
      yellow-green, their lowest secondary lobes 1–2 times as long as
      the basal entire portion of the lf.   2. sylvaticum

9 Stems with spreading hairs; lobes of lvs lobed or deeply toothed;
      petals 15–18 mm., white or pale lilac with violet veins. Intro-
      duced and naturalized on hedge-banks mainly in S. England
      and Wales, common in Cornwall.   *G. versicolor L.
   Stems nearly glabrous; lobes of lvs shallowly crenate-dentate;
      petals 15–18 mm., lilac with violet veins. Introduced and rarely
      naturalized.   *G. nodosum L.

10 Sepals ±spreading; lvs dull- or grey-green.   11
   Sepals erect, somewhat connivent near their tips; lvs bright green.   15

11 Lvs divided nearly to base (at least $\frac{5}{6}$) into linear lobes; lobes usually pinnatifid with linear segments; sepals awned (the point 1–2 mm.).   *12*

Lvs divided to $\frac{3}{4}$ or less, lobes widened above, trifid or 3-lobed; sepals mucronate (the point less than $\frac{1}{2}$ mm.).   *13*

12 Stalks of individual fls 2 cm. or more; carpels glabrous.   **5. columbinum**

Fl.-stalks not more than 1·5 cm.; carpels pubescent.   **6. dissectum**

13 Petals entire; seeds pitted.   **7. rotundifolium**

Petals emarginate; seeds smooth.   *14*

14 All 10 stamens with anthers.   *15*

Five stamens without anthers.   **9. pusillum**

15 Perennial; petals 7–10 mm.; carpels pubescent.   **4. pyrenaicum**

Annual; petals 3–7 mm.; carpels glabrous.   **8. molle**

16 Lvs 5–7-lobed to half-way or less.   **10. lucidum**

Lvs ternate or palmate.   *17*

17 Petals (8–)9–12 mm.; pollen orange.   **11. robertianum**

Petals 6–9 mm.; pollen yellow; otherwise much like *G. robertianum*. Hedge-banks and rocks near the sea and on shingle. Cornwall to Sussex and Gloucester, very local; very rare in Wales and S. Ireland.   *G. purpureum* Vill.

1. G. pratense L.   Meadow Cranesbill.

Perennial, 30–100 cm.; rhizome stout, oblique. Stems ± erect, with short deflexed hairs below and densely clothed with hairs and glands above (like the fl.-stalks and sepals). Basal lvs 7–15 cm. diam., polygonal in outline; dark green, appressed-hairy on both sides, long-stalked, deeply 5–7-lobed; lobes 7–10 mm. across at base, bipinnately lobed, the 2 basal lobes often touching; stem-lvs smaller, short stalked. Fls in pairs on common axillary stalks. Stalks of individual fls reflexed after fl., becoming erect in fr. Petals 15–22 mm., of a beautiful bluish-violet, rounded at the apex, very shortly clawed; corolla ± cup-shaped. Carpels glandular, smooth; beak c. 2·5 cm. in fr. Fl. 6–9. Meadows and roadsides to 1750 ft., on base-rich soils; widespread but rather local in Great Britain northwards to C. Scotland, but rare to Orkney; in Ireland native only in Antrim.

2. G. sylvaticum L.   Wood Cranesbill.

Like *G. pratense* but somewhat less robust. Basal lvs pale green, deeply (5–)7-lobed, the 2 basal lobes separated; lobes 8–15 mm. across at base, pinnately lobed. Fls in pairs, more numerous than in *G. pratense*

and forming a loose cyme. Stalks of individual fls erect after fl.
Petals c. 15 mm., white at base, narrower and redder than in *G. pratense*.
Fl. 6–7. Meadows, hedge-banks, damp woods and mountain rock-ledges,
to 3300 ft. on fairly base-rich soils, northwards from Glamorgan,
Gloucester, Derby and Yorks, but rare in Wales and common only in
N. England and Scotland; in Ireland only in Antrim.

3. G. sanguineum L.   Bloody Cranesbill.

Perennial, 10–40 cm., often bushy in habit; rhizome stout, horizontal,
creeping. Stems procumbent to erect, often branching from the base and
geniculate at the nodes, with long ± spreading hairs. Lvs 2–6 cm. diam.,
orbicular or somewhat polygonal in outline, appressed-hairy on both
sides, stalked, deeply 5–7-lobed; lobes 1·5–6 mm. across at the base, not
widened in the middle, mostly trifid, the secondary lobes entire to trifid,
ultimate lobes oblong or linear. Fls solitary (or rarely in pairs) on long
stalks. Petals 12–18 mm., bright purplish-crimson, rarely pink or white,
emarginate. Carpels sparsely hairy, smooth; beak c. 3 cm. in fr. Fl.
7–8. Grassland, woods and among rocks on basic soils and on fixed
dunes, etc., to 1200 ft. Over the greater part of the British Isles but
local, and absent from S.E. England and S. Ireland.

4. G. pyrenaicum Burm. fil.   Mountain Cranesbill.

Perennial, 25–60 cm. Stem erect, glandular-hairy. Basal lvs 5–8 cm.,
orbicular in outline, hairy on both sides, 5–9-lobed, the lobes 1–1½ times
as long as the entire portion of the lf, obovate-cuneate in outline, shortly
and irregularly 3-lobed at apex; upper stem-lvs often 3-lobed. Fls in
pairs forming a loose cyme. Stalks of individual fls deflexed after fl. but
curving upwards near their ends so that the fr. is erect. Petals 7–10 mm.,
purplish (or purplish-white), deeply emarginate. Carpels pubescent,
smooth; beak c. 1 cm. in fr. Fl. 6–8. Doubtfully native. Hedge-banks
waste places and field margins. Common in S. and E. England, rarer
elsewhere but reaching E. Inverness; scattered over much of Ireland.

5. G. columbinum L.   Long-stalked Cranesbill.

Annual, 10–60 cm. Branches ascending or erect, ± appressed-hairy,
Lower lvs 2–5 cm. diam., ± polygonal in outline, appressed-hairy on
both sides, long-stalked, divided nearly or quite to the base into 5–7
pinnately-lobed lobes; secondary lobes c. 5 times as long as the entire
portion of the lf, linear or oblong-linear; upper lvs smaller and more
shortly stalked. Fls in pairs on common stalks 2–12 cm., and with
individual stalks (2–6 cm.) which spread and often curve upwards after

fl. Sepals appressed-hairy but not glandular. Petals 7–9 mm., purplish-pink, rounded or truncate and often with 1 or a few small teeth at the apex. Carpels glabrous or glabrescent; beak 1·5–2 cm. in fr. Fl. 6–7. Open habitats in dry grassland and scrub, mainly on basic soils; throughout England but rather local, and northwards to Moray and S. Inner Hebrides; local in Ireland.

6. G. dissectum L. Cut-leaved Cranesbill.

Annual, 10–60 cm. Branches ascending, densely clothed with reflexed but not appressed hairs. Lower lvs 2–7 cm. diam., orbicular or reniform in outline, long-stalked, ± hairy especially on the veins beneath, deeply divided into (5–)7 pinnately-lobed lobes 2–4 mm. across at the base, 5–10 times as long as the basal entire portion of the lf; secondary lobes oblong or linear, often again lobed; upper lvs smaller, more shortly stalked. Fls in pairs on common stalks 0·5–2(–3) cm.; stalks of individual fls 0·5–1·5 cm., spreading or ascending after fl. Sepals glandular-hairy. Petals c. 5 mm., reddish-pink, obovate, emarginate. Carpels pubescent, smooth; beak 7–12 mm. in fr. Fl. 5–8. Cultivated and waste ground, grassland, hedge-banks, etc., to 1250 ft. Common throughout the British Is.

7. G. rotundifolium L. Round-leaved Cranesbill.

Annual, 10–40 cm. Branches many, erect or ascending, pubescent and glandular-hairy. Lower lvs 3–6 cm. diam., reniform in outline, long-stalked, pubescent on both sides, 5–7 lobed to half-way or less; lobes c. 1 cm. across at base, wedge-shaped, touching, 3-lobed at apex; secondary lobes short, blunt, often 2–3-toothed or shortly lobed; upper lvs smaller, more shortly stalked and often more deeply lobed. Fls numerous, in pairs on common stalks 0·5–3 cm.; stalks of individual fls 0·5–1·5 cm., spreading or somewhat deflexed after fl., turning upwards at apex. Sepals glandular-hairy. Petals 5–7 mm., pink, obovate-cuneate, rounded or slightly retuse at apex. Carpels pubescent, smooth; beak 10–15 mm. in fr. Fl. 6–7. Hedge-banks, wall-tops, etc. From Cornwall and Kent to N. Wales and Lincoln, local; casual further north; scattered in S. Ireland

8. G. molle L. Dove's-foot Cranesbill.

Annual, 10–40 cm., branched from base, densely clothed with long soft white hairs, glandular above. Branches decumbent or ascending. Lower lvs 1–5 cm. diam., orbicular or reniform in outline, long-stalked, irregularly 5–9-lobed; the lobes up to twice as long as the entire portion of the lf, wedge-shaped, contiguous or nearly so, 2–5 mm. across at base, 3-lobed at apex; upper lvs smaller, more shortly stalked and more deeply lobed. Fls numerous, in pairs on common stalks 0·5–3 cm.; stalks of individual fls 0·5–1·5 cm., often curved upwards after fl. Sepals densely hairy and glandular. Petals 3–6(–7) mm., bright rose-purple (or white), deeply emarginate. Stamens 10, all with anthers. Carpels glabrous, usually wrinkled; beak 5–8 mm. in fr. Fl. 4–9. Dry grassland, dunes, waste places and cultivated ground, to c. 1750 ft. Common throughout the British Is.

9. G. pusillum L.   Small-flowered Cranesbill.

Like *G. molle* but shortly and softly pubescent (glandular above), with lower lvs 1–4 cm. diam., 7–9-lobed; the lobes 2–3 times as long as the entire portion of the lf, separated, 2–5 mm. across at base, trifid towards apex; petals 2–4 mm., pale dingy lilac, deeply emarginate; stamens 5 with and 5 without anthers; carpels pubescent, smooth. Fl. 6–9. Cultivated and waste ground and open habitats in dry grassland. Widespread and rather common in England and Wales; local elsewhere and often only casual.

10. G. lucidum L.   Shining Cranesbill.

Annual, 10–40 cm., usually branched from the base, bright green and shining, often red-tinged, brittle. Stems erect or ascending, nearly glabrous. Lower lvs 2–6 cm. diam., orbicular in outline, nearly glabrous, long-stalked, palmately 5-lobed to ½ or ¾; lobes 5–12 mm. across at the base, wedge-shaped, shortly 3-lobed at apex, the secondary lobes also 2–3-lobed or toothed; upper lvs smaller, more shortly stalked. Stalks of individual fls spreading or ascending after fl. Sepals oblong-ovate, awned, erect and somewhat connivent near apex, forming an angled calyx, transversely ridged towards the margins. Petals 8–9 mm. (including the long claw), pink, rounded at apex. Carpels 5-ridged at the shortly downy apex and obscurely ridged down the back, reticulate, glabrous; beak c. 1 cm. in fr., the carpel not remaining attached at the apex. Fl. 5–8. Shady rocks, walls and hedge-banks, especially on limestone, to 2500 ft. Locally common throughout Great Britain and Ireland.

11. G. robertianum L.   Herb Robert.

Annual, 10–50 cm., usually branched from the base, usually red-tinged, brittle, with a strong 'geranium' smell. Stems decumbent or ascending, densely hairy below. Lvs palmate, lower mostly with 5 lflets, polygonal in outline, sparsely appressed-hairy on both sides, long-stalked; lflets

1·5–6·5 cm., deeply pinnately cut, the segments pinnately lobed; upper lvs mostly ternate, smaller. Stalks of individual fls ascending after fl., mostly straight. Sepals oblong-ovate, awned, hairy and glandular, erect and somewhat connivent. Petals (8–)9–12 mm. (including the long claw), bright pink or rarely white, rounded at apex. Pollen orange. Carpels reticulately ridged, with a white hair-tuft near the apex; beak 1–2 cm. in fr. Fl. 5–9. Woods, hedge-banks, among rocks and on shingle, etc. to 2300 ft. Common throughout the British Is.

2. ERODIUM L'Hér.

Herbs. Lvs usually pinnate or pinnately lobed. Fls solitary or in pairs, regular or slightly zygomorphic, without spur. Stamens 10, those opposite the petals devoid of anthers. Seeds 1 per carpel; beak of carpel twisting spirally when ripe, the seeds coming off enclosed in the carpel walls, the beaks remaining attached to them as awns.

1 Lvs simple, lobed.   **1. maritimum**
   Lvs pinnate.   2
2 Stipules and bracts blunt; filaments of fertile stamens with a tooth
      on each side at the base; plant densely glandular, musk-scented;
      lflets deeply toothed, but rarely more than half-way to the midrib.
                                                        2. moschatum
   Stipules and bracts acute or acuminate; filaments of fertile stamens
      enlarged at the base but not toothed; plant glandular or not,
      never musk-scented; lflets once or twice pinnately lobed or cut.
                                                        3. cicutarium

1. E. maritimum (L.) L'Hér.   Sea Storksbill.

Annual, with decumbent stems (rarely to 30 cm.), rather stiffly hairy. Lvs 5–15 mm., ovate, pinnately lobed about half-way to the midrib, long-stalked; lobes coarsely toothed, with white appressed hairs on both sides. Fls 1–2 on common axillary stalks. Sepals c. 4 mm., oblong, mucronate, hairy. Petals pink, not longer than sepals, often 0. Carpels hairy; beak 1 cm. or less in fr. Fl. 5–9. Fixed dunes and open habitats in short dry grassland, mainly near the sea. Local on the south and west coast of Great Britain from Wigtown to Kent and in a few places on the east coast and inland; local on the coast of Ireland; Channel Is.

2. E. moschatum (L.) L'Hér.   Musk Storksbill.

Annual to 60 cm., decumbent or ascending, usually robust, branching from the base and covered all over with long white hairs and stalked glands; musk-scented. Lvs 5–15 cm., pinnate; lflets ovate, toothed, the lower teeth rarely reaching half-way to the midrib; stipules scarious, whitish, blunt. Fls in terminal umbels of 2–8 on common stalks longer than the lvs. Sepals 4–7 mm., oblong, mucronate, enlarging in fr. Petals rosy-purple, rather longer than the calyx. Filaments of fertile stamens with a tooth each side of the base; pollen yellow. Carpels densely clothed with long hairs; beak 25–40 mm. in fr. Fl. 5–7. Waste places, etc., mainly near the sea. Local on the south and west coasts of England and most of the coast of Ireland; in some inland localities but usually only casual.

3. E. cicutarium (L.) L'Hér.   Common Storksbill.

Annual to 60 cm., very variable in habit and hairiness, glandular or not, not musk-scented. Lvs 2–20 cm., pinnate; lflets ovate, pinnately (or bipinnately) lobed or cut; stipules scarious, whitish, acute or acuminate. Fls c. 7–16 mm. diam., in terminal umbels of 2–9 on common stalks equalling or exceeding the lvs. Sepals 3–7 mm., enlarging in fr. Petals bright rosy-purple, rarely white, often with a blackish spot at the base of each of the two upper petals. Filaments of fertile stamens broadened below but not toothed. Fr. 5–6·5 mm., darkish brown, gradually tapering below; beak 22–40 mm. Fls 6–9. Dunes, dry grassland, arable fields and waste places, mainly on sandy soils, to 1200 ft.; common near the sea all round the British Is. and widespread but local inland (rare inland in Ireland).

Ssp. **cicutarium**: Plant eglandular or somewhat glandular, usually robust. Fls zygomorphic; upper petals often with a blackish spot. Mericarp 5–7 mm., with a large apical pit with a furrow at its base. Common.

Ssp. **bipinnatum** (Willd.) Tourlet (*E. glutinosum* Dumort.): Plant densely glandular, usually rather slender. Fls almost actinomorphic; upper petals without a blackish spot. Mericarp 4–5 mm., with a small apical pit which is without or with a very faint furrow at its base. Sandy places near the sea, mainly on the west coast, local.

48. OXALIDACEAE

Herbs, often with fleshy stock, rarely woody. Lvs spirally arranged, pinnate or palmate, often showing sleep movements, with or without stipules; lflets entire, jointed to the lf-stalk or lf-axis. Fls hermaphrodite, actinomorphic, hypogynous. Sepals 5. Petals 5, contorted in bud.

Stamens 10, connate at base. Ovary usually 5-celled; ovules 1 to many; styles usually 5. Fr. a loculicidal capsule, rarely a berry; seeds endospermic.

1. OXALIS L.

Lvs palmate; lflets 1 to many (always 3 in our spp.). Carpels 5, completely united. Fr. a capsule.

1   Fls white or pink   *2*
     Fls yellow   *5*

2   Lvs all basal.   *3*
     Plant with lfy aerial stem; fls pinkish white. Introduced and
       naturalized on walls in S.W. England, casual elsewhere.
                                              *O. incarnata L.

3   Stem a slender creeping rhizome bearing the lvs and fls; fls white,
       rarely lilac; native plant of woods, etc.   **1. acetosella**
     Lvs and fls from the apex of a thick erect rhizome or direct from a
       bulb; fls bright pink; aliens.   *4*

4   Lvs from apex of thick erect rhizome. Introduced. Garden escape,
       sometimes naturalized, especially in S.W. England.
                                         *O. articulata Savigny
     Plant with bulbs. Introduced. Horticultural weeds (see larger
       Flora, ed. 2).

5   Plant with lfy aerial stem.   *6*
     Lvs all basal, from an erect rhizome or bulb. Introduced. Horticultural weeds or on walls in Scilly Is (see larger *Flora*, ed. 2).

6   Lvs spirally arranged; stipules obvious, oblong.   **2. corniculata**
     Lvs whorled or clustered; stipules minute or 0.   *7*

7   Peduncles not reflexed after fl.; infl. of 4 or more fls; stipules awl-
       shaped or 0.   **3. europaea**
     Peduncles reflexed after fl.; infl. of 2–3 fls; stipules narrowly oblong,
       adnate. Introduced. Naturalized on arable land in Sussex.
                                  *O. stricta L. (*O. dillenii* Jacq.)

1. O. acetosella L.   Wood-sorrel.

Perennial herb with slender creeping rhizome clothed at intervals with swollen fleshy scale-like bases of lf-stalks and bearing lvs and fls on long stalks; aerial stem 0. Lvs ternate; lf-stalks 5–15 cm.; lflets 1–2 cm., obcordate, broader than long, entire except at apex, ciliate and with scattered appressed hairs, bright yellow-green. Fls solitary; fl-stalks with 2 small bracteoles in the middle. Petals 10–16 mm., white, veined with lilac, rarely purplish. Capsule 3–4 mm., 5-angled, glabrous. Fl. 4–5(–9). Woods, hedge-banks, shady rocks, etc. Common throughout much of the British Is. on light, usually acid, soils.

*2. O. corniculata L.   Procumbent Yellow Sorrel.

Usually annual herb 5–15 cm.; rhizome 0. Stems many, weak, procumbent, sometimes ascending, ± densely clothed with long spreading hairs. Lvs ternate; lflets 8–15 mm., broader than long, bilobed at apex to ¼ or ⅓, otherwise entire, hairy on margins and upper surface; lf-stalks 1–8 cm.; stipules c. 2 mm., adnate. Fls 1–6 in axillary umbel-like infls; pedicels deflexed after fl. Petals 8–10 mm., yellow. Capsule 12–15 mm., cylindric, pubescent. Fl. 6–9. Introduced. Waste places and as a garden weed; naturalized and rather common in S.W. England but casual elsewhere.

*3. O. europaea Jord. (*O. stricta* auct.)   Upright Yellow Sorrel.

Perennial, with erect solitary stems 10–40 cm., nearly glabrous, lfy; forming slender underground stolons but probably often annual. Lvs ternate; lflets broader than long, bilobed at apex to ⅓, otherwise entire, glabrous or nearly so; lf-stalks 1–12 cm.; stipules minute or 0. Fls 2–5 in axillary umbel-like infls; pedicels erect to spreading after fl. Petals 4–8 mm., yellow. Capsule 12–15 mm., cylindric, sparsely pubescent. Fl. 6–9. Introduced. Waste places and as a garden weed; naturalized only in S.W. England but casual elsewhere.

49. BALSAMINACEAE

Herbs with succulent stems. Lvs simple, exstipulate. Fls hermaphrodite, strongly zygomorphic. Sepals 5 or 3, the lowest large and spurred, the lateral small. Petals 5, the upper large, the 4 lower usually united in pairs on each side of the fl. Stamens 5, alternate with the petals; filaments broad, short, joined above; anthers jointed round the ovary. Ovary superior, 5-celled; ovules numerous, axile. Fr. a loculicidal capsule with the valves dehiscing elastically and coiling; seeds non-endospermic.

1. IMPATIENS L.

Sepals usually 3. Petals apparently 3.

1   Lvs spirally arranged; fls yellow or orange.   2
     Lvs opposite or whorled; fls purplish-pink, rarely white.
                                                        3. glandulifera
2   Fls 1·5 cm. or less, pale yellow; lvs with 20 or more teeth on each
         side.   **2. parviflora**
     Fls 2 cm. or more; lvs with 15 or fewer teeth on each side.   3
3   Fls orange; spur suddenly contracted and bent; teeth of lvs 1–2 mm.
         deep, rarely more than 10 on each side.   **1. capensis**

Fls bright yellow; spur gradually contracted and curved; teeth of lvs
c. 2–3 mm. deep, usually 10–15 on each side. An erect annual
herb 20–60 cm., with lvs 5–12 cm. and fls c. 3·5 cm. Very local by
streams and in wet woods in N.W. England and N. Wales. Touch-
me-not.   *I. noli-tangere* L.

*1. I. capensis Meerb.   Orange Balsam.

An erect glabrous annual 20–60 cm., with stem swollen at the nodes.
Lvs 3–8 cm., spirally arranged, stalked, ovate-oblong, cuneate at the
base, coarsely crenate-serrate with 10 or fewer blunt teeth 1–2 mm.
deep, on each side. Fls often all cleistogamous, the normal ones 2–3 cm.
overall, orange, strongly blotched or spotted with reddish-brown inside.
Lower sepal rather suddenly contracted into a spur which is suddenly
bent or curved upwards or downwards at a right angle. Fl. 6–8. Introduced and naturalized on river-banks, etc. in S. and C. England and S.
Wales.

*2. I. parviflora DC.   Small Balsam.

An erect glabrous annual 30–100 cm. Lvs 5–15 cm., spirally arranged,
ovate, acuminate, tapered at the base into a winged stalk, serrate, with
20 or more forwardly-directed acute teeth on each side. Fls 5–15 mm.
overall, in axillary 4–10-fld racemes, pale yellow, not spotted. Lower
sepal ± conical; spur ± straight. Fl. 7–11. Introduced and naturalized
in woods and waste shady places, mainly in S. and E. England.

*3. I. glandulifera Royle   Policeman's Helmet.

An erect glabrous annual 1–2 m., stem reddish, stout. Lvs 6–15 cm.,
opposite or in whorls of 3, lanceolate or elliptic, acuminate, rounded or
cuneate at the base, stalked, sharply and closely serrate. Fls 2·5–4 cm.,
purplish-pink or white, in long-stalked axillary 5–10-fld racemes. Lower
sepal very broad, blunt, with a short tail-like spur. Fl. 7–10. Introduced
and naturalized on river-banks and in waste places; locally common in
N. and W. England and Wales, less so in S.E. England, Scotland and
Ireland, but increasing.

50. ACERACEAE

Trees or shrubs with opposite simple (often palmately lobed) or pinnate
exstipulate lvs. Infl. racemose. Fls regular, with varying distributions of
sexes. Sepals 4–5. Petals 4–5 or 0. Stamens usually 8, in two whorls,
inserted either within or upon a nectar-secreting disk and so either hypo-

gynous or perigynous. Ovary 2-celled and 2-lobed, compressed, with the septum narrow; ovules 2 per cell; styles 2. Fr. separating into 2 indehiscent 1-seeded winged halves. Seeds non-endospermic.

1. ACER L.

Lvs, in British spp., palmately lobed. Infl. erect or pendulous and catkin-like. Monoecious. Sepals and petals 5; stamens 8, the anthers failing to open in female fls. Each half-fr. winged on one side only.

1   Twigs pubescent; lvs small (mostly 4–7 cm.), pubescent beneath, with
     obtuse lobes; wings of fr. spreading horizontally.   **2. campestre**
   Twigs glabrous; lvs large (5–16 cm.), glabrous except in the axils of
     the veins beneath, their lobes acute or acuminate; wings of fr.
     spreading at an acute angle.   2

2   Infl. pendulous, narrow; lf-lobes acute, coarsely and irregularly
     toothed; juice not milky.   **1. pseudoplatanus**
   Infl. an erect corymb; lf-lobes with few large acuminate teeth; juice
     milky. A commonly planted non-native tree, readily naturalizing.
     Norway Maple.   **A. platanoides* L.

*1. A. pseudoplatanus L.   Sycamore, Great Maple.

A large deciduous tree to 30 m., with broad spreading crown. Bark grey, smooth for a long time, finally scaling. Buds to 12 × 8 mm., ovoid, acute; scales green with blackish margins. Twigs soon light brown, glabrous. Lvs 5-lobed to about half-way, cordate at the base, dark green above, pale and somewhat glaucous beneath; lobes ovate, acute, coarsely and irregularly toothed and often lobed; stalk often red, without milky juice. Infl. 5–20 cm., narrow, terminal on a short lfy branch, with 60–100 fls. Fls c. 6 mm. diam., yellowish-green; stamens hypogynous; ovary downy. Fr. glabrous; wings spreading at an acute angle or incurved. Fl. 4–6. Introduced in the fifteenth or sixteenth century but now very common throughout the British Is; preferring deep, moist, well-drained rich soils and very tolerant of exposure and salt-spray.

2. A. campestre L.   Field Maple.

A small deciduous tree, 9–15(–25) m., with an ovoid crown, or a shrub. Bark light grey with shallow fissures. Buds c. 5 mm., brown. Twigs brown, pubescent, often developing corky wings. Lvs (3–)5-lobed to about half-way, cordate at the base, dull green above, paler green and persistently pubescent beneath especially on the veins; lobes ovate or obovate, ± obtuse, entire or with a few broad shallow teeth or lobes; stalk pubescent, with milky juice. Infl. an erect corymbose panicle terminal on a short lfy branch, with 10–20 fls. Fls c. 6 mm. diam., pale green; stamens perigynous. Fr. usually pubescent, their wings spreading horizontally. Fl. 5–6. Woods, hedges and old scrub, mainly on basic soils; frequently coppiced. Northwards to Westmorland and Durham, but common only in S., E. and C. England. Rare and introduced in Scotland and Ireland.

51. HIPPOCASTANACEAE

Trees or shrubs. Lvs opposite, palmate; stipules 0. Fls in large panicles of scorpioid cymes, zygomorphic, hermaphrodite and male in the same infl. Sepals 5. Petals 5 or 4, clawed. Stamens 5–8, hypogynous. Ovary superior, 3-celled; ovules 2 in each cell; style long, simple. Fr. a large leathery usually 1-seeded loculicidal capsule opening by 2 valves; seed large, nut-like, shining, non-endospermic.

1. AESCULUS L.

Deciduous trees. Buds large. Sepals united.

*1. A. hippocastanum L. Horse-chestnut.

To 25 m., with broad crown. Bark dark grey-brown, eventually shed in scales. Buds to 3·5 cm., ovoid, deep red-brown, very sticky. Twigs glabrous. Lvs long-stalked, with 5–7 lflets each 8–20 cm., obovate-lanceolate, acuminate, long-cuneate at base, irregularly toothed, glabrous above. Infl. 20–30 cm. Fls c. 2 cm. across. Petals 4, white with yellow spots turning pink. Stamens arched downwards; pollen red. Fr. 6 cm., subglobose, prickly. Fl. 5–6. Introduced. Commonly planted and often self-sown.

52. AQUIFOLIACEAE

Trees or shrubs. Lvs spirally arranged, simple, exstipulate. Fls in few-fld axillary cymes, regular, hermaphrodite or dioecious. Sepals and petals imbricate; sepals small; petals free or united at the base. Stamens equalling in number and alternating with the petals, free and hypogynous or inserted on the extreme base of the petals; nectar-secreting disk 0. Ovary usually 4-celled with 1–2 ovules per cell; style very short or 0. Fr. a drupe with 3 or more stones; seeds endospermic.

1. ILEX L.

Fls dioecious, rarely polygamous. Petals united below. Ovary 4–6(–8)-celled.

1. I. aquifolium L.   Holly.

An evergreen small tree or shrub, 3–15 m.; crown cylindric or conic. Bark grey, smooth for a long time, eventually finely fissured. Buds 2–3 mm., ovoid. Twigs green, ± glabrous. Lvs 3–10 cm., thick and leathery, ovate, elliptic or oblong, sinuate-dentate with large triangular spine-pointed teeth or on old trees largely entire, dark green and glossy above, paler beneath, glabrous, with a narrow cartilaginous border; short-stalked. Infls cluster-like, axillary on the old wood. Fls c. 6 mm. diam., white, 4-merous. Fr. 12 mm., scarlet, globose. Fl. 5–8. Woods, scrub, hedges and among rocks, on all but wet soils, ascending to 1800 ft., sometimes dominant in the lower tree- or shrub-layer of woods. Common throughout the British Is except Caithness, Orkney and Shetland.

53. CELASTRACEAE

Woody plants with simple lvs; stipules small or 0. Fls small, regular, hermaphrodite or unisexual, 4–5-merous, with a well-marked fleshy disk. Stamens usually equalling the petals in number and alternate with them, inserted on the face or edge of the disk. Ovary free or partly sunk in the disk, 2–5-celled, usually with 2 ovules per cell; style very short. Fr. various; seeds endospermic.

1. Euonymus L.

Trees or shrubs. Lvs usually opposite. Stamens short, inserted on the wide, angular, fleshy disk. Ovary sunk in the disk, with as many cells as other floral parts; ovules 1–2 per cell. Fr. a fleshy often brightly coloured loculicidal capsule; seed enclosed in a fleshy aril.

1. E. europaeus L.   Spindle-tree.

Deciduous glabrous shrub or small tree, 2–6 m. Bark grey, smooth. Buds 2–4 mm., ovoid, greenish, Twigs green, 4-angled. Lvs 3–13 cm., opposite, broadly lanceolate or elliptic, acute or acuminate, crenate-serrate; stalk 6–12 mm. Fls 8–10 mm. diam., 4-merous, hermaphrodite or polygamous, 3–10 together in axillary cymes. Petals greenish, widely separated. Fr. 4-lobed, deep pink, 10–15 mm. across, exposing the bright orange aril on opening. Fl. 5–6. Woods and scrub, mostly on calcareous soil. Throughout England, Wales and Ireland, rather common, and extending north to the Clyde and Forth.

54. BUXACEAE

Evergreen shrubs or small trees, rarely herbs, with simple exstipulate lvs. Fls regular, small, usually monoecious. Sepals usually 4 or 0. Petals 0. Stamens 4 opposite the sepals, or more numerous, hypogynous; no disk. Ovary superior, usually 3-celled, with 1–2 ovules per cell; styles free. Fr. a loculicidal capsule or drupe. Seeds endospermic.

Buxus sempervirens L., Box, is the only British representative. An evergreen shrub or small tree, 2–5(–10) m., with opposite, entire, oblong lvs, 1–2·5 cm., and whitish-green fls in axillary clusters each having a terminal female and several male fls. Fr. a capsule with 2-horned valves and black seeds. Very local in beech woods and scrub on chalk or limestone in S. England.

55. RHAMNACEAE

Trees or shrubs with simple lvs. Fls small, perigynous, green, yellow or blue, sometimes unisexual. Calyx tubular, 4-5-lobed, Petals 4–5, sometimes 0, small, often hooded over the stamens. Stamens 4–5, opposite the petals. Ovary superior. Fr. 2–4-celled, with 1 seed in each cell, often fleshy.

Thorny; buds with scales; lvs toothed.	1. RHAMNUS
Not thorny; buds without scales; lvs entire.	2. FRANGULA

1. RHAMNUS L.

Lvs alternate or subopposite. Fls usually 4-merous, polygamous or dioecious, greenish. Style cleft. Germination epigeal.

1. R. catharticus L.   Buckthorn.

Often dioecious bush or small tree 4–6(–10) m. Branches opposite. spreading almost at right angles to main stem, many of the laterals forming short lfy spurs or ending in thorns. Bark of old branches cracked and scaling, blaze orange; young wood whitish. Lvs 3–6 cm., ovate to nearly elliptic, dull green turning yellow or brownish in autumn; large lateral veins 2–3 pairs, curving upwards and running nearly to tip of lf. Fls c. 4 mm., diam., on the previous year's wood of the short shoots. Fr. 6–10 mm. diam., fleshy, black when ripe. Fl. 5–6. In woods, scrub and hedges on calcareous soils and on fen peat, locally common but chiefly in England.

2. FRANGULA Miller

Similar to *Rhamnus* but fls usually 5-merous and hermaphrodite, style entire, and germination hypogeal.

1. F. alnus Miller   Alder Buckthorn, Black Dogwood.

Deciduous shrub or small tree, commonly 4–5 m. Branches sub-opposite, ascending at an acute angle to main stem, without marked distinction into long and short shoots. Bark of old branches smooth (except in very old trees), blaze lemon-yellow; young wood dark brown. Lvs 2–7 cm., obovate, with short brownish hairs when young, particularly beneath, shiny green, turning clear yellow and red in autumn; large lateral veins about 7 pairs. Fls c. 3 mm., on the young wood. Fr. 6–10 mm. diam., fleshy, changing from green to red and then violet-black on ripening. Fl. 5–6(–9). Sometimes found with the last but usually on damp somewhat acid and ± peaty soils; local but widely distributed, rare in Scotland.

56. LEGUMINOSAE

Herbs, less frequently shrubs or trees. Lvs usually pinnate or trifoliolate, sometimes ending in a tendril. Fls with a large adaxial petal (*standard*), 2 lateral petals (*wings*) and 2 lower petals usually ± joined by their lower margins (*keel*); standard outside and enclosing the petals in bud. Calyx usually tubular with 5 teeth. Stamens 10, either all joined by their filaments, or upper free, rarely all free. Fr. a pod, usually opening by 2 valves.

1	Trees or shrubs, sometimes small and prostrate.	2
	Herbs.	8
2	Trees with pendulous racemes.	3
	Shrubs; racemes erect or fls solitary in lf-axils.	4
3	Fls white; lvs pinnate. A more or less naturalized tree. False Acacia, Locust. *Robinia pseudoacacia L.	
	Fls yellow; lvs palmate. A more or less naturalized small tree. Laburnum. *Laburnum anagyroides Medicus	
4	Fls solitary in lf-axils.	5
	Fls in erect racemes.	7
5	Spines branched.	3. ULEX
	Spines simple or 0.	6
6	Twigs terete; style curved but not spirally coiled; valves of pod not coiled after dehiscence.	2. GENISTA

Twigs angled; style spirally coiled; valves of pod coiled after dehiscence.   4. CYTISUS

7 Fls white; lvs palmate; pod not inflated.   1. LUPINUS
Fls yellow; lvs pinnate; pod strongly inflated. A naturalized shrub. Bladder Senna.   *Colutea arborescens* L.

8 Lvs grass-like.   16. LATHYRUS
Lvs with lflets.   9

9 Lvs palmate, with 6 or more lflets.   1. LUPINUS
Lvs pinnate with 2 or more lflets, or trifoliolate.   10

10 Lvs trifoliolate.   11
Lvs pinnate with 2 or more lflets.   14

11 Fls solitary in lf axils.   5. ONONIS
Fls 2–many together in stalked heads or racemes.   12

12 Pod sickle-shaped or spirally coiled.   6. MEDICAGO
Pod straight, usually shorter or little longer than calyx.   13

13 Infl. a lax raceme with many fls.   7. MELILOTUS
Infl. a dense head or else few-fld.   8. TRIFOLIUM

14 Lvs without a terminal lflet, usually ending in tendrils.   15
Lvs with a terminal lflet; tendrils 0.   18

15 Lflets 0 or 1 pair.   16. LATHYRUS (part)
Lflets 2 or more pairs.   16

16 Lflets (2–)3–7 cm., normally 2–4 pairs.   17
Lflets rarely as much as 2 cm., and then more than 4 pairs on most lvs.   15. VICIA (part)

17 Standard purple, wings white, or fls less than 10 mm.
    15. VICIA (part)
Not as above.   16. LATHYRUS (part)

18 Fl. 25–30 mm., solitary, pale yellow. Naturalized locally in S. England.   *Tetragonolobus maritimus* (L.) Roth
Not as above.   19

19 Fls in umbels.   20
Fls in racemes.   24

20 Umbels in pairs with a lf-like involucre; calyx woolly, strongly inflated.   9. ANTHYLLIS
Umbels solitary; if closely subtended by a lf-like bract then calyx not as above.   21

21 Umbels 10–20-fld; fls usually pink, rarely white or purple. Naturalized in a few scattered localities. Crown Vetch.
    Coronilla varia L.
Umbels 1–8(–12)-fld; fls yellow or if whitish closely subtended by a pinnate bract.   22

22 Lflets 5, two lower resembling lf-like stipules; pod straight, not
        jointed.   10. LOTUS
     Lflets usually more than 5; if 5 then lowest pair not resembling
        stipules; pod curved or jointed.   23

23 Perennial; fls yellow; stalks of umbels longer than subtending lvs;
        segments of pod horseshoe-shaped.   13. HIPPOCREPIS
     Annual; fls whitish or, if yellow then stalks of umbels shorter than
        subtending lvs; segments of pod not horseshoe-shaped.
                                                       12. ORNITHOPUS

24 Stipules scarious; fls bright pink or red.   14. ONOBRYCHIS
     Stipules green; fls not bright pink or red.   25

25 Plant 60 cm. or more, erect, glabrous; infl. as long as or longer than
        subtending lf. Locally naturalized in waste places. Goat's Rue,
        French Lilac.   *Galega officinalis L.
     Plant 35 cm. or less, hairy; if larger and glabrous then not erect
        and infl. much shorter than subtending lf.   26

26 Keel blunt.   11. ASTRAGALUS
     Keel sharply pointed.   OXYTROPIS (see p. 162)

1. LUPINUS L.

Usually herbs. Lvs palmate; stipules joined to base of petiole. Infl. a
terminal raceme of showy fls. Calyx deeply 2-lipped; stamens all joined
by the filaments. Pod flattened and often constricted between the seeds.
The 2 following are locally naturalized.

Herb; lflets 6–8; fls blue or purple. River gravels in Scotland.
                                                       *L. nootkatensis Sims
Shrub; lflets 7–11; fls yellow or white, rarely bluish-tinged. Near the
     coast, mainly in the south. Tree Lupin.   *L. arboreus Sims

The garden lupin, *L. polyphyllus Lindley, a perennial herb with 9–16
lflets, persists for a time on railway banks and waste ground.

2. GENISTA L.

Small shrubs, sometimes with simple spines. Lvs 1-foliolate; stipules
small or 0. Fls yellow, solitary in the axils of lvs. Calyx shortly 2-lipped,
upper lip deeply bifid, lower 3-toothed. Wings oblong, deflexed after fl.;
keel-petals separating and not resilient after deflection. Stamens all
joined by the filaments. Pod opening explosively.

1 Nearly always spiny; lvs quite glabrous.   **2. anglica**
    Not spiny; lvs hairy, at least on margins.   2

2   Lvs ciliate; fls glabrous or nearly so, borne on long branches.
                                                            1. tinctoria
    Lvs hairy on both surfaces; fls hairy, borne on short lateral branches.
       Very local in south and west England and in Wales.
                                                            G. pilosa L.

1. G. tinctoria L.   Dyer's Greenweed.

30–70 cm., erect or ascending, not spiny. Young twigs green, not striate.
Lvs up to 30 mm., subsessile, oblong-lanceolate, ciliate. Fls c. 15 mm.,
towards the ends of the main branches; standard about as long as keel.
Pedicels 2–3 mm. Calyx and corolla glabrous or nearly so. Pod 25–
30 mm., glabrous, flat, tapering but blunt at both ends. Fl. 7–9. In
rough pastures, local in England and Wales, rare or absent elsewhere.

2. G. anglica L.   Needle Furze, Petty Whin.

10–50(–100) cm., erect or ascending, spiny. Young twigs terete; spines
1–2 cm., lfy when young. Lvs 2–8 mm., ovate, glabrous and glaucous.
Fls c. 8 mm., towards the ends of the main branches; standard shorter
than keel. Pedicels c. 2 mm. Calyx-lobes fringed; corolla glabrous.
Pod 12–15(–20) mm., glabrous, inflated, obliquely narrowed and pointed
at both ends. Fl 5–6. On heaths and moors scattered throughout
Great Britain.

3. ULEX L.

Densely spiny shrubs; spines green and branched. Lvs 3-foliolate on
young plants, spinous or scale-like on old ones. Fls yellow, in the lf-
axils. Calyx yellow, 2-lipped; lower lip minutely 3-toothed, upper
minutely 2-toothed. Keel hairy. Stamens and pod like *Genista*.

1 · Spines deeply furrowed; bracteoles 3–5 mm.; fl. winter and spring.
                                                            1. europaeus
    Spines faintly furrowed or only striate; bracteoles 0·5 mm.; fl. late
       summer and autumn.   2

2   Spines stiff; calyx at least 9.5 mm.; wings curved,
       longer than keel when straightened.   **2. gallii**
    Spines weak; calyx not exceeding 9.5 mm.; wings straight, shorter
       than keel.   **3. minor**

1. U. europaeus L.   Furze, Gorse, Whin.

60–200 cm., sometimes more, forming a dense, spiny, dark green bush.
Fls c. 15 mm. Pedicels 2–3 mm., densely velvety; bracteoles (2–)3–5 ×
1·5–4 mm., much wider than pedicels. Calyx ⅔ length of corolla, hairy;

teeth curving towards each other. Wings rather longer than keel. Pod c.
15 mm., ripening in summer. Fl. 3–6 and sporadically during a mild
winter. In rough grassy places and on heaths, usually on light soils,
widely distributed and often abundant throughout the British Is.

2. U. gallii Planchon   Dwarf Furze.

30–90 cm., resembling *U. europaeus.* Fls 10–12 mm. Pedicels 3–5 mm.,
with appressed hairs; bracteoles 0·5 mm., about as wide as pedicels.
Calyx $\frac{2}{3}$–$\frac{3}{4}$ length of corolla, hairy; teeth curving towards each other.
Wings curved, longer than keel when straightened. Pod c. 10 mm.,
ripening in spring. Fl. 7–9. On acid heaths, etc., throughout England,
Wales and S. Scotland, mainly in the west.

3. U. minor Roth   Dwarf Furze.

30–90 cm., main branches usually procumbent, spines rather weak.
Fls 8–10 mm. Pedicels and bracteoles as in *U. gallii.* Calyx nearly as
long as corolla; teeth divergent. Wings rather shorter than keel, straight.
Pod c. 7 mm., persistent for nearly a year. Fl. 7–9. On heaths in
S. England and Wales and S. Scotland, chiefly in the east.

4. CYTISUS L.

Shrubs with green not spiny twigs. Lvs 1–3-foliolate, soon deciduous.
Fls yellow, in lf-axils. Calyx green, 2-lipped, minutely toothed. Stamens
as in *Genista.* Style long, spirally coiled. Valves of pod coiled after dehiscence. Incl. *Sarothamnus* Wimmer.

1. C. scoparius (L.) Link   Broom.

60–200 cm., much-branched, erect. Twigs glabrous, with 4 raised angles.
Lflets narrowly elliptic to obovate, acute, appressed-hairy. Fls c.
20 mm.; pedicels c. 10 mm., glabrous; bracteoles minute. Calyx $\frac{1}{4}$ length
of corolla, glabrous. Pod 2·5–4 cm., black with brown hairs on margins.
Fl. 5–6. On heaths, in open woods, etc. on acid soils; locally abundant
throughout the British Is.

5. ONONIS L.

Herbs or small shrubs. Lvs pinnately 3-foliolate, sometimes 1-foliolate,
nerves ending in teeth. Stipules joined to the petioles. Fls pink (in our
spp.), in lf-axils. Standard broad, wings oblong, keel pointed. Stamens
joined in a tube by their filaments, 5 or all broadened above. Pod dehiscent, seeds 1–many.

1 Stout perennials with ±woody stems; fls 10 mm. or more; pod
   erect.   *2*
   Slender annual; fls 7 mm. or less; pod deflexed. Very local in the
   south-west. Small Restharrow.   *O. reclinata* L.

2 Stem hairy all round; wings as long as keel.   **1. repens**
   Stem with 2 lines of hairs; wings shorter than keel.   **2. spinosa**

1. O. repens L.   Restharrow.

30–60 cm., usually procumbent. Rhizomatous. Stems usually without
spines, rooting at base, hairy all round. Petioles 3–5 mm., hairy; stipules
clasping the stem, toothed and ±glandular-hairy. Lvs up to 20 mm.,
hairy and ±glandular; lflets obovate, toothed. Fls 10–15 mm., wings
equalling keel. Pod shorter than the enlarged calyx. Fl. 6–9. Widely
distributed; common in calcareous places.

2. O. spinosa L.   Restharrow.

Like **1** but no rhizomes; stems usually spiny, not rooting at base, hairs
in 2 lines; lflets generally narrower; wings shorter than keel; pod longer
than calyx. Fl. 6–9. In similar places to *O. repens*, but absent from
much of Scotland. Intermediates (probably hybrids) between these
2 spp. are found.

6. MEDICAGO L.

Herbs. Lvs pinnately trifoliolate, nerves ending in teeth. Stipules joined
to the petioles. Fls yellow or purple, in racemes. Calyx-teeth 5, nearly
equal. Petals falling after fl.; keel blunt, shorter than wings. Upper
stamen free, 9 joined by their filaments into a tube, not broadened above.
Pod spirally coiled or curved, often spiny, usually indehiscent, 1–many-
seeded.

1 Fls 7–9 mm., yellow or purple.   **1. sativa**
   Fls 2–5 mm., yellow.   *2*

2 Racemes many-fld; pod not spiny, 1-seeded, black when ripe.
                                                             2. lupulina
   Racemes 1–5-fld; pod spiny or with tubercles, several-seeded, brown
      when ripe.   *3*

3 Plant downy; stipules nearly entire. S.E. England; local. Small
      Medick.   *M. minima* (L.) Bartal.
   Plant glabrous or nearly so; stipules distinctly toothed or laciniate.   *5*

5 Lflets not blotched; stipules laciniate; pod flat. S. and E. England,
      local. Hairy Medick.   *M. polymorpha* L.
   Lflets blotched; stipules toothed; pod subglobose.   **3. arabica**

*1. M. sativa L.   Lucerne, Alfalfa.

30–90 cm., deep-rooted, ± erect, perennial. Lflets up to 30 mm., narrowly obovate, toothed in upper third. Stipules linear-lanceolate, acuminate, ± toothed. Racemes up to 4 cm. Fls c. 8 mm., purple or yellow; pedicels shorter than calyx-tube. Pod nearly straight or in a spiral of 1–3 turns, 10–20-seeded. Fl. 8–9. Ssp. *sativa* has purple fls and spiral pods and is planted for fodder and ± naturalized on waste ground. Ssp. *falcata* (L.) Arcangeli has yellow fls and almost straight or sickle-shaped pods. It is local on sandy soils in E. Anglia and sometimes naturalized elsewhere. Hybrids between these sspp. may have green or blackish fls (*M.* × *varia* Martyn).

2. M. lupulina L.   Black Medick, Black Hay.

5–50 cm., procumbent or ascending, usually downy and annual. Lflets 3–20 mm., obovate, apiculate, finely toothed in upper half. Stipules lanceolate, acuminate, ½-cordate at base, shortly toothed. Racemes 3–8 mm., compact; peduncles longer than petioles of subtending lvs. Fls 2–3 mm., bright yellow. Pod 2 mm. diam., reniform, reticulate, coiled in almost 1 complete turn, 1-seeded, black when ripe. Fl. 4–8. Generally distributed and common in grassy places, except on the poorest soils.

3. M. arabica (L.) Hudson   Spotted Medick.

10–60 cm., procumbent, nearly glabrous, annual. Lflets up to 25 mm., obovate or obcordate, toothed towards the top, usually with a dark blotch. Stipules ½-cordate, toothed, acuminate. Racemes 5–7 mm.; peduncles shorter than petioles of subtending lvs. Fls 4–6 mm., yellow. Pod 4–6 mm. diam., subglobose. faintly reticulate, in a spiral of 3–5 turns, with a double row of curved or hooked spines. Fl. 4–8. In grassy places, especially on light soils near the sea.

7. MELILOTUS Miller

Annual or short-lived perennial herbs. Lvs pinnately 3-foliolate; nerves ending in teeth. Stipules joined to the petioles. Fls in racemes, small, yellow or white. Racemes usually long and rather lax; pod short, straight, thick, never spiny and usually indehiscent. Many spp. smell strongly of new-mown hay, especially when drying.

1   Wings and standard equal.   2
     Standard longer than wings.   3

2   Wings, standard and keel all equal; pod hairy, pointed, black when
     ripe.   **1. altissima**
    Keel shorter than other petals; pod glabrous, blunt, brown when
     ripe.   **2. officinalis**

3   Fls 4–5 mm., white; pod 4–5 mm., brown when ripe.   **3. alba**
    Fls c. 2 mm., pale yellow; pod 2–3 mm., olive-green when ripe.
                                          4. indica

*1. M. altissima Thuill.   Tall Melilot.

60–120 cm., erect, branched. Lflets 15–30 mm., oblong or obovate,
toothed. Stipules narrow. Racemes 2–5 cm., lengthening in fr. Fls
5–6 mm., yellow; wings, standard and keel all equal. Pod 5–6 mm.,
hairy, reticulate, flattened-ovoid, pointed, black when ripe. Fl. 6–8. In
grassy places and open woods; widely distributed but rarer in the north.

*2. M. officinalis (L.) Desr.   Common Melilot.

Like 1 but wings and standard equal but longer than keel; pod glabrous,
transversely wrinkled, little flattened, blunt, mucronate, brown when
ripe. Fl. 7–9. Naturalized in fields and waste places, mainly in the
south.

*3. M. alba Desr.   White Melilot

Like 1 but fls 4–5 mm., white; wings and keel nearly equal, somewhat
shorter than standard. Pod 4–5 mm., glabrous, flattened, mucronate,
brown when ripe. Fl. 7–8. Naturalized in fields and waste places, mainly
in the south.

*4. M. indica (L.) All.   Small-flowered Melilot.

Like 1 but smaller in all its parts; fls c. 2 mm., pale yellow; wings and
keel equal, shorter than standard; pod 2–3 mm., glabrous, globular-
ovoid, olive-green when ripe. Fl. 6–10. Introduced; waste places, mainly
in the south.

8. TRIFOLIUM L.

Annual or perennial herbs. Lvs trifoliolate; stipules joined to the petioles.
Fls sessile or shortly stalked in usually dense racemose heads. Calyx-
teeth 5, usually subequal. Petals usually persistent after fl.; wings longer
than keel. Upper stamen free, 9 joined by their filaments into a tube.
Pod 1–8-seeded, indehiscent, opening by 2 valves or by the top falling
off, ± straight, enclosed in the calyx or slightly protruding, often covered
by the persistent standard

1   Fls yellow, not more than 7 mm.   2
    Fls white, pink or purple, if yellowish then more than 10 mm.   5

2   Standard not folded; fls 5–7 mm.; heads about 40-fld.   3
    Standard folded; fls 2–4 mm.; heads up to 26-fld.   4

3   Stipules half-ovate.   **4. campestre**
    Stipules linear-oblong. Occasionally naturalized.   **T. aureum* Pollich

4   Pedicels shorter than calyx-tube; standard entire; heads (4–)10–20-
    fld.   **3. dubium**
    Pedicels equalling calyx-tube; standard notched; heads 2–6-fld.
                                                            2. micranthum

5   Heads 2–6-fld.   6
    Heads with numerous fls.   7

6   Hairy; fls sessile; petals soon falling; pods burrowing in the earth
        and covered by the reflexed calyces of sterile fls.
                                                          14. subterraneum
    Glabrous; fls shortly stalked; petals persistent; pods above ground,
        longer than calyx.   **1. ornithopodioides**

7   Stems creeping and rooting at nodes; peduncles longer than
        petioles of subtending lvs.   8
    Not as above.   14

8   Calyx not 2-lipped, not inflated in fr.; lateral veins of lflets not
        thickened towards the margin.   **6. repens**
    Calyx ± 2-lipped, much inflated and reticulate in fr.; lateral veins of
        lflets thickened towards the margin.   **7. fragiferum**

9   Heads all at ends of branches.   10
    Some heads in the axils of lvs.   14

10 Heads with a pair of ± opposite lvs below them.   11
    One lf at base of peduncle.   **12. incarnatum**

11 Corolla much longer than calyx; fls up to 15 mm.   12
    Corolla little-longer than calyx; fls up to 7 mm. Salt-marshes, rare.
        Sea Clover.   *T. squamosum* L.

12 Fls whitish-yellow; lflets never spotted. E. England; very local.
        Sulphur Clover.   *T. ochroleucon* Hudson
    Fls pink or red, rarely whitish; lflets often with a whitish spot.   13

13 Free part of stipules triangular, with a bristle-like point; calyx-
        tube hairy.   **13. pratense**
    Free part of stipules awl-shaped; calyx-tube glabrous, at least in
        the lower part.   **8. medium**

14 Heads distinctly stalked.   15
    Heads sessile.   17

15 Heads softly downy, cylindrical.   **9. arvense**
    Heads not downy, ovoid or globular.   16

16 Stipules oblong, entire with triangular acuminate tip; fls 8–10 mm.

           5. hybridum

    Stipules ovate, toothed; fls 4–5 mm. Lizard and Channel Is; very local.   *T. strictum* L.

17 Lateral veins of lflets thickened and backward-curved near the margin.   **10. scabrum**

    Lateral veins of lflets thin and ±straight near the margin.   *18*

18 Lflets glabrous.   *19*

    Lflets pubescent, at least on the veins beneath.   *20*

19 Calyx-teeth broad, abruptly narrowed to a short fine point, corolla distinctly longer than calyx, purplish. On light soils near the sea, rare. Clustered Clover.   *T. glomeratum* L.

    Calyx-teeth narrow, gradually tapering; corolla shorter than calyx, whitish. On light soils near the sea, rare.   *T. suffocatum* L.

20 Lflets hairy above.   **11. striatum**

    Lflets glabrous above. Lizard cliffs, very rare.   *T. bocconei* Savi

1. T. ornithopodioides L.   Birdsfoot Fenugreek.

Glabrous prostrate annual 2–20 cm. Lflets 2–8 mm., obovate to obcordate, Stipules lanceolate, acuminate. Heads 1–5-fld. Fls 5–8 mm., white and pink; calyx-teeth narrow. Pod longer than calyx, 5–8-seeded. Fl. 6–7. In dry shallow soils, chiefly near the coast.

2. T. micranthum Viv.   Slender Trefoil.

± Erect and hairy annual 2–10(–20) cm. Lvs shortly stalked; obovate or obcordate; stipules ovate, acute. Heads 2–6-fld. Fls 2·5–3 mm., deep yellow; standard deeply notched, folded over the stipitate pod; pedicels about as long as calyx-tube. Pod 2-seeded. Fl. 6–7. In grassy places on light soils; rather local and absent from most of Scotland.

3. T. dubium Sibth.   Lesser Yellow Trefoil.

Like **2** but rather larger. Heads (4–)10–26-fld. Fls c. 3 mm., pale yellow, turning dark brown; standard narrow, folded over pod; pedicels shorter than calyx-tube. Fl. 5–10. In grassy places; common and widespread.

4. T. campestre Schreber   Hop Trefoil.

Like **2** but up to 35(–50) cm. Stipules ½-ovate, tips triangular, acute. Heads many-fld. Fls c. 5 mm., yellow turning rather light brown; standard broad, not folded, much longer than pod. Pod 2–2·5 mm., several times as long as the persistent style. Fl. 6–9. In grassy places and waste ground; common and widely distributed.

*5. T. hybridum L.   Alsike Clover.

± Erect nearly glabrous perennial up to 60 cm. Lflets 30–35 mm.,
obovate to elliptic. Stipules ovate to oblong with long acuminate tips.
Heads up to c. 2 cm., globular. Fls 7–8 mm., white or pink, inner with
pedicels 2–3 times as long as calyx-tube. Calyx-tube white, half as long
as the narrow green teeth. Standard twice as long as calyx. Pod 3–
4 mm., 2-seeded. Fl. 6–9. Naturalized by roads, etc., and grown for
fodder throughout the British Is.

6. T. repens L.   White Clover, Dutch Clover.

Perennial up to 50 cm., creeping and rooting at nodes, glabrous or
hairy. Lflets 10–30 mm., obovate or obcordate, usually with a whitish
angled band. Stipules oblong with short narrow points. Heads up to
2 cm., globular. Fls 8–10 mm., white or pink, rarely purple; pedicels c.
3(–6) mm., about as long as calyx-tube. Calyx-tube white with green
veins, teeth narrowly triangular, c. half as long as tube. Standard twice
as long as calyx. Pod 4–5 mm., oblong, 3–6-seeded. Fl. 6–9. In grassy
places throughout the British Is.

T. occidentale D. E. Coombe differs from *T. repens* in its very small unmarked almost circular lflets with non-translucent lateral veins and
small heads of scentless fls. Coasts of W. Cornwall, Scilly and Channel Is.

7. T. fragiferum L.   Strawberry Clover.

Perennial up to 30 cm., creeping and rooting at nodes, glabrous or
hairy. Lflets 10–15 mm., obovate or obcordate. Stipules long-acuminate.
Heads 10–15 mm. diam. Fls c. 6 mm., pinkish to purplish. Calyx ± 2-
lipped, often reddish on outer side; teeth very narrow. Petals longer than
calyx. Pod and persistent petals enclosed in the greatly inflated reticulate and sometimes woolly upper lip of calyx. Fl. 7–9. In grassy places
particularly on heavy soils; widely scattered and locally common.

8. T. medium L.   Zigzag Clover.

Nearly glabrous perennial with flexuous stems up to 50 cm. Lflets 15–
40 mm., usually narrower and darker green than in the foregoing, often
with a faint whitish spot. Free part of stipules awl-shaped, spreading.
Heads 1–3 cm., depressed-globose, shortly stalked with a pair of lvs at
base. Fls c. 15 mm., reddish-purple, almost sessile. Calyx-tube glabrous,
at least in the lower part. Petals 2–3 times as long as calyx. Pod splitting
longitudinally. Fl. 6–9. Widely distributed in grassy places, commoner
in the north.

9. T. arvense L. Hare's-foot.

Softly hairy annual up to 40 cm. Lflets 10–25 mm., narrowly obovate-oblong. Stipules ovate with a long bristle-like point. Heads up to 2 cm., cylindrical; stalks as long as or longer than lvs. Fls c. 5 mm., white or pink, nearly sessile. Petals much shorter than calyx. Pod 1·5 mm., 1-seeded. Fl. 6–9. In rather open habitats on light soils; widely distributed and locally common.

10. T. scabrum L. Rough Trefoil.

Erect or decumbent, ± hairy annual up to 20 cm. Lflets 5–8 mm., obovate, hairy on both surfaces; lateral veins backward-curving, thickened towards the margin. Stipules rather stiff, triangular, with a narrow tip. Heads up to 10 mm., sessile, ovoid. Fls c. 5 mm., white. Calyx-tube ribbed, widening slightly upwards; teeth triangular, erect in fl., rigid and recurved in fr., as long as or longer than tube. Petals little longer than calyx. Pod c. 1·7 mm., obovoid. Fl. 5–7. On dry shallow soils; scattered throughout most of the British Is.

11. T. striatum L. Soft Trefoil.

Softly hairy, often procumbent annual 5–30 cm. Lflets 5–15 mm., obovate; lateral veins nearly straight, not thickened towards the margin. Stipules ovate or triangular with long narrow points. Heads up to 15 mm., sessile, ovoid, ± enfolded in the large stipules of the subtending lvs when young, becoming bluntly conical in fr. Fls c. 5 mm., pink. Calyx-tube swollen in fr., ribbed; teeth narrow, stiff, suberect in fr. and shorter than tube. Standard longer than calyx. Pod c. 3 mm., obovoid. Fl. 6–7. In dry and usually calcareous places. Widely scattered but very local.

12. T. incarnatum L. Crimson Clover.

Hairy annual up to 40 cm. Lflets 5–30 mm., broadly obovate. Stipules ovate, obtuse. Heads 3–4(–7) cm., ovoid or cylindrical, subtended by 1 lf. Fls c. 10 mm., crimson, sessile. Petals longer than calyx. Pod 2·5 mm., 1-seeded. Fl. 6–9. Sown for fodder and ± naturalized in the south. Ssp. **molinerii** (Balbis) Hooker fil., with pale fls, is native near the Lizard.

13. T. pratense L.   Red Clover.

Up to 60 cm., ± hairy. Lflets commonly 10–30 mm., oblong or obovate, often with a whitish crescentic spot. Free part of stipules triangular, with a bristle-like point usually pressed to petiole. Heads up to 3 cm., globose, becoming ovoid, sessile, with a pair of short-petioled lvs at base. Fls c. 18 mm., pink-purple or whitish, sessile. Calyx-tube hairy. Petals up to 3 times as long as calyx. Pod opening by the top falling off. Fl. 5–9. Generally distributed in grassy places.

14. T. subterraneum L.   Subterranean Trefoil.

Prostrate hairy annual 3–20(–60) cm. Lflets 5–21 mm., broadly obcordate. Stipules broadly ovate, acute. Heads few-fld. Fertile fls 8–12 mm., cream-coloured; calyx-tube cylindrical, as long as the bristle-like teeth; petals soon falling, longer than calyx. Sterile fls surrounded by the fertile and consisting of rigid, palmately-lobed calyces. Pod c. 3 mm., orbicular, 1-seeded. Fl. 5–6. In pastures on light soils; local and mainly in the south.

9. ANTHYLLIS L.

Herbs or shrubs with imparipinnate lvs, lower sometimes of 1 lflet. Fls in cymose heads, yellow or red and surrounded by an involucre (in our sp.). Calyx inflated, mouth oblique, shortly 5-toothed. Petals with long claws. Stamens all united by their filaments or the upper free. Pod enclosed in calyx, 1–3-seeded.

1. A. vulneraria L.   Kidney-vetch, Ladies' Fingers.

Up to 60 cm., erect or decumbent, hairy. Lvs up to 14 cm., lower often with large terminal lflet only. Lflets of lower lvs ovate to elliptic, terminal much the largest; of upper linear-oblong, all similar. Petioles short or 0. Heads in pairs, rarely 1, up to 4 cm. across, sessile or nearly so within a lf-like involucre. Fls 12–15 mm., yellow, rarely red. Calyx ± woolly, contracted at mouth. Petals longer than calyx. Pod c. 3 mm., semiorbicular, glabrous, reticulate, 1-seeded. Fl. 6–9. Widely distributed in dry places on shallow soils. Very variable.

10. LOTUS L.

Herbs or undershrubs. Lvs 5-foliolate, margins entire; stipules minute but lowest pair of lflets resembling stipules. Fls in cymose heads on stalks arising in lf-axils, yellow or reddish (in our spp.); bracts 3-foliolate. Calyx 5-toothed. Upper stamen free, 9 joined by their filaments. Pod terete, elongate, many-seeded, septate between the seeds, dehiscing by 2 valves.

1   Fls 10 mm. or more   **2**
    Fls not more than 8 mm.   **4**
2   Calyx-teeth erect in bud; stem solid or nearly so.   **3**
    Calyx-teeth spreading in bud; stems with a wide hollow.
                                                3. uliginosus
3   Lflets obovate, not acuminate; fls c. 15 mm.   **1. corniculatus**
    Lflets linear-lanceolate, acuminate; fls c. 10 mm.   **2. tenuis**
4   Pods 6–12 mm., peduncles equalling or longer than lvs. Coast of
    S. England and S. Wales; local. Hairy Birdsfoot-trefoil.
                         L. subbiflorus Lag. (*L. hispidus* Desf.)
    Pods 20–30 mm.; peduncles usually shorter than lvs. Chiefly by the
    coast in S. England. Slender Birdsfoot-trefoil. *L. angustissimus* L.

1. L. corniculatus L.   Birdsfoot-trefoil, Eggs-and-Bacon.

10–40 cm., usually glabrous and spreading. Stolons 0; stem solid or nearly so. Lflets 3–10 mm., obovate, obtuse or apiculate, lower pair broadly ovate or lanceolate; petioles short. Heads (1–)2–6(–8)-fld. Fls c. 15 mm., yellow, often streaked or tipped with red. Calyx-teeth triangular, erect in bud, 2 upper with an obtuse sinus between them. Pod up to 3 cm. Fl. 6–9. Generally distributed in grassy places.

2. L. tenuis Waldst. & Kit.   Slender Birdsfoot-trefoil.

Like 1 but stems more slender, often taller (to 90 cm.), much more wiry; lflets linear-lanceolate, acuminate or rarely narrowly obovate; heads rarely more than 4-fld; fls c. 10 mm.; calyx-teeth narrow. Fl. 6–8. In dry grassy places; much less common than *L. corniculatus*.

3. L. uliginosus Schkuhr   Large Birdsfoot-trefoil.

Like 1 but stoloniferous; stems hollow, soft, up to 1 m; lflets 15–20 mm. obovate, obtuse or mucronate; heads 5–12-fld; calyx-teeth spreading in bud, 2 upper with an acute sinus between them. Fl. 6–8. In damp grassy places; widely distributed.

11. ASTRAGALUS L.

Herbs or shrubs. Lvs imparipinnate, lflets entire, stipules present. Fls in racemes on stalks arising in lf-axils. Calyx tubular, teeth subequal. Keel blunt. Upper stamen free, 9 joined by their filaments. Pod

2-valved, often longitudinally 2-celled, septum developed from suture next the keel; seeds 2–many.

1 Plant nearly or quite glabrous; fls creamy with a greenish-grey tinge.
                                                            2. glycyphyllos
   Plant hairy; fls blue or purplish.   *2*
2 Stipules joined in lower part; fls ±erect; fr. erect.   **1. danicus**
   Stipules free to base; fls spreading or deflexed; fr. pendulous.
      Scottish mountains; rare. Alpine Milk-vetch.   *A. alpinus* L.

1. A. danicus Retz.   Purple Milk-vetch.

5–35 cm., slender, ascending, softly hairy. Lvs 3–7 cm.; lflets 5–12 mm. Infl.-stalks usually longer than the subtending lf. Fls 15 mm., blue-purple; bracteoles oblong to triangular. Calyx with appressed hairs, some black, some white. Pod 7–10 mm., 2-seeded, covered with crisped white hairs. Fl. 5–7. In short turf on calcareous soils; local and mainly in the east.

2. A. glycyphyllos L.   Milk-vetch.

60–100 cm., glabrous, prostrate or ascending. Lvs 10–20 cm.; lflets 15–40 mm., oblong-elliptic; stipules c. 20 mm., free. Infl.-stalks much shorter than the subtending lf. Fls 10–15 mm., creamy-white with a greenish-grey tinge; bracteoles very narrow. Calyx glabrous. Pod 25–35 mm., slightly curved, acuminate, many-seeded. Fl. 7–8. In rough grass; rather uncommon but widely distributed.

OXYTROPIS DC.

Like *Astragalus* but fls with beaked keel.
Two spp. in Scotland; rare.

Infl.-stalks longer than subtending lvs at fl.; fls pale purple. Purple
   Oxytropis.   *O. halleri* Bunge
Infl.-stalks shorter than subtending lvs at fl.; fls yellowish-white tinged
   with purple. Yellow Oxytropis.   *O. campestris* (L.) DC.

12. ORNITHOPUS L.

Herbs. Lvs imparipinnate; stipules small. Fls small, in few-fld umbels. Calyx 5-toothed. Keel blunt, sometimes very short; upper stamen free, 9 joined by their filaments. Pod curved, ± constricted between the seeds and breaking into 1-seeded portions when ripe.

Plant hairy; fls whitish; bracts present, lf-like; pods strongly jointed.

1. perpusillus

Plant nearly glabrous; fls yellow; bracts 0; pods slightly jointed. Very
   local in Scilly and Channel Is.   *O. pinnatus* (Miller) Druce

1. O. perpusillus L.   Birdsfoot.

Finely hairy ± prostrate annual 2–45 cm. Lvs 15–30 mm.; lflets up to
4 mm., usually 4–7 pairs, elliptic to narrow oblong, lowest pair often at
base of petiole, distant from the others and recurved. Umbel 3–6-fld,
subtended by a sessile bract. Fls 3–4 mm., white, veined with red. Pod
10–20 mm., curved, ± resembling the claw of a bird. Fl. 5–8. On well-
drained soils; rather local but widely distributed.

13. HIPPOCREPIS L.

Herbs. Lvs imparipinnate; lflets entire; stipules small or 0. Fls yellow,
umbellate; calyx 5-toothed, 2 upper teeth joined in the lower part. Pod
breaking up into 3–6 horseshoe-shaped 1-seeded segments at maturity.

1. H. comosa L.   Horse-shoe Vetch.

Almost glabrous perennial 10–40 cm. Lvs 3–5 cm.; lflets 5–8 mm.,
usually 4–5 pairs, obovate to oblong; stipules lanceolate, spreading.
Umbels usually 5–8-fld; infl. stalk longer than the subtending lf. Fls c.
10 mm.; pedicels and upper part of infl.-stalk with few appressed hairs.
Pod c. 3 mm., flattened, rough; segments horseshoe-shaped. Fl. 5–7.
In dry calcareous pastures and on cliffs,; rather local.

14. ONOBRYCHIS Scop.

Herbs or shrubs. Lvs imparipinnate; lflets entire; stipules scarious.
Infl. racemose. Wings short, keel obliquely truncate, equalling or exceeding standard. Pod indehiscent, not jointed, 1–2-seeded, often spiny or
tubercled.

1. O. viciifolia Scop.   Sainfoin.

30–60 cm., slightly hairy. Lflets 1–3 cm., 6–12 pairs, obovate to narrowly
oblong, shortly stalked; stipules ovate-acuminate. Infl.-stalk longer
than the subtending lf; racemes dense, c. 50-fld. Fls 10–12 mm., light
pink or red; calyx-teeth narrow, much longer than the often woolly tube.
Pod 6–8 mm., 1-seeded, strongly reticulate and tubercled on the lower
margin. Fl. 6–8. In chalk and limestone grassland, often elsewhere as a
relic of cultivation.

15. VICIA L.   Vetch, Tare.

Climbing or scrambling herbs. Lvs pinnate, terminal lflet 0; tendrils usually present. Lflets in 2–many pairs. Fls in racemes in the axils of lvs. Upper stamen free, 9 joined by their filaments into an obliquely truncate tube. Style glabrous, or equally downy all round, or bearded below the stigma. Pod flattened, 2-valved, dehiscent, several-seeded.

1 Fls more than 10 mm.   *2*
    Fls less than 8 mm.   *6*

2 Racemes long-stalked.   *3*
    Racemes with stalks very short or 0.   *9*

3 Raceme 1–2-fld; fls with purple standard and white wings. Very
    local, mainly near coast.   *V. bithynica* (L.) L.
    Racemes 6–40-fld.   *4*

4 Tendrils 0; plant erect, 30–60 cm.; fls white tinged with purple.
    Very local. Bitter Vetch.   *V. orobus* DC.
    Tendrils present; plant trailing or climbing.   *5*

5 Fls purplish; upper calyx-teeth very small.   **3. cracca**
    Fls white with blue or purple veins; upper calyx-teeth ½ as long as
    lower.   **4. sylvatica**

6 Stalks of racemes very short or 0.   **7. lathyroides**
    Stalks of racemes long.   *7*

7 Calyx-teeth subequal, somewhat longer than tube; pod hairy.
                                                        1. hirsuta
    Calyx-teeth unequal, upper 2 shorter than tube; pod glabrous.

8 Fls c. 4 mm.; pod 4-seeded; hilum oblong.   **2. tetrasperma**
    Fls c. 8 mm.; pod 5–8-seeded; hilum almost circular. Local in
    south. Slender Tare.   *V. tenuissima* (Bieb.) Schinz & Thell.

9 Fls pale dirty yellow or brownish-violet; pod densely hairy. Very
    local, near coast. Yellow Vetch.   *V. lutea* L.
    Fls bluish or purplish; pod glabrous or sparsely hairy.   *10*

10 Standard purple, wings white; stipules c. 10 mm. Very local,
    mainly near coast.   *V. bithynica* (L.) L.
    Standard and wings purplish or bluish; stipules smaller.   *11*

11 Calyx-teeth unequal, the smaller ones shorter than tube.
                                                         5. sepium
    Calyx-teeth subequal, as long as tube.   **6. sativa**

1. V. hirsuta (L.) S. F. Gray   Hairy Tare.

Slender trailing nearly glabrous annual 20–30(–70) cm. Lflets 5–12 mm., usually 4–8 pairs, oblong, truncate to emarginate; tendrils usually branched; stipules often 4-lobed. Racemes 1–9-fld; 4–5 mm., dirty white or purplish; calyx-teeth subequal, rather longer than tube. Pod c. 10 mm., hairy; seeds usually 2. Fl. 6–8. In grassy places and cultivated land; generally distributed.

2. V. tetrasperma (L.) Schreber   Smooth Tare.

Glabrous slender annual (15–)30–60 cm. Lflets 10–20 mm., usually 4–6 pairs, narrowly oblong, blunt; tendrils usually simple; stipules ½-arrow-shaped. Racemes 1–2-fld; fls c. 4 mm., pale blue; calyx-teeth unequal, upper 2 shorter than tube. Pod (9–)12–15 mm., shortly stalked, glabrous; seeds 4. Fl. 5–8. In grassy places; widespread but rather local.

3. V. cracca L.   Tufted Vetch.

± Hairy perennial 60–200 cm. Lflets 10–25 mm., 6–12(–15) pairs, oblong-lanceolate to linear-lanceolate, ± acute; tendrils branched; stipules ½-arrow-shaped, entire. Racemes 10–40-fld; fls 10–12 mm., purplish; calyx-teeth very unequal, upper minute. Pod 10–20 mm., ovate, obliquely truncate, glabrous; seeds 2–6. Fl. 6–8. In hedges etc.; common and generally distributed.

4. V. sylvatica L.   Wood Vetch.

Glabrous perennial 60–130 cm. Lflets 5–20 mm., 6–9(–12) pairs, oblong-elliptic, mucronate; tendrils much-branched; stipules lanceolate, semicircular with many narrow teeth at base. Racemes up to 18-fld; fls 15–20 mm., white with blue or purple veins; calyx-teeth very narrow, upper ½ as long as lower. Pod c. 30 mm., oblong-lanceolate, acuminate at both ends, glabrous; seeds few. Fl. 6–8. In bushy places, woods, shingle and cliffs by the sea; local but widespread.

5. V. sepium L.   Bush Vetch.

Nearly glabrous perennial 30–100 cm. Lflets 10–30 mm., 5–9 pairs, ovate to elliptic; tendrils branched; stipules ½-arrow-shaped, sometimes toothed. Racemes 2–6-fld, subsessile; fls 12–15 mm., pale purplish; calyx-teeth unequal, lower much shorter than tube. Pod 20–25 mm., oblong-lanceolate, beaked, glabrous; seeds 6–10. Fl. 5–8. Hedges etc.; common and widely distributed.

6. V. sativa L.   Common Vetch.

Slightly hairy annual 15–120 cm. Lflets 10–20 mm., 4–8 pairs, linear to obovate. Stipules ½-arrow-shaped, toothed or entire, often with a dark blotch. Fls solitary or in pairs (–4), 10–30 mm., purple; calyx-teeth subequal, as long as tube. Pod 25–80 mm., narrowly oblong, beaked, slightly hairy to glabrous, 4–12-seeded. Fl. 5–9.

Ssp. **sativa** with stout stems, broad lflets, fls 20–30 mm., and pods 40–80 mm., escapes from cultivation. Ssp. **angustifolia** (L.) Gaudin with slender stems, narrow lflets, fls 10–16 mm., and pods 35–50 mm., is widely distributed in grassy places.

7. V. lathyroides L.   Spring Vetch.

5–20 cm., slender, spreading. Lflets 4–10 mm., 2–3 pairs, narrowly oblong to obovate, obtuse or emarginate and mucronate; tendrils small and unbranched or 0; stipules small, ½-arrow-shaped. Fls 5–7 mm., solitary, lilac; calyx-teeth equal, nearly as long as tube. Pod 15–20 mm., tapering at both ends, glabrous; seeds 8–12. Fl. 5–6. In well-drained grassy places; rather local but widely distributed.

Several other spp. occur as introductions (see larger *Flora*).

16. LATHYRUS L.

Like *Vicia* but usually with fewer lflets and winged or angled stems. Staminal tube transversely truncate. Style bearded on its upper side only.

1	Fls yellow.	2
	Fls crimson, purple or bluish.	3
2	Lvs with 1 pair of lflets and a pair of large lf-like stipules.	**2. pratensis**
	Lvs consisting of a simple tendril with a pair of large lf-like stipules at base; lflets 0, except in seedlings. Rare. Yellow Vetchling.	*L. aphaca* L.
3	Tendrils 0; plant ± erect.	4
	Tendrils present on most lvs; plant usually climbing.	6
4	Lvs (phyllodes) grass-like; stipules very small.	**1. nissolia**
	Lvs pinnate, lflets 2–6 pairs; stipules conspicuous.	5
5	Stem winged; lower calyx-teeth as long as tube.	**4. montanus**
	Stem angled; lower calyx-teeth longer than tube. Rare and probably always introduced.	*L. niger* (L.) Bernh.
6	Lflets 1 pair with a pair of stipules at base.	7
	Lflets 2 or more pairs with a pair of stipules at base.	9
7	Stem flattened and winged.	8
	Stem ± square and sharply angled; lflets obovate; fls crimson. Naturalized; very local. Earthnut Pea.	**L. tuberosus* L.
8	Fls 15–17 mm., usually 5–7 together, pink; pod glabrous.	**3. sylvestris**

Fls 10–12 mm., 1–2 together; standard crimson, wings pale blue;
pod densely silky. Rare. Hairy Vetchling.   *L. hirsutus* L.

9 Stem winged; stipules lanceolate; pod flattened. Fens and damp
places; local. Marsh Pea.   *L. palustris* L.
Stem angled; stipules broadly triangular; pod turgid. Shingle
beaches; very local. Sea Pea.   *L. japonicus* Willd.

1. L. nissolia L.   Grass Vetchling.

30–90 cm., nearly glabrous. Phyllodes up to 15 cm. Racemes 1–2-fld;
fls c. 15 mm., crimson; calyx-teeth triangular. Pod 3–6 cm. × 2–3 mm.,
straight; seeds 15–20, rough. Fl. 5–7. Usually amongst long grass; local.

2. L. pratensis L.   Meadow Vetchling.

30–120 cm., ± hairy. Lflets 1–3 cm., 1 pair, lanceolate, acute, veins
parallel; stipules sagittate. Racemes (2–)5–12-fld; fls 15–18 mm.; calyx-
teeth narrowly triangular. Pod 25–35 mm., glabrous or finely hairy,
flattened; seeds 5–10, smooth. Fl. 5–8. In hedges and long grass; common and widely distributed.

3. L. sylvestris L.   Narrow-leaved Everlasting Pea.

1–2 m., glabrous, often glaucous. Lflets 7–15 cm., sword-shaped;
stipules up to 2 cm., less than half as wide as stem, narrowly sword-
shaped with a spreading basal lobe. Fls 15–17 mm.; calyx-teeth
triangular, shorter than tube. Pod 5–7 cm., glabrous, flattened and
narrowly winged along the upper side; seeds 8–14. Fl. 6–8. In thickets
and hedges, on railway banks, etc ; local but widely distributed.

*L. latifolius L., Everlasting Pea, is sometimes naturalized. It is like
L. sylvestris but the lflets are often ovate and the stipules more than
½ as wide as stem.

4. L. montanus Bernh.   Bitter Vetch.

15–40 cm., glabrous. Lvs narrowly lanceolate to elliptic, acute or blunt;
stipules lanceolate. Racemes 2–6-fld; fls c. 12 mm., lurid crimson becoming green or blue; calyx-teeth unequal, lower about as long as tube.
Pod 3–4 cm., subcylindric, glabrous; seeds 4–6. Fl. 4–7. In hedge-
banks, meadows, heaths, etc., in hilly country; widespread but commoner in the west and north.

57. ROSACEAE

Trees, shrubs or herbs. Lvs nearly always spirally arranged and stipulate. Fls regular, perigynous and sometimes epigynous, usually
hermaphrodite. Epicalyx sometimes present Sepals and petals usually
5; petals free, sometimes 0. Stamens usually 2, 3 or 4 times as many as
the sepals, rarely more or fewer. Carpels 1–many, free or sometimes
united, superior, half-inferior or inferior; ovules usually 2; styles usually
free. Fr. of 1 or more achenes, drupes or follicles or a pome (very rarely
a capsule), the receptacle sometimes becoming coloured and fleshy;
seeds non-endospermic.

Liable to be confused with Ranunculaceae by beginners, but apart from the
perigyny or epigyny the presence of stipules (and sometimes of an epicalyx) will
separate any member likely to be taken for a member of the Ranunculaceae.

1	Trees or upright shrubs	*2*
	Herbs or sometimes slightly woody but prostrate.	*14*
2	Fls bright yellow; epicalyx present; lvs pinnate; a low unarmed shrub.	3. POTENTILLA
	Fls white or pink; epicalyx 0.	*3*
3	Receptacle strongly convex; lvs compound.	2. RUBUS
	Receptacle concave.	*4*
4	Receptacle slightly concave; carpels 5; fls c. 8 mm. diam., pink, very numerous in a dense panicle; Shrub 1–2 m., with narrow simple sharply serrate lvs 3–7 cm., exstipulate. Introduced and often naturalized.	*Spiraea salicifolia* L.
	Receptacle strongly concave (carpels 1 or numerous), or ovary truly inferior (carpels 1–5).	*5*
5	Carpels and styles numerous (styles occasionally united into a column); prickly shrubs with pinnate lvs.	11. ROSA
	Carpels and styles 1–5; unarmed or thorny shrubs or trees (if lvs pinnate, then unarmed).	*6*
6	Carpel 1, free from the receptacle; lvs simple, not lobed.	12. PRUNUS
	Carpels 1–5 (if 1, then lvs lobed), united to the receptacle so that the ovary is at least partly inferior.	*7*
7	Fls in racemes. Shrub or small tree to 12 m, with simple serrate lvs 3–7 cm. and a slender spreading or drooping infl. of many white fls; fr. globose, blackish-purple. Introduced and locally naturalized. *Amelanchier grandiflora* Rehder (*A. laevis* auct.)	
	Fls in corymbs or in clusters of 1–3.	*8*
8	Fls in compound corymbs.	*9*
	Fls in clusters of 1–3 or in short simple umbel-like corymbs.	*11*

9   Lvs entire.   13. COTONEASTER
     Lvs toothed, lobed or pinnate.   10

10 Thorny; carpel-wall hard in fr.   14. CRATAEGUS
     Unarmed; carpel-wall cartilaginous in fr.   15. SORBUS

11 Fls less than 1 cm. diam.; fr. c. 1 cm. diam., red.
                                         13. COTONEASTER
     Fls c. 3 cm. diam. or more; fr. larger, brown or green (sometimes
        tinged red).   12

12 Fls solitary; sepals long, often lf-like; carpel-wall hard in fr.
     Thorny shrub 2–3 m., with lanceolate ±entire lvs 5–12 cm.,
     white fls and subglobose fr. 2–3 cm., crowned by the calyx.
     Introduced and occasionally naturalized. Medlar.
                                      *Mespilus germanica L.
     Fls several; carpel-wall papery in fr.; sepals short, not lf-like.   13

13 Styles free; anthers purple; fr. gritty.   16. PYRUS
     Styles connate below; anthers yellow; fr. not gritty.   17. MALUS

14 Carpels 5 or more, on the surface of a convex to weakly concave
     receptacle.   15
     Carpels 1–2(–4), enclosed in the strongly concave receptacle.   21

15 Epicalyx 0.   16
     Epicalyx present.   18

16 Lvs pinnate; fls in many-fld panicles.   1. FILIPENDULA
     Lvs not pinnate; fls not in many-fld panicles.   17

17 Petals c. 8; plant prostrate; fr. a group of achenes with long
     feathery awns. Undershrub with lvs 0·5–2 cm., ovate-oblong,
     blunt, deeply crenate, dark green above, white-tomentose
     beneath; stipules brownish. Fls 2·5–4 cm. diam., white. On
     ledges and in crevices of basic rocks, chiefly on mountains;
     local from N. Wales and Yorks northwards to Orkney; N. and
     W. Ireland. Mountain Avens.   *Dryas octopetala* L.
     Petals (4–)5; plant not prostrate; fr. a group of druplets.
                                           2. RUBUS

18 Style terminal, persistent on the fr. as a long jointed awn; basal lvs
     pinnate, with the terminal lflet much larger than any of the
     lateral ones.   5. GEUM
     Style lateral, not persistent; if lvs pinnate, then terminal lflet not or
     scarcely larger than the lateral ones (though these are sometimes
     unequal).   19

19 Stamens and carpels not more than 10; petals tiny, narrow,
     yellow, or 0; lvs ternate, blue-green, lflets three-toothed. Small
     perennial herb with woody stock and axillary fl.-stems 1–2 cm.
     Mountain tops and grassland; rock crevices, etc., to over
     4000 ft.; from Westmorland northwards to Sutherland and
     Shetland.   *Sibbaldia procumbens* L.

Stamens and carpels usually numerous; petals usually con-
    spicuous; if lvs ternate, then lflets with more than 3 teeth.   *20*

20 Receptacle fleshy in fr.; fls white; lvs ternate; receptacle glabrous.
                                        4. FRAGARIA
    Receptacle dry in fr.; if fls white and lvs ternate, then receptacle
    hairy.   3. POTENTILLA

21 Petals present; fls in racemes or few-fld cymes.   *22*
    Petals 0; fls in heads or many-fld cymes, rarely spikes.   *23*

22 Epicalyx present; receptacle without spines. Perennial herb 20–
    40 cm.; lvs pinnate; fls yellow, each surrounded by an involucre
    of 6–10 sepal-like lobes. Introduced and naturalized in woods
    in C. Scotland.   *Aremonia agrimonoides* (L.) DC.
    Epicalyx 0; receptacle with a crown of spines.   6. AGRIMONIA

23 Fls in cymes (sometimes dense and head-like, then lf-opposed);
    epicalyx present.   *26*
    Fls in terminal heads or spikes; epicalyx 0.   *24*

24 Plant prostrate; fr.-receptacle with 4 spines c. 1 cm., barbed at the
    tip. Creeping undershrub with pinnate lvs and globose heads
    5–10 mm. diam., of greenish fls. A wool-alien, locally natura-
    lized.   *Acaena anserinifolia* (J. R. & G. Forst). Druce
    Plant erect; fr.-receptacle not spiny.   *25*

25 Fls hermaphrodite; stamens 4; infl. oblong, fls crimson.
                                    9. SANGUISORBA
    Fls polygamous; stamens numerous, infl. globose, fls greenish.
                                    10. POTERIUM

26 Perennial; fls in terminal cymes; stamens 4.   7. ALCHEMILLA
    Annual; fls in dense lf-opposed clusters; stamen 1.   8. APHANES

1. FILIPENDULA Miller

Perennial herbs with short rhizome. Lvs spirally arranged, pinnate with
small pinnae between the large ones, stipulate. Infl. a cymose panicle.
Receptacle flat or slightly concave. Sepals and petals 5–8. Stamens
20–40. Carpels 5–15, free, superior; ovules 2 per carpel. Fr. a group
of 1-seeded achenes.

Basal lvs with 8 or more pairs of the larger lflets; lflets less than
    1·5 cm.; achenes straight, erect.   **1. vulgaris**
Basal lvs with 5 or fewer pairs of the larger lflets; lflets 2 cm. or more;
    achenes twisted together in fr.   **2. ulmaria**

1. F. vulgaris Moench   Dropwort.

Perennial herb (7–)15–80 cm., nearly glabrous. Roots bearing ovoid
tubers. Basal lvs 2–25 cm., with 8–20 pairs of main lflets; these 5–

15 mm., oblong in outline, pinnately lobed, with acute dentate lobes, green on both sides; terminal lflets trifid; stem-lvs very few, the upper simple, lobed, very small. Fls in an irregular cymose panicle much broader than high. Sepals usually 6, triangular-ovate, spreading, then reflexed. Petals 5–9 mm., usually 6, spathulate, cream-white, tinged reddish outside. Carpels c. 4 mm. in fr., 6–12, erect, pubescent. Fl. 5–8. Calcareous grassland; widespread but local, northwards to Caithness; W. Ireland; Jersey.

2. F. ulmaria (L.) Maxim.   Meadow-sweet.

Perennial herb 60–120 cm., with a pink rhizome smelling of thymol. Roots not tuberous. Basal lvs 30–60 cm. with 2–5 pairs of main lflets, these 2–8 cm., ovate, acute, sharply doubly serrate, dark green above, usually white-tomentose beneath but sometimes green; terminal lflet 3–(5)-lobed; uppermost stem-lvs simple or with only small lflets apart from the terminal one. Fl.-stems lfy, glabrous or nearly so. Fls very numerous in an irregular cymose panicle, usually rather longer than broad. Sepals usually 5, triangular-ovate, reflexed, pubescent. Petals 2–5 mm., usually 5, obovate, clawed, cream-white. Carpels 6–10, twisted together spirally and c. 2 mm. in fr., glabrous. Fl. 6–9. Swamps, marshes, fens, wet woods and meadows, wet rock-ledges and by rivers, etc., but not on very acid peat; to nearly 3000 ft. Common throughout the British Is and locally dominant in fens and wet woods.

2. RUBUS L.

Herbs, or shrubs with biennial stems dying after flowering. Epicalyx 0. Sepals 5. Petals 5(–8). Stamens numerous. Carpels few to numerous; ovules 2 in each carpel. Fr. of few to numerous 1-seeded druplets aggregated together into a compound fr.

1 Lvs simple, palmately lobed; dioecious herb with creeping rhizome
     and solitary terminal fls.   **1. chamaemorus**
     Lvs pinnately or palmately compound; fls hermaphrodite, rarely
     solitary.   *2*
2 Stems annual, herbaceous, the flowering branches arising from
     ground level; lvs ternate; stipules attached to the stem.   *3*
     Stems usually biennial, woody, bearing axillary flowering branches
     in the 2nd year; stipules attached only to lf-stalk.   *4*
3 Plant with long above-ground stolons; fls 2–8, white; fr. scarlet when
     ripe.   **2. saxatilis**
     Plant without above-ground stolons; fls 1–3, bright rose-pink;
     fr. dark purple when ripe. Very rare and local on Scottish moun-
     tains; ? extinct. Arctic Bramble.   *R. arcticus* L.

4 Lvs pinnate with 3-7 lflets; fr. bright red, coming away from the
   receptacle when ripe.   **3. idaeus**
   Lvs palmate with 3-5 lflets (rarely with 7 and then the 4 lower ones
   from the same point), or all ternate; fr. black or blue-black
   (rarely deep red), coming away with the receptacle when ripe.   5

5 Fr. very pruinose, with 2-5 or 14-20 druplets; stems very pruinose,
   terete, weak; lvs ternate; stipules lanceolate.   **4. caesius**
   Fr. not or slightly pruinose; druplets usually more numerous.
   Stipules linear or linear-lanceolate (if the latter, then lvs not all
   ternate).   **5. fruticosus**

1. R. chamaemorus L.   Cloudberry.

Non-spiny somewhat pubescent dioecious herb with far-creeping
rhizome. Fl-stems 5-20 cm., annual, erect. Lvs few, 1·5-8 cm., simple,
roundish, palmately 5-7-lobed (female more shallowly than male),
deeply cordate at base, somewhat rugose; lobes toothed; stalk 1-7 cm.;
stipules ovate, scarious, attached to stem. Fl. solitary, terminal.
Sepals ovate, acuminate. Petals 8-12 mm., white, much longer than
sepals. Fr. orange when ripe (red earlier), with few large druplets. Fl.
6-8. Mountain moors and blanket bogs. Locally abundant from N.
Wales, Lancashire and Derby northwards to Caithness; very rare
in Ireland.

2. R. saxatilis L.   Stone Bramble.

Herb with long above-ground stolons rooting at their tips and ± erect
annual fl-stems, 8-40 cm., pubescent and with weak or no prickles. Lvs
ternate; terminal lflets 2·5-8 cm., ovate, ± acute, irregularly toothed,
green and almost glabrous above, paler and pubescent beneath; lateral
lflets subsessile; stalk 2-7 cm.; stipules ovate, green, attached to stem.
Fls 2-8 in a compact cyme. Sepals triangular-ovate, shortly pubescent.
Petals 3-5 mm., narrow, white, often shorter than sepals. Fr. scarlet,
translucent, of 2-6 large separate druplets. Fl. 6-8. Stony woods and
shady rocks, especially basic, in hilly districts. Rather local in W.
England and Wales and northwards from Lancashire and N. Lincoln;
widespread in Ireland.

3. R. idaeus L.   Raspberry.

Plant suckering from root-buds. Stems 100-160 cm., biennial, woody,
erect, terete, somewhat pruinose, armed with slender straight prickles.
Lvs pinnate with 3-5(-7) lflets each 5-12 cm., terminal largest, ovate or
ovate-lanceolate, acuminate, rounded or subcordate at base, irregularly
toothed, green and somewhat pubescent above, densely white-tomentose

beneath; stalk 2–7 cm.; stipules very slender, attached only to lf-stalk. Fls 1–10 in dense axillary and terminal cymes on short lateral branches with ternate lvs. Sepals with long acuminate tips. Petals narrow, erect, white, about as long as sepals. Fr. red (rarely pale yellow), opaque; druplets numerous, pubescent. Fl. 6–8. Woods and heaths, especially in hilly districts. Common almost throughout the British Is.

4. R. caesius L.   Dewberry.

Stem procumbent, rooting, weak, terete, glabrous, very pruinose; prickles very weak, straight or curved, scattered. Lflets 3, sparsely hairy above, green beneath, irregularly, coarsely and doubly toothed or shallowly lobed; stipules lanceolate, attached only to the lf-stalk. Infl. short, lax, few-fld. Sepals grey-green, white-margined, cuspidate with long points clasping the fr. Petals white or pinkish. Fr. with very few (rarely to 20 or more) druplets, densely pruinose and so appearing bluish, acid. Fl. 6–9. Dry grassland and scrub, mainly on basic soils, also common in fen carr. Widespread and common in England and Wales, local in Scotland and rather local in Ireland.

Forms believed to have originated as hybrids between *R. caesius* and *R. fruti-cosus* are widespread on basic soils (*R. corylifolius* agg.). They have stems terete or nearly so with scattered prickles, lflets 3–5, the basal ones sessile or nearly so, ± lanceolate stipules and fr. with few large druplets.

5. R. fruticosus L., *sensu lato*   Blackberry, Bramble.

Stems biennial, woody, terete to sharply angled, clothed more or less densely, and in very varying proportions, with prickles, bristles and stalked glands, the prickles being stout or slender, straight or curved, scattered or confined to the angles, and the various types of emergence being distinct or intergrading. Lvs ternate or palmate; the basal pair of lflets usually being distinctly (though often very shortly) stalked; stipules very slender, attached only to the lf-stalk. Infl. usually compound. Fls hermaphrodite. Petals white to deep pink. Fr. black (occasionally deep red or bluish with bloom), coming away with the fleshy receptacle when ripe. Fl. 6–9. Common, in various forms, throughout the British Is.

The forms of this aggregate, commonly treated as separate species, are bewilderingly numerous and difficult to determine. This arises from the facts that most of them usually set seed without fertilization by a pollen-derived male nucleus, but can on occasion reproduce sexually and give rise to hybrids. There is one form, the commonest British bramble, which is normally sexual:

R. ulmifolius Schott: stem arching, often climbing or forming dense tangled bushes, furrowed, finely pubescent; prickles strong; glands 0. Lflets 3–5, small, dull dark green and somewhat rugose and convex above, densely and closely white-tomentose beneath; terminal obovate, shortly acuminate, rounded at base. Infl. cylindrical; branches tomentose and with strong curved prickles. Petals crumpled, bright purplish-pink. Stamens about equalling styles. Fl. late. The commonest bramble in general and, unlike most spp., growing on chalky and heavy clay soils.

3. POTENTILLA L.   Cinquefoil.

Usually perennial herbs or small shrubs. Lvs palmate, pinnate or ternate; stipulate. Epicalyx present. Stamens 10–30, inserted at the outer edge of the narrow nectar-secreting ring. Carpels (4–)10–80, free, superior, borne on the convex receptacle; ovule 1; style ± lateral, withered or not persistent in fr. Fr. an aggregate of achenes borne on the dry or spongy receptacle.

1	Lvs pinnate.	2
	Lvs palmate or ternate.	5
2	Shrub c. 1 m.; lflets (3–)5(–7), entire; fls 5-merous. A very local plant of damp rocky ground in N. England and W. Ireland. Shrubby Cinquefoil.	*P. fruticosa* L.
	Herbs, sometimes woody at the base; lflets toothed.	3
3	Calyx and corolla purple; woody at base; bogs and fens.	**1. palustris**
	Calyx green, corolla white or yellow; not woody; plants of dryish ground.	4
4	Fls white, in terminal cymes. A pubescent herb 20–50 cm. E. Wales and Sutherland; extremely rare. Rock Cinquefoil.	*P. rupestris* L.
	Fls yellow, solitary.	**3. anserina**
5	Fls white; lvs ternate; carpels hairy.	**2. sterilis**
	Fls yellow; carpels glabrous.	6
6	Lvs densely white-tomentose beneath.	**4. argentea**
	Lvs green on both sides.	7
7	Fls axillary and terminal, forming a terminal cyme, 5-merous.	8
	Fls solitary, or, if forming a cyme, all or mostly 4-merous.	10
8	Corolla not exceeding calyx; fl.-stems terminal, many-fld. Herb 20–50 cm., with ternate lvs and coarsely toothed lflets; fls 7–8 mm. diam.; calyx increasing to 15–20 mm. diam. in fr. Introduced and locally naturalized on waste ground.	**P. norvegica* L.
	Corolla larger than calyx; fl.-stems axillary from a basal rosette; fls few.	9

9 Stems prostrate, rooting, forming mats; free part of stipules of basal lvs long and linear; fls 10–15 mm. diam.
             5. tabernaemontani
 Without long prostrate rooting stems and never forming mats; free part of stipules of basal lvs ovate; fls 15–25 mm. diam. A very local plant chiefly of mountain rock-ledges and crevices on basic substrata in N. Wales, N. England and Scotland.
           P. crantzii (Crantz) G. Beck

10 Stems decumbent to erect, never rooting; lvs all ternate; fls nearly all 4-merous, c. 10 mm. diam.   **6. erecta**
 Stems decumbent or prostrate, rooting, at least late in the season; at least some of the lower lvs with 4–5 lflets; fls 14–25 mm. diam., 4–5-merous.   *11*

11 Stems decumbent; basal lvs with 3–5 lflets, stem-lvs mostly with 3, short-stalked; some fls 4-, some 5-merous, 14–18 mm. diam.; carpels 20–50 (see also hybrids).   **7. anglica**
 Stems prostrate; almost all lvs with 5 lflets; stem-lvs long-stalked; all fls 5-merous, 17–25 mm. diam.; carpels 60–120. **8. reptans**

1. P. palustris (L.) Scop.   Marsh Cinquefoil.

Rhizome woody, long-creeping. Stems 15–45 cm., ascending, dying back in winter to the short persistent woody base. Lvs pinnate, the lower long-stalked, with 5–7 lflets each 3–6 cm., oblong, coarsely and sharply toothed, subglaucous and glabrous to shaggy beneath; upper lvs smaller and shorter-stalked; bracts ternate. Fls in a loose terminal cyme, 5-merous, their stalks glandular-pubescent. Sepals purplish, enlarging in fr. Petals ovate-lanceolate, deep purple, shorter than sepals. Stamens, carpels and styles deep purple. Achenes glabrous; receptacle hairy, spongy. Fl. 5–7. Fens, marshes, bogs, wet heaths and moors to 3000 ft., throughout the British Is.

2. P. sterilis (L.) Garcke   Barren Strawberry.

Perennial herb 5–15 cm., softly hairy. Stock thick, oblique, somewhat woody, ending in a rosette of lvs, frequently forming prostrate stolons. Rosette-lvs ternate, long-stalked; lflets 0·5–2·5 cm., short-stalked, orbicular or broadly obovate ± truncate, with 5–7 teeth on each side and a terminal tooth much smaller than its neighbours, bluish-green, dull and hairy above, paler and more densely covered with spreading hairs beneath; stem-lvs smaller. Fl-stems axillary, decumbent, 1–3-fld. Fls 10–15 mm. diam., 5-merous. Petals white, widely separated, about equalling the sepals. Carpels hairy only near the tip; receptacle hairy but not spongy. Fl. 2–5. Scrub, wood-margins, open woods, etc., to over 2000 ft. Throughout the British Is, common.

The spreading hairs on the underside of the lflets distinguish this plant from *Fragaria vesca*.

3. P. anserina L.   Silverweed.

Perennial ± silky herb with short thick stock ending in a rosette of lvs and producing long creeping, rooting and flowering stolons to 80 cm. Basal lvs 5–25 cm., pinnate, with 7–12 pairs of main lflets alternating with smaller ones; lflets 1–6 cm., oval or oblong, deeply, narrowly and regularly toothed; usually silver-silky on both sides or only beneath; lvs of stolons smaller. Fls 5-merous, solitary, axillary, long-stalked. Petals c. 1 cm., yellow, obovate. Achenes glabrous; receptacle hairy. Fl. 6–8. Waste places, roadsides, damp pastures, dunes, etc., to 1400 ft.; common throughout the British Is.

4. P. argentea L.

Perennial herb 15–50 cm., with short thick stock. Basal lvs palmate, with 5 lflets; lflets 1–3 cm., long-stalked, obovate-cuneiform, pinnately lobed, with 2–5 oblong ± entire lobes on each side, densely white-tomentose beneath, margins narrowly recurved; stem-lvs similar but shorter-stalked and the upper sometimes ternate. Fl.-stems terminal, decumbent or ascending, ± tomentose. Fls 10–15 mm. diam., 5-merous, several in a dichotomous cyme. Sepals ovate, tomentose. Petals yellow, slightly longer than sepals. Fl. 6–9. Dry sandy grassland from Moray southwards; local, especially in the west; Channel Is.

5. P. tabernaemontani Ascherson   Spring Cinquefoil.

Perennial mat-forming herb 5–20 cm. Stock thick, much branched, emitting prostrate usually rooting branches. Basal lvs long-stalked, palmate with 5 lflets; lflets 0·5–2 cm., obovate-cuneiform with 2–9 teeth on each side, the terminal tooth markedly smaller than its neighbours, green and ± appressed-hairy on both sides or only beneath; free part of stipules long and linear-lanceolate; stem-lvs few, the upper smaller, ternate, subsessile. Fl.-stems axillary, decumbent, pubescent and shaggy. Infl. a lax few-fld cyme. Fls 10–15 mm. diam., 5-merous. Petals yellow, obovate, deeply emarginate, longer than sepals. Receptacle hairy. Fl. 4–6. Dry basic grassland, especially on the edge of rocky outcrops, from Angus and Cumberland southwards, very local.

6. P. erecta (L.) Raüschel   Common Tormentil.

Perennial herb. Stock very thick, woody, vertical or oblique, red-fleshed, bearing a terminal rosette of lvs which often wither before fl. Lvs usually all ternate; basal lvs long-stalked, their lflets 5–10 mm., broadly obovate-cuneiform, with 3–4 coarse teeth on each side of the truncate apex; stem-lvs ± sessile, their lflets 1–2 cm., narrower, toothed above the middle, ± glabrous above, appressed silky-hairy on the margins and veins beneath; stipules of stem-lvs large, palmately lobed, like additional lflets. Fl.-stem 10–30 cm., axillary, slender, decumbent to ascending, never rooting. Fls many, in loose terminal cymes, almost all 4-merous, 7–11(–15) mm. diam., long-stalked. Petals yellow, emarginate, rather longer than sepals. Stamens usually 16. Carpels 4–8(–20); receptacle hairy. Fl. 6–9. Grassland, heaths, bogs, fens, open woods, mountain-tops, etc., especially on light acid soils, to nearly 3500 ft. Throughout the British Is.

7. P. anglica Laicharding   Trailing Tormentil.

Differs from *P. erecta* in the persistent rosette of lvs; the high proportion (c. 50 %) of the basal lvs with 5 lflets; the distinctly stalked stem-lvs (1–2 cm. on lower lvs), which are mostly ternate; the broader lflets and the mostly entire stipules of the stem-lvs; the longer fl.-stems (15–80 cm.), usually rooting at the nodes in late summer and producing new plants; the larger fls (14–18 mm. diam.) which are c. 75 % 4-merous and c. 25 % 5-merous, and the more numerous carpels (20–50). Fl. 6–9. Wood-margins and clearings, heaths, hedge-banks, etc., especially on light acid soils, to 1350 ft.; widespread but local, and rare in N. Scotland; Ireland. Hybrids of **6 & 8, 6 & 7** and **7 & 8** are usually sterile.

8. P. reptans L.   Creeping Cinquefoil.

Perennial herb with stock bearing a persistent terminal rosette of lvs. Basal lvs long-stalked, mostly palmate with 5(–7) lflets; lflets 0·5–3 cm., obovate, toothed in the upper part or all round, with 6–10 teeth on each side, hairy or almost glabrous above and beneath. Stem-lvs hardly different but short-stalked. Fl.-stems 30–100 cm. or more, prostrate, quickly rooting at the nodes and forming new plants. Fls 17–25 mm. diam., solitary, long-stalked, 5-merous. Petals yellow, obovate, emarginate, up to twice as long as sepals. Stamens c. 20. Carpels 60–120; receptacle hairy. Fl. 6–9. Hedge-banks, waste places and sometimes grassland, mainly on basic and neutral soils to 1400 ft. Common throughout the British Is.

4. FRAGARIA L.   Strawberry.

Perennial stoloniferous herbs. Lvs ternate, stipulate, in a basal rosette. Fls 5-merous, in few-fld cymes on axillary scapes. Epicalyx present. Stamens numerous (c. 20), inserted on the outer edge of the narrow nectar-secreting ring. Receptacle glabrous. Carpels numerous (10–80), free, superior; style lateral, not persistent in fr. Fr. a group of achenes on the surface of a much enlarged, fleshy, juicy, brightly coloured receptacle.

1   Lflets hairy above, the terminal one cuneate at base; fls 12–20(–25) mm. diam.; fr. 2 cm. diam. or less, with projecting achenes.   *2*

    Lflets usually glabrous above, the terminal one rounded at base; fls 20–35 mm. diam.; fr. c. 3 cm., with the achenes sunk in the flesh. Introduced and often ± naturalized. Garden Strawberry.

                                             *F. × ananassa Duchesne

2   Fl.-stalks (at least the upper) with appressed hairs; lateral lflets ± sessile; fls 12–18 mm. diam.; fr. red, with achenes all over.

                                                   1. vesca

    Fl.-stalks with spreading hairs; lateral lflets stalked; fls 15–25 mm. diam.; fr. purplish-red or partly greenish, without achenes at base. Introduced and perhaps naturalized in a few places. Hautbois Strawberry.   **F. moschata* Duchesne

1. F. vesca L.   **Wild Strawberry.**

Stock rather thick and woody, producing very long arching runners which root at the nodes and form fresh plants. Lflets 1–6 cm., ± ovate, bright green and somewhat hairy above, pale and glaucous beneath and clothed with silky appressed hairs, coarsely toothed, the terminal tooth not or scarcely shorter than its neighbours; lateral lflets ± sessile, terminal sessile or shortly stalked, cuneate at base; lf-stalks long, with spreading white hairs. Fl.-stems 5–30 cm., ± erect, with spreading hairs; stem-lvs 0 but lowest bracts usually lf-like; fl.-stalks, at least the upper, with ± appressed hairs. Fls 12–18 mm. diam. Calyx spreading or reflexed in fr.; sepals ovate, acuminate; epicalyx about as long as calyx. Petals white, nearly contiguous to overlapping. Fr.-receptacle 1–2 cm. ovoid, red (rarely white) covered all over with achenes which project from the surface. Fl. 4–7. Woods and scrub on base-rich soil and in basic grassland; sometimes locally dominant in wood-clearings on calcareous soils; to 2400 ft. Common throughout the British Is.

5. Geum L.

Perennial herbs. Lvs unequally pinnate, stipulate. Fls 5-merous. Epicalyx present. Stamens 20 or more, inserted at the outer edge of the narrow nectar-secreting ring. Carpels 20 or more, free, superior; ovule 1; style terminal, enlarged and persistent as an awn on the fr. Fr. a group of achenes on a dry receptacle.

Fls erect; petals 5–9 mm., yellow, spreading, neither emarginate nor
   clawed.   **1. urbanum**
Fls nodding; petals 10–15 mm., reddish, erect, emarginate, clawed.
                                                       2. rivale

1. G. urbanum L.   Herb Bennet, Wood Avens.

Stems 20–60 cm., ± erect, ± pubescent; rhizome short, thick. Basal
lvs pinnate, with 2–3 pairs of unequal lateral lflets 5–10 mm. long and a
large suborbicular lobed terminal lflet 5–8 cm., or with the upper pair
of laterals also large; lflets all crenate or dentate; upper stem-lvs usually
simple; stipules 1–3 cm., lf-like. Fls erect, few, on long stalks in very
open cymes. Sepals triangular-lanceolate, green. Petals 5–9 mm.,
yellow, spreading, about equalling the sepals, neither clawed nor
emarginate. Carpels hairy, remaining as a sessile head in fr.; awn
purplish, jointed near the tip, the lower part hooked and persistent,
glabrous throughout. Fl. 6–8. Woods, scrub, hedge-banks and shady
places on good damp soils. Common throughout the British Is.

2. G. rivale L.   Water Avens

Stem 20–60 cm., pubescent; rihzome short, thick. Basal lvs pinnate,
with 3–6 pairs of unequal lateral lflets 2–20 mm. long and a large suborbicular lobed terminal lflet 2–5 cm.; lflets all ± dentate; stem-lvs few,
small, simple or with a few very small lateral lflets; stipules c. 5 mm.,
green but scarcely lf-like. Fls nodding, few, in a narrow cyme. Sepals
triangular-lanceolate, purple. Petals 1–1·5 cm., dull orange-pink, erect,
long-clawed, retuse or emarginate, about equalling the sepals. Carpels
hairy, head becoming stalked in fr.; awn jointed rather above the middle,
the lower part glabrous hooked and persistent, the upper joint hairy.
Fl. 5–9. Marshes, streamsides, wet rock-ledges, etc., usually in shade;
reaching 3200 ft. Widespread through most of the British Is and common in the north and west but local in S. England.

   The hybrid between *G. urbanum* and *G. rivale* (*G.* × *intermedium* Ehrh.) is
commonly found where the parents grow together. It is fertile and forms
hybrid swarms with much variation.

6. AGRIMONIA L.

Perennial herbs. Lvs unequally pinnate, stipulate. Fls in terminal spike-
like racemes, 5-merous, hermaphrodite. Epicalyx 0. Stamens 10–20.
Carpels 2, free, superior, at the base of a deeply concave receptacle
which is covered with small spines and becomes hard in fr.; style

terminal; ovule 1. Fr. of 2 achenes enclosed in the persistent receptacle and shed with it.

Lvs not (rarely slightly) glandular beneath; fr.-receptacle obconic, deeply grooved throughout; basal spines spreading horizontally.

1. eupatoria

Lvs with numerous glands beneath; fr.-receptacle bell-shaped, without grooves at the base; basal spines deflexed.   **2. procera**

1. A. eupatoria L.   Common Agrimony.

Stems 30–60 cm., erect, often reddish, ± shaggy, not or sparingly glandular, densely lfy below. Lower lvs pinnate, with 3–6 pairs of main lflets becoming larger upwards, and 2–3 pairs of smaller lflets between successive main pairs; largest lflets 2–6 cm., elliptic, deeply and coarsely toothed, shaggy and greyish, but not glandular, beneath; upper lvs with few lflets; stipules lf-like. Fls c. 5–8 mm. diam., numerous, yellow. Fr.-receptacle obconic, deeply grooved almost throughout its length, covered above with hooked spines; lowest spines ascending or spreading horizontally. Fl. 6–8. Hedge-banks, roadsides, edges of fields, etc.; to 1600 ft. Common throughout the British Is except N. Scotland, where it is rare (absent from Orkney and Shetland).

2. A. procera Wallr. (*A. odorata* auct.)   Fragrant Agrimony.

Like 1 but more robust (often 1 m. high) and sweet smelling; lvs larger; lflets larger and relatively narrower, with narrower more acute teeth, greener and less hairy beneath and always with numerous small sessile shining glands beneath; fr.-receptacle bell-shaped, more shallowly grooved, the grooves ceasing well above the base (or almost without grooves); lowest spines deflexed. Fl. 6–8. Grows in similar places to *A. eupatoria* but usually absent from calcareous soils. From Kintyre and Kincardine southwards; local in Ireland; Channel Is.

7. ALCHEMILLA L.   Lady's-Mantle.

Perennial herbs. Lvs palmate or palmately lobed, stipulate. Fls 4-merous, small, green, in cymes. Epicalyx present. Petals 0. Stamens 4(–5), alternate with the sepals, inserted on the outer margin of the nectar-secreting disk which forms the rim of the deeply concave receptacle. Carpel 1, with basal style and 1 ovule, borne at the base of the receptacle, which is small, dry and non-spiny in fr.; fr. a single achene enclosed in the receptacle and shed with it.

The following key includes only the very distinct *A. alpina* and the three species of *A. vulgaris* agg. which grow in lowland Britain. For the remaining species (mostly mountain plants) the larger *Flora* should be consulted

1 Lvs divided to the base or nearly so into oblong-lanceolate seg-
       ments, green above, silvery-silky beneath.   **1. alpina**
   Lvs divided to less than halfway, green on both sides.

                                                                 (**vulgaris** agg.)   2
2 Stem (at least the lower part) and lf-stalks ± densely covered with
       spreading hairs.   3
   Stem and lf-stalks with appressed hairs or glabrous.   **4. glabra**
3 Lvs glabrous above.   **3. xanthochlora**
   Lvs hairy above.   **2. vestita**

1. A. alpina L.   Alpine Lady's-Mantle.
Stock somewhat woody, branched, shortly creeping; lvs mostly basal; stems 10–20 cm., ascending. Basal lvs 2·5–3·5 cm. diam., orbicular or reniform in outline, palmately divided almost or quite to the base into 5–7 oblanceolate-oblong segments, green and glabrous above, densely silvery-silky beneath; segments 1–2 × 3–6 mm., sharply toothed at the extreme tip; stalk long; stipules brown. Stem-lvs few, small, short-stalked. Stems, fl.-stalks, receptacle and sepals appressed-silky. Fls c. 3 mm. diam., in rather dense clusters forming a terminal cyme. Fl. 6–8. Mountain grassland, rock-crevices, screes and mountain-tops, from sea-level (in Skye) to over 4000 ft. N.W. England (local) and from Stirling and Arran northwards (widespread and locally abundant); very rare in Ireland.

2. A. vestita (Buser) Raunk. (*A. filicaulis* Buser ssp. *vestita* (Buser)
                                                              M. E. Bradshaw).
Stems, lf- and fl.-stalks and receptacle rather densely covered with spreading hairs. Lvs usually 7-lobed, ± reniform in outline with a wide basal sinus, hairy all over both surfaces; lobes not much broader than long, not overlapping, toothed all round, the teeth acute, somewhat curved towards the tip of the lobe, all ± equal; stipules purplish. Fls 3–4 mm. diam., in a compound terminal cyme. Fl. 6–9. Damp grass-land, open woods, rock-ledges, etc., especially on basic and neutral soils. Throughout most of the British Is, but rare in S.E. England and absent from Channel Is.

3. A. xanthochlora Rothm.
Robust. Stems and lf-stalks densely clothed with spreading hairs; infl. and receptacle glabrous or nearly so. Lvs usually 9-lobed, reniform in outline with a wide basal sinus, glabrous above (rarely with a few hairs

on the folds), hairy on the veins and sometimes thinly on the surface beneath; lobes rounded, rather broad, not overlapping, toothed all round, the teeth acute, somewhat curved towards the tip of the lobe, all ± equal; stipules brown. Fls c. 3 mm. diam., in dense clusters in a compound terminal cyme. Fl. 6–9. Grassland, etc. at low altitudes. Throughout most of the British Is and generally common except in S.E. England and S. Ireland.

4. A. glabra Neygenfind

Stems sparsely and closely appressed-hairy on the lowest 1–2 internodes, or almost glabrous; upper part of plant glabrous; lf-stalks appressed-hairy or glabrous. Lvs 7–9-lobed, reniform in outline with a wide basal sinus, glabrous except for the apical portion of the veins beneath; lobes rounded or straight-sided, rather broad, not or rarely overlapping, toothed all round, the teeth somewhat curved towards the tip of the lobe, those in the middle of the lobe larger and broader than the upper and lower ones, and the apical tooth conspicuously narrower and usually shorter than its neighbours. Fls 3–4 mm. diam., in loose clusters (or scarcely clustered) in compound terminal cymes. Fl. 6–9. Grassland, open woods and rock-ledges, to nearly 4000 ft. Almost throughout the British Is and the commonest sp. on mountains but rare in S. England and absent from S.E. England and S.E. Ireland.

8. APHANES L.   Parsley Piert.

Like *Alchemilla* but annuals with fls in dense lf-opposed cluster-like cymes and 1(–2) stamen opposite a sepal and inserted on the inner margin of the nectar-secreting disk.

Lobes of stipules surrounding infl. triangular-ovate; fr. 2·2–2·6 mm., bottle-shaped.   **1. arvensis**
Lobes of stipules surrounding infl. oblong; fr. 1·4–1·8 mm. calyx-teeth converging.   **2. microcarpa**

1. A. arvensis L.

Small rather inconspicuous annual 2–20 cm., usually much branched from the base, hairy, pale greyish-green; branches decumbent or ascending. Lvs 2–10 mm., fan-shaped, short-stalked, cut into 3 segments each divided at the tip into 3–5 oblong lobes; stipules united into a lf-like cup with 5–7 blunt lobes. Fls in sessile clusters which are half-enclosed by the stipular cup; lobes of the enclosing cup triangular-ovate, little longer than broad, c. half as long as the entire portion. Epicalyx-segments very small. Fr. (including sepals) 2·2–2·6 mm., the sepals ascending so that the whole appears bottle-shaped. Fl. 4–10. Arable land and bare places in grassland, mainly on dry soils (both acid and basic). Common throughout the British Is.

2. A. microcarpa (Boiss. & Reuter) Rothm.

Like **1** but usually more slender and not greyish. Lobes of stipular cup surrounding the infl. oblong, c. twice as long as broad, nearly as long as the entire portion. Fr. (including sepals) 1·4–1·8 mm., the sepals converging so that the whole appears ovoid. Arable land and bare places in grassland, chiefly on acid sandy soils; much less common than *A. arvensis.*

9. SANGUISORBA L.

Erect perennial herbs. Lvs pinnate, stipulate. Fls in dense terminal spikes or heads, hermaphrodite. Epicalyx 0. Sepals 4, coloured. Petals 0. Stamens 4, perigynous. Carpel 1, superior, at the base of the deeply concave receptacle; style terminal, stigma simple; ovule 1. Receptacle becoming corky in fr., enclosing the single achene; not spiny. Nectar secreted by a ring round the style (insect-pollinated).

1. S. officinalis L. Great Burnet.

Stock thick. Stems erect, 30–100 cm., branched above, glabrous. Basal and lower stem-lvs pinnate with 3–7 pairs of lflets which increase in size upwards; larger lflets 2–4 cm., ovate, stalked, mostly cordate at base, blunt, toothed; upper stem-lvs small and few. Fls dull crimson, in oblong heads 1–2 cm. Stamens equalling or somewhat longer than sepals. Fr.-receptacle 4-winged, smooth between the wings. Fl. 6–9. Damp grassland, to 1500 ft. Locally common from Ayr and Berwick southwards, but not in S.E. England; local in N. and W. Ireland.

10. POTERIUM L.

Like *Sanguisorba* but fls monoecious or polygamous; sepals green; stamens numerous, exserted; carpels 2(–3), free; stigmas feathery; no nectar (wind-pollinated).

Fr. receptacle with 4 longitudinal entire ridges, the surface between them with a fine raised network. **1. sanguisorba**
Fr. receptacle with 4 longitudinal wavy wings, the surface between them strongly and irregularly pitted and ridged, the ridges toothed. **2. polygamum**

1. P. sanguisorba L.   Salad Burnet.

Stock somewhat woody. Stems erect, 15–40(–60) cm., glabrous except for some long flexuous hairs below; smelling of cucumber when crushed. Basal lvs pinnate with 4–12 pairs of lflets which increase in size upwards; larger lflets 0·5–2 cm., orbicular or shortly oval, short-stalked, rounded at base, ± truncate at tip, deeply toothed with the terminal tooth smaller than the lateral; stem-lvs few or 0, the upper small. Fls green, often purple-tinged, in globose heads 7–12 mm. diam., the lower fls male, the middle hermaphrodite, the upper female, Styles purple-red. Fr.-receptacle c. 4 mm., ovoid, 4-angled with entire ridges down the angles, the surface between them with a fine raised entire network, not pitted. Fl. 5–8. Calcareous grassland, to 1650 ft. Widespread and common in England and Wales; very local in Scotland northwards to Dunbarton and Angus; widespread in C. and S.E. Ireland, rare elsewhwere.

*2. P. polygamum Waldst. & Kit.

Like *P. sanguisorba* but taller (30–80 cm.) and more robust; several stem-lvs with narrow lflets; fr.-receptacle c. 6 mm., ovoid, 4-angled with wavy wings down the angles, the surface between them with coarse raised toothed ridges. Fl. 6–8. Introduced and formerly grown for fodder; now naturalized on field-borders, etc., especially in S. England and Wales.

11. Rosa L.

Shrubs, sometimes trailing or scrambling, usually deciduous. Lvs pinnate, stipulate. Stems usually prickly. Fls terminal, solitary or in corymbs, hermaphrodite, 5-merous. Stamens numerous, perigynous, round the disk-bearing rim of the deeply concave urn-shaped receptacle. Carpels numerous, free, superior, borne on the inner wall of the receptacle; style terminal; ovule 1. Fr. consisting of numerous achenes enclosed in the coloured fleshy receptacle ('hip').

No attempt is made in this book to describe the several species and numerous varieties included in the section *Caninae*. It should usually be possible to place a plant in its appropriate 'aggregate species', though some intermediate forms may be encountered. A fuller treatment will be found in the larger *Flora*.

1   Styles united into a column which equals at least the shorter stamens;
    shrub with weak trailing stems.   **1. arvensis**
  Styles free or united into a short column; stems not trailing.   *2*

2   Fls always solitary, without bracts; lflets small, 3–5 pairs; stems densely prickly and bristly, not tomentose; sepals entire.
                                    2. pimpinellifolia
     Fls 1 or more, bracts resembling the stipules; lflets 2–3(–4) pairs; stems not bristly or with a few bristles or, if with numerous bristles, then tomentose.   *3*

3   Stems prickly and bristly, tomentose; sepals entire; fls 6–8 cm. diam.; fr. large (2–2·5 cm.), red, crowned by the erect sepals. An introduced rose with rugose lflets, often used as a stock and locally naturalized.   ***R. rugosa** Thunb.
     Stems with few or no bristles, not tomentose; outer sepals pinnate-lobed; fls 5 cm. diam. or less; fr. usually less than 2 cm.   *4*

4   Styles united into a column at fl., becoming free in fr.; disk conical, prominent.   **3. stylosa**
     Styles free throughout; disk flat or nearly so.   *5*

5   Lvs ± densely covered over the whole lower surface with conspicuous brownish sticky fruity-scented glands, glabrous or somewhat pubescent but not tomentose; prickles hooked.
                                  6. rubiginosa agg.
     Lvs not covered beneath with conspicuous fruity-scented glands (though small glands may occur on the veins or scattered over the whole surface, which is then usually densely tomentose).   *6*

6   Prickles hooked or strongly curved; lvs glabrous or pubescent, simply or doubly serrate.   **4. canina** agg.
     Prickles straight or slightly curved; lvs always pubescent, usually very tomentose, always doubly serrate.   **5. villosa** agg.

1. R. arvensis Hudson   Field Rose.

Deciduous shrub, glabrous or nearly so, with weak trailing often purple-tinted stems, either decumbent and forming low bushes 50–100 cm. high, or climbing over other shrubs; rarely more erect, to 2 m. Prickles hooked, all ± equal. Lflets 2–3 pairs, 1–3·5 cm., ± ovate, usually simply serrate, glabrous on both sides or pubescent on the veins beneath, rather thin. Fls 1–6, white, 3–5 cm. diam. Sepals soon falling, less than 1 cm., ovate-acuminate, the 2 outer with very few pinnately-arranged lobes. Styles glabrous, united into a column equalling the stamens. Fr. small, red, smooth. Fl. 6–7. Woods, hedge-banks and scrub, to 1250 ft. Rather common in S. England and Wales but becoming very rare in Scotland northwards to Stirling; rather local in Ireland.

2. R. pimpinellifolia L. (*R. spinosissima* L.)   Burnet Rose.

Low erect deciduous shrub 10–40(–100) cm., spreading by suckers and forming large patches; prickles numerous, straight, mixed with numerous stiff bristles and passing into them. Lflets 3–5 pairs, small, 0·5–1·5(–2) cm., suborbicular or oval, blunt, usually simply serrate and ± glabrous on both sides. Fl. solitary, cream-white, rarely pink, 2–4 cm. diam. Sepals entire. Styles woolly. Fr. 1–1·5 cm., subglobose, purplish-black. Fl. 5–7. Dunes, sandy heaths, limestone pavement and scree, etc., especially near the sea but reaching 1700 ft. Northwards to Caithness and Outer Hebrides, but rather local; throughout Ireland.

3. R. stylosa Desv.

Shrub 1–4 m.; stems arching; prickles hooked, some with very stout bases. Lflets 2–3 pairs, usually simply serrate, usually pubescent beneath, eglandular. Fls 3–5 cm. diam., 1–8 or more, white or pale pink. Sepals reflexed after fl., falling before fr. is ripe. Styles at first fused into a glabrous exserted column shorter than the stamens, becoming free later; stigmas in an ovoid head; disk conical, prominent. Fr. red, smooth. Fl. 6–7. Hedges, etc.; local, from N. Wales, Worcester, Lancaster and Suffolk southwards and in the southern half of Ireland; Channel Is.

4. R. canina agg.   Dog Rose.

Shrubs 1–3 m.; stems arching; prickles curved or hooked, equal, usually stout. Lflets 2–3 pairs, glabrous or pubescent (but not tomentose on both sides); usually eglandular or with a few glands on the main veins beneath (rarely with numerous glands which are not strongly scented). Fls 1–4 or more, 3–5 cm. diam., pink or white. Styles free, glabrous to villous. Fr. red or scarlet, usually smooth. Fl. 5–7. Woods, hedges. scrub, etc., throughout the British Is. Very variable.

5. R. villosa agg.   Downy Rose.

Shrubs 40 cm.–2 m.; stems erect or arching; prickles straight or curved, usually equal, rather slender. Lflets 2–3(–4) pairs, doubly glandular-serrate, usually densely tomentose on both sides and usually glandular all over beneath, the glands small and ± scentless. Fls 1–4 or more, 3–5 cm. diam., commonly deep pink but sometimes paler or white. Styles free. Disk usually broad and conspicuous. Fr. red, usually glandular-hispid. Fl. 6–7. Woods, hedges, scrub, etc.; throughout the British Is.

6. R. rubiginosa agg.   Sweet-briar.

Shrubs 1–2 m.; stems erect or arching; prickles usually hooked, equal or unequal, sometimes mixed with a few stout bristles mainly on the fl. branches. Lflets 2–3(–4) pairs, usually rounded at the base, doubly glandular-serrate, glabrous or somewhat pubescent (never tomentose), ± thickly clothed beneath with sweet-scented sticky brownish glands. Fls 1–4, usually pink. Styles free. Fr. scarlet, smooth or glandular-hispid. Fl. 6–7. Woods, hedges, scrub, etc., mainly on calcareous soils; widespread and locally common in England and Wales, rarer in Scotland; local in Ireland. There are two common and readily distinguishable species in this group: **R. rubiginosa** L., with erect stems, sepals erect or spreading, persistent at least until the fr. reddens, and with hairy styles; and **R. micrantha** Borrer ex Sm., with arching stems, sepals reflexed and falling early, and glabrous styles.

12. PRUNUS L.

Trees or shrubs with simple stipulate lvs. Fls hermaphrodite. Petals and sepals 5. Stamens usually 20, perigynous, borne on the rim of the concave nectar-secreting receptacle. Carpel 1, superior; style terminal; ovules 2. Fr. a 1-seeded drupe.

1	Fls in clusters or umbels or solitary.	2
	Fls in racemes.	6
2	Fls 1–3 from axillary buds; bracts 0 or small and few.	3
	Fls in few-fld umbels; infl. surrounded at its base by the persistent bud-scales which form an involucre.	5
3	Twigs and lvs dull; plant usually ± hairy on twigs, lvs and fl.-stalks; lvs nearly always broadest above the middle.	4
	Twigs and upper surface of lvs somewhat glossy; plant glabrous except for lf-stalks and lower part of midrib; lvs 3–7 cm., broadest above or below the middle. Deciduous shrub, rarely thorny, with usually solitary white fls (petals 7–11 mm.) and a globose yellowish or reddish fr., 2–2·5 cm. Introduced. Cherry Plum.	

P. cerasifera Ehrh.

4	Very thorny shrub; fls usually before the lvs, mostly solitary, rarely 2; petals 5–8 mm.; fr. 10–15 mm.	**1. spinosa**
	Unarmed or somewhat thorny shrub or small tree; fls with the lvs, 1–3; petals 7 mm. or more; fr. usually 2 cm. or more.	

2. domestica

5	Lvs light dull green, somewhat pubescent beneath; most of the infls without lf-like scales; petals obovate; usually a tree.	**3. avium**
	Lvs dark green, soon nearly glabrous beneath; infl. with the inner scales lf-like; petals orbicular; usually a shrub, suckering freely. Introduced. Hedges, etc., very local. Sour Cherry.	

P. cerasus L.

6	Lvs 5–10 cm., deciduous, rather thin, closely and finely toothed; infl.-stalks with 1 or 2 lvs near base.	**4. padus**

Lvs 5–18 cm., evergreen, glossy above, thick and leathery, distantly
    toothed or subentire; infl.-stalks lfless; racemes erect; petals
    white, 4 mm.; fr. c. 8 mm., blackish. Introduced. Cherry-Laurel.
                                                    P. laurocerasus L.

1. P. spinosa L.   Blackthorn, Sloe.

Rigid, much-branched, deciduous shrub, 1–4 m., suckering and often
forming dense thickets. Twigs dull, usually becoming blackish in the 1st
winter and losing their initial pubescence after about 1 year; short lateral
shoots numerous, becoming thorns. Buds hairy. Lvs 2–4 cm., stalked,
cuneate at base, broadest above the middle, crenate-serrate, ± pubescent
beneath, dull above. Fls usually before the lvs, solitary (–2); fl.-stalks
usually glabrous. Petals 5–8 mm., pure white. Fr. 10–15 mm., globose,
blue-black with a whitish bloom, very astringent. Fl. 3–5. Scrub, woods
and hedges on a great variety of soils; to 1360 ft. Throughout Great
Britain and Ireland.

*2. P. domestica L.   Plum, Bullace, Greengage, etc.

Deciduous shrub or small tree, 2–6(–12) m., often suckering. Twigs dull,
becoming grey or brown in the 1st winter; thorns few or 0. Lvs 4–
10 cm., stalked, obovate or elliptic, cuneate at base, crenate-serrate,
± pubescent beneath, dull above. Fls with the lvs, 1–3 together. Petals
7–12(–15) mm., white. Fr. variable in size, shape and colour. Fl. 4–5.
Introduced. Several different cultivated forms may be found in
hedges, etc.

3. P. avium (L.) L.   Gean, Wild Cherry.

Deciduous tree 5–25 m., suckering freely. Bark smooth, reddish-brown,
peeling off in thin strips. Branches spreading or ascending. Lvs 6–
15 cm., obovate-elliptic, acuminate, light dull green and glabrous above,
sparingly but persistently appressed-pubescent beneath, rather thin,
drooping; stalk with 2 large glands near the top. Fls 2–6, cup-shaped,
in ± sessile umbels; bud-scales greenish, usually not lf-like, persisting as
an involucre below the infl. Receptacle constricted above. Petals
8–15 mm., white, obovate, gradually narrowed at the base. Fr. c. 1 cm.,
subglobose, bright or dark red, sweet or bitter. Fl. 4–5. Woods and
hedges on better soils. Rather common in England, Wales and Ireland,
less so in Scotland but reaching Caithness.

4. P. padus L.   Bird-Cherry.

Deciduous tree 3–15 m. Bark brown, strong-smelling, peeling. Twigs
brown or grey. Lvs 5–10 cm., elliptic or obovate, acuminate, rounded

or cordate at base, closely and sharply serrate, glabrous except sometimes for tufts in the vein-axils beneath; stalk with a gland on each side
at the top. Fls 10–40, in long, loose, ascending to drooping racemes,
7–15 cm.; infl.-stalk with 1–2 lvs. Petals 4–6(–10) mm., white, irregularly
toothed. Fr. 6–8 mm., ovoid, black, astringent. Fl. 5. Woods, etc., to
2000 ft.; widespread and rather common in Scotland, N. England and
Wales and reaching Gloucester, Derby and E. Yorks.

13. COTONEASTER Medicus

Shrubs, rarely small trees, not thorny. Lvs simple, entire, short-stalked,
stipulate. Sepals 5, short, ± triangular. Petals 5. Stamens c. 20, perigynous. Carpels 2–5, free on the inner side, ½-inferior, the wall stony
in fr.; ovules 2 per carpel; style free. Fr. with mealy flesh and 2–5
stones.

1   Petals erect, pink; lvs 10 mm. or more, broadest near the middle.   2
     Petals spreading, white; lvs 5–8 mm., broadest near the tip, cuneate
       at base, blunt to retuse, glaucous and appressed-hairy beneath.
       Introduced. A low evergreen shrub to 1 m., with crimson fr. c.
       6 mm., naturalized in many places and especially on limestone
       near the sea.   **C. microphyllus* Wall. ex Lindley

2   Lvs 1·5–4 cm., grey-tomentose beneath, ovate to suborbicular,
       rounded at base; fr. c. 6 mm., red. A very rare deciduous shrub of
       limestone cliffs near the sea in N. Wales.   *C. integerrimus* Medicus
     Lvs 1–3 cm., sparsely hairy beneath, ovate, cuneate at base; fr. c.
       8 mm., scarlet. Introduced. A ± deciduous shrub, 1–4 m.,
       naturalized in many places.   **C. simonsii* Baker

14. CRATAEGUS L.   Hawthorn.

Deciduous trees or shrubs with long and short shoots, the latter usually
thorny. Lvs simple, lobed or toothed, stipulate. Fls in corymbs. Sepals
short, ± triangular. Stamens 5–25, perigynous, on the outer edge of the
nectar-secreting ring. Carpels 1–5, free at the apex, united at least at the
base of the inner side, so that the ovary is ± inferior; wall stony in fr.;
ovules 2 per carpel, the upper sterile; styles free. Fr. with mealy flesh
and 1–2(–3) stones.

Lvs of short-shoots lobed less than half-way to the midrib, the lobes
    broader than long and usually rounded; styles usually 2 (1 or 3 in
       some fls).   **1. laevigata**
Lvs of short-shoots lobed more than half-way to the midrib, the lobes
    longer than broad, ± triangular; style usually 1.   **2. monogyna**

1. C. laevigata (Poiret) DC. (*C. oxyacanthoides* Thuill.)

Thorny much-branched shrub or small tree, 2–10 m. Adult twigs glabrous. Lvs of short-shoots 1·5–5 cm., obovate in outline, glabrous when mature, 3–5-lobed, the lobes shallow (less than half-way to the midrib), usually rounded, broader than long, serrate nearly all round; those of the long shoots usually more deeply lobed, with conspicuous lf-like stipules. Fls usually 10 or fewer. Petals 5–8 mm., white. Anthers pink or purple. Styles mostly 2 (often 1 or 3 on some fls). Fr. usually 8–10 mm., deep red, usually with 2 stones. Fl. 5–6, about a week earlier than *C. monogyna*. Woods, less frequently scrub or hedges; much less common than *C. monogyna* and more shade-tolerant; mostly on clay or loam. In England local and mainly in the east but extending to Somerset, Carmarthen, Denbigh, and Cumberland; rare in N. England and very rare in Scotland; scattered over Ireland.

2. C. monogyna Jacq.

Habit of *C. laevigata*. Lvs of short-shoots 1·5–3·5 cm., ovate or obovate in outline, glabrous when mature except for tufts of hairs in the vein-axils beneath, 3–7-lobed, the lobes deep (reaching more than half-way to the midrib), straight-sided and tapering to a ± acute apex, longer than broad, serrate only at the apex; those of long shoots more deeply lobed, with conspicuous lf-like stipules. Fls often up to 16 or more. Petals 4–6 mm., white or pink. Anthers pink or purple. Style 1 (or 2 in a few fls.). Fr usually 8–10 mm., deep red, stone usually 1. Fl. 5–6. Scrub, woods and hedges to 1800 ft.; the commonest shrub of most soils, rare only on wet peat and acid sands. Throughout the British Is.

15. SORBUS L.

Deciduous trees or shrubs, not thorny. Lvs pinnate or simple and lobed or toothed, stipulate. Fls in compound corymbs. Sepals ± triangular. Petals white. Stamens 15–25, perigynous. Carpels 2–5, united at least to the middle, inferior or ½-inferior, walls cartilaginous in fr.; ovules 2 per carpel, with no false septum; styles free or joined below. Fr. a pome, variously coloured, with 1–2 seeds per cell.

Sorbus includes the whitebeams, a group of closely related forms which are difficult to separate. The following key and descriptions deal only with the few most widespread species. For a more detailed treatment the larger *Flora* should be consulted.

1   Lvs pinnate with 4 or more pairs of lflets, the terminal lflet ± equalling the lateral ones.   **1. aucuparia**

Lvs simple or with up to 3 pairs of free lflets at the base and with the
terminal part several times as large. 2

2 Lvs 7–11 cm., mostly with 1–3 pairs of free lflets at the base; styles
2–3; fr. subglobose, scarlet. A rare hybrid tree, frequently planted
and sometimes occurring naturally. *S. aria* × *aucuparia*
Lvs without free lflets. 3

3 Lvs green on both sides, subglabrous beneath (except when very
young), deeply lobed; lobes acuminate; fr. brown. **5. torminalis**
Lvs persistently grey- or white-tomentose beneath. 4

4 Lvs yellowish-grey-tomentose beneath (whitish-grey in allied spp.),
with broad acute ascending lobes, the deepest reaching $\frac{1}{4}-\frac{1}{3}(-\frac{1}{2})$ of
the way to the midrib. **2. intermedia**
Lvs white or grey beneath, not lobed or with lobes not reaching
more than $\frac{1}{4}$ of the way to the midrib. 5

5 Lvs greenish-grey-tomentose beneath, ± ovate, acute, rounded at
base, with broadly triangular acute or acuminate lobes or at least
with very prominent straight teeth terminating the main veins; fr.
9–15 mm., orange or brown. Woods in S.W. England and S.E.
Ireland; occasionally naturalized elsewhere.
S. latifolia (Lam.) Pers., *sensu lato*
Lvs whitish-tomentose beneath; fr. red. 6

6 Lvs ovate to oval, blunt or acute, rounded or cuneate at base,
entire for the basal $\frac{1}{3}$ or less of the otherwise toothed margin; fr.
usually longer than broad. **3. aria**
Lvs obovate or oblanceolate, usually rounded at apex and cuneate
at base, not toothed in the basal $\frac{1}{3}-\frac{1}{2}$; fr. broader than long.
4. rupicola

1. S. aucuparia L. Rowan, Mountain Ash.

Slender tree to 15(–20) m. with narrow crown and ± ascending branches.
Bark greyish, smooth, shining. Twigs becoming glabrous and greyish-
brown. Buds 10–15 mm., ovoid, dark brown. Lvs 10–25 cm., pinnate;
lflets usually 6–7 pairs (the terminal lflet never larger than the laterals),
3–6 cm., elliptic-oblong, ± acute, serrate, dark green and glabrous above,
subglaucous and initially pubescent beneath. Petals c. 3·5 mm. Styles
3–4. Fr. 6–9 mm., subglobose, scarlet. Fl. 5–6. Woods, scrub, mountain rocks, etc., to 3200 ft., mainly on light soils; throughout the British
Is, but rare in much of lowland England.

*2. S. intermedia (Ehrh.) Pers.

Tree to 10 m. with spreading branches and broad crown. Twigs stout.
Lvs 7–12 cm., elliptic or oblong-elliptic, rounded or broadly cuneate at
base, lobed, with broad acute ascending lobes, the deepest reaching $\frac{1}{4}-\frac{1}{3}$

or more of the way to the midrib, serrate; dark yellowish-green and glabrous above, yellowish-grey-tomentose beneath. Petals c. 6 mm. Fr. 12–15 mm., oblong, much longer than broad, scarlet. Fl. 5. Introduced. Rather commonly planted and naturalized in a few places.

Allied native plants with lvs whitish-grey beneath occur very locally in S.W. England, Wales, Kerry and Arran.

3. S. aria (L.) Crantz   White Beam.

Tree to 15(–25) m., with wide dense crown, or a large shrub. Bark dark grey, shallowly fissured. Twigs becoming chestnut brown, then dark grey. Buds up to 2 cm., ovoid, greenish. Lvs 5–12 cm., variable in shape from ovate to obovate, blunt or acute, rounded or cuneate at base, the sides mostly gradually curved (i.e. not triangular at apex or base), lobes broader than long, not reaching more than $\frac{1}{6}$ of the way to the midrib, or 0; margin toothed except in the basal $\frac{1}{3}$ or less, the teeth usually curved towards the apex; surface dull yellow-green above, densely and evenly pure white-tomentose beneath. Petals c. 6 mm., oval. Styles 2. Fr. 8–15 mm., usually longer than broad, scarlet. Fl. 5–6. Woods and scrub, usually on chalk and limestone, local elsewhere.

4. S. rupicola (Syme) T. Hedl.)   Rock White Beam.

Shrub to c. 2 m., rarely a small tree. Lvs usually 8–14·5 cm., obovate or oblanceolate, usually $1\frac{1}{2}$–2 times as long as broad, rounded at apex, usually tapering uniformly to a cuneate base from the middle of the lf or above; coarsely and unequally serrate except in the basal $\frac{1}{3}$–$\frac{1}{2}$, the teeth mostly curved towards the apex; dark green above, rather thickly white-tomentose beneath. Petals c. 7 mm. Styles 2. Fr. 12–15 mm., broader than long, carmine (green on one side, reddish on the other, when ripening). Fl. 5–6. On limestone or other basic rocks and crags, to 1500 ft. Local northwards from Stafford and Derby and in Wales, S. Devon, N. and W. Ireland.

Several other spp. allied to *S. aria* and *S. rupicola* occur very locally in W. England and Ireland.

5. S. torminalis (L.) Crantz   Wild Service Tree.

Tree to 25 m. with wide crown and spreading branches. Bark dark grey, shallowly fissured. Twigs becoming glabrous and dark brown. Buds greenish. Lvs 7–10 cm., ovate in outline, acuminate, usually rounded or cordate at base, with deep triangular-acuminate lobes, the lowest pair spreading and reaching half-way or more to the midrib, the remainder ascending, all finely toothed; lf green on both sides and becoming sub-glabrous. Petals c. 6 mm. Fr. 12–16 mm., considerably longer than broad, brown. Fl. 5–6. Woods, usually on clay, sometimes on lime-stone, local. England, from Westmorland and Lincoln southwards.

16. PYRUS L.

Deciduous trees or shrubs, sometimes thorny. Lvs simple. Fls in short umbel-like simple corymbs. Stamens 20–30. Carpels 2–5, completely united with each other and with the receptacle, their walls cartilaginous in fr.; ovules 2 in each carpel; styles free. Fr. a brownish or greenish pome, not deeply indented at base; flesh with numerous pockets of hard 'stone-cells'.

Petals 10–15 mm.; fr. 25 mm. or more, the calyx persistent; infl.-axis
  rarely longer than 1 cm. **1. pyraster**
Petals 8–10 mm.; fr. 12–18 mm., the calyx eventually falling; infl.-axis
  1 cm. or more. Very rare shrub. 3–4 m., with lvs 1–4 cm. Hedges
  near Plymouth. *P. cordata* Desv.

1. P. pyraster Burgsd. (*P. communis* auct.) Pear.

Tree or large shrub, 5–15 m., sometimes thorny. Bark deeply fissured, scaly. Twigs and buds glabrous or nearly so, yellowish-brown; short shoots numerous. Lf-blades 2·5–6 cm., ovate or roundish, rounded or subcordate at base, acute to cuspidate, finely crenate-toothed or almost entire, woolly when young, becoming glabrous or nearly so; stalk almost equalling blade. Infl.-axis to 1 cm.; fls c. 5–9. Petals 10–15 mm., white. Anthers purple. Fr. 2–4 cm., rounded or tapered at base, brownish, with many lenticels; calyx persistent. Fl. 4–5. Hedges, etc. Widespread (as isolated trees) in England and Wales; Channel Is.

17. MALUS Miller

Deciduous trees or shrubs, occasionally thorny. Lvs simple. Fls in short umbel-like corymbs. Stamens 15–50. Carpels 3–5, completely united with each other and with the receptacle, the walls cartilaginous in fr.; ovules 2 in each carpel; styles united below. Fr. a green to red pome, usually deeply indented at base; flesh without pockets of stone-cells.

1. M. sylvestris Miller Crab Apple.

Small tree with dense round crown, or shrub, 2–10 m., pubescent or nearly glabrous. Bark grey-brown, irregularly fissured and scaly. Twigs red-brown; short shoots numerous. Lf-blades 3–4 cm., ovate or oval, broadly wedge-shaped or rounded at base, crenate-toothed; mature lvs glabrous on both sides (tomentose at least beneath in cultivated apples). Infl. almost an umbel; fls c. 4–7. Sepals tomentose within. Petals 1·3–2·8 cm., white suffused with pink. Anthers yellow. Fr. 2 cm. or more, subglobose with a depression at each end, yellowish-green often speckled or flushed with red; calyx persistent. Fl. 5. Woods, hedges and scrub. Rather common throughout England, Wales and Ireland and reaching Ross but rare in C. and N. Scotland. Descendants of cultivated apples have fl.-stalk, receptacle and outside of calyx tomentose instead of glabrous or nearly so.

58. CRASSULACEAE

Herbs or undershrubs, mostly succulent. Lvs usually simple and entire; Stipules 0. Fls regular, hypogynous or slightly perigynous, hermaphrodite or rarely dioecious, 3–18-merous; usually 5-merous. Petals free or united. Stamens as many as petals and alternate with them or twice as many. Carpels as many as petals, free or united at the base, each with a nectariferous scale at the base; ovules usually numerous. Fr. a bunch of follicles or a capsule; seeds with no or little endosperm.

1 Petals united into a cylindrical tube for more than half their length; lvs peltate.   2. UMBILICUS
  Petals free or united only at their base; lvs not peltate.   2

2 Stamens as many as petals; a small decumbent or ascending annual with tiny opposite lvs and small solitary axillary whitish 3-merous fls. Very local on bare sandy or gravelly ground in S. and E. England and Channel Is.   *Crassula tillaea* Lester-Garland
  Stamens twice as many as petals; fls conspicuous.   3

3 Fls 8–18-merous; lvs (except on the fl.-stem) densely crowded into a basal rosette 5–14 cm. diam.; infl. branched, often a panicle. Introduced and often planted on roofs and walls. Houseleek.
                                               *Sempervivum tectorum L.
  Fls 4–5-merous; plants with elongated lfy stems.   1. SEDUM

1. SEDUM L.

Herbs or undershrubs, with usually spirally arranged succulent lvs. Infl. cymose. Sepals shortly united at the base. Petals free. Stamens twice as many as petals, free or adnate to petals below. Nectar-scales small, entire or slightly toothed. Carpels free or shortly united at the base; ovules usually many. Fr. a group of follicles.

1 Lvs broad, flat, usually toothed.   2
     Lvs terete, or, if flat above, linear, entire.   4

2 Stems erect, several from a central stock.   3
     Stems numerous, creeping and rooting; no central stock. A mat-
     forming perennial with ascending branches c. 15 cm., terminat-
     ing in dense flat-topped cymes of pink (or white) fls. Introduced
     and often escaping.   *S. spurium* Bieb.

3 Stock very thick and fleshy, scaly; fls 4-merous, dioecious,
     greenish.   **1. rosea**
     Stock neither fleshy nor scaly, short, bearing carrot-like tubers; fls
     5-merous, hermaphrodite, reddish-purple.   **2. telephium**

4 Commonly biennial 5–15 cm.; petals pink; lvs spirally arranged,
     glandular-pubescent. Streamsides, wet stony ground and base-
     rich flushes in the mountains, from Yorks and Lancashire to
     Inverness and Argyll. Hairy Stonecrop.   *S. villosum* L.
     Perennial; petals yellow or white (sometimes tinged pink); lvs
     glabrous or, if glandular-pubescent, opposite.   5

5 Petals white.   6
     Petals yellow.   8

6 Lvs spirally arranged, glabrous.   7
     Lvs 3–5 mm., mostly opposite, ± glandular-pubescent. Probably
     introduced and naturalized on old walls in S. England; perhaps
     native in Cork.   *S. dasyphyllum* L.

7 Lvs 3–5 mm., glaucous; infl. with 2(–3) main branches; fls c.
     12 mm. diam.; nectar-scales red.   **3. anglicum**
     Lvs 6–12 mm., green; infl. with several branches; fls 6–9 mm.
     diam.; nectar-scales yellow. Possibly native in a few places in W.
     England but naturalized elsewhere. White Stonecrop.

                                                         S. album L.

8 Lvs 7 mm. or less, ovoid or cylindric, blunt.   9
     Lvs 8 mm. or more, linear, terete or flat above, acute or apiculate.   10

9 Lvs ovoid, imbricate, hot and acrid to the taste.   **4. acre**
     Lvs linear-cylindric, spreading, dense but not imbricate, not hot to
     the taste. Introduced and naturalized on walls in a few places.
                                                   S. sexangulare L.

10 Lvs flat above; sterile shoots clothed with withered lvs below and
     with rosettes of crowded lvs above. Very local on rocks and screes
     in S.W. England and Wales; naturalized elsewhere.
                                                   S. forsteranum Sm.
     Lvs terete; sterile shoots equally lfy for a considerable distance,
     their tips not rosette-like and dead lvs not persistent. Introduced
     and often naturalized on old walls and rocks.   *S. reflexum* L.

1. S. rosea (L.) Scop.   Rose-root.

Glabrous glaucous perennial 15–30 cm. Stock thick, fleshy, branched, projecting above ground-level and crowned with brownish chaffy scales, flesh-pink, rose-scented; roots not tuberous. Stems erect, simple, usually several from each branch of the stock. Lvs 1–4 cm., spirally arranged, numerous, dense, obovate-oblong, usually acute, sessile and rounded at the base, ± dentate at the tip, decreasing in size downwards. Infl. terminal and compact. Fls dioecious, 4-merous, greenish-yellow; sepals and petals linear; nectar-scales emarginate, conspicuous. Male fls with petals half as long again as sepals, often red-tinged. Female fls with petals equalling sepals; stamens 0; carpels 4–6, erect, greenish. Fl. 5–8. Common in crevices of mountain-rocks (to 3850 ft.) from S. Wales and Yorks northwards and of sea-cliffs in W. Scotland and Ireland.

2. S. telephium L.   Orpine, Livelong.

Glabrous subglaucous perennial 20–60 cm., often tinged with red. Stock short, stout; roots tuberous, carrot-like. Stems erect, simple, clustered. Lvs 2–8 cm., spirally arranged, numerous, ± oblong, sessile or shortly stalked, rounded or cuneate at the base. Infl. terminal, of compact subglobose terminal and axillary cymes. Fls 9–12 mm. diam., reddish-purple. Nectar-scales yellow. Carpels erect, purple. Fl. 7–9. Woods and hedge-banks over the greater part of the British Isles but rather local.

Two sspp. occur:

Ssp. **telephium**: Upper lvs truncate at base, sessile; follicles grooved on back.

Ssp. **fabaria** (Koch) Kirschleger: Upper lvs cuneate at base, sometimes stalked; follicles not grooved on back.

3. S. anglicum Hudson   English Stonecrop.

Glabrous glaucous evergreen perennial 2–5 cm., often tinged red, with numerous slender creeping and rooting stems forming mats and ascending flowering and non-flowering branches. Lvs 3–5 mm., spirally arranged, ovoid or shortly ellipsoid, terete, entire, spreading, clasping the stem and with a small spur at the base. Infl. with 2(–3) main branches each with 3–6 fls and with a fl. at the fork. Fls c. 12 mm. diam., on short stout stalks c. 1 mm. Petals white, tinged pink on back. Nectar-scales red. Carpels erect in fr., becoming red. Fl. 6–8(–9). Rocks, less frequently dry grassland, to 3500 ft.; also on dunes and shingle; absent from strongly basic soils; common in the west and in Ireland, and also in a few places near the south and east coasts.

4. S. acre L.   Wall-pepper, Biting Stonecrop.

Glabrous green evergreen perennial 2–10 cm., with numerous creeping stems forming mats, and ascending or erect non-flowering and fl.-branches; taste hot and acrid. Lvs 3–5 mm., spirally arranged, ascending, imbricate (except sometimes on the fl.-stems), ovoid-trigonous, blunt, entire, sessile, spurred at the base. Infl. with 2–3 main branches, each with 2–4 fls at the fork. Fls c. 12 mm. diam., bright yellow. Nectar-scales whitish. Carpels spreading in fr. Fl. 6–7. Dry grassland, dunes, shingle and walls, especially on basic soils. Common throughout the British Is, except Shetland.

2. UMBILICUS DC.

Perennial herbs with tuberous stock. Lvs spirally arranged, stalked, peltate or cordate, fleshy. Infl. a narrow terminal spike or raceme. Fls 5-merous. Petals connate for most of their length into a cylindric or nearly bell-shaped tube. Stamens usually 10. Nectar-scales small. Carpels free; styles short; ovules numerous. Fr. a group of follicles; seeds very small.

1. U. rupestris (Salisb.) Dandy   Wall Pennywort, Navelwort.

Glabrous, 10–40 cm. Stock subglobose. Stem usually solitary, simple. Lvs mostly basal, blades 1·5–7 cm. diam., orbicular, peltate, depressed at the junction with the long stalk, crenate with very broad teeth, or sinuate; stem-lvs few, becoming smaller upwards and with shorter stalks, the uppermost often not peltate. Infl. long, usually more than half the total length of the stem, many-fld; fls drooping. Corolla 8–10 mm., whitish-green, tubular; lobes ovate, short. Stamens inserted on corolla-tube, included. Fl. 6–8(–9). Crevices of rocks and walls, especially acid, to 1800 ft. Rather common in W. England and Wales, rarer eastwards but extending to Kent and Leicester and northwards to Argyll and to mid Inner Hebrides; throughout Ireland, but local.

59. SAXIFRAGACEAE

Herbs. Lvs usually spirally arranged. Fls hermaphrodite, hypogynous to epigynous, regular or rarely somewhat zygomorphic. Petals sometimes 0. Stamens usually twice as many as the petals, sometimes as many and alternating with them. Carpels 2(–5), commonly united at the base and free above, tapering gradually into the free styles, superior to inferior; ovules numerous. Fr. a capsule; seeds numerous, endospermic.

Fls 5-merous; petals present.   1. SAXIFRAGA
Fls 4-merous; petals absent.   2. CHRYSOSPLENIUM

1. SAXIFRAGA L.

Usually perennial herbs. Lvs simple. Fls usually perigynous, 5-merous in British spp. Petals present. Stamens 10 (rarely 8). Carpels usually 2, ± united below, forming a superior or partly inferior ovary with axile placentation.

1	Lvs spirally arranged; fls white or yellow.	*2*
	Lvs opposite, small, crowded and appearing 4-rowed; fls purple.	
		6. oppositifolia
2	Fls yellow.	*3*
	Fls white, sometimes spotted with yellow or purple.	*4*
3	Fl.-stems erect; fls 1–3; petals 10–13 mm., yellow, often with orange spots near the tubercled base; ovary superior. Very local and rare in wet grassy ground on moors in N. England, Scotland and Ireland. Yellow Marsh Saxifrage. *S. hirculus* L.	
	Fl.-stems ascending; fls 1–10 in a loose terminal cyme; petals 4–7 mm.; ovary partly inferior. **5. aizoides**	
4	Lvs confined to a basal rosette; stem-lvs 0 (apart from bracts).	*5*
	Stem-lvs present.	*9*
5	Fls 3–12 in a dense head-like infl.; petals greenish-white, unspotted; lvs spathulate, toothed, ± purple beneath. Wet rocks on mountains, very local and rare. *S. nivalis* L.	
	Fls in an open panicle; petals spotted or blotched with yellow or crimson or both.	*6*
6	Lvs nearly sessile, distantly toothed; capsule inflated. **1. stellaris**	
	Lvs distinctly stalked and with contiguous teeth; capsule not inflated.	*7*
7	Lvs cuneate at base, glabrous; lf-stalk 2 mm. broad or more; petals spotted with red.	*8*
	Lvs orbicular or reniform, cordate at base, with long hairs on both sides; lf-stalk scarcely 1 mm. broad; petals with yellow basal blotch but often without red spots. S.W. Ireland, very local.	
		S. hirsuta L.
8	Teeth of lvs blunt, the terminal tooth shorter than its neighbours; lf-stalk shorter than blade, densely ciliate throughout; styles ascending in fr. Introduced in Yorks. **S. umbrosa* L.	

('London Pride' is *S. umbrosa* × *spathularis*.)

Teeth of lvs acute, the terminal tooth equalling or longer than its neighbours; lf-stalk longer than blade, sparsely ciliate only at the base; styles spreading widely in fr. Local among rocks in south and west Ireland. St Patrick's Cabbage.   *S. spathularis* Brot.

9 Annual erect herb 2–15 cm.; petals 2–3 mm.   **2. tridactylites**
   Perennial herbs of varying habit; petals usually more than 3 mm. long.   10

10 Plants with numerous non-flowering lf-rosettes or elongated creeping lfy non-flowering shoots; basal lvs palmately lobed or cut with segments longer than wide, usually narrow.   11
   Plants without numerous non-flowering lf-rosettes or elongated creeping non-flowering shoots; basal lvs reniform or ovate, long-stalked, with broad crenations or palmate, ovate lobes.   13

11 Lobes of lvs ± oblong, the largest always more than 1 mm. wide, blunt or subacute, sometimes mucronate, never awned; sterile shoots erect or ascending, without axillary bulbils.   12
   Lobes of lvs ± linear, rarely more than 1 mm. wide, acute or acuminate, awned; non-flowering shoots procumbent, usually with axillary spindle-shaped lfy bulbils.   **4. hypnoides**

12 Plant very compact; lobes of lvs very blunt; glandular hairs on lvs short and dense. Very rare on high mountain-rocks.
                                                                   S. cespitosa L.
   Plant rather loose; lobes of lvs subacute, glandular hairs on lvs short and dense. Donegal only.   *S. hartii* D. A. Webb
   Plant compact to rather loose; lobes of lvs subacute; glandular hairs on lvs few or 0. Mountains and limestone hills in Ireland, local.   *S. rosacea* Moench

13 Basal lvs hairy, with 7 or more broad crenations or with short broad palmate lobes; fls 2–12; lowland.   **3. granulata**
   Basal lvs subglabrous, 3–7-lobed with ovate lobes; fls 1–3 (or all replaced by bulbils); high-alpine.   14

14 Plant with numerous red bulbils in the axils of lvs and bracts; fl.-stems solitary, erect; petals 8–13 mm.; fls often 0, replaced by bulbils. A very rare alpine.   *S. cernua* L.
   Plant lacking bulbils; fl.-stems usually several, ascending; petals 3–5 mm. A very rare alpine.   *S. rivularis* L.

1. S. stellaris L.   Starry Saxifrage.

Perennial, with short stock and 1 to several basal lf-rosettes which often elongate; not purple-tinged. Lvs 0·5–3 cm., ± spathulate, scarcely stalked, ± acute, distantly toothed, with scattered long hairs, especially above. Fl.-stems devoid of lvs (apart from the very narrow bracts), with scattered long hairs below and glandular hairs above. Fls up to 12 or more in an open panicle, their stalks slender. Sepals reflexed. Petals

4–5 mm., pure white with 2 yellow spots near the base. Anthers orange-red. Ovary superior. Fr. c. 6 mm., ± inflated. Fl. 6–8. Common on mountains by streams, in springs, on wet rock-ledges and wet stony ground, to 4400 ft.; Wales, N. England, Scotland and Ireland.

2. S. tridactylites L.   Rue-leaved Saxifrage.

Erect annual herb 2–15 cm., ± glandular-hairy all over. Basal lvs crowded, ± spathulate, broadly stalked, blade 1 cm. or less, palmately 3–5-lobed or the smaller lvs entire; upper lvs smaller, passing into the entire bracts. Fls in cymes or solitary and terminal. Sepals erect, ovate. Petals 2–3 mm., white. Ovary almost completely inferior. Fr. sub-globose Fl. 4–6. Dry open habitats in grassland, on walls, rocks, etc., and on sand-dunes, mainly on basic substrata; chiefly lowland; throughout much of the British Is.

3. S. granulata L.   Meadow Saxifrage.

Perennial 10–50 cm., overwintering by bulbils in the axils of basal lvs. Basal lvs in a rosette, 0·5–3 cm., reniform, cordate at base, with 7 or more broad crenations (or sometimes shortly palmate-lobed) and with scattered long white hairs, glandular-ciliate; stalk many times longer than blade; stem-lvs few, smaller upwards, short-stalked, cuneate at base. Fl.-stems hairy below, glandular especially above. Fls 2–12 in a loose terminal cyme. Sepals ovate. Petals 10–17 mm., white. Ovary about ¾-inferior. Fr. 6–9 mm., ovoid. Fl. 4–6. Basic and neutral grass-land, to 1600 ft.; rather local, commoner in the east and extending northwards to Moray and Renfrew; very local in E. Ireland.

4. S. hypnoides L.   Dovedale Moss.

Perennial 5–20 cm., forming mats with flowering rosettes and numerous procumbent or decumbent non-flowering shoots (up to 15 cm), usually bearing axillary spindle-shaped lfy bulbils. Lvs of flowering rosettes to 1 cm., 3–5(–9)-lobed, subglabrous and non-glandular, narrowed into a sparsely ciliate stalk longer than the blade; lf-lobes linear to linear-oblong, acute or acuminate, awned, c. 1(–1·5) mm. wide. Lvs of non-flowering shoots mostly entire, like the upper stem-lvs. Stem glabrous or sparsely glandular; fl.-stalks and calyx glandular. Fls 1–5. Sepals triangular, awned. Petals 4–10 mm., pure white. Ovary almost com-pletely inferior. Fl. 5–7. Rock-ledges, scree and stony grassland on base-rich substrata, to 4000 ft. Locally common in hilly and moun-tainous districts from Somerset northwards; local in Ireland.

5. S. aizoides L.   Yellow Mountain Saxifrage.

Perennial 5–20 cm., with numerous decumbent non-flowering shoots and ascending flowering stems. Lvs 1–2 cm., numerous, oblong-linear, acute, sessile, usually distantly ciliate with stiff hairs, otherwise glabrous, rather thick. Fl.-stems pubescent; fls 1–10 in a loose terminal cyme; bracts like the lvs but smaller. Sepals triangular-ovate, blunt, spreading. Petals 4–7 mm., distant, yellow, often spotted with red. Ovary less than ½-inferior. Fr. c. 7 mm., ovoid. Stream-sides and wet stony ground on base-rich substrata in mountain districts to 3850 ft.; locally common. Wales, N. England, Scotland and N. Ireland.

6. S. oppositifolia L.   Purple Saxifrage.

Perennial, with numerous long prostrate branching stems. Lvs 2–6 mm., in opposite pairs but so crowded as to appear 4-rowed, ± oblong, concave, thickened and flattened at the tip, dark-green, glabrous except for small marginal bristles, sessile. Fls solitary on less densely lfy stems 1–2 cm. Sepals ovate, suberect. Petals 6–10 mm., rosy-purple, rarely white. Ovary about ½-inferior. Fr. 3–6 mm., ovoid. Fl. 3–5. Damp rocks and stony ground on base-rich substrata in mountain districts, to 4000 ft.; local but often abundant. Wales, N. England, Scotland, N.W. Ireland.

2. CHRYSOSPLENIUM L.   Golden Saxifrage.

Herbs with simple stalked lvs. Fls small, perigynous, yellowish, in flattened dichasial corymbs made more conspicuous by the broad yellow-green bracts. Sepals 4–5. Petals 0. Stamens 8. Ovary partly inferior; placentae 2, parietal; styles 2, free. Fr. a capsule opening along the inner edges of the carpels.

Lvs opposite; basal lvs truncate or broadly cuneate at base.
                                                          1. oppositifolium
Stem-lvs not opposite (usually only 1); basal lvs cordate at base.
                                                          2. alternifolium

1. C. oppositifolium L.

Perennial 5–15 cm., with numerous decumbent lfy rooting stems which form large patches; fl.-stems ascending. Lvs in opposite pairs, with stalks not longer than the blades; lower 1–2 cm., roundish, ± crenate, truncate or broadly cuneate at base, with scattered appressed hairs above; lvs of fl.-stem 1–3 pairs, smaller, shorter-stalked, usually glabrous. Bracts bright greenish-yellow. Calyx 3–4 mm. diam., greenish-yellow. Fl. 5–7. Stream-sides, springs, wet rocks, wet ground in woods, to 3400 ft., usually in shade. Common throughout the British Is, except E. and S.E. England, where it is local.

2. C. alternifolium L.

Like **1** but more robust and spreading by stolons bearing only scale-lvs; basal lvs 2–4, much longer-stalked, with blade 1–2·5 cm., reniform, crenate, cordate at base; stem-lvs 1(–3), not opposite; bracts rather larger and more deeply crenate. Calyx 5–6 mm., diam. Fl. 4–7. In similar places to *C. oppositifolium* but much more local; not in Ireland or in extreme west of Great Britain.

60. PARNASSIACEAE

Perennial herbs. Lvs simple, spirally arranged, exstipulate, chiefly basal. Fl. solitary, terminal on a stem from the axil of a basal lf; regular, hermaphrodite, 5-merous. A ring of 5 large fringed staminodes, with pseudonectaries on the upper surface, lies inside and opposite the petals. Fertile stamens 5, alternating with the staminodes. Ovary superior, usually 4-celled below, 1-celled above; placentae axile below, parietal above, with numerous ovules; stigmas usually 4, subsessile. Fr. a loculicidal capsule.

1. PARNASSIA L.

The only genus.

1. P. palustris L.   Grass of Parnassus.

Glabrous 10–30 cm., with short vertical stock. Basal lvs 1–5 cm., ovate, cordate at base, subacute, entire, shining beneath, stalk longer than blade. Fl.-stem erect, straight, with 1 sessile deeply cordate lf near the base. Petals 7–12 mm., white with conspicuous greyish veins. Staminodes c. ⅓ as long as petals, broadened upwards, with 7–15 long fine processes tipped with yellowish glands. Ovary ovoid. Fl. 7–10. Marshes and moors, to 2600 ft. Widespread throughout the British Is, but rather local in lowland England.

61. ESCALLONIACEAE (p. xxiii)

62. GROSSULARIACEAE

Shrubs. Lvs spirally arranged, simple, often palmately lobed; stipules 0 or small and adnate to lf-stalk. Fls in racemes, hermaphrodite or dioecious, actinomorphic, 4–5-merous, epigynous and with the conspicuous and often coloured receptacle prolonged upwards as a perigynous tube, cup or rim. Sepals coloured like the receptacle. Petals usually shorter than sepals. Stamens as many as and alternating with petals. Ovary inferior, 1-celled, with parietal placentae; styles 2; ovules numerous. Fr. a berry crowned by the persistent calyx; seeds endospermic.

1. Ribes L.

The only genus.

1	Non-spiny; fls 5 or more in racemes.	2
	Spiny; fls 1–2(–3).	**3. uva-crispa**
2	Fls hermaphrodite, usually in spreading or drooping racemes; bracts less than half as long as fl.-stalks.	3

2 Fls dioecious, 4–6 mm. diam., yellow-green; racemes erect in fl. and fr.; bracts longer than fl.-stalks; fr. red. Cliffs and rocky woods on limestone in Wales and in England from Staffs northwards to Cumberland; very local. Mountain Currant.

R. alpinum L.

3 Lvs and ovary without glands, odourless; fr. red, rarely white. 4
Lvs and ovary with sessile glands, strong-smelling; fr. black.

2. nigrum

4 Receptacle obscurely 5-angled with raised rim round style, saucer-shaped; connective broad; lvs cordate at base with narrow sinus.

1. rubrum

Receptacle circular, without raised rim, cup-shaped; connective very narrow on inner side, broad on outer; lvs truncate at base or cordate with wide sinus; fls 7 mm. diam., pale green. Woods on limestone from Lancashire and Yorks to Caithness, very local. Erect-spiked Red Currant. *R. spicatum* Robson

1. R. rubrum L. (*R. sylvestre* (Lam.) Mert. & Koch) Red Currant.

Deciduous shrub, 1–2 m. Lf-blades to 6 cm., 3–5-lobed, deeply cordate at base with narrow sinus, glabrous or nearly so at maturity, not glandular, odourless; lobes triangular-ovate, toothed; lf-stalk usually glandular. Fls 6–20 in drooping racemes with bracts less than half as long as fl-stalks. Fls c. 5 mm. diam., pale green or slightly purplish, not glandular, hermaphrodite; sepals broader than long; petals very small; stamens with a broad connective; receptacle saucer-shaped, somewhat

5-angled with a raised 5-angled rim round the style. Fr. 6–10 mm., red, globose, not contracted above when young. Fl. 4–5. Woods and hedges, especially by streams and in fen carr. Widespread throughout Great Britain but probably introduced in Scotland and in Ireland where it is very local. Most cultivated red and white currants belong to this sp.

2. R. nigrum L.   Black Currant.

Deciduous shrub, 1–2 m. Lf-blades 3–10 cm., 3–5-lobed, cordate at base, glabrous above, ± pubescent on veins beneath, with scattered sessile glands, strong-smelling; lobes ± deltate, acute, coarsely and irregularly toothed; lf-stalk broad, pubescent beneath and with sessile glands. Racemes lax, drooping, 5–10-fld, pubescent, often glandular; bracts less than half as long as fl.-stalks. Fls c. 8 mm. diam., dull purplish-green, hermaphrodite; sepals recurved at tip; petals whitish, about half as long as sepals; receptacle broadly bell-shaped, pubescent outside; ovary glandular. Fr. 12–15 mm., black, globose. Fl. 4–5. Woods and hedges, throughout Great Britain especially by streams and in fen carr; introduced and very local in Ireland.

3. R. uva-crispa L.   Gooseberry.

Deciduous much-branched shrub, c. 1 m. Spines 1–3 at each node. Twigs with many short axillary shoots. Lf-blades 2–5 cm., 3–5-lobed; lobes ovate-rhombic, blunt, deeply blunt-toothed in upper part. Fls 1–2(–3) on each stalk, drooping; bracts small. Fls c. 1 cm. diam., greenish, purple-tinged, hermaphrodite; sepals reflexed; petals small, whitish; receptacle broadly bell-shaped, hairy like the calyx on both sides; ovary glandular-bristly or pubescent. Fr. 10–20 mm. (more in cultivated forms), green, yellowish or purple, smooth to bristly. Fl. 3–5. Woods and hedges, often an escape but probably native in damp woods; introduced and local in Ireland.

63. DROSERACEAE

Perennial glandular insectivorous herbs. Fls 4–8-merous, actinomorphic, hermaphrodite, often in unbranched circinate cymes. Stamens 4–20, often 5. Ovary superior; styles 3–5. Fr. a capsule with numerous small seeds.

1. DROSERA L.

Lvs often reddish, densely glandular and fringed with long glandular hairs ('tentacles'), spirally arranged and often crowded in a rosette. Fls in circinate cymes, rarely solitary. Stamens as many as petals.

1 Lvs orbicular, spreading horizontally, abruptly narrowed at base.
                                                  1. rotundifolia
  Lvs distinctly longer than broad, obovate to narrowly oblong,
    ± erect, gradually narrowed at base.   2

2 Scape in fl. up to twice as long as lvs, straight, apparently arising
    from centre of rosette.   **2. anglica**
  Scape in fl. little longer than lvs, curved or decumbent at base,
    arising laterally below the rosette.   **3. intermedia**

1. D. rotundifolia L.   Sundew.

6–25 cm., reddish. Lvs long-stalked, in a rosette; blade up to 1 cm. diam.; stalks hairy. Scape erect, usually unbranched, 2–4 times as long as lvs, arising from centre of rosette. Fls 5 mm., white, short-stalked, usually 6-merous, arranged in 2 rows. Capsule acute, as long as or longer than sepals. Fl. 6–8. In bogs, etc.; widely spread in suitable habitats.

2. D. anglica Hudson   Great Sundew.

10–30 cm., reddish. Lvs long-stalked, in a rosette; blade up to 3 cm., narrowly obovate to narrowly oblong; stalks nearly or quite glabrous. Fls similar to the preceding. Capsule blunt, longer than sepals. Fl. 7–8. In the wetter parts of bogs, chiefly in the north and west.

3. D. intermedia Hayne   Long-leaved Sundew.

5–10 cm., reddish. Lvs long-stalked, in a rosette; blade c. 1 cm., obovate; stalks glabrous. Fls similar to the preceding. Capsule pear-shaped, as long as or slightly longer than sepals. Fl. 6–8. In bog pools and damp peaty places on heaths and moors, mainly in the west.

    D. anglica × rotundifolia (*D. × obovata* Mert. & Koch) occurs with the parents. It resembles *D. intermedia* but has a straight scape 2–3 times as long as the lvs and is sterile.

64. SARRACENIACEAE (p. xx)

65. LYTHRACEAE

Herbs, shrubs or sometimes trees. Lvs usually opposite or in whorls. Fls usually actinomorphic, hermaphrodite, with a perigynous 'calyx-tube' which often bears appendages alternating with the calyx-teeth. Petals present or 0, inserted at or near the top of the calyx-tube. Stamens

4 or 8(–12), inserted below the petals. Ovary superior. Fr. usually a capsule; seeds small.

Stems erect or ascending, not rooting at nodes; lvs linear to ovate; fls
    5 mm. or more; calyx-tube cylindrical.   1. LYTHRUM
Stem prostrate, rooting freely at nodes; lvs obovate; fls c. 1 mm.;
    calyx-tube obconical.   2. PEPLIS

1. LYTHRUM L.

Herbs or small shrubs. Stems 4-angled, at least when young. Lvs quite entire. Fls solitary or in small cymes in axils of lvs, purple or pink. Petals 4–6, rarely 0. Stamens 6–12, sometimes unequal in length.

Perennial; fls 10–15 mm., magenta, in cymes.   **1. salicaria**
Annual; fls c. 5 mm. pale pinkish-purple, solitary. Very local. Grass
    Poly.   *L. hyssopifolia* L.

1. L. salicaria L.   Purple Loosestrife.

60–120 cm., ± hairy, branches few or 0. Lvs 4–7 cm., sessile, lanceolate to ovate, acute, ± cordate at base, the lower often in whorls of 3, the upper sometimes alternate. Infl. up to 30 cm. or more, spike-like, many fld. Calyx-tube c. 6 mm., cylindrical, ribbed. Petals 8–10 mm., obovate. Capsule 3–4 mm., enclosed in calyx. Fl. 6–8. By rivers, lakes, etc. in reedswamp, often forming large stands, locally abundant.

2. PEPLIS L.

Small annual herbs. Lvs entire. Fls small, solitary, usually 6-merous. Calyx-tube obconical. Petals withering or dropping soon after the fl. opens.

1. P. portula L.   Water Purslane.

4–25 cm., glabrous. Stems bluntly 4-angled, often pink. Lvs 1(–2) cm., obovate-spathulate, somewhat fleshy. Fls c. 1 mm., subsessile, solitary in the axils of all but the lower lvs. Calyx-teeth triangular, about as long as tube, alternating with bristle-like appendages. Capsule sub-globose. Fl. 6–10. At muddy margins of pools or puddles in open communities, locally common on non-calcareous soils.

66. THYMELAEACEAE

Trees or shrubs. Lvs usually spirally arranged, exstipulate. Fls regular, usually hermaphrodite, with an elongated, often coloured, perigynous 'calyx-tube'. Calyx often coloured. Petals scale-like or 0. Stamens usually twice as many as sepals, sometimes fewer, perigynous. Ovary superior, 1-celled; ovule 1; style simple or 0; stigma capitate. Fr. various; seeds endospermic or not.

1. DAPHNE L.

Shrubs. Lvs short-stalked, spirally arranged. Fls hermaphrodite. Sepals 4, coloured. Petals 0. Stamens 8; filaments short. Stigma large, sessile or nearly so. Fr. a drupe; endosperm sparse.

Fls purple, appearing before the lvs in 2–4-fld clusters in the axils of fallen lvs of the previous year; lvs thin, light green; fr. red. Deciduous shrub, 50–100 cm., fl. 2–4. Woods and scrub on calcareous soils; very local and rare. Mezereon. *D. mezereum* L.
Fls green, in short racemes in the axils of the evergreen leathery dark-green lvs; fr. black. **1. laureola**

1. D. laureola L.
Spurge Laurel.

Evergreen shrub, 40–100 cm., with erect stems, the lvs often all near the top of the plant. Lvs 5–12 cm., obovate-lanceolate, ± acute, dark glossy green, leathery, glabrous. Fls 8–12 mm., in short axillary 5–10-fld racemes, green, short-stalked. Sepals up to half as long as perigynous tube. Fr. c. 12 mm., ovoid, black. Fl. 2–4. Woods, mainly on calcareous soils, widespread and rather common in S. England and extending north to Lancashire and S. Northumberland; local in Wales; Channel Is.

67. ELAEAGNACEAE

Trees or shrubs densely covered with peltate or stellate silvery-brown scale-hairs. Lvs entire; stipules 0. Fls regular. Sepals 2 (as in the British sp.) or 4, valvate. Petals 0. Stamens equal in number to sepals and alternate with them or twice as many (4 in the British sp.); perigynous. Ovary superior, 1-celled, at the base of the perigynous tube; ovule 1; style long; stigma simple. Fr. drupe-like, consisting of the true dry fr. surrounded by the lower half of the perigynous tube which becomes fleshy. Seed non-endospermic.

The only British sp. is Sea Buckthorn, **Hippophaë rhamnoides** L., a much-branched thorny deciduous shrub, 1–3 m., with linear-lanceolate silver-scaly lvs, 1–8 cm., greenish dioecious fls in short spikes before the lvs, and ovoid orange fr., 6–8 mm. On fixed dunes and occasionally on sea-cliffs from Yorks to Sussex, local but sometimes dominant; often planted elsewhere.

68. ONAGRACEAE

Usually perennial herbs with simple lvs; stipules 0. Fls usually actinomorphic, hermaphrodite, epigynous and commonly also with a perigynous 'calyx-tube'; 4-merous, rarely 2-merous. Petals free. Ovary inferior, (1–)2–4-celled with 1 to many ovules in each cell; style single. Fr. usually a loculicidal capsule.

1 Shrubs with pendulous crimson and violet fls.   4. FUCHSIA
  Herbs.   2
2 Petals 0; stamens 4. A creeping aquatic herb with opposite ovate to
  elliptical lvs, resembling *Peplis*; fls 3 mm. diam., solitary, axillary.
  Shallow pools in acid fen. Only in New Forest (Hants.) and Jersey.
                         Ludwigia palustris (L.) Elliott
  Petals 2 or 4.   3
3 Petals 2, white; stamens 2; fr. indehiscent, covered with barbed
  bristles.   5. CIRCAEA
  Petals 4; stamens 4+4; fr. a capsule.   4
4 Fls yellow; seeds not plumed.   3. OENOTHERA
  Fls rose or white; seeds plumed.   5
5 Lvs all spirally arranged; fls held horizontally and slightly zygomorphic; calyx-tube 0.   2. CHAMAENERION
  At least the lower lvs in opposite pairs; fls ± erect and actinomorphic; calyx-tube short.   1. EPILOBIUM

1. EPILOBIUM L.

Perennial herbs with at least the lower lvs in opposite pairs. Fls 4-merous, rose or white, ± erect, actinomorphic. Fr. a long 4-valved capsule with many plumed seeds.

1 Stem quite prostrate, slender, rooting at the nodes; lvs 2·5–8 mm.,
  almost circular; fls solitary, axillary.   11. nerterioides
  Stem not wholly prostrate; fls in terminal and axillary racemes,
  rarely solitary and then terminal.   2
2 Stem with spreading non-glandular and glandular hairs; stigma
  4-lobed.   3
  Stem glabrous or with ± appressed non-glandular hairs (spreading
  hairs, if present, glandular only); stigma either 4-lobed or entire
  and club-shaped.   4
3 Lvs usually 6–12 cm., somewhat clasping and slightly decurrent;
  petals more than 1 cm. long, deep rose.   1. hirsutum
  Lvs usually 3–7 cm., neither clasping nor decurrent; petals less
  than 1 cm. long, pale rose.   2. parviflorum

4 Stigma 4-lobed.   *5*
  Stigma entire, club-shaped.   *6*

5 All lvs opposite, ovate-lanceolate, short-stalked, toothed through-
    out, including the rounded base.   **3. montanum**
  Only the lower lvs opposite; lvs lanceolate, stalked, toothed except
    for the wedge-shaped base.   **4. lanceolatum**

6 Lvs usually with stalks 3–20 mm.   **5. roseum**
  Lvs sessile or very short-stalked.   *7*

7 Stem with no ridges or raised lines decurrent from the lvs, though
    sometimes with 2 rows of crisped hairs; lvs narrowly lanceolate-
    elliptical; slender below-ground stolons, ending in bulbils, pro-
    duced in summer.   **9. palustre**
  Stem with 2–4 ridges or raised lines, decurrent from the lvs; lvs
    usually lanceolate to ovate, or, if narrowly lanceolate, then not
    markedly narrowed at base.   *8*

8 Herbs with erect or ascending stems, usually 25–80 cm.; top of infl.
    and buds erect.   *9*
  Small alpine herbs with procumbent or ascending stems, 5–
    20(–30) cm.; top of infl. nodding in fl. and young fr.   *11*

9 Stem with numerous slender spreading glandular hairs as well as
    appressed crisped hairs; fls 4–6 mm. diam.   **6. adenocaulon**
  Spreading glandular hairs 0 or only on calyx-tube; fls 6–12 mm.
    diam.   *10*

10 Glandular hairs present on calyx-tube; capsule 4–6 cm.; elongat-
    ing above-ground stolons formed in summer.   **8. obscurum**
  Plant wholly without glandular hairs; fr. 7–10 cm.; almost sessile
    rosettes formed at base of stem in autumn.   **7. tetragonum**

11 Stem very slender (1–2 mm. diam.); lvs 1–2 cm., yellowish-green,
    ±lanceolate; stolons above-ground with small green lvs; fls
    4–5 mm. diam. By streams and springs on mountains from N.W.
    Yorks and Durham northwards. Alpine Willow-herb.
                                                    E. anagallidifolium Lam.
  Stem 2–3 mm. diam.; lvs 1·5–4 cm., bluish-green, ±ovate; stolons
    below-ground with distant yellowish scale-lvs; fls 8–9 mm. diam.
    By streams and springs on mountains; Caernarvon and north-
    wards from Cumberland, N.W. Yorks and Durham; Leitrim.
    Chickweed Willow-herb.   *E. alsinifolium* Vill.

1. E. hirsutum L.   Great Hairy Willow-herb.

Stem 80–150 cm., stout, erect, densely glandular and with numerous
soft spreading hairs. White fleshy underground stolons produced in
summer. Lvs mostly opposite, oblong-lanceolate, sessile, half-clasping
and slightly decurrent, ±hairy on both sides, toothed. Fls 15–23 mm.

diam., deep purplish-rose, in ± corymbose racemes; buds erect. Stigma of 4 revolute lobes exceeding the stamens. Fr. 5–8 cm., downy. Fl. 7–8. Stream-banks, marshes, drier parts of fens, etc., throughout Great Britain and Ireland.

2. E. parviflorum Schreber   Lesser Hairy-Willow herb.

Stem 30–60(–90) cm., erect, glandular above and usually with short soft spreading hairs throughout. Lf-rosettes, at first sessile but later terminating short above-ground stolons, formed at the base of the stem in autumn. Only the lower lvs opposite; all oblong-lanceolate, rounded and sessile at the base but not clasping or decurrent, softly hairy on both sides, distantly toothed. Fls 6–9 mm. diam., pale purplish-rose, in ± corymbose racemes; buds erect. Stigma of 4 non-revolute lobes about equalling the stamens. Fr. 3·5–6·5 cm., subglabrous or downy. Fl. 7–8. Stream-banks, marshes and fens throughout most of the British Isles.

3. E. montanum L.   Broad-leaved Willow-herb.

Stem (5–)20–60 cm., slender, erect, subglabrous or sparsely hairy. Short stolons, above or below ground, produced in late autumn. Lvs mostly opposite, ovate-lanceolate to ovate, rounded at the base and with short narrowly winged stalks, sharply and irregularly toothed, subglabrous. Fls 6–9 mm. diam., pale rose, in a terminal raceme; buds ± drooping. Stigma of 4 short non-revolute lobes. Fr. 4–8 cm., downy. Fl. 6–8. In woods on the more base-rich soils, in hedge-rows, on walls and rocks and as a weed in gardens; common throughout the British Isles.

4. E. lanceolatum Sebastiani & Mauri   Spear-leaved Willow-herb.

Stem 20–60(–90) cm., erect, with 4 hardly raised lines; subglabrous below and downy with short crisped hairs above. Subsessile lf-rosettes produced at the base of the stem in late autumn. Lvs, of which only the lower are opposite, elliptical to elliptical-lanceolate, blunt, narrowing into a stalk 4–8 mm.; hairy on the veins and margins and with small distant teeth except at the cuneate base. Fls 6–7 mm. diam., pale pink becoming deeper, in a terminal raceme drooping in bud. Stigma of 4 short spreading lobes. Fr. 5–7 cm., downy. Fl. 7–9. Roadsides, railway-banks, walls, dry waste places, etc. in S. and S.W. England and Glamorgan.

5. E. roseum Schreber   Small-flowered Willow-herb.

Stem 25–60(–80) cm., erect, fragile, usually with 2 distinct and 2 indistinct raised lines; glabrous below but with crisped and glandular hairs

above. Very short stolons terminating in lax subsessile lf-rosettes produced in late autumn. Lvs, of which only the lower are usually opposite,
elliptical, narrowed both to the acute apex and to the cuneate base and
with a longish stalk, 3–20 mm.; glabrous or hairy only on the veins, the
margin finely and sharply toothed. Fls 4–6 mm. diam., at first white
then streaked with rose, their buds cuspidate and drooping. Stigma
entire. Fr. 5–7 cm., downy with crisped and glandular hairs. Fl. 7–8.
Damp places, woods and copses, railway banks and cultivated ground;
throughout lowland Great Britain northwards to Perth and Angus;
N.E. Ireland, extending to Sligo and Dublin.

*6. E. adenocaulon Hausskn.   American Willow-herb.

Stem 30–90(–150) cm., much-branched, stiffly erect, with 4 raised lines
(only 2 near base); subglabrous below, ± densely clothed with appressed
crisped hairs and short spreading glandular hairs above. Subsessile
basal rosettes, at first with small fleshy rounded lvs but later developing
normal lvs, formed in late summer. Lvs all opposite except uppermost,
oblong-lanceolate, gradually tapering to an acute tip and suddenly
narrowed at the rounded base into a short stalk, 1·5–3 mm., subglabrous,
with numerous small irregular forwardly-directed marginal teeth. Fls
4–6 mm. diam., pale pink, erect in bud, numerous in terminal racemes.
Stigma entire. Fr. 4–6·5 cm. Fl. 6–8. Introduced. Damp woods, copses,
stream-sides, railway-banks, gardens, waste places, etc. Common in
S.E. England and rapidly spreading northwards.

7. E. tetragonum L. (*E. adnatum* Griseb.)

Square-stemmed Willow-herb.

Stem 25–60(–80) cm., erect, firm, with 4 conspicuously raised lines or
wings; subglabrous below but downy above with ± appressed whitish
hairs. Lax lf-rosettes, terminating very short stolons, produced at the
base of the stem in late autumn. Lvs, of which the lower are opposite,
strap-shaped to narrowly oblong-lanceolate, blunt at the tip, narrowed
into a ± sessile base and decurrent into the raised lines of the stem,
shining and greasy-looking above, subglabrous, the margins strongly
and irregularly toothed. Stigma entire, equalling the style. Fr. 7–
9(–11) cm., downy. Fl. 7–8. Damp woodland clearings and hedge-
banks, stream- and ditch-sides and cultivated ground; throughout lowland Great Britain northwards to Inverness and Argyll; rare in Ireland.

Plants like *E. tetragonum* but with broader lvs of which at least the
upper are narrowed into a short stalk, and petals 10–12 mm. (5–7 mm.
in *tetragonum*) have been named *E. lamyi* F. Schultz. S. England and
northwards to Brecon, Leicester and Huntingdon.

8. E. obscurum Schreber　　　　　　　　Dull-leaved Willow-herb.

Stem 30–60(–80) cm., erect from a curving base, with 4 distinctly raised lines; glabrous below but downy above with appressed hairs. Slender ± elongated above- or below-ground stolons bearing distinct pairs of small lvs are produced in late summer but do not form distinct rosettes. Lvs (of which the lower are opposite) lanceolate or ovate-lanceolate, blunt at the apex, rounded at the sessile base and contracted suddenly to become decurrent into raised lines of the stem; dull above, subglabrous, with a few distant irregular small marginal teeth. Fls 7–9 mm. diam., deep rose, their buds erect, acute. Calyx-tube with a few glandular hairs (absent in *E. tetragonum*). Stigma entire, equalling the style. Fr. 4–6 cm., downy. Fl. 7–8. Marshes, stream- and ditch-banks, moist woods and cultivated ground; locally common throughout most of the British Isles.

9. E. palustre L.　　　　　　　　　　　　Marsh Willow-herb.

Stem 15–60 cm., erect from a curving base, terete though often with 2 rows of crisped hairs; subglabrous or downy with short crisped hairs. Filiform below-ground stolons, bearing distant pairs of yellowish scale-lvs and terminating in a bulbil-like bud with fleshy scales, produced in summer. Lvs mostly opposite, lanceolate to linear-lanceolate, blunt, cuneate at the base, ± sessile or the uppermost very shortly stalked, subglabrous, entire or very obscurely toothed. Fls 4–6 mm. diam., pale rose or whitish; buds blunt, initially erect but soon drooping so that the top of the raceme hangs to one side. Stigma entire, shorter than the style. Fr. 5–8 cm., downy. Fl. 7–8. In marshes and acid fens, flushes, etc.; calcifuge; locally common throughout the British Isles.

*10. E. nerterioides A. Cunningham

Prostrate, with slender creeping stems to 20 cm., rooting at the nodes. Lvs 2·5–8 mm., opposite, broadly ovate to suborbicular, very short stalked, entire or faintly toothed, subglabrous. Fls 3–4 mm., pink, solitary in the axils of lvs, on stalks 0·5–4 cm. Fr. 2–4 cm., glabrous, on an erect stalk to 6 cm. Fl. 6–7. Introduced from New Zealand. On moist stony ground, rocky sides and beds of streams, etc.; spreading rapidly, especially in N. and N.W. England, N. Wales, Scotland and Ireland.

2. CHAMAENERION Seguier

Lvs all spirally arranged. Fls 4-merous, rose, held horizontally and somewhat zygomorphic. Stamens, like the style, exserted and ultimately bending downwards. Fr. a long 4-valved capsule with many plumed seeds.

1. C. angustifolium (L.) Scop.   Rose-bay Willow-herb.

Stems 30–120 cm., erect, glabrous below, ± downy above. Long stout perennial horizontally spreading and branching roots give rise to new lfy shoots. Lvs narrowly oblong-elliptical, entire or with small distant teeth, margins often waved, glaucous beneath, lateral veins joining to a continuous wavy marginal vein. Fls 2–3 cm. diam., rose-purple, in a long dense raceme. Upper petals broader than the lower. Fr. 2·5–8 cm., downy. Fl. 7–9. Rocky places, scree-slopes, wood-margins and woodland clearings (especially after fire), disturbed ground, gardens, bombed sites, etc., to 2300 ft. in Scotland; throughout the British Isles, but commonest in the south.

3. OENOTHERA L.   Evening Primrose.

Lvs all spirally arranged. Fls opening and scented in the evening, 4-merous, yellow, with a long narrow calyx-tube. Fr. a 4-valved capsule with many non-plumed seeds.

1   Petals less than 30 mm.   *2*
    Petals more than 30 mm.   *3*

2   Stem decumbent, bearing long hairs whose bases become enlarged
     and reddish; stem-lvs narrowly lanceolate, rather fleshy and thick,
     white-hairy; tips of sepals in bud standing apart down to their
     bases; petals 11–16 mm. Local on sand-dunes.
                               *O. ammophila Focke
    Stem usually without red-based hairs; stem-lvs ± broadly lanceo-
     late, hairy; tips of sepals in bud closely connivent at their bases;
     petals 20–30 mm.   **1. biennis**

3   Stem with many red bulbous-based hairs; lvs broadly oblong-
     lanceolate, strongly crinkled, ± pubescent; petals 3·5–5(–6) cm.,
     remaining yellow throughout.   **2. erythrosepala**
    Stem without red-based hairs; lvs linear-lanceolate with waved
     margins; petals 3–4 cm., yellow at first, then turning wine-red.
                                     3. stricta

*1. O. biennis L.

Usually biennial with erect robust stems, 50–100 cm., ± pubescent but lacking hairs with red bulbous bases. Lvs lanceolate, basal stalked, the rest subsessile; all hairy, finely toothed, with midrib eventually turning reddish. Sepals green, their tips in bud closely connivent into a narrow tube. Petals 20–30 mm. Stigmas at ± same level as anthers.

Fr. ± cylindrical, downy but lacking red bulbous-based hairs. Fl. 6–9. Introduced. Dunes, roadsides, railway-banks and waste places. A frequent casual, locally naturalized, throughout the British Is northwards to Perth and Angus.

*2. O. erythrosepala Borbás (*O. lamarckiana* de Vries, non Ser.)

Usually biennial with erect robust stems, 50–120 cm., ± pubescent and with numerous longer hairs having enlarged red bases. Rosette-lvs 8–12 × 3–5 cm., broadly elliptic-oblanceolate, bluntish, stalked, pubescent; stem-lvs broadly oblong-lanceolate, blunt to acute, subsessile, rather sparsely hairy; all strongly crinkled and with shallow distant marginal teeth; midrib usually white. Sepals red-striped. Petals 35–50(–60) mm., golden-yellow. Stigmas at much higher level than anthers. Fr. hairy, with red stripes and some red-based hairs. Fl. 6–9. Introduced. Dunes, railway-banks, roadsides and waste places. Naturalized in various parts of the British Is.

*3. O. stricta Ledeb. ex Link (*O. odorata* Jacq., p.p.)

Annual or biennial with ± erect slender stem, 50–90 cm., pubescent and glandular but lacking red-based hairs. Basal lvs linear-lanceolate, narrowed into a stalk-like base; stem-lvs broader, subsessile; all with the margins waved and ciliate, distantly but sharply toothed. Sepals hairy, becoming reddish. Petals 30–45 mm., yellow at first then turning wine-red. Fr. 1·8–2·5 cm., clavate, glandular. Fl. 6–9. Introduced and naturalized on sand-dunes, especially in S.W. England and Channel Is.

Other *Oenothera* spp. are established locally; see larger *Flora*.

4. Fuchsia L.

Shrubs with opposite subglabrous stalked lvs with deciduous stipules. Fls axillary, pendulous, 4-merous. Stamens 4 + 4, exserted, like the style. Fr. a 4-celled berry with many seeds.

1. F. magellanica Lam. var. **macrostema** (Ruiz & Pavon) Munz

Small shrub to 3 m. Lvs ovate-oblong, acuminate, short-stalked, often in whorls of 3. Fls solitary. Calyx-tube red. Sepals spreading, red. Petals erect, violet. Berry black. Fl. 6–9. Introduced; of S. American ancestry. Much cultivated as a hedge plant and established in S.W. England and W. Ireland.

5. CIRCAEA L.

Perennial herbs usually with slender far-creeping rhizomes, stalked opposite lvs and terminal racemes of small white 2-merous fls. Sepals falling early; petals 2-lobed; stamens 2; ovary inferior, 1–2-celled, with 1 ovule in each cell; style 1. Fr. indehiscent, 1–2-seeded, covered with ± hooked bristles.

1 Open fls in a terminal cluster; petals less than 2 mm., shallowly
    notched; ovary 1-celled.   **3. alpina**
   Open fls spaced along the infl.-axis; petals 2–4 mm., notched to at
    least half-way; ovary with 2 equal or unequal cells.   2

2 Fls with a conspicuous dark nectar-secreting ring; lvs not or
   slightly cordate, sinuate-toothed or with distant small teeth,
   gradually acuminate; ovary equally 2-celled; fr. ripening.
                                      1. lutetiana
   Nectar-secreting ring inconspicuous or 0; lvs strongly cordate,
   dentate, shortly acuminate; ovary unequally 2-celled; fr. rarely
   ripening.   **2. × intermedia**

1. C. lutetiana L.   Enchanter's Nightshade.

Stem usually 20–70 cm., swollen at the nodes, sparsely glandular-pubescent. Lvs 4–10 cm., ovate, gradually acuminate, rounded or slightly cordate at base, minutely and distantly toothed, thin, dull above, paler and shining beneath, subglabrous; stalk furrowed above, not winged, pubescent all round. Infl.-axis elongating before the fls drop. Fl.-stalks glandular-pubescent, reflexed in fr. Sepals glandular-pubescent. Petals 2–4 mm., rounded at base. Stigma deeply 2-lobed. Fr. c. 3 mm. diam., obovoid, equally 2-celled, densely covered with stiff hooked white bristles. Fl. 6–8. Woods and shady places on moist base-rich soils; locally common throughout most of the British Is.

2. C. × intermedia Ehrh.   Intermediate Enchanter's Nightshade.

Stem 15–40 cm., sparsely glandular to glabrous. Lvs 3–8 cm., ovate, abruptly acuminate, distinctly cordate at base, strongly but distantly toothed, thin, ± translucent, subglabrous; stalk furrowed above, rarely winged, pubescent only above. Infl.-axis elongating before the fls drop. Fl.-stalks and sepals sparsely glandular to glabrous, sepals somewhat reflexed in fr. Petals 2–4 mm., cuneate at base. Stigma deeply 2-lobed. Fr. 1·5–3 mm. diam., obovoid, 2-celled but with 1 cell small and aborting, densely covered with softer bristles than in *C. lutetiana*; very rarely setting viable seed. Fl. 7–8. Shady rocky places and mountain woods in the west and north of Great Britain; N. and E. Ireland. Very variable; often mistaken for *C. alpina*. A hybrid of *C. alpina* and *C. lutetiana*.

3. C. alpina L.   Alpine Enchanter's Nightshade.
Stem (5–)10–30(–40) cm., subglabrous. Lvs 2–6(–8) cm., ovate, acute or acuminate, cordate at base, strongly but distantly toothed, thin, usually translucent, somewhat shining above, glabrous; stalk flat above, winged, ± glabrous. Infl.-axis not elongating until after the fls have dropped. Fl.-stalks glabrous, little reflexed in fr. Sepals glabrous. Petals 0·6–1·5 mm., cuneate at base. Stigma entire or notched. Fr. 1–1·5 mm. diam., narrowly obovoid, 1-celled, covered with soft bristles which are not consistently hooked. Fl. 7–8. Woods in mountain districts and shady rocky places in the W. and N. of Great Britain; Hebrides; Shetland.

69. HALORAGACEAE

Herbs, usually aquatic or subaquatic, often very large. Lvs variously arranged; stipules 0. Fls inconspicuous, actinomorphic, hermaphrodite or unisexual. Sepals 4, 2 or 0, free, small; petals free, as many as sepals but often much larger, sometimes 0; stamens 4+4, 4 or 2, free; ovary inferior, usually 4-celled with 1 ovule in each cell; styles 1–4, often very short; stigmas 1–4, feathery or coarsely papillose. Fr. a nut or drupe or separating into 1-seeded nutlets; seeds endospermic.

Water plants, submerged except for the infl.; lvs whorled, pinnate with
    capillary segments.   1. MYRIOPHYLLUM
Gigantic rhubarb-like marsh plant with orbicular-cordate palmately-
    lobed lvs up to 2 m. diam. and very large panicles of small incon-
    spicuous mostly unisexual fls. Introduced and established in a few
    places.   *Gunnera tinctoria* (Molina) Mirbel (*G. chilensis* Lam.)

1. MYRIOPHYLLUM L.

Perennial aquatic herbs, free-floating or with rhizomes in the substratum and the lfy shoots submerged apart from the infl. Lvs in whorls of 3–6, pinnately divided into unbranched hair-like segments; aerial lvs and bracts sometimes simple, toothed or entire. Fls in lfy or bracteate terminal spikes, sessile in whorls in the axils of lvs or bracts; upper fls commonly male, lower female. Calyx inconspicuous, of 4 small lobes in the male fl., minute in the female, corolla in the male fl. of 4 boat-shaped petals, soon falling, minute or 0 in the female; stamens usually

8; ovary 4-celled; style very short or 0; stigmas oblong, recurved. Fr. separating into 4 or fewer 1-seeded nutlets.

1 Lvs usually 5 in a whorl, much exceeding the internodes; fls in the axils of lf-like bracts or of bracts which are pectinate or toothed and as long as the fls even near the tip of the spike.

1. verticillatum

Lvs usually 4 in a whorl, about equalling the internodes; uppermost bracts entire and shorter than the fls.   *2*

2 Lf with 6–18 segments; spike short, at first drooping at the tip; lowest bracts lf-like; upper fls often alternate, not whorled; petals yellow with red streaks.   **3. alterniflorum**

Lf with 13–35 segments; spike erect throughout; all bracts equalling or falling short of the fls; all fls whorled; petals dull red.

2. spicatum

1. M. verticillatum L.   Whorled Water-milfoil.

Rhizomatous. Lfy shoots 50–300 cm. Lvs 2·5–4·5 cm., usually 5 in a whorl, commonly much exceeding the internodes, each with 25–35 rather distant segments. Spike 7–25 cm. Fls usually in whorls of 5 in the axils of shortly pinnate or pectinate bracts of very variable length, but never entire and never shorter than the fls even at the tip of the spike. Usually a few hermaphrodite fls between male and female fls. Petals of male fls c. 2·5 mm., greenish-yellow, rarely reddish, soon falling. Fr. at first subglobular. Fl. 7–8. Forms clavate detachable winter-buds (turions). Ponds, lakes and slow streams of lowland districts in England, Wales and Ireland, especially in base-rich water; not common.

2. M. spicatum L.   Spiked Water-milfoil.

Rhizomatous. Lfy shoots 50–250 cm., naked below through decay of older lvs. Lvs 1·5–3 cm., usually 4 in a whorl, about equalling the internodes, each with 13–35 segments. Spike 5–15 cm. Fls usually in whorls of 4 in the axils of bracts all but the lowest of which are entire and shorter than the fls. About 4 basal whorls of female fls, then 1 of hermaphrodite, the rest of male fls with dull red petals, c. 3 mm., soon falling. Fr. at first subglobular. Fl. 6–7. Turions 0. Lakes, ponds, ditches, etc. Locally common, especially in calcareous water, throughout most of the British Is.

3. M. alterniflorum DC.   Alternate-flowered Water-milfoil.

Rhizomatous. Lfy shoots 20–120 cm., naked below through decay of older lvs. Lvs 1–2·5 cm. (rarely shorter), usually 4 in a whorl, sometimes 3, about equalling the internodes, each with 6–18 segments. Spike

1–2(–3) cm., its tip drooping in bud. Basal whorl usually of 3 female fls in the axils of lf-like pinnate bracts, then other female fls in the axils of short pectinate bracts, next hermaphrodite fls and in the upper half usually c. 6 male fls, singly or in opposite pairs in the axils of entire bracts shorter than the fls; petals 2·5 mm., yellow with red streaks. Fr. longer than wide. Fl. 5–8. Turions 0. Lakes, streams, ditches, etc. Locally common through most of the British Is, but especially in the west and north and in base-poor and peaty water.

70. HIPPURIDACEAE

Aquatic herb with whorled linear lvs; stipules 0. Fls solitary axillary, hermaphrodite or unisexual. Perianth a rim round the top of the ovary. Stamens 1. Ovary inferior, 1-celled, with 1 ovule; style 1, long and slender with stigmatic papillae throughout its length. Fr. an achene.

1. Hippuris L.

The only genus.

1. H. vulgaris L.   Mare's-tail.

Perennial usually aquatic herb with stout creeping rhizome and lfy shoots, 25–75(–150) cm. if wholly or partly submerged, but only 7–20 cm. in terrestrial forms. Lvs 1–7·5 cm. × 1–3·5 mm., 6–12 in a whorl, linear, sessile, entire, glabrous, with a hard acute tip; submerged shoots have longer, thinner, more flaccid and more translucent lvs than those of emergent shoots, and the internodes are longer and less rigid. Fls small, greenish, in the axils of emergent lvs only. Stamen with reddish anther, sometimes 0. Fr. ovoid, smooth, greenish. Fl. 6–7. Lakes, ponds and slow streams, especially in base-rich water. Local throughout the British Is.

71. CALLITRICHACEAE

Annual or perennial, aquatic or subaquatic herbs with thread-like stems and opposite, entire, linear to ovate, exstipulate lvs. Monoecious. Fls axillary, either solitary or a male and female fl. in the same axil. Bracteoles 2, crescent-shaped, or 0; perianth 0; stamen 1, with a long slender filament and reniform anther; ovary 4-lobed, at first 2-celled, becoming 4-celled by secondary septation; styles 2, long; ovules 1 per cell. Fr. 4-lobed, the lobes ± keeled or winged, separating at maturity into 4 druplets; seeds endospermic.

1. CALLITRICHE L.   Star-wort.
The only genus.

The shapes of lvs, the presence or absence of floating lf-rosettes and the capacity to produce fls and fr. vary greatly with the depth and movement of the water in which the plants are growing. The following key should be used only when healthy plants with fls and fr. are available.

1   Lvs all ±linear, submerged.   2
    Lvs not all ±linear; upper lvs ovate or spathulate, forming a float-
        ing rosette.   4
2   Lvs dark blue-green, mostly less than 10 × 1 mm., very slightly taper-
        ing to a truncate, feebly emarginate tip; stems reddish; fr. (rarely
        seen) c. 1 mm., its lobes not keeled. Pools and ditches in S.
        England and Channel Is, local, very rare in Ireland.
                                            C. truncata Guss.
    Lvs pale or yellowish green, mostly more than 1 cm. long, with a
        distinctly emarginate tip: lobes of fr. keeled or winged.   3
3   Lvs commonly 15–25 × 0·5–1 mm., ±parallel-sided then abruptly
        expanded at the emarginate tip and shaped like a bicycle spanner;
        fr. 0·8–1·3 mm., its lobes keeled but not winged.   4. intermedia
    Lvs commonly 10–20 × 1–2 mm., widest at the base and distinctly
        tapering above to the narrow emarginate tip; fr. 2 mm., its lobes
        broadly winged.   5. hermaphroditica
4   Rosette-lvs ±rhombic; submerged lvs narrower, sometimes linear;
        fr. c. 1·5 × 1 mm., its lobes with rounded margins and with
        a very shallow groove between members of a parallel pair.
                                                        3. obtusangula
    Rosette-lvs broadly ovate to narrowly elliptical, not rhombic;
        submerged lvs linear or not; lobes of fr. keeled, with a distinct
        groove between members of a parallel pair.   5
5   Rosette-lvs broadly ovate to almost circular; submerged lvs often
        not very different in shape and never linear; fr. 1·6–2 mm., some-
        what broader than long, distinctly winged, its lateral grooves
        deep.   1. stagnalis
    Rosette-lvs broadly to narrowly elliptical; submerged lvs usually
        narrowly elliptical or linear.   6
6   Rosette concave above; stamens and stigmas submerged; lower lvs
        linear, parallel-sided but with an expanded and deeply emarginate
        (bicycle spanner) tip; fr. almost circular, c. 1·2 mm. diam.
                                                         4. hamulata
    Rosette ±convex above; stamens and stigmas not submerged; lower
        lvs linear or narrowly elliptical, emarginate but not expanded at
        the tip; fr. somewhat longer than broad, c. 1·5 mm. diam.
                                                        2. platycarpa

1. C. stagnalis Scop.

Annual to perennial, usually with floating lf-rosettes; sometimes prostrate on mud. Lowest lvs elliptical or spathulate, slightly emarginate; rosette-lvs c. 10–20 mm., with a broadly elliptical or almost circular blade narrowed rather abruptly into the stalk; blade usually 5-veined, rounded or slightly emarginate at the tip. Fls in the axils of rosette-lvs only. Stamen c. 2 mm. Stigmas 2–3 mm., soon curving downwards, persisting in fr. Fr. 1·6–2 × 2 mm., suborbicular, conspicuously keeled, with deep lateral grooves; seeds broadly winged, divergent. Fl. 5–9. Ponds, ditches, streams, wet mud, etc., to 3000 ft.; common throughout the British Is.

2. C. platycarpa Kütz.

Perennial, usually with ± convex floating lf-rosettes; sometimes prostrate on mud. Lowest lvs linear or narrowly elliptical, emarginate but not expanded at the tip; rosette-lvs c. 10–20 mm., with a broadly elliptical blade ± gradually narrowed into the stalk; blade 3(–5)-veined, slightly emarginate at the tip. Fls in the axils of rosette-lvs. Stamen c. 4 mm. Stigmas 3–5 mm., their erect bases sometimes persistent in fr. Fr. 1·5–1·8 × 1·3–1·5 mm., suborbicular to shortly elliptical, distinctly keeled with fairly deep lateral grooves; seeds winged, parallel. Fl. 5–9. Ponds, ditches, streams, etc.; probably common throughout lowland Britain.

3. C. obtusangula Le Gall

Perennial, usually with dense convex floating lf-rosettes, sometimes prostrate on mud. Lowest lvs (sometimes not persisting until fl.) 10–40 × 0·5–2 mm., linear, deeply emarginate but not expanded at the tip. Rosette-lvs with a rhombic blade rather abruptly narrowed into the stalk; blade 3–5-veined, slightly emarginate at the tip. Fls in the axils of rosette-lvs. Stamen c. 5 mm. Stigmas c. 4 mm., their erect bases persisting in fr. Fr. c. 1·5 × 1 mm., with broadly rounded lobes and barely discernible grooves; seeds unwinged. Fl. 5–9. Ponds, ditches, lakes; locally frequent in S. England, rarer northwards to Inverness, Ireland.

4. C. hamulata Kütz. (*C. intermedia* Hoffm.)

Commonly a winter annual, with or without concave floating lf-rosettes; sometimes prostrate on mud. Lower lvs 10–25(–40) × 0·5–1·3 mm., linear, ± parallel-sided for most of their length then widened and deeply emarginate (like a bicycle spanner) at the tip; rosette-lvs c. 1 cm., their blades c. 6 × 3 mm., elliptical-obovate, 3-veined, gradually narrowed into the stalk, rounded to slightly emarginate at the tip. Fls in the axils of submerged lvs. Stamen c. 1 mm. Stigmas 1·5–2 mm., reflexed close to the fr. Fr. 0·8–1·3 mm., sometimes stalked, suborbicular, with keeled lobes and distinct but shallow lateral grooves; seed narrowly winged. Fl. 4–9. Lakes, pools, ditches and slow streams throughout the British Is.

5. C. hermaphroditica L. (*C. autumnalis* L.)

Stems 15–50 cm., yellowish, entirely submerged. Lvs 10–20 × 1–2 mm., 1-veined, those in the middle of the stem longest, linear-lanceolate, widest at the base and distinctly tapering to an emarginate tip, pale green, darkening on drying. Bracteoles 0. Stamens with filaments little longer than anthers. Stigmas soon falling. Fr. (1·5–)2 mm. diam., ± orbicular, of 4 readily separating, broadly and acutely winged lobes. Fl. 5–9. Lakes and streams in N. Britain from Anglesey and Yorks northwards; Ireland.

72. LORANTHACEAE

Mostly shrubs partially parasitic on trees. Lvs usually opposite or whorled, entire. Fls actinomorphic, hermaphrodite or unisexual. Perianth of 2 whorls, the outer sometimes suppressed; petals free or united. Stamens inserted on the inner per. segs. Ovary inferior. Fr. a berry with 1 seed.

1. Viscum L.

Fls unisexual. Outer per. segs much reduced, inner sepal-like, usually 4. Stamens sessile, opening by pores. Berries white, viscid.

1. V. album L.   Mistletoe.

A somewhat woody evergreen. Stems up to c. 1 m., green, much-branched, branching apparently dichotomous. Lvs 5–8 cm., narrow-obovate and often somewhat curved, blunt, thick and leathery, yellow-green, narrowed at base into a short stalk. Infl. of 3–5 subsessile fls, usually dioecious. Fr. c. 1 cm. diam. Fl. 2–4. Parasitic on a great variety of deciduous trees, most commonly apple, rarely on evergreens and very rarely on conifers; mainly in S. England and W. Midlands.

73. SANTALACEAE

Herbs or woody plants, mostly root-hemiparasites, with entire lvs; stipules 0. Fls actinomorphic, hermaphrodite or unisexual. Perianth of 1 whorl, sepaloid or petaloid, often fleshy; lobes 3–6, valvate. Stamens the same in number as the per. segs and opposite them. Ovary inferior or half-inferior, 1-celled; ovules 1–3. Fr. a nut or drupe; seeds endospermic, without testa.

1. THESIUM L.

Perennial hemiparasitic herbs attached by their roots to roots of host-plants. Lvs spirally arranged, narrow. Fls small, greenish, hermaphrodite. Perianth funnel- or bell-shaped; lobes 5(–4). Ovules 3.

1. T. humifusum DC.   Bastard Toadflax.

Stock woody. Stems spreading or prostrate, slender, yellow-green, glabrous, rough-angled. Lvs 5–15(–25) mm., linear, 1-veined; lower distant, scale-like. Infl. terminal, cymose; 1 bract and 2 narrow bracteoles at base of fl.-stalk. Fls 3 mm. diam. Fr. 3 mm., ovoid, ribbed, green, crowned by persistent perianth. Fl. 6–8. Parasitic on roots of various herbs of chalk and limestone grassland. Local in S. and C. England.

74. CORNACEAE

Trees or shrubs, rarely herbs. Lvs simple, usually stipulate. Fls small, usually in panicles, sometimes in umbels or heads, actinomorphic, usually hermaphrodite. Sepals 4(–5), small; petals 4(–5), free, rarely 0; stamens 4–(5), alternating with petals; ovary inferior, usually 2-celled with 1 ovule in each cell. Fr. a drupe, rarely a berry; seeds endospermic.

Shrub; fls cream-coloured in corymbose panicles without involucral
    bracts; fr. black.   1. THELYCRANIA
Herb with lvs 3–5-veined from the base; fls tiny, purplish-black, in a
    terminal fl.-like umbel with 4 large white involucral bracts; fr. red.
    A local plant of mountain moors, very rare in N. England, more
    frequent in Scotland. Dwarf Cornel.
       Chamaepericlymenum suecicum (L.) Ascherson & Graebner

1. THELYCRANIA (Dumort.) Fourr.

Trees or shrubs, usually deciduous. Lvs usually opposite, entire; stipules 0. Fls in corymbose cymes; bracts 0. Ovary 2-celled. Fr. a drupe with a 2-celled stone.

1. T. sanguinea (L.) Fourr. (*Cornus sanguinea* L.) Dogwood.

Deciduous shrub ¼–4 m. Twigs purplish-red at least on the sunny side. Lvs opposite, blade 4–8 cm., ovate or oval, cuspidate or acuminate, rounded at base, appressed-pubescent on both sides, paler below, usually becoming purplish-red in autumn; main lateral veins curving round towards the apex; stalk 8–15 mm., grooved above. Infl. many-fld, its stalk 2·5–3·5 cm. Calyx-teeth very small. Petals 4–6 mm., creamy-white. Fr. 6–8 mm. diam., subglobose, black. Fl. 6–7. Woods and scrub, especially on calcareous soils. Common locally in England and Wales northwards to Durham and Cumberland; in Ireland native only in the south midlands and the north-west; introduced in Scotland and elsewhere in Ireland.

*T. sericea (L.) Dandy (*Cornus stolonifera* Michx), with deep blood-red twigs, numerous decumbent suckering branches and white fr., is often planted and occasionally naturalized.

75. ARALIACEAE

Trees, shrubs or woody climbers, occasionally herbs. Lvs usually spirally arranged, simple or compound, usually stipulate. Fls small, regular, hermaphrodite or unisexual, usually 5-merous. Sepals small or obsolete. Petals free or united, Stamens usually equal in number and alternate with the petals, epigynous. Ovary inferior; cells usually as many as the petals; ovules 1 per cell. Fr. a drupe or berry; seeds endospermic.

1. HEDERA L.

Evergreen woody root-climbers. Lvs spirally arranged, simple, leathery, glabrous, exstipulate. Fls in terminal umbels. Calyx with 5 small teeth. Petals 5, free, valvate. Stamens 5. Ovary 5-celled; styles joined into a column. Fr. a berry.

1. H. helix L. Ivy.

Stems climbing to 30 m. or creeping along the ground, up to 25 cm. diam., densely clothed with adhesive roots. Young twigs stellate-pubescent. Lvs dark green above, often with pale veins, paler beneath; blades of those on creeping or climbing stems 4–10 cm., palmately 3–5-lobed with ± triangular entire lobes; those of the fl. branches (which are found only where sun reaches the tops of climbing stems) entire, ovate or rhombic. Fls in subglobose umbels, these often grouped into a terminal panicle; infl.-branches, fl.-stalks and receptacle stellate-tomentose.

Petals yellowish-green, 3–4 mm. Fr. black, globose, 6–9 mm. Fl. 9–11. Climbing in woods, hedges or on rocks and walls or creeping in woods on all but very acid, very dry or water-logged soils; reaching 2000 ft. Common throughout the British Isles.

76. UMBELLIFERAE

Herbs, very rarely shrubs. Stems often furrowed, pith wide and soft or internodes hollow. Lvs alternate, exstipulate, usually much divided. Infl. an umbel, rarely a head; bracts and bracteoles usually present, whorled. Calyx-teeth 5, usually small, often 0. Petals 5, generally notched with an inflexed point, sometimes very unequal. Stamens 5, alternating with petals. Ovary inferior, 2(–1)-celled, crowned by the enlarged base of the two styles (*stylopodium*). Carpels indehiscent, usually separating from each other in fr.

1	Lvs peltate.	1. HYDROCOTYLE
	Lvs not peltate.	2
2	Lvs entire.	BUPLEURUM (see p. 233)
	Lvs variously toothed or divided.	3
3	Lvs spiny.	3. ERYNGIUM
	Lvs not spiny.	4
4	Lvs palmately lobed.	5
	Lvs pinnate or ternate.	6
5	Bracts small, inconspicuous.	2. SANICULA
	Bracts forming a conspicuous coloured involucre. Naturalized in a few places.	*Astrantia major* L.
6	Aerial lvs few or 0; aquatics with finely divided submerged lvs.	7
	Aerial lvs numerous and well-developed at fl.	8
7	Plant slender; rays 1–2(–3).	11. *Apium inundatum*
	Plant stout; rays 4 or more.	19. OENANTHE
8	Lower lvs or lowest aerial lvs simply pinnate or pinnately lobed.	9
	Lower lvs 2–4-pinnate or ternate.	22
9	Plant ± hairy, sometimes minutely so (use a lens).	10
	Plant quite glabrous.	14
10	Calyx-teeth conspicuous, about as long as petals.	11
	Calyx-teeth minute or 0.	12
11	Lvs stiffly hairy beneath, glabrous above; inner fls long-stalked, outer subsessile. A rare casual.	

Turgenia latifolia (L.) Hoffm. (*Caucalis latifolia* L.)

Lvs stiffly hairy on both sides; all fls subsessile; fr. hispid, strongly compressed with much thickened margins. Doubtfully native. Hedge-banks and riversides in S.E. England; rare.

Tordylium maximum L.

12 Fls yellow.   23. PASTINACA
     Fls white or pinkish.   13

13 Bracteoles 0.   16. PIMPINELLA
     Bracteoles several, narrow, reflexed.   24. HERACLEUM

14 Segments of all lvs very narrow; partial umbels ± globose in fr.
                               19. OENANTHE
     Segments of at least the lower lvs lanceolate to suborbicular; partial umbels ± flat or irregular.   15

15 Bracts present, at least some ½ as long as the shorter rays. 16
     Bracts 0, or if present then all much shorter than rays. 18

16 Bracts awl-shaped, stem solid.   12. PETROSELINUM
     Bracts broader, often divided; stem hollow.   17

17 Lf-segments 4–6 pairs. Ditches, etc.; local.   *Sium latifolium* L.
     Lf-segments 7–10 pairs.   18. BERULA

18 Bracts 0 or if present then umbels lf-opposed.   19
     Bracts always present; umbels not lf-opposed.   21

19 At least some umbels sessile and lf-opposed.   11. APIUM
     Umbels all distinctly stalked, terminal.   20

20 Bracteoles 0.   16. PIMPINELLA
     Bracteoles present. Erect annual; fr. hard, red-brown. Introduced. A rare casual. Coriander.   *Coriandrum sativum* L.

21 Lf-segments palmately divided into very fine lobes. 14. CARUM
     Lf-segments not as above; plant with a smell of petrol and nutmeg when crushed.   13. SISON

22 Plant ± hairy, often minutely so, hairs sometimes confined to stalks or rays of umbels or margins and nerves of upper surface of lf (use a lens).   23
     Plant glabrous; lf margins sometimes finely toothed.   37

23 Stems with a large hollow in the middle.   24
     Stems solid or nearly so.   30

24 Only rays and fl.-stalks hairy.   PEUCEDANUM (see p. 240)
     Plant hairy elsewhere.   25

25 Segments of lower lvs 3 cm. or more, not or slightly pinnatifid. 26
     Segments of lower lvs smaller, frequently and deeply pinnatifid. 28

26 Bracteoles 0.   16. PIMPINELLA
     Bracteoles several.   27

27 Lvs ternately divided; stems terete, nearly smooth. 22. ANGELICA
     Lvs pinnately divided; stems ridged.   24. HERACLEUM

28 Plant not aromatic.   29
    Plant strongly aromatic.   7. MYRRHIS

29 Rays 3 or more; bracteoles entire or 0.   5. ANTHRISCUS
    Ray 1(–2); some bracteoles 2–3-fid.   6. SCANDIX

30 Stems, at least in the younger parts, with short straight appressed
        downward-pointing hairs, not purple-spotted.   8. TORILIS
    Stems purple-spotted or glabrous, or hairs short and crisped or
        longer and pointing in various directions, not appressed.   31

31 Stems and sheaths glabrous; lvs sparsely and minutely hairy on nerves
        and near margins on upper surfaces. S.W. England and Bucks.,
        very local. Bladder-seed.   *Physospermum cornubiense* (L.) DC.
    Stems hairy, at least near nodes.   32

32 Hairs minute, crisped.   33
    Hairs spreading or pointing in various directions, or stems purple-
        spotted.   34

33 Bracts 0; stock without fibres.   16. PIMPINELLA
    Bracts present; stock crowned with fibres Chalk grassland in S. and
        E. England; very local.   *Seseli libanotis* (L.) Koch

34 Bracts 7–13, large, ternate or pinnatifid.   25. DAUCUS
    Bracts 0–5, small, simple.   35

35 Bracteoles large, green, some 2–3-fid.   6. SCANDIX
    Bracteoles all entire, often ± scarious.   36

36 Stem purple-spotted or entirely purple; rays (4–)8–25.
                                                4. CHAEROPHYLLUM

    Stem not purple-spotted; rays 2–3(–5). A rare casual.
                                              Caucalis platycarpos L.

37 Fls yellow, yellowish or pinkish.   38
    Fls white or only slightly tinged with green, pink or purple.   43

38 Lf-segments linear, entire.   39
    Lf-segments not linear, but pinnatifid, lobed or toothed.   41

39 Lvs ternate, segments long-acuminate; pedicels long, filiform.
                                                PEUCEDANUM (see p. 240)
    Lvs pinnate, segments filiform or acute; pedicels not exceptionally
        long or slender.   40

40 Lf-segments fleshy; bracts and bracteoles numerous; bushy plant
        up to c. 30 cm. Sea cliffs; local. Rock Samphire.
                                              Crithmum maritimum L.
    Lf-segments very narrow, not fleshy; bracts and bracteoles 0–1;
        erect plant up to c. 130 cm., smelling of aniseed. Waste places,
        often near the sea; locally naturalized. Fennel.
                                        Foeniculum vulgare Miller

41   Segments of lower lvs 3 cm. or more, crenate or toothed, blunt.
                                        9. SMYRNIUM
     Segments of lower lvs less than 3 cm., pinnatifid or lobed, acute.   *42*

42   Sheathing bases of petioles with broad hyaline margins; lf-margins
     not minutely toothed.   12. PETROSELINUM
     Sheathing bases of petioles with green or only narrowly hyaline
     margins; lf-margins minutely toothed.   21. SILAUM

43   Stems ± glaucous, purple-spotted.   10. CONIUM
     Stems not purple-spotted.   *44*

44   Bracteoles 0 or 1.   *45*
     Bracteoles 2 or more.   *47*

45   Segments of lower lvs 3 cm. or more, toothed but not pinnatifid;
     plant rhizomatous.   17. AEGOPODIUM
     Segments of lower lvs smaller, pinnatifid; plant not rhizomatous.   *46*

46   Basal lvs soon withering; petioles long and slender, largely under-
     ground; plant perennial, tuberous.   15. CONOPODIUM
     Basal lvs persistent; petioles short, stout, sheathing, arising above
     ground; plant biennial, root spindle-shaped.   14. CARUM

47   Umbels simple or of 2 stout rays; bracteoles ciliate, usually bifid or
     pinnatifid; fr. 3–7 cm.   6. SCANDIX
     Umbels with (2–)3 or more rays; bracteoles not as above; fr. not
     more than 1 cm.   *48*

48   Lf-segments narrow, up to 30 cm. regularly and sharply toothed,
     ± sickle-shaped; bracts and bracteoles numerous, narrow.
     Naturalized in a few places.   **Falcaria vulgaris* Bernh.
     Not as above.   *49*

49   Bracts ½–⅔ length of rays, at least some pinnatifid with narrow
     segments. Uncommon casual.   **Ammi majus* L.
     Bracts smaller or 0, never pinnatifid, rarely 3-cleft.   *50*

50   Bracteoles 3–4 on outside of each partial umbel, conspicuous,
     green, pointing downwards.   20. AETHUSA
     Not as above.   *51*

51   Lower lvs soon withering; stem tapering downwards from surface
     of earth to junction with tuber, underground portion flexuous.   *52*
     Not as above.   *53*

52   Stem hollow after fl.; stylopodium gradually narrowed into the
     styles; styles suberect in fr. Common, but almost absent from
     chalky soils.   15. CONOPODIUM
     Stem solid after fl.; stylopodium abruptly contracted into the
     styles; styles reflexed in fr. Rare; on chalk in south-east
     England.   *Bunium bulbocastanum* L.

53   Lf-segments very narrow; abundant persistent fibrous remains of
     old lvs at base of stem.   *54*
     Segments of at least the lower lvs broader; fibres 0 or very scanty.   *55*

54 Stem 3–20 cm., solid; plant glaucous, branched from base; dioe-
    cious. Very local on limestone in S.W. England. Honewort.
                                          Trinia glauca (L.) Dumort.
    Stem 20–60 cm., hollow; plant neither glaucous nor dioecious.
    Mountain districts; local. Spignel, Meu.
                                          Meum athamanticum Jacq.

55 Segments of lower lvs ovate, obovate or suborbicular, little lobed,
    at least some 2 cm. or more.   56
    Segments of lower lvs smaller and usually narrower, often deeply
    pinnatifid.   58

56 Lower lvs 3–4-pinnate, segments many.   19. OENANTHE
    Lower lvs 1–2-ternate, segments few.   57

57 Lf-segments toothed in upper half only; rays 8–12. Rocky
    northern coasts. Lovage.   *Ligusticum scoticum* L.
    Lf-segments toothed nearly to base; rays 20–25.
                                          PEUCEDANUM (see p. 240)

58 Lf-segments narrow-lanceolate, sharply toothed but not lobed;
    bracts 0; bracteoles strap-shaped, some longer than the partial
    umbels. In shallow water; very local. Cowbane.
                                          Cicuta virosa L.
    Lf-segments broader and variously lobed or else not toothed.   59

59 Pedicels not longer than the oblong fr.; roots with several tubers or
    plant aquatic often with translucent submerged lvs.
                                          19. OENANTHE
    Pedicels longer than fr. or else fr. globose; roots not tuberous;
    translucent submerged lvs 0.   60

60 Bracts 4 or more.   61
    Bracts fewer than 4.   62

61 Stems hollow, strongly ridged and ±angled.
                                          PEUCEDANUM (see p. 240)
    Stems solid or nearly so, striate, terete. S.W. England and Bucks.;
    very local. Bladder-seed.
                                          Physospermum cornubiense (L.) DC.

62 Annual; lf-segments not toothed; rays (2–)3–5(–10), smooth.
    Rare casual. Coriander.   *Coriandrum sativum* L.
    Perennial; lf-segments minutely toothed; rays 10–20, rough on the
    ridges. Fens in E. England; rare.   *Selinum carvifolia* (L.) L.

1. HYDROCOTYLE L.

1. H. vulgaris L.   Pennywort, White-rot.

Slender, creeping or sometimes floating, rooting freely at the nodes.
Petioles 1–25 cm., erect, slightly hairy; blade 1·5 cm. diam., peltate,

orbicular, crenate. Stalks of infl. shorter than petioles; whorls of fls 2–5 mm. diam., of 2–5 fls, solitary or sometimes several. Bracts triangular. Fls c. 1 mm. diam., pinkish-green. Fr. 2 mm. diam., covered with brownish resinous dots. Fl. 6–8. In damp or wet places, usually on acid soils; common in suitable habitats.

2. SANICULA L.

Perennial. Lvs palmately lobed. Umbels small, irregular, compound; partial umbels subglobose; bracts few, small, lf-like; bracteoles few. Calyx-teeth longer than petals. Fr. ovoid, with hooked bristles.

1. S. europaea L.   Sanicle.

20–60 cm., glabrous. Basal lvs 2–6 cm., 3–5-lobed, lobes cuneate, coarsely toothed. Umbels of few (often 3) few-fld partial umbels; bracts 3–5 mm., 2–5, simple or pinnatifid. Fls pink or white, outer shortly stalked, inner subsessile. Fl. 5–9. In woods, locally common except on poor soils.

3. ERYNGIUM L.

Perennial, rigid, often glaucous and spiny. Fls bracteolate, sessile in dense heads, surrounded by rigid lf-like bracts; calyx-teeth longer than petals. Fr. ovoid.

Plant glaucous; basal lvs 3-lobed; bracts oblong-cuneate, spiny-toothed.   **1. maritimum**
Plant pale green; basal lvs pinnate; bracts narrowly lanceolate, nearly or quite entire. Very rare; in S. England.   *E. campestre* L.

1. E. maritimum L.   Sea Holly.

30–60 cm. Basal lvs 5–12 cm. diam., stalked, suborbicular; cauline sessile, palmate; both spinous-toothed with a thickened cartilaginous border. Heads 1·5–2·5 cm. Bracteoles narrow, spiny, 3-fid, often purplish-blue and rather longer than fls. Fls c. 8 mm. diam., bluish. Fr. covered with hooked projections. Fl. 7–8. On sandy and shingly shores around the coasts.

4. CHAEROPHYLLUM L.

Lvs usually 2–3-pinnate. Umbels compound; bracts few or 0; bracteoles several. Fls white (British spp.); calyx-teeth small or 0. Fr. oblong or oblong-ovoid, scarcely beaked, laterally flattened.

Biennial; lvs dark green; styles not longer than stylopodium; ripe fr. 5–7 mm.   **1. temulentum**

Perennial; lvs yellow-green; styles twice as long as stylopodium; ripe fr. 8–12 mm. Naturalized in a few places in Scotland.   *C. aureum L.

1. C. temulentum L.   Rough Chervil.

30–100 cm., rough. Stem solid, swollen below the nodes, purple-spotted or entirely purple. Lvs up to 20 cm.; segments ovate, pinnately lobed, lobes toothed. Umbels 3–6 cm. diam., rather irregular, nodding in bud, rays (4–)8–10(–15), rough-hairy, slender; bracts 0–2, bracteoles 5–8, lanceolate to ovate, aristate, fringed, spreading in fl., deflexed in fr. Fr. 5–7 mm., oblong-ovoid, narrowed upwards. Fl. 6–7. In hedge-banks, etc., common.

5. ANTHRISCUS Pers.

Lvs usually 2–3-pinnate. Umbels compound; bracts 0(–1); bracteoles several. Fls white; calyx-teeth minute or 0. Fr. ovoid or oblong, beaked and with ridges only on the beak.

1 Some umbels shortly stalked; rays usually not more than 5; stem
     nearly smooth.   *2*
     Umbels all long-stalked; rays usually 8–16; stem furrowed.
                                          2. sylvestris

2 Rays glabrous; pedicels thicker than rays in fr.; stem glabrous.
                                          1. caucalis
     Rays hairy; pedicels not thicker than rays; stem hairy above nodes.
     Introduced, very local.   *A. cerefolium* (L.) Hoffm.

1. A. caucalis Bieb. (*A. neglecta* Boiss. & Reuter)   Bur Chervil.
Stems 25–50(–100) cm., hollow, glabrous, somewhat thickened below the nodes. Lvs up to 10 cm., usually less, glabrous above, with scattered hairs beneath; segments pinnatifid, lobes blunt. Umbels 2–4 cm. diam., rays 3–6, often widely spreading or recurved in fr. Bracteoles c. 2 mm., ovate, aristate, fringed. Fr. 3 mm., ovate, prickly, beak short, glabrous; pedicels with a ring of hairs at top. Fl. 5–6. In hedge-banks, etc., chiefly on light soils, local and absent from the north.

2. A. sylvestris (L.) Hoffm.   Cow Parsley, Keck.
Stems 60–100 cm., hollow, downy below, glabrous above. Lvs up to 30 cm., ± hairy; segments ovate, pinnatifid and coarsely toothed. Umbels 2–6 cm. diam.; rays (4–)8–10(–16), glabrous. Bracteoles 2–5 mm., ovate, aristate, fringed, spreading or deflexed, often pink. Fr. 5 mm., oblong-ovoid, smooth, beak very short. Fl. 4–6. In hedges, waste places and wood-margins. By far the commonest of the early-flowering umbellifers in the southern half of England.

6. SCANDIX L.

Lvs 2–3-pinnate. Umbels simple or compound; bracts 1 or 0; bracteoles several, entire or variously lobed. Fls white; calyx-teeth minute or 0; petals often unequal. Fr. subcylindric, with a very long beak.

1. S. pecten-veneris L.   Shepherd's Needle.

Nearly glabrous annual 15–50 cm. Stems hollow when old, striate. Lvs oblong or narrowly triangular; segments up to 5 mm., spathulate, scarcely toothed. Umbels simple or of 2 stout rays; bracteoles 5–10 mm., bifid or pinnatifid, sometimes entire, margins coarsely ciliate. Fr. 30–70 mm., rough. Fl. 4–7. In arable fields, widely distributed but rare in the north and west.

7. MYRRHIS Miller

Lvs 2–3-pinnate. Umbels compound; bracts few or 0; bracteoles several, whitish. Fls white; calyx-teeth minute or 0. Fr. narrowly oblong, beaked.

1. M. odorata (L.) Scop.   Sweet Cicely.

Strongly aromatic slightly hairy perennial 60–100 cm. Stems hollow, somewhat grooved. Lvs up to c. 30 cm., pale beneath; segments oblong-ovate, pinnatisect, lobes coarsely toothed; petioles of stem-lvs sheathing. Umbels 1–5 cm. diam.; rays 5–10; bracteoles c. 5, lanceolate, aristate. Fr. 20–25 mm., strongly and sharply ridged, ridges often rough. Fl. 5–6. Grassy roadsides, etc.; widely spread but common only in N. England and S. Scotland.

8. TORILIS Adanson

Annual. Lvs 1–3-pinnate. Umbels compound; bracts several or 0; bracteoles several, narrow. Fls white or pinkish; calyx-teeth small, triangular-lanceolate. Fr. ovoid; carpels with slender ciliate ridges, the furrows between thickly beset with tubercles or spines.

1	Umbels nearly sessile, with 2–3 very short rays.	**3. nodosa**
	Umbels long stalked, with 3–12 rays.	2
2	Bracts 4–12; rays 5–12	**1. japonica**
	Bracts 0–1; rays (2–)3–5(–8)	**2. arvensis**

1. T. japonica (Houtt.) DC.   Upright Hedge-parsley.

5–125 cm., appressed hairy. Stems solid, hairs deflexed. Lf-segments 1–2 cm., ovate to lanceolate, pinnatifid, toothed. Umbels 1·5–4 cm.

diam. Fls pinkish or purplish-white, outer petals longer than inner. Fr. 3–4 mm. Fl. 7–8. Abundant in hedges etc.

2. T. arvensis (Hudson) Link   Spreading Hedge-parsley.
10–40(–100) cm., ± appressed hairy. Stems solid, hairs deflexed, or sometimes almost glabrous. Lf-segments 0·5–3 cm., lanceolate, pinnatifid or coarsely toothed. Umbels 1–2·5 cm. diam. Fls white or pinkish, outer petals longer than inner. Fr. 4–5 mm. Fl. 7–9. In arable fields, rather uncommon.

3. T. nodosa (L.) Gaertner   Knotted Hedge-parsley.
5–35 cm., usually prostrate. Stems solid, hairs sparse, spreading, deflexed. Lf-segments ovate, deeply pinnatifid, lobes linear-lanceolate. Umbels 0·5–1 cm. Fls pinkish, petals subequal. Fr. 2–3 mm., outer carpels with spreading spines, inner tubercled. Fl. 5–7. On dry banks and in arable fields, locally common in the south, rarer elsewhere.

9. SMYRNIUM L.

Basal lvs ternately compound, segments broad. Umbels compound; bracts and bracteoles small and few or 0. Fls yellow; calyx-teeth small or 0. Fr. ovoid, laterally compressed and narrowed where the 2 carpels join; carpels with sharp ridges.

*1. S. olusatrum L.   Alexanders.
Stout, glabrous biennial 50–150 cm. Stem solid, becoming hollow when old, furrowed. Lvs dark green and shiny, the basal c. 30 cm.; segments rhombic, bluntly toothed or lobed, stalked; upper stem-lvs often opposite, petiole base sheathing. Umbels subglobose; rays (3–)7–15. Fls yellow-green. Fr. c. 8 mm., nearly black. Fl. 4–6. Widely naturalized in hedge-banks, on cliffs, etc., especially near the sea.

10. CONIUM L.

Glabrous biennials. Lvs 2–3-pinnate. Umbels compound; bracts and bracteoles few, small. Fls white; calyx-teeth 0. Fr. broadly ovoid to suborbicular; carpels with prominent ridges. Very poisonous.

1. C. maculatum L.   Hemlock.
Up to 2 m. Stems smooth, purple-spotted, glaucous. Lvs up to 30 cm., ovate to deltate; segments 1–2 cm., oblong-lanceolate to deltate, coarsely toothed. Umbels 2–5 cm. diam.; rays 10–20; bracts 2–5 mm.,

few, reflexed; bracteoles similar but smaller and only on one side of partial umbel. Fr. c. 3 mm., suborbicular; ridges wavy. Fl. 6–7. Damp places throughout most of the British Is.

BUPLEURUM L. Hare's Ear.

All spp. are rare or local.

1 Shrub, evergreen. Malvern Hills; introduced. **B. fruticosum* L.
   Herbs. 2
2 Upper lvs perfoliate; bracts 0. 3
   Upper lvs not perfoliate; bracts present. 4
3 Rays usually 2–3; bracteoles suborbicular; fruit covered with
   tubercles. Cornfields and waste places, rare.
   B. lancifolium Hornem.
   Rays usually 5–10; bracteoles oblanceolate to ovate; fruit without
   tubercles. Cornfields and waste places, very rare.
   B. rotundifolium L.
4 Perennial; stem hollow; bracteoles shorter than fls. Very rare.
   S.E. England. *B. falcatum* L.
   Annual; stem solid; bracteoles longer than fls. 5
5 Bracteoles ovate, concealing fls; fr. not granulate. Very rare.
   S. England. *B. baldense* Turra
   Bracteoles awl-shaped, not concealing fls; fr. granulate. Salt-
   marshes; local. *B. tenuissimum* L.

11. APIUM L.

Glabrous marsh or water plants. Lvs pinnate or ternate. Bracts few or 0; bracteoles several or 0. Fls white; calyx-teeth 0 or minute. Fr. broadly ovoid or elliptic-oblong, laterally compressed; carpels with equal or subequal ridges.

1 Lower lvs with deltate to rhombic segments, lower segments
   stalked; bracteoles 0. **1. graveolens**
   Lower lvs with sessile or very narrow segments; bracteoles present. 2
2 Lower lvs with deeply lobed segments, lobes very narrow; segments
   of upper lvs cuneate, often 3-lobed; fr. elliptic-oblong.
   3. inundatum
   All lvs with toothed, sometimes slightly lobed segments, never very
   narrow; fr. broadly ovoid. **2. nodiflorum**

1. A. graveolens L. Wild Celery.

Erect strong-smelling biennial 30–60 cm. Stem grooved. Basal lvs 1-pinnate; segments 0·5–3 cm., lobed and toothed; upper stem-lvs ternate, subentire. Umbels short-stalked or sessile; rays unequal. Fls greenish-white. Fr. c. 1·5 mm., broadly ovoid. Fl. 6–8. Native in damp places, especially near the sea.

2. A. nodiflorum (L.) Lag.   Fool's Watercress.

Prostrate or ascending perennial 30–100 cm. Stems finely furrowed. Lvs 1-pinnate, bright green and shiny; segments 1–3·5 cm., 4–6 pairs, lanceolate to ovate, toothed, often slightly lobed. Umbels lf-opposed, sessile or short-stalked; rays unequal, spreading or recurved; bracteoles c. 5, narrowly lanceolate. Fr. c. 2 mm., ovoid. Fl. 7–8. Ditches and shallow ponds; common and generally distributed.

3. A. inundatum (L.) Reichenb. fil.

Often submerged or floating. Stems 10–50 cm., nearly smooth. Lvs pinnate. Umbels lf-opposed, stalked, 1–3·5 cm.; rays 2–3(–4), somewhat unequal; bracteoles 3–6, lanceolate, blunt, unequal. Fr. 2 mm., elliptic or oblong. Fl. 6–8. Lakes, ponds and ditches; local but widely distributed.

12. PETROSELINUM Hill

Annual or biennial, glabrous, smelling like parsley. Lvs pinnate, segments broad. Umbels compound; bracts and bracteoles several. Fls white or yellowish; calyx-teeth 0. Fr. ovoid or subspherical, laterally compressed; carpels with 5 slender ridges.

Lvs 3-pinnate; fls yellowish.   **1. crispum**
Lvs 1-pinnate; fls white.   **2. segetum**

*1. P. crispum (Miller) Airy-Shaw   Parsley.

Stem 30–75 cm., stout, erect, solid, striate; branches ascending at a narrow angle. Lvs deltate, shiny; segments 1–2 cm., cuneate, lobed, much crisped in cultivars. Umbels 2–5 cm. diam., flat-topped; rays 8–15; bracts 2–3, erect, entire or 3-lobed; bracteoles c. 5, narrowly oblong to ovate-cuspidate. Fr. 2·5 mm., ovoid. Fl. 6–8. Naturalized in many scattered localities.

2. P. segetum (L.) Koch   Corn Caraway.

Stem 30–100 cm., slender, dark-green, ± glaucous, solid, striate; branches spreading at a wide angle, lower prostrate or nearly so. Lvs narrowly oblong; segments 0·5–1 cm., ovate, toothed or sometimes lobed, margins thickened. Umbels 1–5 cm. diam., very irregular; rays 2–5; bracts and bracteoles 2–4, awl-shaped. Fr. 3–4 mm., ovoid. Fl. 8–9. Hedgerows and grassy places, mainly in the south.

13. SISON L.

Biennial. Lvs pinnate, segments broad. Umbels compound, bracts and bracteoles few, rarely 0. Fls white, calyx-teeth 0. Fr. broadly ovoid or subglobose, laterally compressed; carpels with slender ridges.

1. S. amomum L.   Stone Parsley.

Erect, glabrous, with a nauseous smell when crushed. Stem 50–100 cm., solid, finely striate, branches ascending. Lvs 1-pinnate, lower 10–20 cm., long-stalked; segments 2–7 cm., oblong-ovate, toothed and often lobed, margins thickened; upper stem-lvs with very narrow segments. Umbels 1–4 cm. diam.; rays 3–6; bracts and bracteoles (0–)2–4, awl-shaped. Fr. 3 mm., subglobose. Fl. 7–9. Hedgerows, etc., mainly in the south.

14. CARUM L.

Perennial or biennial, glabrous, Lvs pinnate, segments usually narrow. Umbels compound; bracts and bracteoles several, few or 0. Fls white or pinkish; calyx-teeth minute. Fr. ovoid or oblong, laterally compressed.

Stem solid; lvs 1-pinnate, segments palmately divided into numerous
    very narrow lobes appearing as if whorled; bracts and bracteoles
    several. Very local. Chiefly in the west. Whorled Caraway.
                                  C. verticillatum (L.) Koch
Stem hollow; lvs 2-pinnate, segments pinnately divided into rather
    narrow lobes; bracts and bracteoles 0 or 1.   **1. carvi**

*1. C. carvi L.   Caraway.

Stem 25–60 cm., erect, much-branched, striate. Lvs all basal in 1st year, narrowly triangular to oblong. Umbels 2–4 cm. diam., rather irregular; rays 5–10; bracts and bracteoles very narrow. Fr. 3–4 mm., oblong, strong-smelling when crushed; ridges low, blunt. Fl. 6–7. Naturalized by roads, etc.; widely distributed but not common.

15. CONOPODIUM Koch

Perennial, with tubers. Lvs ternately divided, segments narrow. Umbels compound; bracts and bracteoles few or 0, thin. Fls white; calyx-teeth 0; petals of outer fls often unequal. Fr. ovoid or oblong, often shortly beaked, narrowed at junction of 2 carpels; ridges slender, inconspicuous.

1. C. majus (Gouan) Loret   Pignut, Earthnut.

Slender erect glabrous perennial 30–50(–90) cm. Tuber 1–3·5 cm., dark brown, irregular. Stem hollow after fl., greatly narrowed near base. Basal lvs 5–15 cm., broadly deltate, soon withering; petioles slender, flexuous, mostly underground; primary divisions stalked, 2-pinnate; segments cut into narrow lobes. Umbels 3–7 cm., diam. nodding in bud; rays 6–12; bracts and bracteoles 0–5, narrow. Fr. 4 mm., narrowly ovoid, beaked. Fl. 5–6. Fields and woods; widely distributed but absent from chalk.

16. PIMPINELLA L.

Perennial, rarely annual. Lvs usually pinnate. Umbels compound; bracts 0, bracteoles 0 or few. Fls white or pink; calyx-teeth small or 0. Fr. ovoid or oblong, laterally compressed; carpels with slender ridges.

Stem with small hairs, subterete, tough.   **1. saxifraga**
Stem glabrous, strongly ridged and angled, brittle.   **2. major**

1. P. saxifraga L.   Burnet Saxifrage.

30–100 cm. Basal lvs and upper stem-lvs usually 1-pinnate, lower stem-lvs usually 2-pinnate, very variable; segments 1–2·5 cm., ovate to narrowly lanceolate, toothed or cut, sessile or nearly so; petioles of upper stem-lvs sheath-like, often purplish. Umbels 2–5 cm. diam.; rays 10–20. Styles much shorter than petals. Fr. 3 mm., broadly ovoid. Fl. 7–8. Dry grassy usually calcareous places; widely distributed.

2. P. major (L.) Hudson   Greater Burnet Saxifrage.

Stem 50–120 cm. often magenta towards base. Lvs all 1-pinnate (lower rarely 2-pinnate); segments 2–8 cm., those of basal lvs ovate, sub-cordate, shortly stalked, those of stem-lvs narrower, sessile, all toothed and sometimes lobed; petioles of upper stem-lvs sheath-like, green. Umbels 3–6 cm. diam.; rays 10–20. Styles about as long as petals. Fr. 4 mm., ovoid. Fl. 6–7. Grassy places at margins of woods and on hedge-banks; widely scattered but local.

17. AEGOPODIUM L.

Glabrous rhizomatous perennials. Lvs 1–2-ternate, segments broad. Umbels compound; bracts and bracteoles few or 0. Fls white; calyx-teeth 0. Fr. ovoid, laterally compressed; carpels with slender ridges.

1. A. podagraria L.

Goutweed, Bishop's Weed, Ground Elder, Herb Gerard.

40–100 cm., stout, erect. Rhizomes far-creeping, white when young. Stem hollow, grooved. Lvs 10–20 cm., deltate; segments 4–8 cm.,

sessile or shortly stalked, lanceolate to ovate, acuminate, often oblique
at base, irregularly toothed; petioles much longer than blade, triangular
in section. Umbels 2–6 cm. diam., rays 15–20; bracts and bracteoles
usually 0. Fr. 4 mm., ovoid. Fl. 5–7. In waste places and a persistent
weed in gardens; generally distributed.

18. BERULA Koch

Glabrous perennials. Lvs pinnate, segments broad. Umbels compound;
bracts and bracteoles many. Fls white; calyx-teeth small, acute. Fr.
orbicular, deeply constricted where the 2 carpels join; carpels with
prominent blunt or thickened ridges, the lateral not marginal.

1. B. erecta (Hudson) Coville   Narrow-leaved Water-parsnip.

Stoloniferous. Stem 30–100 cm., hollow, striate. Lvs up to 30 cm.,
long-stalked, 1-pinnate, dull bluish-green; segments 2–5 cm., sessile,
(5–)7–10 pairs, oblong, lanceolate to ovate, toothed, or slightly lobed;
stem-lvs small, usually very irregularly toothed. Umbels 3–6 cm. diam.,
lf-opposed, rather irregular; rays usually 10–15; bracts and bracteoles
lf-like, often 3-fid. Fr. 2 mm., almost suborbicular, broader than long.
Fl. 7–9. Ditches, marshes, etc.; widely distributed.

19. OENANTHE L.

Glabrous, usually marsh or water plants. Lvs 1–3-pinnate. Umbels
compound; bracts variable, bracteoles (British spp.) many. Fls white;
calyx-teeth acute. Fr. ovoid, cylindrical or globose, ± terete.

1	Segments of upper lvs narrow.	*2*
	Segments of upper lvs lanceolate to ovate.	*4*
2	Blade of stem-lvs shorter than hollow stalk.	**1. fistulosa**
	Blade of stem-lvs longer than solid or flattened stalk.	*3*
3	Roots with rounded tubers towards their ends; stem solid; bracteoles narrow; pedicels thickened at top. Rare. *O. pimpinelloides* L.	
	Roots tuberous for most of their length; stem hollow; bracteoles ovate-acuminate; a strong constriction at junction of fr. and stalk. Rare. *O. silaifolia* Bieb.	
	Roots cylindrical or somewhat spindle-shaped; stem solid (rarely hollow); bracteoles oblong-lanceolate; pedicels neither thickened nor constricted at top. **2. lachenalii**	
4	Roots tuberous; umbels not lf-opposed, most with stalks longer than rays. **3. crocata**	

Roots fibrous; at least some umbels lf-opposed, with stalks shorter
    than rays.   5
5 Segments of submerged lvs (if present) very narrow, cylindric; fr.
    3–4 mm., about twice as long as styles.   **4. aquatica**
   Segments of submerged lvs cuneate, cut at tip; fr. 5–6 mm., at least
    3 times as long as styles. Rivers in the south; local.
                                         O. fluviatilis (Bab.) Coleman

1. O. fistulosa L.   Water Dropwort.

30–60 cm., erect. Roots with spindle-shaped tubers. Stems hollow,
often constricted at nodes and rooting at lower nodes. Lvs 1(–2)-pin-
nate; segments of lower lvs shortly stalked, ovate, lobed; of upper very
narrow, entire, distant. Umbels stalked; rays 2–4, stout; partial umbels
c. 1 cm. diam., dense, flat-topped in fl., spherical in fr.; bracteoles shorter
than pedicels. Fr. 3–4 mm., angular; pedicels not thickened at top.
Fl. 7–9. Marshes and shallow water, widely distributed but rather
local.

2. O. lachenalii C. C. Gmelin   Parsley Water Dropwort.

30–100 cm., erect. Roots cylindrical or somewhat spindle-shaped. Stems
solid or with a small cavity when old. Lvs 2-pinnate, lower soon
withering; lobes narrow, ± obtuse. Rays 5–15, slender (c. 0·2 mm.
diam.) in fr. Fr. 2 mm., ovoid. Fl. 6–9. Brackish and freshwater marshes
and fens; widely distributed but rather local.

3. O. crocata L.   Hemlock Water Dropwort.

50–150 cm., stout, erect. Root tubers 2–3 cm. diam., spindle-shaped,
sweetish-tasting but poisonous. Stems hollow, grooved. Lvs 30 cm. or
more, deltate, 3–4-pinnate; segments ovate or suborbicular, cuneate at
base, 1–2-lobed and toothed; segments of stem-lvs narrower; petioles
entirely sheathing. Umbels 5–10 cm. diam.; stalks usually longer than
rays; rays 12–40; bracts and bracteoles many, soon falling. Fr. 4–6 mm.,
cylindrical. Fl. 6–7. Wet places, mainly in the south and west.

4. O. aquatica (L.) Poiret   Fine-leaved Water Dropwort.

30–150 cm., stoloniferous. Root fibrous. Stem hollow, striate, often
very stout. Lvs 3-pinnate, all aerial or lower submerged; submerged lvs
with very narrow segments; aerial lvs with deeply lobed segments, lobes
lanceolate to ovate, acute. Umbels 2–5 cm. diam., some lf-opposed;
stalks usually shorter than rays; rays 4–10; bracts 0–1; bracteoles
several. Fr. 3–4 mm., ovoid or oblong-ovoid. Fl. 6–9. Slow-flowing or
stagnant water; local.

20. AETHUSA L.

Glabrous annual. Lvs ternately 2-pinnate. Umbels compound; bracts 0–1, bracteoles 1–5, deflexed. Fls white; calyx-teeth small or 0. Fr. broadly ovoid; carpels dorsally compressed, ridges very broad, lateral narrowly winged.

1. A. cynapium L. Fool's Parsley.

Stem 5–120 cm., hollow, finely striate, somewhat glaucous. Lvs deltate, segments ovate, pinnatifid. Umbels 2–6 cm. diam., some lf-opposed; rays (4–)10–20; bracts usually 0; bracteoles usually 3–4 on the outer side of the partial umbels, narrow. Petals unequal. Fr. 3–4 mm. Fl. 7–8. A weed of cultivated ground; common and widely distributed.

21. SILAUM Miller

Glabrous perennials. Lvs 1–3-pinnate. Umbels compound; bracts few or 0, bracteoles several. Fls yellowish; calyx-teeth minute. Fr. ovoid or oblong; carpels with slender ridges, the lateral ones winged.

1. S. silaus (L.) Schinz & Thell. Pepper Saxifrage.

Stem 30–100 cm., erect, solid, striate. Lower lvs deltate, 2–3-pinnate, segments 1–1·5 cm., entire or pinnatisect, exceedingly finely serrate; upper lvs few, small, sometimes reduced to petioles. Umbels 2–6 cm. diam., long-stalked; rays 5–10, rather unequal; bracteoles narrowly lanceolate with scarious margins. Fr. 4–4·5 mm., oblong-ovoid. Fl. 6–8. Meadows, etc.; rather local and becoming rare in the north.

22. ANGELICA L.

Tall perennials. Lvs 2–3-pinnate, segments broad and large. Umbels compound; bracts few or 0, bracteoles usually many. Fls greenish, white or pinkish; calyx-teeth minute or 0. Fr. ovoid, dorsally compressed; carpels flat with broad marginal wings and 3 dorsal ridges.

Stems usually purplish; lf-segments not decurrent; fls white or pinkish.
                                                              1. sylvestris
Stems usually green; lf-segments somewhat decurrent; fls greenish.
    Naturalized locally.   *A. archangelica* L.

1. A. sylvestris L. Wild Angelica.

Stem (7–)30–200 cm. or more, stout, hollow, striate and pruinose, hairy towards base. Lvs 30–60 cm., deltate, lower primary divisions of basal lvs long-stalked; segments 2–8 cm., obliquely oblong-ovate, sharply toothed; petioles laterally flattened, deeply channelled on upper side, sheathing at base; upper lvs reduced to inflated sheathing petioles which ± enclose the fl. buds. Umbels 3–15 cm. diam., stalks shortly hairy; rays many, shortly hairy; bracteoles few, narrow. Fr. 5 mm. Fl. 7–9. Fens, damp meadows and woods; common.

PEUCEDANUM L.

All spp. are rare or local; they may distinguished as follows:

Stem solid; lf-segments 4–10 cm., narrowly linear, quite entire. By the sea; E. Kent and Essex. Sulphur-weed. *P. officinale* L.

Stem hollow; lf-segments pinnatifid, lobes c. 0·5 cm., finely toothed. Fens and marshes; very local. Milk Parsley, Hog's Fennel.
　　　　　　　　　　　　　　　　　　　　　　　P. palustre (L.) Moench

Stem hollow; lf-segments 2–10 cm., broad, lobed and coarsely toothed. Naturalized in a few places. Master-wort.　　*P. ostruthium* (L.) Koch

23. PASTINACA L.

Lvs pinnate. Umbels compound; bracts and bracteoles 0 or 1–2 and soon falling. Fls yellow; calyx-teeth small or 0. Fr. broadly ovoid or orbicular, strongly dorsally flattened; carpels with a broad flattened rather narrowly winged margin, dorsal ridges narrow, distant from lateral.

1. P. sativa L.　　　　　　　　　　　　　　　Wild Parsnip.

30–150 cm., strongly-smelling, ± hairy. Stems hollow, furrowed and ± angled. Lvs 1-pinnate, segments ovate, lobed and toothed. Umbels 3–10 cm. diam.; rays 5–15. Fr. 5–8 mm. Fl. 7–8. Grassy places, particularly on calcareous soils in the south and east.

24. HERACLEUM L.

Annual or perennial, sometimes very large. Lvs 1–3-pinnate, segments broad. Umbels compound; bracts 0 or soon falling, bracteoles several. Fls white or pinkish; calyx-teeth small, unequal. Fr. orbicular, obovate or oblong, strongly dorsally flattened; carpels nearly flat, marginal ridges forming a broad wing, dorsal ones slender.

1. H. sphondylium L.　　　　　　　　Cow Parsnip, Hogweed, Keck.

Stems 50–200 cm., stout, hispid, hollow, ridged, hairs deflexed. Lvs 15–60 cm., 1-pinnate, segments 5–15 cm., variously divided, toothed, ovate to narrowly lanceolate, lower stalked. Umbels 5–15 cm. diam.; rays 7–20; bracteoles very narrow, reflexed. Fr. 7–8 mm., suborbicular, whitish. Fl. 6–9. Common in grassy places and open woods.

*H. mantegazzianum Sommier & Levier is naturalized in some places and is easily recognized by its enormous size (stem up to 3·5 m. and 10 cm. in diam.).

25. DAUCUS L.

Annual or biennial, hispid. Lvs pinnate. Umbels compound; bracts and bracteoles numerous, persistent. Fls white; calyx-teeth small or 0. Fr. ovoid to oblong; carpels convex; primary ridges 5, narrow; secondary 4, stouter, bearing a single row of spines.

1. D. carota L.   Wild Carrot.

Biennial. Stem 30–100 cm., solid, striate or ridged. Lvs 3-pinnate; segments pinnatifid, lobes c. 5 mm. Umbels 3–7 cm. diam., flat or convex in fl.; rays numerous, usually nearly or quite glabrous; bracts 7–13, ternate or pinnatifid, conspicuous, segments linear. Central fl. of each umbel usually red or purple. Fr. 3–4 mm., spines separate at base. Fl. 7–8. Grassy places, particularly on chalky soils and near the sea.

Ssp. **carota**, with a tough not fleshy root, lvs ovate in outline, and umbels strongly concave in fr., is the common plant. Ssp. **gummifer** (Lam.) Thell. with the lvs narrower in outline and the umbels flat in fr., occurs locally on the south coast. Ssp. *sativus* (Hoffm.) Arcangeli, with a thick and fleshy root in the first year, is the cultivated carrot.

77. CUCURBITACEAE

Usually herbs climbing by tendrils and with very sappy often roughly hairy stems. Tendrils spirally coiled, each arising at the side of a lf-axil. Lvs spirally arranged, palmately veined and often palmately lobed; stipules 0. Fls usually unisexual, actinomorphic. Sepals 5, narrow; petals 5, free or united below; stamens 5, rarely all free, more usually of 2 pairs of ± completely united members with 1 free stamen; anthers often borne on a common column and variously curved and twisted; ovary inferior, 1-celled (or 3-celled when the 3 parietal placentae meet in the centre); ovules many; style usually 1; stigmas 3. Fr. a berry or berry-like fruit with a hard wall (pepo); seeds non-endospermic.

1. Bryonia L.

Perennial tendril-climbing herbs with palmately-lobed lvs and unisexual fls in axillary cymes or clusters. Stamens in 2 pairs of members with common filament and 2 pollen-sacs, and 1 free stamen with 1 pollen-sac. Stigmas 3, each bifid. Fr. a small smooth globular berry.

1. B. dioica Jacq.   White (or Red) Bryony.

Stock erect, massively tuberous, branched. Stem very long, branching especially from near the base, brittle, angled, rough with swollen-based hairs. Lvs palmately (3–)5-lobed with the lobes sinuate-toothed, cordate at base, with short curved stalk. Dioecious. Fls in cymes; those of male plant stalked, corymbose, of 3–8 pale greenish fls 12–18 mm. diam. with triangular spreading sepals, oblong hairy net-veined petals 2–3 times as long as sepals, and yellow anthers; those of female plant in a ± sessile umbel of 2–5 greenish fls 10–12 mm. diam. with a short-stalked ovary and prominent bifid stigmas. Berry 5–8 mm. diam., red when ripe; seeds 3–6, yellow and black. Fl. 5–9. Hedgerows, copses, scrub; avoided by rabbits and common in warrens. Locally common in England and Wales northwards to N.W. Yorks and Northumberland.

78. ARISTOLOCHIACEAE

Herbs or woody climbers. Lvs spirally arranged, simple, exstipulate. Fls hermaphrodite, usually 3-merous. Perianth usually in 1 whorl, petaloid, often brown or lurid purple, united below into a tube and either regularly 3–6-lobed or strongly zygomorphic, entire or lobed. Stamens usually 6 or 12. Ovary inferior, 4–6-celled with numerous axile ovules. Fr. a capsule.

Fls 15 mm., regular, with brownish 3-lobed perianth and 12 ± free stamens in 2 whorls. A rhizomatous woodland herb 2–5 cm., with reniform lvs and solitary fls. Perhaps native in a few scattered localities. Asarabacca.   *Asarum europaeum* L.

Fls 2–3 cm., strongly zygomorphic, the dull yellow perianth with a long somewhat curved tube swollen at the base and an oblong entire limb. Stamens 6, in 1 whorl, joined to the style. An erect perennial herb 20–80 cm., with cordate lvs. Introduced as a medicinal plant and naturalized mainly in E. England. Birthwort. *Aristolochia clematitis* L.

79. EUPHORBIACEAE

Trees, shrubs or (as in all British spp.) herbs. Infl. often compound. Fls regular or nearly so, unisexual. Perianth of one whorl or 0. Stamens 1 to many, ovary superior, usually 3-celled with 1 or 2 ovules per cell. Fr. usually a capsule separating into 3 parts and leaving the persistent axis, rarely a drupe; seeds endospermic.

Perianth present, 3-merous; fls in clusters, often spike-like; without
milky juice. 1. MERCURIALIS
Perianth 0; fls very small, one female and several male fls borne in a 4–5-lobed perianth-like involucre and simulating a single fl.; juice milky.
2. EUPHORBIA

1. MERCURIALIS L.

Herbs with watery juice. Lvs opposite; stipules small. Male fls in clusters arranged in long-stalked spikes; per. segs 3, sepaloid; stamens 8–15. Female fls axillary, solitary or in clusters; per. segs 3, sepaloid; stamens represented by 2–3 sterile filaments; ovary of two 1-seeded cells; styles 2, free. Fr. dehiscing by 2 valves.

Hairy perennial with creeping rhizome; stems simple. **1. perennis**
± Glabrous annual; stems branched. **2. annua**

1. M. perennis L. Dog's Mercury.

Hairy perennial with long creeping rhizome. Stems 15–40 cm., erect, simple. Lvs 3–8 cm., ± broadly elliptic, crenate-serrate; stalk 3–10 mm. Dioecious. Fls 4–5 mm. diam.; female 1(–3), long-stalked. Fr. hairy, 6–8 mm. wide. Fl. 2–4. Woods on good soils and shady mountain-rocks to 3400 ft.; especially abundant on calcareous soils. Common over most of Great Britain, but not in Orkney or Shetland and very rare in Ireland.

2. M. annua L. Annual Mercury.

Annual, nearly glabrous. Stems 10–50 cm., erect, branched. Lvs 1·5–5 cm., ± broadly elliptic, crenate-serrate; stalk 2–15 mm. Usually dioecious. Female fls subsessile in the lf-axils. Fr. hispid, 3–4 mm. wide. Fl. 7–10. Waste places and gardens, often only a casual. Widespread in S. and E. England, but rare in Wales, N. England and E. Scotland; S. Ireland.

2. EUPHORBIA L.   Spurge.

Herbs with milky juice. Lvs usually spirally arranged, usually exstipulate. Monoecious. Fls very small, perianth 0, a number of male fls and
a single female fl. being grouped within a cup-shaped perianth-like
involucre (cyathium) with 4–5 small marginal teeth alternating with
conspicuous glands. Cyathia in compound cymes, the primary branches
forming an umbel; the subsequent branching dichasial with conspicuous
opposite bracts which usually differ from the lvs. Male fl. a single
stamen on a jointed stalk; female fl. a 3-celled ovary on a stalk which
elongates in fr.; ovules 1 per cell; styles 3. Fr. a 3-valved capsule.

1   Lvs and bracts not differing, very unequal on the two sides at the
     base; stipules present, very narrow; an annual procumbent
     maritime plant, glaucous and often purplish, very rare on sandy
     shores in S.W. England and Channel Is. and sporadic elsewhere.
                                    E. peplis L.
     Lvs equal at base; stipules 0; bracts often markedly different from
     lvs; plant ±erect.   2

2   Lvs opposite. A robust glabrous glaucous biennial 30–120 cm.
     Doubtfully native in woods; locally common as a garden weed.
     Caper Spurge.   *E. lathyrus* L.
     Lvs spirally arranged.   3

3   Involucral glands rounded on the outer edge.   4
     Involucral glands with concave outer edge, the tips ±prolonged
     into horns.   10

4   Perennials with ±numerous stems; lvs oblong or lanceolate,
     cuneate at base.   5
     Annuals with single stems; lvs either obovate and very blunt or
     narrower and deeply cordate at base.   8

5   Capsule smooth or minutely tuberculate; lvs usually hairy on both
     sides.   6
     Capsule strongly warty; lvs glabrous above, sometimes hairy
     beneath.   7

6   Bracts of the umbel c. 2 cm., oval, resembling the yellowish partial
     bracts; capsule glabrous or sparsely hairy; lvs 4–10 cm., oblong.
     In and round a wood near Bath. Hairy Spurge.
             E. villosa Waldst. & Kit. ex Willd. (*E. pilosa* auct.)
     Bracts of the umbel c. 7–8 cm., oblong-lanceolate, resembling the
     lvs; partial bracts green or red-tinged; capsule woolly. Intro-
     duced; very local in Sussex. Coral spurge.   **E. corallioides* L.

7   Partial bracts subcordate or rounded at base, yellowish; glands
     yellowish, finally brown; lvs 5–10 cm., ±oblong; capsule

5–6 mm. Woods, hedges and pastures chiefly in S.W. England
and S.W. Ireland. Irish Spurge.   *E. hyberna* L.
Partial bracts truncate at base, green; glands at first green, soon
   purple; lvs 3–5 cm., ±oblong; capsule 2–3 mm. Introduced
   and naturalized in a few places, chiefly in Scotland.

                                                          *E. dulcis L.

8 Lvs and bracts acute or subacute; lvs cordate at base; capsule warty.   9
   Lvs and bracts very blunt; lvs tapered to base; capsule smooth.
                                                          2. helioscopia

9 Capsule shallowly grooved, almost globose, its warts ±hemi-
   spherical; lower partial bracts similar in shape to upper,
   markedly different from the bracts of the umbel.
                                                          1. platyphyllos
   Capsule 3-angled; its warts cylindrical or conical, as long as or
   longer than broad; lower partial bracts intermediate in shape
   between the upper and the bracts of the umbel. Like *E. platyphyllos* but more slender and always glabrous. Limestone woods
   in W., very local. Upright Spurge.
                                      E. serrulata Thuill. (*E. stricta* L.)

10 Bracts all free; plant glabrous.   11
    Bracts (except those of the umbel) connate in pairs; plant hairy.
                                                          8. amygdaloides

11 Annuals with single stems; lvs few.   12
    Biennials or perennials with ±numerous stems, some of them
    usually non-flowering, tufted or from a creeping rhizome;
    lvs many.   13

12 Lvs oval or obovate, stalked, green.   **3. peplus**
    Lvs linear, sessile, glaucous.   **4. exigua**

13 Lvs ovate to oblanceolate, all less than 2 cm., ±leathery; maritime
    plants without creeping rhizome; umbel with 6 rays or fewer.   14
    Lvs linear-lanceolate or linear, mostly over 2 cm. (or, if less, then
    narrowly linear), thin; inland plants with creeping rhizome;
    umbel with 6 or more rays.   15

14 Lvs obovate or oblanceolate, slightly leathery, midrib prominent
    below; seeds pitted.   **5. portlandica**
    Lvs ovate or oblong, very leathery, midrib obscure; seeds smooth.
                                                          6. paralias

15 Lvs of fl.-stem 4 mm. broad or more.   **7. esula**
    Lvs of fl.-stem 2 mm. broad or less; umbel 9–15-rayed; glands
    lunate with very short horns; forming large patches. Possibly
    native in calcareous grassland and scrub in a few places, casual
    elsewhere; widespread but local.   *E. cyparissias* L.

1. E. platyphyllos L.   Broad Spurge.

Annual 15–80 cm., glabrous or pubescent. Lvs 1–4·5 cm., spirally
arranged, obovate- or oblong-lanceolate, acute, deeply cordate at base,
sessile, finely toothed except at base. Umbels usually 5-rayed; bracts
2–3 cm., elliptic-oblong. Partial bracts 5–15 mm., deltate, mucronate,
the lowest not differing from the upper and markedly different from the
bracts of the umbel. Glands suborbicular, entire. Capsule 2–3 mm.,
subglobose, shallowly grooved, with hemispherical warts. Fl. 6–10.
Arable land and waste places, widespread but local in S. England, rare
in Wales and N. England.

2. E. helioscopia L.   Sun Spurge.

Glabrous annual 10–50 cm. Lvs 1·5–3 cm., spirally arranged, obovate,
very blunt, tapered from near apex to a narrow base, finely toothed
above. Umbels 5-rayed. Bracts all ± similar to lvs but yellowish.
Glands transversely oval, entire, green. Capsule 3–5 mm., subglobose,
smooth. Fl. 5–10. Common in cultivated ground throughout the
British Is.

3. E. peplus L.   Petty Spurge.

Glabrous green annual 10–30 cm. Lvs 0·5–3 cm., spirally arranged,
oval or obovate, blunt, short-stalked, entire. Umbels 3-rayed. Bracts
like the lvs but subsessile. Glands lunate, with long slender horns.
Capsule c. 2 mm., 3-angled, each valve with 2 narrow wings on the back.
Fl. 4–11. Very common in cultivated and waste ground throughout the
British Is.

4. E. exigua L.   Dwarf Spurge.

Glabrous glaucous annual 5–30 cm. Lvs 0·5–3 cm., spirally arranged,
linear, usually acute, mucronate, sessile, entire. Umbel 3(–5)-rayed.
Lower bracts triangular-lanceolate, upper triangular-ovate, all subcordate at base, Glands lunate, with long slender horns. Capsule c.
2 mm., 3-angled, smooth or slightly rough on back. Fl. 6–10. Common
in arable land in Great Britain northwards to Lanark and Angus and in
E. Ireland; rare elsewhere.

5. E. portlandica L.   Portland Spurge.

Glabrous glaucous biennial or perennial 5–40 cm., with several stems
from the base, all flowering or some not. Lvs 0·5–2 cm., spirally
arranged, dense, numerous, often falling from fl.-stems, somewhat
leathery, obovate to oblanceolate, acute, mucronate, tapered to base,

± sessile, entire, midrib prominent below. Umbel 3–6-rayed, its bracts oval or obovate. Partial bracts 5–8 mm., triangular-rhombic, broader than long, mucronate. Glands lunate, with long horns. Capsule c. 3 mm., 3-angled, granulate on the back of the valves. Seeds pitted. Fl. 5–9. Sea-sands, young dunes and cliff-ledges on the S. and W. coasts from Sussex to Wigtown, very local; all round the Irish coast; Channel Is.

6. E. paralias L. Sea Spurge.

Glabrous glaucous perennial 20–40 cm., with several fertile and sterile stems from a short woody stock. Lvs 0·5–2 cm., spirally arranged, numerous, dense, often imbricate, very thick and fleshy, entire, somewhat concave, ovate or oblong, ± blunt, not mucronate, with broad sessile base; midrib obscure. Umbel 3–6(–8)-rayed, its bracts ovate. Glands lunate, with short horns. Partial bracts 5–10 mm., orbicular-rhombic, thick and fleshy, mucronate. Capsule c. 4 mm., 3-angled, granulate. Seeds smooth. Fl. 6–10. Sea-sands and mobile dunes from Wigtown and Norfolk southwards, rather local; all round the Irish coast; Channel Is.

*7. E. esula L. (*E. uralensis* auct.)

Glabrous perennial 30–80 cm., with numerous erect flowering and non-flowering stems, mostly root-borne and often forming large patches. Flowering stems with axillary, usually non-flowering branches above. Lvs spirally arranged, numerous, linear-lanceolate, on the main stems 2–8 × 4–6 mm., sessile, entire. Umbel 6–12-rayed; its bracts 5–35 mm., lanceolate or linear-oblong. Partial bracts 5–10 mm., yellowish. Glands lunate, with rather long horns. Capsule 2–3 mm., 3-angled, slightly rough on the back of the valves. Fl. 5–7. A variable sp., widespread in Europe and introduced and naturalized in grassland and waste ground in many places in England and E. Scotland. The commonest plant in Britain is probably the hybrid between ssp. *esula* and ssp. *tommasiniana* (Bertol.) Nyman, and it is this which has been called *E. uralensis*.

8. E. amygdaloides L. Wood Spurge.

Pubescent perennial 30–80 cm., with thick stock and tufted stems which are sterile the first year and elongate and flower in the second, the first-year lvs persisting. Lvs of first year 3–8 cm., oblanceolate, ± blunt, tapering into a short stalk, entire, dark green; of the fl.-stem oblong, not tapering, subsessile. Umbel with 5–10 rays, its bracts oval. Partial bracts 5–10 mm., reniform, members of pairs joined for about half the width of

their bases, yellowish. Glands lunate, with converging horns. Capsule
c. 4 mm., somewhat 3-angled, granulate. Fl. 3–5. Damp woods on rich
light soils, sometimes very abundant. Throughout England and Wales,
rather common in the south, very local in the north; rare in Ireland and
probably introduced.

80. POLYGONACEAE

Herbs, shrubs, climbers or rarely trees. Lvs usually spirally arranged
and usually with sheathing stipules. Fls hermaphrodite or unisexual.
Per. segs 3–6, sepaloid or petaloid, free or joined, persistent. Stamens
usually 6–9, hypogynous. Ovary superior, 1-locular. Fr. indehiscent,
hard, 1-seeded, usually enveloped in the perianth.

1 Tiny annual; some lvs opposite; stipules minute; fls small, yellowish-
   green, nearly sessile in the lf-axils. Very local in Skye and Mull.
                               Koenigia islandica L.
   Not as above.   2

2 Lvs kidney-shaped; per. segs. 4 Mountains: local. Mountain
   Sorrel.   *Oxyria digyna* (L.) Hill
   Lvs various but not as above; per. segs usually 5 or 6.   3

3 Infl. unbranched or with a few short branches; perianth petaloid,
   usually pink or white; per. segs usually 5, not enlarging in fr.; lvs
   usually less than 10 cm.   4
   Infl. usually with many long branches; perianth sepaloid; per. segs
   6, in 2 whorls, the inner enlarging in fr. or else fls dioecious; lvs
   usually more than 10 cm.   2. RUMEX

4 Fls in short-branched panicles; lvs triangular, hastate or cordate at
   base; fr. much longer than perianth. An annual c 50 cm., grown
   as a crop and casual. Buckwheat.
                       Fagopyrum esculentum Moench
   Fls in unbranched or short-branched spikes; lvs various but if at all
   like the above then the plant a stout perennial, 1–2 m.; fr. shorter
   or little longer than perianth.   1. POLYGONUM

1. POLYGONUM L.

Annual or perennial herbs, rarely shrubby. Stipules joined in a sheathing tube. Perianth of 3–6, usually 5, spirally arranged segments, the
outer sometimes enlarging slightly in fr., but never tubercled. Stamens
4–8. Styles 2–3. Fr. triangular or biconvex in section, ± enclosed in the
perianth.

1   Stems slender, twining; lvs cordate-sagittate.   2
    Stems not twining; lvs not cordate-sagittate.   3

2   Stalk of fr. 1–2 mm., jointed above the middle; fr. not shiny.
                                                        10. convolvulus
    Stalk of fr. up to 8 mm., jointed at or below the middle; fr. shiny.
        Very local in thickets and hedges in S. England.
                                                        P. dumetorum L.

3   Infl. 1–8-fld. all in lf-axils.   4
    Infl. usually many-fld, some or all at the ends of branches.   6

4   Fr. enclosed by or slightly longer than the perianth.   **1. aviculare**
    Fr. distinctly longer than the perianth.   5

5   Lvs flat; stipules with few, unbranched veins.   **2. oxyspermum**
    Lf-margins revolute; stipules with more numerous branched veins.
        Sea-shores in S.W. England; very rare. Sea Knot grass.
                                                        P. maritimum L.

6   Stout erect herbs 1–4 m. high, with somewhat woody stems.   14
    Plant rarely more than 50 cm. high, or, if more, then with soft stems
        ± decumbent at base and with nodes often swollen.   7

7   Stems unbranched, never floating.   8
    Stems usually branched, or, if simple, then generally floating.   9

8   Basal lvs truncate or subcordate at base; lf-stalk winged in
        upper half; infl. stout; bulbils 0.   **3. bistorta**
    Basal lvs tapering at base; lf-stalk not winged; infl. slender,
        bearing bulbils as well as fls. Mountain districts; local.
                                                        P. viviparum L.

9   Lf-base cordate or rounded; stamens longer than perianth.
                                                        4. amphibium
    Lvs narrowed to base; stamens shorter than perianth.   10

10 Plant with glands (sometimes sparse) on infl. or its stalk.   11
    Infl. and its stalk entirely without glands.   12

11 Infl. slender, nodding; per. segs conspicuously glandular; no
        glands on stalk of infl.   **7. hydropiper**
    Infl. stout, erect; per. segs sparsely glandular; glands on stalk of
        infl.   **6. lapathifolium**

12 Infl. dense, blunt; lvs often with a dark blotch.   **5. persicaria**
    Infl. few-fld, acute; lvs never blotched.   13

13 Lvs usually 10–25 mm. wide; infl. slightly nodding; fr. 3–4 mm.
                                                            8. mite
    Lvs usually 5–8 mm. wide; infl. erect.; fr 2–2·5 mm.   **9. minus**

14 Lvs lanceolate, c. 3 times as long as broad. Naturalized; local.
                                                        *P. polystachyum Wall.
    Lvs ovate, usually less than 1½ times as long as broad.   15

15   Lvs truncate at base, suddenly narrowed to a long sharp apex.
                                                      11. cuspidatum
        Lvs weakly cordate at base, gradually narrowed at top. Natura-
        lized; local.   **P. sachalinense* Friedrich Schmidt Petrop.

1. P. aviculare agg.   Knotgrass.

5–100 cm., spreading, often prostrate, glabrous. Lvs up to c. 5 cm.,
elliptic or narrower; stipules silvery. Fls usually pink and white. Fr.
2–4 mm., triangular in section, usually dotted or striate. Fl. 7–10.
Common and generally distributed in waste places, arable land and on
the sea-shore.

4 spp. are recognized in this country within this agg.

P. aviculare L. Robust, up to 1 m. Lvs of main stems 2–3 times as long
as those of fl. branches. Petioles included in the stipules. Per. segs
joined at base only. Fr. 2·5–3·5 mm. Widely distributed.

P. boreale (Lange) Small. Like *P. aviculare* but petioles well exserted
from stipules and fr. 3·5–4·5 mm. Shetland.

P. arenastrum Boreau (*P. aequale* Lindm.). 10–30 cm. Lvs of main stems
and of fl. branches about the same size. Per. segs joined for at least half
their length. Fr. 2–2·5 mm. Widely distributed.

P. rurivagum Boreau. Lvs of main stem 2–3 times as long as those of fl.
branches, very narrow. Stipules up to 10 mm., brownish below. Per.
segs joined at base only. Fr. 2·5–3·5 mm. Very local, in arable fields on
chalky soils.

2. oxyspermum Lecleb. (*P. raii* Bab.)   Ray's Knotgrass.

10–80 cm. Stems ± woody at base. Lvs 1–2·5 cm. Fls pink or greenish-
white. Per. segs 3 mm. Fr. 5–6 mm., ovoid, acute, reddish-brown,
shiny. Fl. 6–9. Sandy shores or fine shingle above high-water spring
tides, local round the coasts.

3. P. bistorta L.   Snake-root, Easter-ledges, Bistort.

25–50 cm., erect. Rhizome very stout and usually shaped like a flattened
S. Basal lvs 5–15 cm., broadly ovate, blunt, ± hairy beneath, folded
longitudinally in bud and showing 'creases' when mature; upper lvs
triangular-acuminate. Infl. 10–15 mm. diam. Fls 4–5 mm., pink, rarely
white. Fl. 6–8. Meadows and grassy roadsides, commonest on siliceous
soils in the north.

4. P. amphibium L. Amphibious Bistort.

Commonly 30–75 cm. Aquatic and terrestrial forms show considerable vegetative differences. Aquatic: glabrous; stems floating and rooting at nodes; lvs floating, 5–15 cm., ovate-oblong, truncate or subcordate at base. Terrestrial: ± hairy; stems ascending or erect, rooting only at lower nodes; lvs 5–12(–15) cm., oblong-lanceolate, rounded at base. Infl. dense, stalked. Fls pink or red. Glands 0 on infl. Stamens 5. Fl. 7–9. Pools, canals, etc. the terrestrial form on banks by water. Widely distributed.

5. P. persicaria L. Persicaria.

Erect, nearly glabrous annual 25–75 cm. Stems often purplish, swollen above the nodes. Lvs 5–10 cm., lanceolate, sometimes woolly beneath; stipules truncate, fringed. Infl. usually continuous, lfless or with 1 lf at base. Fls pink. Glands 0. Fr. c. 3 mm., bluntly triangular in section, shiny. Fl. 6–10. Cultivated ground, waste places and beside ponds, common.

6. P. lapathifolium L. (incl. P. nodosum Pers.) Pale Persicaria.

Erect ± hairy annual up to 100 cm. Stems usually greenish, swollen above the nodes. Lvs 5–20 cm., lanceolate, sometimes woolly or glandular beneath, often with a dark blotch. Stipules truncate, at least some shortly fringed. Infl. ± continuous, lfless. Fls greenish-white, or pink. Glands rather sparse. Fr. usually biconvex in section, shiny. Fl. 6–10. In similar places to the preceding.

7. P. hydropiper L. Water-pepper.

Nearly glabrous erect annual 25–75 cm. Lvs usually 5–10 cm., lanceolate; stipules truncate, not or shortly fringed. Infl. acute, interrupted and lfy below. Fls greenish. Per. segs covered with yellow glands. Fr. c. 3 mm., biconvex or bluntly triangular in section, not shiny. Fl. 7–9. Damp places or shallow water, widely distributed. Plant acrid and burning to the taste.

8. P. mite Schrank

Like 7 but stipules conspicuously and coarsely fringed and lvs rather abruptly narrowed at base. Infl. nearly erect, slender, interrupted but scarcely lfy. Glands 0. Fls pink, rarely white. Fr. 3–4 mm., biconvex in section, shiny. Fl. 6–9. In similar habitats to P. hydropiper but less common.

9. P. minus Hudson

Spreading ± decumbent annual 10–30(–40) cm. Lvs usually 2–5 × 0·5–
0·8 cm., narrowly lanceolate, ± obtuse; stipules truncate, conspicuously
and coarsely fringed. Infl. ± interrupted. Glands 0. Fls pink or rarely
white. Fr. 2–2·5 mm., biconvex in section, shiny. Fl. 8–9. Wet marshy
places and shores of ponds and lakes, rather local.

10. P. convolvulus L.   Black Bindweed.

Somewhat mealy annual 30–120 cm. Stem angular. Lvs 2–6 cm.,
ovate-acuminate, longer than petioles. Infl. stalked or subsessile, interrupted. Per. segs 5, 3 outer bluntly keeled, rarely narrowly winged
in fr., rough on back. Fr. 4 mm., minutely dotted, black. Fl. 7–10.
Waste places, arable land and gardens, common and widespread.

*11. P. cuspidatum Sieber & Zucc.

Perennial 1–2 m. Lvs broadly ovate, cuspidate, truncate at base,
glabrous; stipules much shorter than internodes. Infl. of many slender
lax spikes, some of the axillary ones much longer than the subtending lf.
Fls pink. Fl. 6–8. Native of Japan now commonly naturalized, particularly by roads and on railway banks in the south.

2. RUMEX L.

Annual to perennial herbs, often with long stout roots. Lvs spirally
arranged; stipules tubular. Fls hermaphrodite or unisexual, in whorls
in simple or branched infls. Per. segs in 2 whorls of 3; outer always small
and thin; inner (fr. per. segs) enlarging and usually becoming hard in fr.
Fr. per. segs with or without spherical or ovoid tubercles on their midribs; tubercles developing as fr. ripens. Stamens in 2 whorls of 3. Fr.
sharply triangular in section, outer wall woody.

Hybrids occur frequently and may usually be recognized by their
fairly high degree of pollen and fr. sterility.

Really ripe fr. is essential for the accurate identification of many spp.

1 Foliage acid; lvs hastate; fls usually unisexual.	2
Foliage not or scarcely acid; lvs not hastate; fls usually herma- phrodite.	4
2 Lvs about as long as broad, upper stalked, Naturalized; rare.	
	R. scutatus L.
Lvs several times as long as broad; if nearly as broad as long then upper subsessile and clasping the stem at their base.	3

3   Lobes of lvs spreading or forward-pointing; upper lvs not clasping
    the stem at their base.   **1. acetosella** agg.
  Lobes of lvs ± downward-directed; upper lvs clasping the stem at
    their base.   **2. acetosa**

4   All fr. per. segs without tubercles.   5
  At least one fr. per. seg with a distinct tubercle.   8

5   Plant slender; up to 6 fls in a whorl; fr. per. segs with 3–5 hooked
    teeth. A rare alien.   ***R. brownii** Campd.
  Plant stout; whorls many-fld; fr. per. segs not toothed.   6

6   Fr. stalks with an almost imperceptible joint. Scotland; very rare.
                        R. aquaticus L.
  Fr. stalks distinctly jointed.   7

7   Rhizomatous; lvs about as long as broad; fr. per. segs ovate,
    truncate at base. Naturalized; very local in the north. Monk's
    Rhubarb.   ***R. alpinus** L.
  Not rhizomatous; lvs distinctly longer than broad; fr. per. segs
    kidney-shaped.   **4. longifolius**

8   Fr. per. segs distinctly toothed (teeth more than 1 mm.).   9
  Fr. per. segs entire or minutely toothed (teeth not more than
    1 mm.).   13

9   Branches making a wide angle with the main stem, forming a
    tangled mass in fr.; lower lvs fiddle-shaped, blade rarely more
    than 10 cm. Local and chiefly near the sea in the south. Fiddle
    Dock.   **R. pulcher** L.
  Branches usually making a narrow angle with the main stem, not
    forming a tangled mass in fr.; lvs very rarely fiddle-shaped, blade
    often more than 10 cm.   10

10   Fr. per. segs 4–7 mm., reddish or brown.   11
  Fr. per. segs 2–3 mm., yellow or golden.   12

11   Fr. per segs. 5–6 × 3 mm., triangular, truncate at base, one (rarely
    all 3) with a prominent tubercle.   **6. obtusifolius**
  Fr. per. segs 4–5 × 4–5 mm., roundish with a short triangular apex,
    all 3 with small tubercles. A rare alien.   ***R. stenophyllus** Ledeb.

12   Anthers c. 1 mm.; fr. per. segs with strap-shaped blunt tips, teeth
    rigid, bristle-like, shorter than width of segs; fr. 2–2·5 mm.
    By lakes, ponds, etc.; local.   **R. palustris** Sm.
  Anthers c. 0·5 mm.; fr. per. segs with triangular ± acute tips, teeth
    very fine, longer than width of segs; fr. 1–1·5 mm.
                      9. maritimus

13   Lvs narrowly obovate, thick and leathery; rhizome long; infl. little
    branched. Naturalized on dunes in the S.W.
                 ***R. frutescens** Thouars
  Not as above.   14

14   Primary infl. overtopped by others arising from the lower lf-axils
     after fl.; tubercles of fr. per. segs rough. A rare alien.
                                  R. triangulivalvis (Danser) Rech. f.
     Not as above.   15

15   Infl. dense, whorls crowded; fr. per. segs (3·5–)4·5(–8) mm.;
     tubercles not more than ½ as long as fr. per. segs.   16
     Infl. lax; fr. per. segs up to 3 mm., or if more then tubercles ⅔–¾
     length of fr. per. segs.   19

16   Fr. per. segs triangular, truncate or subcuneate at base.
                                       3. hydrolapathum
     Fr. per. segs broadly ovate or orbicular, cordate at base.   17

17   Lvs lanceolate, usually narrow, undulate or crisped; plant up to
     1 m.; fr. per. segs 2–5 mm. broad.   **5. crispus**
     Lvs ovate- or oblong-lanceolate, not crisped; plant 1–2 m.; fr. per.
     segs 5–7 mm. broad.   18

18   Veins in middle of lf at an angle of 45–60° with midrib. Locally
     naturalized.   *R. patientia* L.
     Veins in middle of lf at an angle of 60–90° with midrib. Very locally
     naturalized.   *R. cristatus* DC.

19   Lvs glaucous; fr. per. segs c. 4 mm., all with prominent tubercles
     ⅔–½ as long as segs. Very local by the sea in S.W. Shore Dock.
                                     R. rupestris Le Gall
     Lvs not glaucous; fr. per. segs up to 3 mm.   20

20   Stems usually almost straight, branches making an angle of
     20°(–40°); lowest whorls on branches subtended by lvs; tubercle
     (usually 1) globular.   **7. sanguineus**
     Stems usually flexous, branches making an angle of 30–90°;
     whorls usually subtended by lvs for ⅔ length of branches;
     tubercles (usually 3) oblong.   **8. conglomeratus**

1. R. acetosella agg.   Sheep's Sorrel.

Stems to 30 cm., from buds on roots. Lvs to 4 cm., upper ± distinctly
stalked; stipules hyaline. Infl. to c. 15 cm., lfless or nearly so. Fls
dioecious; outer per. segs pressed to inner; fr. per. segs not or scarcely
enlarged. Fl. 5–8. Includes the following:

R. tenuifolius (Wallr.) Á. Löve. Small, ± decumbent. Lower lvs 7–10
times as long as broad (excluding lobes), margins inrolled. Fr. c.
1 × 0·7 mm. Common on sandy heaths, chiefly in the S.E.

R. acetosella L. Variable in size, erect. Lower lvs 3–4 times as long as
broad (excluding lobes), margins flat. Fr. c. 1·5 × 0·8 mm. Widespread
and common but not on the poorest soils.

2. R. acetosa L.   Sorrel.

Up to 100 cm., nearly glabrous. Lvs up to 10 cm., usually less, oblong-lanceolate, ± blunt, upper subsessile. Infl. up to 40 cm., lfless or nearly so. Fls dioecious; outer per. segs reflexed and pressed to stalks after fl.; fr. per. segs 3–4 mm., c. twice as long as fr., orbicular-cordate. Fl. 5–6. Common in grassland and open woods.

3. R. hydrolapathum Hudson   Great Water Dock.

Up to 2 m., stout. Lvs up to 1·1 m., lanceolate to ovate, acute or acuminate, narrowed to the cuneate or rarely slightly cordate base, margins flat or somewhat undulate. Infl. usually with many branches ascending at a narrow angle, somewhat lfy below. Whorls ± crowded. Fr. per. segs c. 6 mm., all with tubercles, margin usually with a few short teeth towards the base. Fr. 4 mm. Fl. 7–9. In wet places and shallow water; widespread but rare in the north.

4. R. longifolius DC.

Up to 120 cm., stout. Lvs up to 80 cm., ovate to lanceolate, ± blunt, ± cordate at base, margins undulate and crisped. Infl. dense, spindle-shaped. Whorls confluent, lfy at base. Fr. per. segs 6 mm., thin, reni-form. Fr. 3–4 mm. Fl. 6–7. By water and in damp grassy places in the north.

5. R. crispus L.   Curled Dock.

50–100 cm. Lvs up to 30 cm., lanceolate or oblong-lanceolate, narrowed or ± rounded at base and tapering from about the middle to a blunt point, margins usually undulate and strongly crisped, rarely flat or nearly so. Infl. usually little-branched, branches ascending at a narrow angle, usually subtended by narrow much-crisped lvs. Whorls close, distinct, or confluent towards the ends of branches. Fr. per. segs (3·5–)4·5(6–) mm., broadly ovate-cordate, usually all 3 with tubercles, one larger than others or, less frequently, all equal; margins entire or minutely toothed. Fr. 2·5–3 mm. Fl. 6–10. A common and variable weed of arable land; also in grassland, dune-slacks, and on shingle beaches, etc.

6. R. obtusifolius L.   Broad-leaved Dock.

50–100 cm. Lvs up to 25 cm., lower ovate-oblong, cordate at base, blunt, margins undulate, upper narrower, all usually hairy on lower surface and veins. Infl. with rather spreading branches, lfy in the lower part. Whorls distinct. Fr. per. segs with 3–5 long teeth. Fr. 3 mm. Fl. 6–10. A common weed, often growing with the preceding.

7. R. sanguineus L. Red-veined Dock.

Up to 100 cm. Lvs ovate-lanceolate, with a rounded or subcordate base, upper narrower, all acute, usually green and thin, occasionally rusty-red (var. *viridis* Sibth.) or rarely thicker with purple veins (var. *sanguineus*). Infl. much-branched. Whorls rather distant. Fr. per. segs c. 3 mm., oblong, blunt; tubercle c. 1·5 mm. diam., margins entire. Fr. 1·25–1·75 mm. Fl. 6–8. In grassy places, open woods and waste ground, common in the south, less frequent in the north.

8. R. conglomeratus Murray Sharp Dock.

Like **7** but lvs oblong or fiddle-shaped, base rounded or subcordate, upper narrower, all acute; fr. per. segs c. 3 mm., ovate to oblong, blunt, all with tubercles 1·25–1·75 mm.; fr. 1·75–2 mm. Fl. 5–6. In similar habitats to the preceding, very variable and sometimes not easily distinguishable from it.

9. R. maritimus L. Golden Dock.

Up to 100 cm., annual or sometimes longer-lived, usually golden-yellow in fr. Lvs oblong-lanceolate, tapering at base, ± acute, upper very narrow. Whorls usually crowded. Anthers 0·45–0·62 mm. Fr.-stalks very slender, mostly longer than fr. per. segs. Outer per. segs thin, shorter than ½ diam. of fr. per. segs, horizontally spreading or weakly reflexed. Fr. per. segs 2–3 mm., tubercles acute in front. Fl. 6–9. Bare muddy ground at margins of lakes and reservoirs, in dried-up ponds and, more rarely, damp grassy places; rather local, rare in the north.

81. URTICACEAE

Herbs, small shrubs or rarely soft-wooded trees, often with stinging hairs. Lvs simple. Fls small, generally unisexual. Perianth 4–5-merous, often enlarged in fr. Stamens opposite the per. segs, inflexed in bud, springing open at maturity. Ovary 1-celled. Fr. an achene (in our spp.).

1 Plant usually with stinging hairs; lvs toothed. 2. URTICA
  Plant without stinging hairs; lvs entire. 2

2 Stems not rooting at nodes; lvs mostly 1 cm. or more, stalked; fls in
    clusters. 1. PARIETARIA
  Stems rooting at nodes; lvs not more than 6 mm., nearly sessile; fls
    solitary. Naturalized on walls, etc., chiefly in S.W. England.
    Mind-your-own-business.
        Soleirolia soleirolii (Req.) Dandy (*Helxine soleirolii* Req.)

1. PARIETARIA L.

Herbs. Lvs alternate. Fls hermaphrodite or unisexual, green. Perianth of female fls tubular, 4-toothed, of male and hermaphrodite fls 4-partite. Fr. enclosed in perianth.

1. P. judaica L. (*P. diffusa* Mert. & Koch)   Pellitory-of-the-Wall.
Softly hairy perennial 30–100 cm. Stems spreading, usually reddish. Lvs up to c. 7 cm., lanceolate to ovate, blunt to acuminate, stalk shorter than blade. Fls usually unisexual, female terminal, male lateral. Fl. 6–10. Cracks in rocks, old walls and hedge-banks, rather local.

2. URTICA L.   Nettle.

Herbs, usually with stinging hairs. Lvs opposite. Infl. lateral, arising from an often very short lfy branch, usually spike-like. Fls green, unisexual. Female fls with unequal per. segs, the larger enclosing the fr.

Annual; blades of lower lvs shorter than stalks.   **1. urens**
Perennial; blades of lower lvs longer than stalks.   **2. dioica**

1. U. urens L.   Small Nettle.
10–60 cm., readily uprooted. Lvs 1·5–4 cm., ovate or elliptic, blunt to acuminate, deeply and sharply toothed. Infl. c. 1 cm., on usually well-developed lateral branches. Fl. 6–9. Cultivated ground, etc., particularly on light soils, common in the S.E., rather local elsewhere.

2. U. dioica L.   Stinging Nettle.
30–150 cm., with very tough yellow roots. Some stems creeping and rooting at nodes. Lvs 4–8 cm., ovate, toothed. Infl. up to c. 10 cm., on usually very short lateral branches. Fl. 6–8. Hedge-banks, woods, fens, etc., especially where the ground is covered with litter or rubble, abundant and generally distributed.

82. CANNABACEAE

Herbs. Lvs usually lobed. Fls dioecious. Male fls stalked, perianth 5-partite; stamens erect in bud. Female fls sessile, perianth entire; ovary sessile; fr. an achene enclosed in the perianth. The infructescences of *Humulus* are used in brewing; **Cannabis sativa* L., a rare casual on rubbish dumps, etc., yields a valuable fibre (hemp) and a narcotic resin.

1. Humulus L.

Climbing perennials rough with deflexed hairs. Lvs opposite, palmately lobed or entire, stalked. Infl. glandular, male much branched, female a stalked cone-like spike (hop).

1. H. lupulus L.   Hop.

3–6 m., stems climbing by twisting in a clockwise direction. Lvs (4–)10–15 cm., broadly ovate, ±cordate at base, usually deeply 3–5-lobed and coarsely toothed, lobes acuminate. Male fls c. 5 mm. diam. Female 'cone' 15–20 mm., enlarging to 5 cm. in fr.; bracts c. 10 mm., ovate, acute, pale yellowish-green. Fls 2–3 in the axils of the bracts. Fl. 7–8. In hedges and thickets, widely distributed but doubtless often an escape from cultivation.

83. ULMACEAE

Trees. Lvs alternate, simple, often asymmetrical at base. Fls clustered on the 1-year-old twigs. Perianth greenish, shortly 4–8-lobed. Stamens the same number as the perianth-lobes and opposite to them. Fr. flattened, dry or slightly fleshy, often winged.

1. Ulmus L.   Elm.

Trees, usually with suckers. Lvs serrate. Fls appearing before the lvs; perianth usually 4–5-lobed; anthers reddish. Fr. with a broad wing, notched at top.

Descriptions of lvs refer to those on slow-growing twigs on mature trees. Those on quickly-growing shoots, and particularly suckers, differ greatly from them and are very variable.

1	Lvs medium-sized to large (5–16 cm.), broadly ovate, elliptic or obovate, very rough above, stalk less than 3 mm.	**1. glabra**
	Not with the above combination of characters.	2
2	Lvs medium-sized to large (6–12 cm.), broad, smooth above.	4
	Lvs small to large (2–10 cm.), if broad less than 5 cm. or else rough above, otherwise rough or smooth above.	3
3	Lvs broadly ovate to suborbicular, rough, dark green.	**2. procera**
	Lvs narrow to broad; if broad smooth and light green.	**3. minor**
4	Main branches straight, ascending; cork not conspicuous on shoots arising from the trunk; not suckering. Commonly planted. Huntingdon Elm.	

U. × hollandica Miller var. *vegeta* (Loudon) Rehder

Main branches crooked, spreading upwards; shoots arising from the trunk with large cork flanges; suckering freely. Sometimes planted. Dutch Elm. *U.* × *hollandica* Miller var. *hollandica*

1. U. glabra Hudson Wych Elm.

Up to c. 40 m., often dividing into 2 or 3 large branches near the base. Suckers few or 0. Twigs stout, hairy, becoming smooth and grey by the third year. Lvs (5–)8–16 cm., broadly ovate, elliptic or obovate, cuspidate, rough above, hairy beneath; base unequal, the long side forming a rounded lobe which overlaps and often hides the short stalk; lateral nerves 12–18 pairs. Ssp. **glabra** has broad lvs with well-developed basal lobes and occurs in the south. Ssp. **montana** (Stokes) Lindquist has narrow lvs with small basal lobes and occurs in the north. Fl. 2–3. Woods, hedges and beside streams, widespread but commoner in the north and west.

2. U. procera Salisb. English Elm.

Up to 30 m., usually with a well-marked trunk and few large branches in the lower part. Suckers usually abundant. Twigs rather stout, hairy. Lvs 4·5–9 cm., suborbicular to ovate, acute, rough above, hairy beneath; base unequal but without a well-marked lobe; lateral nerves 10–12 pairs. Fl. 2–3. In hedges and by roads, widespread in England but uncommon in the north and in East Anglia.

3. U. minor Miller (*U. carpinifolia* G. Suckow) Smooth Elm.

Up to 20(–30) m., very variable in habit and lf-shape. Branches ascending sharply (var. *stricta* Lindl.) to spreading. Twigs often long, slender and ±pendulous. Lvs (2·5–)4–10(–11·5) cm., oblanceolate to suborbicular, nearly or quite smooth and shining above; base ±unequal; lateral nerves (5–)9–12(–14) pairs. Fl. 2–3. Thickets and hedgerows, common in S. and E. England, absent from Scotland and much of Wales.

84. JUGLANDACEAE

Trees with spirally arranged pinnate exstipulate lvs. Fls monoecious, solitary in the axils of bracts. Male fls in axillary usually pendulous catkins; bracteoles usually 2; perianth 0 or small; stamens 3–40; rudimentary ovary sometimes present. Female fls solitary or in spikes, terminal; bracteoles and perianth as in the male; ovary inferior, 1-celled with 1 basal ovule; style short; stigmas 2. Fr. a drupe or small nut; seeds non-endospermic.

1. JUGLANS L.

Deciduous trees. Twigs with septate pith. Buds with few scales. Male catkins pendulous. Female fls in erect terminal few-fld spikes. Bracteoles and perianth present in both sexes; perianth 1–5-lobed in the male, 4-lobed in the female. Bracteoles united with the ovary, not persisting in fr. Stamens 8–40. Stigmas simple. Fr. a large indehiscent drupe with the stone ('walnut') incompletely 2–4-celled and the surface of the cotyledon strongly convoluted.

*1. J. regia L.   Walnut.

Large tree to 30 m. with a spreading crown. Bark grey, smooth for many years, finally fissured, not scaling. Buds c. 6 mm., broadly ovoid, blackish, glabrous (or the terminal one greyish, tomentose). Twigs glabrous; lf-scars Y-shaped. Lflets usually 3–4 pairs, 6–12 cm., obovate or elliptic, acute or acuminate, becoming glabrous except for tufts in vein-axils beneath, fragrant when bruised. Male catkins 5–15 cm. Fr. 4–5 cm., subglobose, green, gland-dotted, aromatic; stone ovoid, 4-celled below. Introduced and sometimes naturalized in S. England.

85. MYRICACEAE

Trees or shrubs. Lvs spirally arranged, simple, dotted with resinous glands, aromatic, exstipulate. Fls unisexual, solitary in the axils of bracts, forming catkins. Perianth 0. Male fl. usually without bracteoles; stamens 2–16. Female fl. with 2 or more small bracteoles; ovary 1-celled with 1 basal ovule; style short; stigmas 2, thread-like. Fr. a drupe or nut; seeds non-endospermic.

1. MYRICA L.

The only genus.

1. M. gale L.   Bog Myrtle, Sweet Gale.

Deciduous shrub, 60–150 cm., spreading by suckers. Twigs red-brown with scattered yellowish glands, ascending at a narrow angle. Buds small, ovoid, reddish-brown, with several scales. Lvs 2–6 cm., oblanceolate, cuneate at base, subsessile, toothed near the apex or subentire, grey-green and usually becoming glabrous above but ± pubescent beneath, with conspicuous shining yellow sessile glands on both sides. Mostly dioecious. Catkins before the lvs, ascending, forming panicles at the ends of shoots; male 7–15(–30) mm., with red-brown bracts, bracteoles 0, stamens c. 4, anthers red; female 5–10 mm. in fr., bracteoles 2, styles red. Fr. dry, flattened, gland-dotted, the bracteoles forming 2 wings. Fl. 4–5. Bogs, wet heaths and fens, to 1800 ft.; often with *Molinia*. Local throughout the British Is.

86. BETULACEAE

Deciduous trees or shrubs. Buds scaly. Lvs alternate, simple; stipules caducous. Fls monoecious, the sexes in different infls. Male fls in drooping catkins, three together in the axil of each bract, with 2–3 bracteoles to each group of fls; perianth present; stamens 2 or 4. Female fls in erect cylindric or ovoid catkins, 2–3 in the axil of each bract, 2–4 bracteoles to each group of fls; perianth 0; ovary 2-celled with 1 ovule per cell; styles 2. Fr. a flattened nutlet, 1–3 on the surface of a scale formed from the enlarged and fused bract and bracteoles; seed 1, nonendospermic.

Stamens 2, bifid beneath the anthers; fruiting catkin cylindric, its scales
   3-lobed, falling with the fr.; fls with the lvs.   1. BETULA
Stamens 4, entire; fruiting catkin cone-like, its scales 5-lobed, persistent;
   fls before the lvs.   2. ALNUS

1. BETULA L.

Buds with several scales. Bracteoles 2 to each group of fls. Stamens 2, bifid below the anthers; perianth minute. Female fls 3 to each bract. Fr. catkins cylindric, the scales 3-lobed, relatively thin, falling with the fr. Fr. winged, styles persistent.

1 Lvs ± ovate, subacute to acuminate, more or less finely toothed; lf-
    stalk 7 mm. or longer.   2
    Lvs orbicular (5–15 mm.), deeply crenate, glabrous when mature;
    lf-stalks 3 mm. or less. A shrub to 1 m., the stems mostly pro-
    cumbent; fr. catkin 5–10 mm.; fr-scales with 3 ± equal narrow
    erect apical lobes. Mountain moors in N. England and Scotland,
    very local. Dwarf Birch.   *B. nana* L.
2 Lvs acuminate, sharply doubly serrate with prominent primary
    teeth, glabrous; trunk black and fissured at the base, white above.
                                           1. pendula
    Lvs acute or subacute, irregularly or evenly serrate without markedly
    projecting primary teeth; trunk ± smooth throughout.
                                          2. pubescens

1. B. pendula Roth   Silver Birch.

Tree to 25 m. with single stem; bark smooth and silvery-white above, peeling, ± suddenly changing near the base of the trunk to black and fissured into rectangular bosses. Branches ± pendulous. Twigs glabrous, brown, somewhat shining, with ± conspicuous pale warts. Buds not sticky. Lvs usually 2·5 cm., glabrous, stalked, ovate-deltate, acuminate, truncate or broadly cuneate at base, sharply doubly serrate with the primary teeth very prominent and curved towards the lf-tip. Male catkins 2–6 cm. Female fr. infl. 1·5–3 × c.1 cm., its scales with short broad cuneate base, lateral lobes broad, spreading and curving downwards, middle lobe deltate, blunt; wings of fr. 2–3 times as broad as achene, the upper edge surpassing the stigmas. Fl. 4–5. Woods, especially on light dry soils (but rare on chalk and limestone), and heaths. Throughout the British Is, but rare in N. Scotland.

2. B. pubescens Ehrh.   Birch.

Tree to 20 m., or shrub with many stems; bark smooth, less silvery than in **1**, often brownish, much the same to base of trunk, there sometimes grooved but not broken up into rectangular bosses. Branches usually not pendulous. Twigs ± pubescent or glabrous, dark brown or blackish, not or scarcely shining, with or without brown resinous warts. Buds sticky or not. Lvs stalked, 1·5–5·5 cm., very variable in shape, subacute or acute, rounded or cuneate at the base, coarsely and sometimes irregularly serrate, the teeth not curved towards the lf-tip, usually pubescent at least on the veins beneath. Male catkins 3–6 cm. Female fr. infl. 1–4 × 0·5–1 cm., its scales with cuneate base, lateral lobes rounded or nearly square, ± spreading or ascending, middle lobe long or short, narrow, oblong- or triangular-lanceolate; wings of fr. 1–1½ times as broad as achene, the upper edge not surpassing the stigmas. Fl. 4–5. In similar places to *B. pendula* but more tolerant of wet and cold, only abundant on wet soils (including peat) in S. England but forming pure woods above the limit of oakwood in the north and west.

Ssp. **pubescens**, mainly in southern and lowland areas, has twigs usually conspicuously pubescent, with warts few or 0, and buds not sticky.

Ssp. **carpatica** (Willd.) Ascherson & Graebner (ssp. *odorata* (Bechst.) E. F. Warburg), of northern and mountain areas, is often shrubby with twigs not or scarcely pubescent, covered when young with brown resinous warts, and sticky buds. The plant has a sweet resinous smell when the buds are unfolding.

2. ALNUS Miller

Bracteoles 4 to each group of fls. Stamens 4, not bifid. Female fls 2 to each bract. Fr. catkins ovoid or ellipsoid, cone-like; the scales 5-lobed, thick and woody, long persistent after the fall of the winged fr.

1. A. glutinosa (L.) Gaertner   Alder.

Tree to 20(–40) m. with ± oblong crown. Bark dark brown, fissured. Twigs glabrous. Buds purplish, blunt, stalked, with a large outer scale almost hiding the inner ones. Lvs stalked, 3–9 cm., suborbicular or broadly obovate, usually truncate or retuse at the apex, cuneate at base, irregularly and often doubly serrate, bright green and glabrous on both sides except for axillary tufts beneath. Male catkins 2–6 cm. Female fr. infl. 1·5–3 cm., ovoid or ellipsoid. Fl. 2–3, before the lvs. Wet places in woods and by lakes and streams, often forming pure woods. Common almost throughout the British Is.

87. CORYLACEAE

Deciduous trees or shrubs. Buds scaly. Lvs alternate, simple; stipules caducous. Fls monoecious, the sexes in different infls. Male fls in drooping catkins, solitary in the axil of each bract. Female fls 2 to each bract; perianth irregularly lobed; ovary inferior, 2-celled, with 1 ovule per cell; styles 2. Fr. a nut, surrounded or subtended by a lf-like involucre formed from the fusion of the accrescent bract and bracteoles; seed 1, endospermic.

Female fls numerous, in drooping catkins; fr. involucre 3-lobed, uni-
    lateral.   1. CARPINUS
Female fls few, in short erect bud-like spikes; fr. involucre irregularly
    lobed, surrounding the nut.   2. CORYLUS

1. CARPINUS L.

Trees. Male fls without bracteoles. Stamens c. 10. Female fls in terminal drooping catkins. Fr. a small nut subtended by a large uni-lateral 3-lobed bract-like involucre.

1. C. betulus L.   Hornbeam.

Tree to 30 m., often coppiced as a shrub. Trunk fluted; bark smooth, grey; branches ascending at an angle of 20–30°. Buds 5–10 mm., narrowly oblong, pointed, pale brown, with numerous scales. Twigs brown, sparsely pubescent. Lvs 3–10(–12) cm., ovate, acute or acuminate, cordate or rounded at the base, sharply and doubly serrate, folded along the veins in bud, glabrous except for long hairs on the main veins beneath; stalk 5–15 mm. Male catkins 2·5–5 cm., bracts ovate, greenish. Female catkins c. 2 cm. in fl., 5–14 cm. in fr.; involucre 2·5–4 cm., the middle lobe much longer than the 2 laterals. Fr. 5–10 mm., greenish, ovoid, compressed, strongly veined, crowned by the persistent perianth. Fl. 4–5. Woods and hedgerows, often dominant as a coppice shrub in oakwoods on sandy or loamy clays in S.E. England; perhaps native northwards and westwards to S. Norfolk and Monmouth.

2. CORYLUS L.

Shrubs, rarely trees. Male fls with 2 bracteoles. Stamens c. 4. Female fls few in erect short bud-like spikes. Fr. a large nut surrounded or enclosed by a lobed involucre.

1. C. avellana L.   Hazel, Cob-nut.

Shrub 1–6 m. with several stems, usually coppiced, rarely a small tree. Bark smooth, coppery-brown, peeling in thin papery strips. Buds c. 4 mm., ovoid, blunt, with ciliate scales. Twigs thickly clothed with reddish glandular hairs. Lvs 5–12 cm., suborbicular, cuspidate, cordate at the base, sharply doubly serrate or lobed, slightly pubescent above, more so beneath, becoming nearly glabrous; stalk 8–15 mm., glandular-hispid. Fls before the lvs. Male catkins 1–4 together, 2–8 cm., bracts ovate, anthers bright yellow. Female spikes 5 mm. or less, styles red. Fr. in clusters of 1–4, 1·5–2 cm., globose or ovoid, brown, with hard woody shell, surrounded by a deeply lobed involucre about as long as or rather longer than itself. Fl. 1–4. Woods, scrub and hedges on damp or dry basic and damp neutral or moderately acid soils; the common shrub-layer dominant (as coppice) of lowland oakwoods and sometimes also of ashwoods, or forming scrub on exposed limestone; to 2000 ft. Common almost throughout the British Is.

88. FAGACEAE

Trees or shrubs with scaly buds and spirally arranged simple lvs; stipules usually falling early. Fls monoecious, usually in different infls. Male fls in catkins or in many-fld tassel-like heads; perianth 4–6-lobed; stamens usually twice as many as perianth-lobes; rudimentary ovary sometimes present. Female fls, singly or in threes, arranged in spikes or at the base of the male infl.; each fl. or triad with a basal scaly involucre; perianth 4–6-lobed; ovary inferior, 3- or 6-celled with 2 ovules per cell; styles 3 or 6. Fr. a 1-seeded nut, 1–3 being surrounded or enclosed by a scaly or spiny 'cupule' formed from the enlarged involucre; seed non-endospermic.

1 Lvs almost entire, deciduous; buds spindle-shaped, acute; male fls in tassel-like heads; fr. completely enclosed by a woody 4-valved cupule.   1. FAGUS
   Lvs toothed or lobed (or, if entire, then evergreen); buds ovoid; male fls in catkins.   2

2 Lvs 10–25 cm., with awned teeth, deciduous, not lobed; male catkins ± erect with female fls near the base; cupule prickly, completely enclosing the fr., splitting into 2–4 valves.   2. CASTANEA
   Lvs lobed or, if not, 2–9 cm., evergreen and densely tomentose beneath; male catkins drooping; female fls in separate spikes; cupule cup-like, enclosing only the lower half of the fr., not prickly or splitting into valves.   3. QUERCUS

1. FAGUS L.

Deciduous trees with acute spindle-shaped buds. Male fls in hanging long-stalked tassel-like heads; perianth bell-shaped, 4–7-lobed; stamens 8–16. Female fls usually in pairs, surrounded at the base by the stalked scaly cupule; styles 3. Nuts sharply 3-angled, enclosed singly or in pairs in the woody regularly 4-valved scaly or prickly cupule. Germination epigeal.

1. F. sylvatica L.   Beech.

Tree to 30(–40) m., with broad dense crown. Bark grey, smooth. Buds 1–2 cm., reddish-brown. Lvs 4–9 cm., stalked, ovate-elliptic, acute, somewhat sinuate, glabrous except for the long-ciliate margins and silky hairs on veins and in axils beneath; veins 5–9 pairs. Stalk of male tassel 5–6 cm. Fr. 12–18 mm., reddish-brown; cupule c. 2·5 cm., pale brown, with narrow spreading scales. Fl. 4–5. Woods, etc., in S.E. England north and west to Hertford, Gloucester, S.E. Wales, Somerset and Dorset; widely planted and naturalized almost throughout the British Is.

2. CASTANEA Miller

Deciduous trees. Buds with 3–4 scales. Fls in suberect terminal catkins, female fls at the base, the rest male. Male fls with 6-lobed perianth and 10–20 stamens. Female fls 3(–7) in each cupule; ovary 6-celled; styles usually 6. Fr. large, brown, 1–3 (or rarely more) enclosed, except for the styles, in a prickly cupule dehiscing rather irregularly by 2–4 valves. Germination hypogeal.

*1. C. sativa Miller   Sweet Chestnut, Spanish Chestnut.

Tree to 30 m., with wide crown. Bark dark brownish-grey, fissured, the longitudinal fissures often spirally curved. Twigs olive-brown, glabrous or slightly pubescent. Buds c. 4–5 mm., ovoid, blunt, yellowish-green, tinged brownish. Lvs 10–25 cm., stalked, oblong-lanceolate, acute, becoming glabrous, with coarse and regular awned teeth. Catkins 12–20 cm., made conspicuous by the yellowish-white anthers of the numerous male fls. Fr. 2–3·5 cm., deep brown, shining; cupule green, with dense long branched spines. Fl. 7. Introduced; extensively naturalized in S.E. England.

3. QUERCUS L.

Deciduous or evergreen trees or sometimes shrubs. Buds with numerous scales. Male and female fls in separate infls. Male fls in drooping catkins; perianth 4–7-lobed; stamens 4–12, usually 6. Female fls solitary or in spikes, each fl. (in British spp.) in the axil of a bract and with a cupule at its base; ovary usually 3-celled; styles variously shaped, the stigma on the ± flat inner surface. Fr. (acorn) large, borne singly in a cup-like indehiscent symmetrical scaly cupule. Germination hypogeal.

1   Lvs evergreen, dark green above, grey-tomentose beneath, entire or distantly toothed; cupule covering less than ½ the fr., grey-tomentose, with small appressed scales. Introduced and sometimes naturalized in S. England. Holm Oak.   ***Q. ilex** L.
    Lvs deciduous, green and glabrous or pubescent beneath, lobed.   2

2   Cupule-scales to 1 cm., very narrow, spreading; buds surrounded by long narrow persistent stipules; lvs rough above, lobes unequal, acute. Introduced. Turkey Oak.   ***Q. cerris** L.
    Cupule-scales ovate, appressed; buds without surrounding narrow stipules; lvs smooth above, lobes blunt.   3

3   Lf-stalk short (less than 1 cm.) or 0; lvs glabrous beneath, with small tightly reflexed auricles at the base; acorns on a stalk 2–8 cm. long.   **1. robur**
    Lf-stalk 1 cm. or more; lvs with stellate hairs along either side of the midrib beneath, without tightly reflexed auricles; acorns sessile or on a stalk 1 cm. long or less.   **2. petraea**

1. Q. robur L.   Common Oak, Pedunculate Oak.

Large deciduous tree to 30(–40) m., with broad crown; bark brownish-grey, fissured. Twigs glabrous, greyish-brown. Buds ovoid, blunt, glabrous. Lvs 5–12 cm., obovate-oblong in outline, blunt, rounded to cordate at base, usually with a small tightly reflexed auricle on either side, pinnately lobed, dull green and glabrous above, paler and, even if sparsely hairy when young, soon glabrous beneath; lobes 3–5(–6) pairs, blunt, rather unequal; stipules linear, soon falling; stalk to 5 (rarely 10) mm., often almost 0. Male catkins 2–4 cm.; stamens commonly 6–8. Female spikes 1–5-fld, stalked; styles 3, obvious. Acorn 1·5–4 cm., brown, 1 or more on a stalk 2–8 cm., cup 1·5–2 cm. diam., covering ⅓–½ of the fr., clothed with small ovate appressed scales. Fl. 4–5. Woods, hedgerows, etc. Throughout Great Britain and Ireland as the characteristic dominant tree of heavy and especially basic soils (clays and loams) and thus of most of S., E. and C. England; sometimes also dominant or co-dominant with *Q. petraea* on the damper acid sands; in other parts of the British Is mainly on alluvium; not thriving on acid peat or shallow limestone soils.

2. Q. petraea (Mattuschka) Liebl.   Durmast Oak, Sessile Oak.

Differs from *Q. robur* as follows: Usually branching higher; crown rather narrower. Buds with ciliate scales. Lvs wedge-shaped to cordate at base, without tightly reflexed auricles, somewhat shining above, persistently stellate-pubescent with large hairs along either side of the base of the midrib and main veins beneath; lobes more regular, 4–6 pairs; stalk 1–2·5 cm. Stigmas subsessile. Fr.-stalk 0–1 cm. Fl. 4–5. Woods throughout the British Is, and the characteristic dominant of the siliceous soils of N. and W. England, Wales, Scotland and Ireland.

89. SALICACEAE

Deciduous trees and shrubs. Lvs spirally arranged, simple, stipulate. Fls dioecious, in catkins, each fl. solitary in the axil of a bract (scale). Bracteoles 0. Perianth 0, but fls usually have a cup-like disk or 1 or 2 small nectaries which may represent a perianth. Stamens 2–many, filaments long and slender, occasionally fused. Ovary of 2 carpels, 1-celled with 2 or 4 parietal placentae; ovules usually numerous; style 1; stigmas 2, often bifid. Fr. a 2-valved capsule; seeds with long silky hairs from the funicle, non-endospermic.

Catkin-scales toothed or laciniate; stamens numerous; fls with a cup-like disk; buds with several outer scales.   1. Populus
Catkin-scales entire; stamens 5 or fewer; fls with 1 or 2 nectaries, without disk; buds with 1 outer scale.   2. Salix

1. POPULUS L.

Trees. Buds with several outer scales; terminal bud usually present. Fls appearing before the lvs in pendulous catkins; catkin-scales toothed or laciniate; each fl. with a cup-like disk; stamens 4–many, their anthers red or purple.

1   Bark smooth, grey (except at the base of old trunks) with con-
       spicuous rhomboid lenticels; lvs of normal shoots blunt, coarsely
       toothed or lobed with up to c. 10 teeth; catkin-scales with long
       marginal hairs.   2
    Bark fissured, dark (or smooth and grey or brown above in *P. nigra*
       var. *italica*); lvs acuminate, crenate or with numerous (c. 20 or
       more) teeth; catkin-scales glabrous.   4

2   Lvs, at least at the ends of long shoots, persistently tomentose
       beneath; lf-stalk scarcely flattened laterally; catkin-scales not
       divided to half-way; stigmas greenish.   3
    Lvs (except of sucker shoots) glabrous; lf-stalk strongly flattened
       laterally; catkin-scales divided to more than half-way; stigmas
       purple.   2. tremula

3   Lvs of sucker shoots and of ends of long shoots 3–5 cm., deeply
       palmately 5-lobed; those of short shoots 3–5 cm., ovate-orbicular,
       dark green above, persistently tomentose beneath (white at first
       but becoming greyer); catkin-scales toothed. Tree to 25 m., freely
       suckering. Introduced and much planted but not naturalized.
       White Poplar.   *P. alba* L.
    Lvs of sucker shoots and of ends of long shoots ± evenly toothed,
       not or shallowly lobed; those of short shoots becoming sub-
       glabrous; catkin-scales laciniate.   1. canescens

4   Lvs with narrow translucent border, not scented when unfolding,
       cuneate to subcordate at base; lf-stalk flattened laterally.   5
    Lvs without translucent border, strongly balsam-scented (like the
       large sticky buds) when unfolding, triangular-ovate, acuminate,
       cordate at base (or later lvs truncate), ciliate, very pale and densely
       pubescent on the veins beneath; lf-stalk not flattened. Female
       tree to 20(–30) m., often suckering. Introduced by streams, ponds,
       woods, etc. Balm of Gilead.   *P. gileadensis* Rouleau

5   Branches suberect, forming a very narrow crown. Male tree with
       lvs like *P. nigra* but smaller (to 6 cm.).   3. nigra var. italica
    Branches forming a wide crown.   6

6   Trunk and larger branches usually with swollen bosses; branches
       spreading and arching downwards; lvs without glands at the
       junction with the stalk.   3. nigra
    Trunk and larger branches without swollen bosses; branches ascend-
       ing and curving upwards to form a wide fan-like crown; lvs 6–

10 cm., deltate, truncate or subcordate at base, at least some on each twig with 1–2 glands at the junction with the stalk. Hybrid male tree (*P. nigra* × *P. deltoidea* Marsh), to 40 m. Introduced and commonly planted on damp soils; commoner than *P. nigra*. Black Italian Poplar.

P. × *canadensis* Moench var. *serotina* (Hartig) Rehder

1. P. canescens (Aiton) Sm.   Grey Poplar.

Tree to 35 m., freely suckering, with wide spreading crown. Bark smooth, grey, with conspicuous rhomboid lenticels; rough only at the base of old trunks. Buds ± thinly tomentose, neither sticky nor fragrant. Lvs of short shoots 3–6 cm., ovate or orbicular-ovate, blunt, rounded or subcordate at base, with 4–6 blunt triangular teeth on each side, white- or grey-tomentose beneath when young, becoming subglabrous in summer; stalk not or scarcely flattened; summer lvs (especially of suckers) larger, persistently tomentose beneath. Stigmas oblong, greenish-yellow. Catkin-scales laciniate, the divisions not reaching half-way. Fl. 2–3. Probably native in damp and wet woods in lowland England, northwards to Shropshire and Derby; Channel Is.

2. P. tremula L.   Aspen.

Tree to 20 m., suckering freely. Bark smooth, grey, with conspicuous rhomboid lenticels. Buds glabrous, somewhat sticky. Lvs of normal shoots 2·5–6 cm., orbicular, rounded at base, with c. 8–10 teeth on each side, silky-pubescent when young, soon glabrous; stalk strongly flattened laterally, causing the lvs to tremble; lvs of sucker shoots up to 15 cm., ovate-cordate, greyish-pubescent beneath. Stigmas broadened above, purple. Catkin-scales deeply laciniate, the divisions reaching more than half-way. Fl. 2–3. Woods, especially on poorer soils where it may be locally dominant. Throughout the British Is; common in the north and west.

3. P. nigra L.   Black Poplar.

Tree to 35 m., rarely suckering. Bark nearly black, with long deep fissures; trunk and larger branches usually with large swollen bosses; branches spreading, arching downwards to form a wide crown. Buds long, reddish-brown, curving outwards at apex. Lvs 5–20 cm., rhombic- to deltate-ovate, truncate or cuneate at base, with numerous teeth and with a narrow translucent border, soon glabrous; stalks laterally flattened, without glands at the junction with the blade. Stigmas greenish, stout. Catkin-scales laciniate, glabrous. Fl. 4. Probably native in wet woods and stream-sides in E. and C. England, but less

common than the introduced hybrid **P. canadensis** var. **serotina** (see key).
Var. **italica** Duroi, Lombardy Poplar, with erect branches and very
narrow crown, is often planted.

2. SALIX L.   Willows.

Trees or shrubs. Buds with 1 outer scale; terminal bud 0. Fls appearing before or after the lvs, in usually erect catkins with entire scales;
each fl. with 1 or 2 small nectaries; stamens 2–5(–12), usually 2.

Willows hybridize readily, are dioecious and often flower before the
lvs expand. For these reasons identification may be a difficult matter.
The following notes and key should make it possible to name the more
commonly encountered forms. For a fuller account the larger *Flora*
should be consulted.

Notes on plants with catkins:

(*a*) Catkins usually open before the lvs in the sallows (*S. caprea, cinerea,
aurita*), the common and purple osiers (*S. viminalis* and *purpurea*) and in most
of the hybrids within and between these two groups; also in *S. repens, S. lapponum* and the introduced *S. daphnoides*, and often in *S. nigricans*.

(*b*) Catkin-scales are uniformly yellowish in *S. pentandra, alba, fragilis,
triandra* and × *lanceolata*; reddish or brownish at the tip in *S. repens, S. triandra* × *viminalis* (and the alpine *S. arbuscula*), and blackish at the tip in the
remaining spp. and hybrids.

(*c*) Stamens are 5 (4–12) in each male fl. in *S. pentandra*; 3 in *S. triandra*;
2 but occasionally 3 in *S. triandra* × *viminalis* and other *triandra* hybrids;
2 united so completely as to appear single in *S. purpurea*; 2 with their filaments
united below in *S. purpurea* × *viminalis* and 2, quite free, in remaining willows.

(*d*) Anthers are yellow in most willows but are usually red-tinged in
S. cinerea and red or purple in *S. purpurea* and some of its hybrids.

(The alpine willows *S. lapponum, arbuscula, myrsinites* and *reticulata* also
have reddish or purplish anthers.)

(*e*) Styles are short or 0 in most willows but are long in *S. viminalis,
nigricans* and *phylicifolia* (and in some alpine spp.) Hybrids of *S. viminalis*
commonly have long or moderate styles.

1	Small prostrate creeping shrubs with lvs rounded or retuse at apex.	2
	Lvs not rounded at apex (or, if so, tall shrubs).	4
2	Lvs densely appressed-silky, at least beneath; catkins lateral on the previous year's wood, appearing with or before the lvs.	
		12. repens
	Lvs glabrous or nearly so; catkins terminal, appearing after the lvs.	3
3	Lvs 6–20 mm., bright green and shining on both sides, crenate-serrate; veins not impressed above. Mountain tops and rock-ledges, mostly at high altitudes.	*S. herbacea* L.

Lvs 1–3(–5) cm., dark green and rugose above, glaucous and prominently net-veined beneath, entire; veins strongly impressed above. Base-rich mountains in Scotland; very local.

S. reticulata L.

4 Lvs, or some of them, in subopposite pairs, ±oblanceolate, glabrous, dull and slightly bluish green; stamens 2, united so as to appear single. **5. purpurea**
Lvs never in subopposite pairs. 5

5 Lvs glabrous or nearly so. 6
Lvs hairy at least beneath. 16

6 Trees or ±erect shrubs 1 m. or more, mainly lowland. 7
Low shrubs 60 cm. or less, branches procumbent to ascending; Scottish mountains. 14

7 Lvs 2–8 cm., acute, not long-tapering to apex, glaucous beneath, 1½–2½(–5) times as long as broad. **11. phylicifolia**
Lvs 5–15 cm., acuminate or gradually tapering to an acute apex (except sometimes in S. purpurea × viminalis, and then 5–12 times as long as broad). 8

8 Twigs dark purple with a whitish waxy bloom; shrub or tree 7–10 m., with oblong to linear-lanceolate lvs, 5–10 cm., shining above, glaucous beneath. Introduced. *S. daphnoides Vill.
Twigs without a white waxy bloom. 9

9 Bark scaling off in patches; stamens 3. **4. triandra**
(Or, if some lvs 8 or more times as long as broad, S. × lanceolata Sm., an osier known only as female.)
Bark not scaling. 10

10 Twigs very brittle at junctions; lvs lanceolate, usually asymmetric at apex, with a glaucous bloom (or less frequently pale green) beneath. **3. fragilis**
Twigs not brittle at junctions; lvs not or scarcely asymmetric at apex, usually green beneath. 11

11 Lvs 2–4 times as long as broad, dark green and very glossy above, green beneath, somewhat leathery; stamens 5. **1. pentandra**
Lvs more than 4 times as long as broad, not especially dark and glossy or leathery. 12

12 Twigs bright yellow. **2. alba var. vitellina**
Twigs not bright yellow. 13

13 Shrub; lvs dull above, 5–12 times as long as broad; stamens 2, their filaments united below, free above. Grown as an osier.

S. purpurea × viminalis

Shrub; lvs somewhat glossy above, 5½–9 times as long as broad; stamens free, 2 or occasionally 3. Midland England.

S. triandra × viminalis

Tree with branches ascending at a narrow angle.

2. alba var. **caerulea**

14   Lvs equally bright green and shining on both sides. Rare and local
      on wet basic mountain rocks.   *S. myrsinites* L.
    Lvs glaucous or pale beneath.   *15*

15   Shrub 30–60 cm., often forming a crown from the base; main
      branches usually ± decumbent; lvs rarely more than 2 cm., often
      sparingly appressed-pubescent. Damp ledges of basic mountain
      rocks; very local.   *S. arbuscula* L.
    Shrub to 4 m., with main branches spreading or erect, not forming
      a crown; lvs 2–8 cm.; style long.   **11. phylicifolia**

16   Lvs lanceolate or linear-lanceolate, more than 5 times as long as
      broad; pubescence of appressed silky hairs; trees or shrubs 3 m.
      or more.   *17*
    Lvs 4 times as long as broad or less (rarely more in *S. repens*); if
      trees or tall shrubs, pubescence not silky.   *18*

17   Lvs 5–10 cm., moderately and ± equally silky on both sides; tree.

2. alba

    Lvs 10–25 cm., usually 7–18 times as long as broad, dark green
      above, densely silvery-silky beneath; styles long.   **6. viminalis**

18   Lvs with closely appressed silky hairs; shrubs 1¼ m. or less; lvs
      5 cm. or less.   *19*
    Hairs not closely appressed.   *20*

19   Lvs densely silky beneath, often silky above.   **12. repens**
    Lvs glabrous above and with a few hairs beneath (see *15* above).

S. arbuscula L.

20   Lvs clothed beneath and usually above with dense white matted
      silky-woolly hairs; low bushy shrubs (to 1½ m.).   *21*
    Lvs not clothed as above.   *22*

21   Lvs ovate or obovate, not more than twice as long as broad;
      branches stout; buds large, woolly; catkin-scales densely clothed
      with golden-yellow silky hairs when young. Very local and rare
      on basic mountain rocks in Scotland.   *S. lanata* L.
    Lvs elliptic or oblong, 2–4 times as long as broad; branches not
      markedly stout; buds small; catkin-scales with long white hairs.
      Local on wet mountain rocks, Westmorland to Sutherland.

S. lapponum L.

22   Lvs thinly pubescent beneath, frequently only on the veins, not or
      rarely glaucous, hairs not rust-coloured, blackening when dried;
      style long.   **10. nigricans**
    Lvs rather densely tomentose or pubescent all over beneath; if
      thinly pubescent then glaucous and usually with some rust-
      coloured hairs; not blackening when dried.   *23*

23 Lvs usually less than 3 times as long as broad; lf-apex blunt or
    shortly acuminate or cuspidate, rarely acute; style short.   **24**
    Lvs more than 3 times as long as broad, gradually tapered to an
    acute or acuminate apex; style usually moderate to long.
    Hybrids of *S. viminalis* with *S. caprea, cinerea* and *aurita*.
    Scattered localities in England, Ireland and S. Scotland.
        (**S. calodendron* Wimmer, with large lvs and old twigs
        blackish-tomentose, may be a species or a hybrid but is
        introduced).

24 Lvs 2–3 cm., rugose above, grey-tomentose beneath; usually a low
    shrub (1–2 m.) with wide-angled branching.   **9. aurita**
    Lvs 3 cm. or more, not rugose; shrubs or small trees 2 m. or more;
    branching not wide-angled.   **25**

25 Lvs oval or ovate-oblong (sometimes obovate), 5–10 cm., thickly
    and softly tomentose beneath, without rust-coloured hairs; twigs
    soon glabrous, without ridges under the bark.   **7. caprea**
    Lvs obovate, oblanceolate or elliptic, 3–7(–10) cm., usually with
    rusty hairs beneath; twigs often pubescent, with long narrow
    ridges under the bark when two years old.   **8. cinerea**

1. S. pentandra L.   Bay Willow

Shrub or small tree, 2–7 m. Twigs glabrous, shining as if varnished, not
fragile. Buds ovoid, sticky, shining. Lvs 5–12 cm., 2–4 times as long as
broad, ± elliptic, acuminate, rounded or broadly cuneate at base,
glandular-serrate, glabrous, dark green and glossy above, paler beneath,
somewhat leathery when mature, sticky and fragrant when young; stalk
with 1–3 pairs of glands near the top; stipules small, soon falling.
Catkins appearing after the lvs, cylindric; scales uniformly yellowish.
Male catkin 2–6 × 1–1·5 cm., dense; male fls with (4–)5(–12) stamens,
anthers golden yellow before dehiscence. Female catkin 2–5 × c. 1 cm.;
female fl. with 2 nectaries. Fl. 5–6. Stream-sides, marshes, fens and wet
woods, to 1500 ft.; from N. Wales, Derby and Yorks northwards and in
N. Ireland, rather uncommon.

2. S. alba L.   White Willow.

Tree 10–25 m., branches ascending at 30–50°, forming a narrow crown,
appearing silvery-grey in lf; often pollarded. Rootlets whitish. Bark
greyish, not peeling, fissured, the ridges forming a closed network. Twigs
silky when young, later glabrous and olive, not fragile. Buds oblong,
pubescent for some time. Lvs 5–10 cm., usually 5½–7½ times as long as
broad, lanceolate, acuminate, not or slightly asymmetric at apex, cuneate
at base, finely serrate, covered with white silky appressed hairs on both
sides; stalks without glands; stipules usually soon falling. Catkins

appearing with the lvs, dense, cylindric; catkin-scales uniformly yellowish. Male catkin 2·5–5 cm. × 6 mm.; male fl. with 2 free stamens with yellow anthers. Female fl. with usually 1 nectary. Fl. 4–5. By streams and rivers, marshes, fens and wet woods on richer soils throughout most of the British Is.

Ssp. **caerulea** (Sm.) Sm., Cricket-bat Willow, has branches ascending at a narrower angle and less hairy bluish-green lvs.

Ssp. **vitellina** (L.) Stokes is a smaller tree with twigs bright yellow or orange in the first year and lvs soon nearly glabrous above.

3. S. fragilis L.   Crack Willow.

Tree 10–25 m., branches spreading widely at 60–90°, forming a broad crown, appearing green in lf; often pollarded. Rootlets red. Bark greyish, not peeling, deeply fissured, the ridges forming an open network. Twigs sometimes slightly pubescent when young, soon glabrous and olive, very fragile at the junctions and separating with a distinct snap. Buds brown, ovate, becoming glabrous, ± sticky. Lvs 6–15 × 1·5–4 cm., usually 4½–9 times as long as broad, lanceolate, long-acuminate, usually asymmetric at apex, cuneate at base, rather coarsely serrate, becoming glabrous and bright above, with a glaucous bloom (or less often paler green) beneath; stalk with 2 glands at the top; stipules usually soon falling. Catkins appearing with the lvs, drooping, rather dense, cylindric; catkin-scales uniformly yellowish. Male catkins 2·5–6 cm., up to 1 cm. broad; male fl. with 2 free stamens with yellow anthers. Female catkin 3–7 cm.; female fl. usually with 2 nectaries. Fl. 4 (earlier than *S. alba*). By streams and rivers, marshes, fens and wet woods; more tolerant of poor soils than *S. alba*; common from Perth southwards; probably introduced in Ireland.

S. decipiens Hoffm. differs from *S. fragilis* in having bark flaking off when old and twigs red on the exposed side when young; stamens occasionally 3.

4. S. triandra L.   Almond Willow.

Shrub, rarely a small tree, 4–10 m. Bark smooth, peeling off in patches. Twigs glabrous, olive or reddish-brown, rather fragile. Buds brown, ovoid, glabrous. Lvs 5–10 cm., usually 3½–7½ times as long as broad, oblong-ovate or oblong-lanceolate, acute or shortly acuminate (not asymmetric), rounded at base, serrate, glabrous, dark green and somewhat shining above, glaucous (or less often pale green) beneath; stalk with 2–3 small glands at the top; stipules rather large, usually persistent. Catkins appearing with the lvs, erect, cylindric. Male catkin 3–5 cm., slender; male fl. with 3 free stamens with yellow anthers. Female catkin rather shorter and denser than male; female fl. with 1 nectary. Fl. 3–5.

Sides of rivers and ponds, marshes, etc., often planted as an osier; widespread and rather common in England; less so in Wales and very local in S. Scotland; E. Ireland.

5. S. purpurea L.   Purple Osier.

Shrub 1½–3 m., rather slender. Bark bitter. Twigs slender, straight, glabrous, shining, usually purplish at first, becoming olive or yellowish-grey. Buds oblong, acute, glabrous. Lvs 4–10 cm., usually sub-opposite at least near the ends of twigs, 3–15 times as long as broad, obovate-oblong to oblanceolate-linear, acute or acuminate, gradually narrowed to the base, very finely serrate, glabrous, dull and slightly bluish-green above, paler and often glaucous beneath; stipules small, quickly falling. Catkins 2–3·5 cm., appearing before the lvs, dense, cylindric, suberect to spreading, subsessile; catkin-scales blackish at the apex. Stamens 2, completely united and appearing single, their anthers reddish or purplish before dehiscence. Female fl. with 1 nectary (like the male). Fl. 3–4. Fens, where it may be locally dominant, marshes and by rivers and ponds, throughout Great Britain and Ireland; sometimes planted as an osier, though not as commonly as *S. viminalis* and its hybrid with *S. purpurea* (whose 2 stamens have their filaments joined below but free above).

6. S. viminalis L.   Common Osier.

Shrub 3–5(–10) m., with long straight flexible branches. Twigs densely pubescent, becoming glabrous later, not ridged under the bark. Buds ovoid-oblong, acuminate, pubescent. Lvs 10–25 cm., usually 7–18 times as long as broad, lanceolate or linear-lanceolate, gradually narrowed to the apex, cuneate at base, dark green and glabrous above, silvery silky-tomentose beneath; margins undulate, revolute when young, entire or nearly so; stipules small, usually falling soon. Catkins appearing before or with the lvs, dense, subsessile; male 2·5–4 cm., ovoid-oblong; female cylindric; catkin-scales blackish at the apex. Stamens 2, free; anthers yellow. Female fl. with 1 nectary (like the male); style long, as long as the ovary and half as long as the fl. By streams and ponds, in marshes and fens, and commonly planted as an osier. Throughout the British Is, and common in lowland areas.

7. S. caprea L.   Great Sallow, Goat Willow.

Shrub or small tree, 3–10 m. Bark coarsely fissured. Twigs rather stout, pubescent at first, becoming ± glabrous by autumn, without ridges under the bark. Buds ovoid-conic, 3-angled, acute, pubescent at first, later glabrous. Lvs 5–10 cm., 1·2–2 times as long as broad, oval to ovate-oblong or obovate, ± blunt or shortly acuminate with the tip obliquely reflexed, subcordate to cuneate at base. Margin somewhat undulate, crenate to entire, dark green and finally ± glabrous above, persistently softly and densely grey-tomentose beneath and strongly net-veined; stipules half-cordate. Catkins appearing before the lvs, dense, subsessile; male 2–3·5 × 1·5–2 cm., oblong-ovoid; female finally lax and 3–7 cm.; catkin-scales blackish at the apex. Stamens 2, free, with yellow anthers. Female fl. with 1 nectary (like the male). Fl. 3–4. Woods, scrub, hedges, etc. to 2800 ft., common throughout the British Is.

8. S. cinerea L.   Common Sallow.

Shrub or small tree, 2–10 m.; branches mostly long, straight and sub-erect. Twigs rather stout, shortly pubescent at first, brown (or reddish in winter), persistently but thinly pubescent or ± glabrous later, their branching not wide-angled, with narrow ridges visible on the surface of the wood when the bark is removed. Buds ovoid, thinly pubescent or glabrous. Lvs 2·5–6 cm., c. 2–3½ times as long as broad, obovate or oblanceolate, apiculate or shortly cuspidate, cuneate at base, margin often somewhat undulate, distinctly crenate-serrate or subentire; soon glabrous and dark grey-green above; thinly pubescent mainly on the veins beneath, but not soft to the touch as in *S. caprea* and *S. aurita*, and glaucous with all or some of the hairs rust-coloured when adult; stipules small, usually persistent. Catkins appearing before the lvs, dense, subsessile; male 2–3 × 1–2 cm., ± ovoid; female finally 3·5–5 cm.; catkin-scales blackish at the apex. Stamens 2, free, with yellow anthers red-tinged when young. Female fl. with 1 nectary (like the male). Fl. 3–4. Woods and heaths, marshes and fens, and by ponds and streams, to 2000 ft; common throughout the British Is. The above description is of the widespread ssp. **atrocinerea** (Brot.) P. Silva & G. Sobr. Ssp. **cinerea** has its twigs thickly tomentose and its lvs softly tomentose beneath with no rust-coloured hairs. It is a local plant mainly in the E. Anglian fens.

9. S. aurita L.   Eared Sallow.

Shrub 1–2(–3) m., with numerous spreading branches. Twigs rather slender, soon glabrous, brown, usually angular and with wide-angled branching, with ridges under the bark. Buds oval, ± glabrous. Lvs 2–3 cm., c. 1½–2½ times as long as broad, obovate, shortly cuspidate with the cusp often obliquely decurved, ± cuneate at base, undulate, toothed to subentire, distinctly wrinkled, dull grey-green and ± pubescent above, grey-tomentose beneath. Stipules large and conspicuous, ± reniform, persistent. Catkins appearing before the lvs, subsessile; male 1–2 cm., ovoid; female 1–2·5 cm., cylindric; catkin-scales blackish at the apex. Stamens 2, free, with yellow anthers. Female fl. with 1 nectary (like the male). Fl. 4. Damp woods, heaths, rocks by streams and on moors, etc., to 2600 ft. Common throughout the British Is.

10. S. nigricans Sm.   Dark-leaved Willow.

Shrub 1–4 m., or spreading and procumbent. Twigs pubescent or ± glabrous, dull, blackish to olive green, ridged under the bark. Buds oval. Lvs c. 2–7 × 1½–3 cm., usually 1½–3 times as long as broad, variable in shape from orbicular-ovate to lanceolate, acute or shortly acuminate, ± rounded at base, toothed to subentire, deep green and rather dull above, paler and ± pubescent beneath at least on the veins, usually turning blackish when dried; stipules usually rather large, half-cordate. Catkins with or before the lvs; male 1·5–2·5 cm., subsessile, ovoid or oblong; female finally 5–7 cm., rather lax, on short lfy stalks; catkin-scales blackish at the apex. Stamens 2, free; anthers yellow. Female fl. with 1 nectary (like the male) and a long style; capsule glabrous. Fl. 4–5. By lakes and streams and on damp rock-ledges, to 3000 ft.; from Lancashire and Yorks northwards and in N. Ireland.

11. S. phylicifolia L.   Tea-Leaved Willow.

Shrub 1–4 m. Twigs glabrous, shining as though varnished, brown at maturity, ridged under the bark. Buds narrow, ovate, acute, yellowish, ± glabrous. Lvs 2–8 cm., c. 1½–2½ times as long as broad, ± ovate, ± acute, rounded or cuneate at base, margin ± crenate or shortly gland-toothed, glabrous and shining above, somewhat glaucous and glabrous beneath, rather leathery, not blackening when dried; stipules small, usually soon falling. Catkins appearing with the lvs, subsessile or shortly stalked; male 1·5–2·5 cm., ovoid; female 2·5–6 cm., rather lax in fr.; catkin-scales blackish at the apex. Stamens 2, free; anthers yellow. Female fl. with 1 nectary (like the male) and a long style. Fl. 4–5. By lakes and streams and among wet rocks, to 2300 ft.; from Lancashire and Yorks northwards, rather local; N. and N.W. Ireland.

S. hibernica Rech. fil. has almost entire very shortly acuminate lvs. Known only from Ben Bulben, Co. Sligo.

12. S. repens L.   Creeping Willow.

Shrub 30–150 cm., with creeping rhizome. Stems prostrate to erect, slender. Twigs silky-pubescent when young, with a few fine ridges under the bark. Buds ovate, soon glabrous. Lvs 0·5–4·5 cm., usually 1½–3½ times as long as broad, very variable in shape, appressed silvery-silky on both sides when young but sometimes becoming glabrous above, prominently net-veined beneath; stipules usually soon falling or 0. Catkins appearing before the lvs, subsessile or short-stalked; male 5–20 mm., slender, ovoid to oblong; female 8–25 mm., globose to oblong; catkin-scales brownish at the apex. Stamens 2, free; anthers yellow. Female fl. with 1 nectary (like the male). Fl. 4–5. Damp and wet heaths and fens (ssp. **repens**, with lvs usually silky beneath only and those of sterile shoots only up to 25 × 10 mm.), and dune-slacks (ssp. **argentea** (Sm.) A. & G. Camus, with lvs usually silky on both sides and those of sterile shoots reaching 45 × 25 mm.); throughout the British Is, but local.

METACHLAMYDEAE

90. ERICACEAE

Shrubs or rarely trees with simple exstipulate lvs. Fls hermaphrodite, regular or nearly so. Calyx persistent. Corolla usually gamopetalous, rarely free, on the edge of a fleshy disk. Stamens usually twice as many as corolla-lobes, hypogynous or epigynous; anthers usually opening by apical pores or slits, often with awn-like appendages; pollen in tetrads. Ovary superior or inferior, usually 4–5-celled with axile placentae; ovules 1–many; style simple; stigma capitate. Fr. usually a capsule or berry; seeds small, endospermic.

1	Ovary superior.	*2*
	Ovary inferior.	6. VACCINIUM
2	Petals free, cream; lvs rusty-tomentose beneath, linear, with revolute margins. A shrub to 1 m., possibly native in bogs in C. Scotland and a rare escape elsewhere. *Ledum palustre* L.	
	Petals ±united; lvs not rusty-tomentose.	*3*
3	Corolla falling after fl.; lvs usually spirally arranged or, if opposite, then not imbricate.	*4*
	Corolla persistent in fr.; lvs whorled or opposite and imbricate.	*12*
4	Lvs opposite; a small creeping evergreen alpine shrub with pink fls 4–5 mm. diam., and a capsular fr.; local on mountains in N. Scotland. Mountain Azalea. *Loiseleuria procumbens* (L.) Desv.	
	Lvs spirally arranged.	*5*
5	Corolla bell-shaped, slightly zygomorphic, c. 5 cm. diam.	1. RHODODENDRON
	Corolla urceolate, regular, smaller.	*6*

6 Fls 4-merous, 8–12 mm., rose-purple, in lax racemes; lvs white-tomentose below. A local plant of W. Ireland. St Dabeoc's Heath. *Daboecia cantabrica* (Hudson) C. Koch.
Fls normally 5-merous; lvs at most glaucous beneath (and then fls in clusters). 7

7 Lvs linear or linear-elliptic, at most 5 mm. broad; fr. dry; fls in clusters. 8
Lvs ovate, obovate or elliptic; fr. fleshy. 9

8 Lvs 5–9 mm., blunt, serrulate; corolla ovoid, purple; calyx and fl.-stalks glandular. A heath-like shrub; on two mountains in Scotland. *Phyllodoce caerulea* (L.) Bab.
Lvs 15–35 mm., acute, entire; corolla subglobose, pink; calyx and fl.-stalks glabrous. 2. ANDROMEDA

9 Lvs 5–12 cm., rounded to cordate at base; fr. a capsule enclosed in the fleshy calyx and resembling a black berry (1 cm.). A suckering shrub to 1 m., forming large patches; fls urceolate, white tinged with pink. Introduced for pheasants and often naturalized. *Gaultheria shallon* Pursh
Lvs cuneate at base; fr. a drupe or berry; plant not suckering. 10

10 Erect shrub or tree to 10 m.; lvs 5–10 cm., ±elliptic, serrate, shining above; fr. a red globose ±warty strawberry-like berry (1·5–2 cm.); fls urceolate, creamy-white, sometimes pink-tinged. Locally abundant in W. Ireland. Strawberry Tree.
Arbutus unedo L.
Prostrate shrubs with conspicuously net-veined obovate lvs. 11

11 Lvs persistent, leathery, entire; fr. red. 3. ARCTOSTAPHYLOS
Lvs deciduous, thinnish, finely toothed; fr. black (6–10 mm.); fls urceolate, white. Very local on mountains in N. Scotland, to 3000 ft. *Arctous alpinus* (L.) Niedenzu

12 Calyx longer than and coloured like the corolla; lvs opposite.
4. CALLUNA
Calyx much shorter than corolla; lvs whorled. 5. ERICA

1. RHODODENDRON L.

Shrubs, rarely trees. Buds scaly. Lvs spirally arranged, usually entire, short-stalked. Fls usually in terminal racemes, usually 5-merous. Corolla bell- or funnel-shaped, slightly zygomorphic, falling after fl. Stamens from as many to twice as many as corolla-lobes, hypogynous, anthers opening by pores, awnless. Ovary superior. Fr. a septicidal capsule.

*1. R. ponticum L.

Evergreen shrub to 3 m. Lvs 6–12 cm., elliptic, dark green above, paler beneath, acute. Corolla widely bell-shaped, c. 5 cm. diam., dull purple spotted with brown. Stamens 10. Fl. 5–6. Introduced and thoroughly naturalized in many places on sandy and peaty soils.

2. Andromeda L.

Low evergreen shrubs with spirally arranged oblong or linear entire lvs. Fls 5-merous, in terminal clusters. Corolla falling after fl., urceolate, lobes short. Stamens 10, hypogynous; anthers awned, opening by pores. Ovary superior. Fr. a loculicidal capsule.

1. A polifolia L.   Marsh Rosemary.

Glabrous shrub to 30 cm., with creeping rhizome and scattered erect stems. Lvs 1·5–3·5 cm., linear, acute, revolute, dark green above, glaucous beneath. Fls 2–8, nodding. Corolla 5–7 mm., subglobose, pink. Capsule subglobose, glaucous. Fl. 5–9. Bogs, rarely wet heaths, to 1750 ft.; from Somerset and Norfolk northwards to Perth; Ireland.

3. Arctostaphylos Adanson

Evergreen shrubs with spirally arranged lvs. Fls usually 5-merous in terminal infls. Corolla falling after fl., urceolate; stamens 10, hypogynous; anthers awned, opening by pores. Ovary superior, 5–10-celled with 1 ovule per cell. Fr. a drupe.

1. A. uva-ursi (L.) Sprengel   Bearberry.

Prostrate mat-forming glabrous shrub with long rooting branches. Lvs 1–2 cm., obovate, bluntish, cuneate at base, dark green above, paler beneath, conspicuously net-veined, entire. Fls 5–12 in short dense racemes. Corolla 4–6 mm., ± globose, white tinged with pink. Fr. 6–8 mm., red, globose, glossy. Fl. 5–7. Moors, rocks and banks, to 3000 ft. From Derby and Westmorland northwards, common in the Scottish Highlands; N. and W. Ireland.

4. Calluna Salisb.

Evergreen shrub with very small opposite lvs. Fls 4-merous. Perianth persistent in fr. Calyx large, of same colour and texture as the shorter bell-shaped corolla. Stamens 8, hypogynous; anthers awned, opening by pores. Ovary superior. Fr. a septicidal capsule, few-seeded.

1. C. vulgaris (L.) Hull Ling, Heather.

Diffuse shrub to 60(–100) cm. with numerous tortuous, decumbent or ascending stems, rooting at the base. Lvs 1–2 mm., linear, sessile, with two short downward projections at the base; margins strongly revolute; glabrous to densely grey-tomentose; those of the main stems distant, those of the short shoots densely imbricate in 4 rows. Fls solitary, axillary, on main and short shoots, forming a loose raceme-like infl., 3–15 cm. Under each fl. 4 ovate bracteoles form a calyx-like involucre. Calyx 4 mm., somewhat scarious, pale purple like the corolla, 2–3 mm., whose free lobes slightly exceed the basal tube. Capsule c. 2 mm., globose, enclosed in the persistent calyx and corolla. Fl. 7–9. Heaths, moors, bogs and open woods on acid soils. Common throughout the British Is.

5. ERICA L.

Evergreen shrubs with small entire whorled lvs. Infl. various. Fls 4-merous. Calyx much shorter than corolla, deeply lobed, not petaloid. Corolla persistent in fr., urceolate, bell-shaped or cylindric; lobes short. Stamens 8, hypogynous; anthers opening by pores. Ovary superior. Capsule loculicidal, many-seeded.

1	Stamens included in corolla-tube.	2
	Stamens exserted, at least partly; lvs glabrous.	6
2	Lvs and sepals ciliate with long usually glandular hairs.	3
	Lvs and sepals glabrous.	5
3	Fls in terminal umbel-like clusters; anthers awned.	4

Fls in 1-sided terminal racemes; anthers not awned; corolla 8–10 mm., deep pink, urceolate, curved above and swollen below, with an oblique mouth; lvs in whorls of 3, ovate, glabrous above. Heaths in S.W. England and W. Ireland. Dorset Heath.

E. ciliaris L.

4 Lvs grey-pubescent above as well as ciliate, revolute nearly to the midrib, shorter and more distant below infl. **1. tetralix**

Lvs dark green and glabrous above, somewhat revolute but leaving much of the white undersurface exposed, not different below infl.; calyx-lobes glabrous except for the cilia; corolla bright deep pink. Very local in W. Ireland. *E. mackaiana* Bab.

5 Corolla urceolate, pink or purple, rarely white (and then white in bud); low shrub; lvs 3 in a whorl. **2. cinerea**

Corolla narrowly bell-shaped, pink in bud, white when open. Erect shrub 1–2 m.; lvs 3–4 in a whorl, very slender, glabrous; fls in long panicles. Introduced and naturalized in S.W. England.

*E. lusitanica J. H. Rudolph

6   Fls longer than their stalks; corolla ±tubular, dull purplish-pink;
       anthers about ½ exserted; fl. 3–5. Shrub to 2 m.; lvs 4 in a whorl.
       Bog-margins and heaths in W. Ireland, local.
                    E. hibernica (Hooker & Arn.) Syme (*E. mediterranea* auct.)
     Fls much shorter than their stalks; corolla widely bell-shaped,
       usually pale lilac; anthers fully exserted; fl. 7–8. Shrub 30–80 cm.
       Heaths round the Lizard (Cornwall). Cornish Heath.
                                                                   E. vagans L.

1. E. tetralix L.   Cross-leaved Heath.

Diffuse shrub to 60 cm., with numerous tortuous ascending branched
stems, rooting at the base, without short axillary shoots. Lvs 4 in a
whorl, 2–4 mm., linear, glandular-ciliate, grey-pubescent above, margins
revolute nearly to midrib, hiding the under-surface; lvs below the infl.
more distant and usually shorter. Fls 4–12 in terminal umbel-like
clusters, nodding in fl., becoming erect in fr. Calyx-lobes c. 2 mm.,
pubescent on the surface and ciliate with long wavy glandular hairs.
Corolla 6–7 mm., urceolate, rose-pink; lobes very short. Anthers
included, awned. Capsule pubescent. Fl. 7–9. Bogs, wet heaths and
moors, to 2400 ft. Common in suitable habitats throughout the
British Is.

2. E. cinerea L.   Bell-heather.

Diffuse shrub to 60 cm., with numerous ascending branched stems rooting at the base and with numerous lfy axillary shoots which often appear
as bunches of lvs. Lvs 3 in a whorl, 5–7 mm., linear, glabrous, dark
green, margins strongly revolute. Fls in short terminal racemes and on
the upper axillary shoots. Calyx-lobes glabrous, usually purple. Corolla
5–6 mm., urceolate, usually crimson-purple; lobes very short, reflexed.
Anthers included, awned. Capsule glabrous. Fl. 7–9. Heaths and
moors, usually dry, to 2200 ft. Common throughout the British Is.

6. VACCINIUM L.

Shrubs with erect or creeping stems and spirally arranged or 2-ranked
short-stalked lvs. Fls 4–5-merous. Calyx-lobes short. Corolla falling
after fl. Stamens 8–10, epigynous; anthers awned, opening by pores.
Ovary inferior. Fr. a berry with persistent calyx-lobes at its apex.

1   Corolla bell-shaped or urceolate with small lobes; aerial stems not
       creeping and thread-like.   2
     Corolla divided nearly to the base, lobes reflexed; stems creeping
       and thread-like.   4

2 Lvs evergreen, dark and glossy; fls in racemes; corolla bell-shaped;
    fr. red.   **1. vitis-idaea**
   Lvs deciduous; fls 1–4, axillary; corolla urceolate; fr. blackish.   *3*

3 Lvs finely toothed, acute, bright green; twigs angled, green; fls
    solitary.   **2. myrtillus**
   Lvs entire, blunt, blue-green; twigs terete, brownish; fls 1–4 in each
    axil, c. 4 mm., subglobose or ovoid, pale pink; fr. black with a
    glaucous bloom. A shrub with obovate lvs, 1–2·5 cm., on moors
    (to 3500 ft.) from Durham and Cumberland northwards, local.
    Bog Whortleberry.   *V. uliginosum* L.

4 Lvs ± ovate, acute, to 8 mm.; infl. terminal.   *5*
   Lvs oblong, blunt, 6–18 mm.; infl.-axis ending in a lfy shoot; fls
    1–10 in a raceme; corolla-lobes 6–10 mm., pink; fr. 10–20 mm.,
    red. Introduced for its fr. and sometimes naturalized. Large
    Cranberry.   ***V. macrocarpon*** Aiton

5 Fl-stalks shortly downy; lvs 4–8 mm., ± oblong-ovate, equally wide
    for some distance at the base.   **3. oxycoccos**
   Fl.-stalks glabrous; lvs 3–5 mm., ± triangular-ovate, widest near the
    base. Fr. lemon- or pear-shaped. Local in bogs in the Scottish
    Highlands.   *V. microcarpum* (Rupr.) Hooker fil.

1. V. vitis-idaea L.   Cowberry, Red Whortleberry.

Evergreen shrub to 30 cm., with creeping rhizome and numerous
± erect often arching stems. Twigs terete. Lvs 1–3 cm., obovate, blunt
or emarginate, dark green and glossy above, paler and gland-dotted
beneath, leathery, obscurely crenulate. Fls c. 4 in short terminal drooping racemes. Corolla c. 6 mm., bell-shaped with revolute lobes; white
tinged with pink. Fr. 8–10 mm., red, globose, edible. Fl. 6–8. Rocky
moors and woods on acid soils, to over 3500 ft. Common in Scotland
and extending to N. and W. England; N. and E. Ireland.

2. V. myrtillus L.   Bilberry, Blaeberry, Whortleberry, Huckleberry.

Glabrous deciduous shrub to 60 cm., with creeping rhizome and
numerous erect stems and branches. Twigs angled, green. Lvs 1–3 cm.,
ovate, acute, finely toothed, bright green, conspicuously net-veined.
Fls usually solitary, axillary, hanging. Corolla c. 6 mm., urceolate-
globose, greenish-pink, with very short reflexed lobes. Fr. c. 8 mm.,
black with a glaucous bloom, globose, edible. Fl. 4–6. Heaths, moors
and woods on acid soil to 4000 ft.; more tolerant of exposure and shade
than *Calluna*. Common throughout most of the British Is, but local in
the E. and S.E.

3. V. oxycoccos L. (*Oxycoccus palustris* Pers.)　　　　Cranberry.

Stems prostrate, rooting, thread-like. Lvs 4–8 mm., distant, oblong-ovate, acute, dark green above, glaucous beneath, strongly revolute. Fls 4-merous, 1–4 in a terminal raceme. Corolla pink, rotate, 4-lobed nearly to the base, the lobes 5–6 mm., reflexed. Fr. 6–8 mm., globose or pear-shaped, red or brown-spotted, edible. Fl. 6–8. Bogs and wet heaths, local, but occurring through most of Great Britain, Ireland and the Inner Hebrides.

91. PYROLACEAE

Evergreen perennial herbs, often far-creeping. Fls hermaphrodite, regular, 5-merous. Stamens twice as many as petals; anthers opening by pores, pollen usually in tetrads. Ovary superior, incompletely 5-celled with thick fleshy axile placentae and numerous ovules; style simple; stigma capitate. Fr. a loculicidal capsule. Seeds very small, numerous, endospermic.

1　Fls 5–12 mm. diam., in terminal racemes; lvs spirally arranged.　　　　2
　　Fls c. 15 mm. diam., solitary, white; lvs opposite. A very rare plant
　　　chiefly of Scottish pinewoods.　　　　*Moneses uniflora* (L.) A. Gray
2　Fls greenish, c. 5 mm. diam., in 1-sided racemes; lvs acute, their
　　　stalks c. 1 cm. A very local plant of woods and damp rock-ledges
　　　in S. Wales, N. England, Scotland and N. Ireland.
　　　　　Orthilia secunda (L.) House (*Ramischia secunda* (L.) Garcke)
　　Fls pure white or pinkish, 6 mm. diam. or more, not in 1-sided
　　　racemes; lvs ± blunt, their stalks 2 cm. or more.　　　1. PYROLA

1. PYROLA L.　　　　　　　　　　　　Wintergreen.

Herbs with a slender creeping rhizome and short, often distant, aerial stems frequently reduced to a basal rosette of lvs. Lvs spirally arranged. Fls in racemes, which are not 1-sided, on scapes usually bearing a few scales. Pollen in tetrads. Valves of the capsule webbed at the edges.

1　Style straight; fls ± globose in outline.　　　　　　　　　　2
　　Style strongly curved; corolla nearly flat.　　　　**3. rotundifolia**
2　Style 1–2 mm., not thickened below the stigma.　　　　**1. minor**
　　Style c. 5 mm., thickened into a ring below the stigma.　　**2. media**

1. P. minor L.　　　　　　　　　　　Common Wintergreen.

Stem very short or lvs all basal. Lvs 2·5–4 cm., ovate or oval, blunt or subacute, crenulate, light green; their stalks 2·5–3 cm., shorter than the blade. Scape 10–30 cm. Fls c. 6 mm., ± globose in outline. Petals pinkish. Style 1–2 mm., straight, included, shorter than stamens, without a ring below the stigma; stigma with 5 large spreading lobes. Fl. 6–8. Woods, moors, damp rock-ledges and dunes, to 3750 ft. Rather local over much of Great Britain and Ireland.

2. P. media Swartz   Intermediate Wintergreen.

Lvs 3–5 cm., all basal, ± orbicular, blunt, obscurely crenulate, dark green; their stalks 2·5–5·5 cm., about equalling the blade or longer. Scape 15–30 cm. Fls c. 10 mm., ± globose in outline. Petals white, tinged pink. Style c. 5 mm., straight, exserted, longer than stamens, expanded into a ring below the stigma, which has 5 erect lobes. Fl. 6–8. Woods and moors, to 1800 ft. Very local through much of N. England and Scotland and with a few localities in S. and W. England; W. and N. Ireland. Commonest in Scottish pinewoods.

3. P. rotundifolia L.   Larger Wintergreen.

Lvs 2·5–5 cm., all basal, orbicular or oval, obscurely crenulate, dark green and glossy; their stalks 3–7 cm., longer than the blade. Scape 10–40 cm. Fls c. 12 mm. diam., ± flat. Petals pure white. Style in a curve, concave upwards, longer than petals and stamens, expanded into a ring below the stigma, which has 5 small erect lobes. Fl. 7–9. Bogs, fens, damp rock-ledges, woods and dune-slacks, to 2500 ft. Sussex and Wales to Orkney, very local and with a distinct eastern tendency; very rare in Ireland; Channel Is.

92. MONOTROPACEAE

Saprophytic herbs without chlorophyll. Lvs scale-like. Fls hermaphrodite, regular. Petals usually free. Stamens twice as many as petals; anthers opening by longitudinal slits; pollen-grains single. Ovary superior, incompletely septate, with numerous ovules on massive placentae. Fr. a capsule with numerous tiny seeds.

1. Monotropa L.

Fls in short racemes, 4–5-merous. Sepals free, large, oblong-spathulate. Petals saccate at the base. Stamens 8–10. Disk of 8–10 glands. Style stout, straight; stigma broad, funnel-shaped; ovary 8–10-lobed, 4–5-celled below. Fr. a loculicidal capsule.

1. M. hypopitys L., *sensu lato*　　　　　　　　Yellow Bird's-nest.

Whole plant yellowish-white, fragrant when dry. Roots short, thick, much-branched. Stems 8–30 cm., stout, all root-borne. Lvs 5–10 mm., ovate-oblong, entire, ± erect, numerous, especially at the base of the stem. Infl. up to 11-fld. drooping, erect in fr. Fls 10–15 mm., narrowly bell-shaped. Petals up to twice as long as sepals. Fl. 6–8. Woods, especially of beech and pine, and in dunes among *Salix repens*, etc.; widespread but rather local in England and Wales northwards to Westmorland and York; in Scotland only near Moray Firth; local in Ireland.

93. EMPETRACEAE

Small evergreen heath-like shrubs. Lvs spirally arranged, entire, with strongly revolute margins; stipules 0. Fls small, regular, 2–3-merous. Per. segs 4–6, in two ± similar whorls. Stamens half as many, hypogynous. Ovary 2–9-celled, superior, with 1 ovule per cell; style short; stigmas as many as ovary-cells. Fr. a drupe with 2–9 stones; seeds endospermic.

1. EMPETRUM L.　　　　　　　　　　Crowberry.

Fls 1–3, axillary. Per. segs 6. Stamens 3. Ovary 6–9-celled; stigmas toothed. Fr. juicy.

Fls dioecious; stems prostrate and rooting round the edge of the tuft; young stems reddish; lvs 3–4 times as long as broad, their margins almost exactly parallel.　　　　　　　　　　　　　　**1. nigrum**

Fls hermaphrodite; stems not prostrate nor rooting round the edge of the tuft; young stems green, becoming brown; lvs 2–3 times as long as broad, their margins somewhat rounded. Mountain tops and high-level moors, c. 2000–3500 ft., in N. Wales, N.W. England and Scotland.　　　　　　　　　　　　*E. hermaphroditum* Hagerup

1. E. nigrum L.

Low shrub 15–45 cm., with relatively long and slender procumbent and ascending stems, prostrate and rooting round the edge of the tuft. Young twigs reddish, becoming red-brown. Lvs dense, 4–6 mm., oblong or linear-oblong, parallel-sided, c. 3–4 times as long as broad, blunt, short-stalked, glandular on the margin when young, otherwise glabrous. Fls 1–2 mm. diam., dioecious, pale pinkish with large red anthers or dark violet stigmas. Fr. c. 5 mm., black, subglobose, with 6–9 stones. Fl. 5–6. Moors and the drier parts of blanket bogs, to c. 2500 ft. Common in Scotland, N. England and Wales, rare in S.W. England and absent from the whole south-eastern part of England; throughout Ireland.

94. DIAPENSIACEAE (p. xxv)

95. PLUMBAGINACEAE

Perennial or rarely annual herbs or shrubs with simple usually spirally arranged lvs often confined to a basal rosette; stipules 0. Infl. usually cymose, the cymes sometimes closely aggregated into heads. Bracts scarious. Fls actinomorphic, hermaphrodite. Calyx tubular below, scarious and often pleated above, persistent. Petals free or slightly joined at the base or with a long basal tube. Stamens 5, free and hypogynous or inserted on the corolla-tube, opposite the petals. Ovary superior, 1-celled, with 1 basal ovule; styles 5, or 1 with 5 stigma-lobes. Fr. dry with a thin papery wall, opening with a lid, or irregularly, or remaining closed; seeds endospermic.

Fls lavender, in terminal panicles; styles glabrous throughout.

1. LIMONIUM

Fls pink, in dense almost globular heads; styles hairy below.

2. ARMERIA

1. LIMONIUM Miller

Usually perennial herbs with woody stock and with lvs confined to a basal rosette. Fls short-stalked in 1–5-fld spikelets each with 3 scale-like bracts, the spikelets further aggregated into spikes. Calyx funnel-shaped, green and 5–10-ribbed below, bluish and scarious above, with or without smaller teeth between the 5 free lobes. Corolla-tube very short. Stamens inserted at base of corolla.

1   Lvs pinnately veined; calyx-lobes with intermediate teeth.   2
     Lvs not pinnately veined; calyx-lobes without intermediate teeth.   3
2   Stem usually not branched below the middle; infl. ± corymbose, the
        spikelets crowded into short spreading spikes.   **1. vulgare**
     Stem usually branched below the middle; infl. not corymbose, the
        spikelets distant in long ± erect spikes.   **2. humile**
3   Infl. with very numerous slender zig-zag barren branches below;
        outermost bract entirely scarious.   **3. bellidifolium**
     Infl. with few or no barren branches below; outermost bract
        scarious only at the margin.   4
4   Lvs broadly obovate-spathulate with 5 or more veins, narrowed
        below into a 5–9-veined stalk. A robust glaucous plant of mari-
        time rocks in the Channel Is.
          L. auriculae-ursifolium (Pourret) Druce (*L. lychnidifolium* auct.)
     Lvs narrowly obovate to oblong-linear, usually 3-veined below and
        with an obscurely veined stalk.   **4. binervosum**

1. L. vulgare Miller   Sea Lavender.

Perennial with deep tap-root and stout branched woody stock; buds
borne on lateral roots. Lvs 4–12(–25) cm., strongly pinnate-veined,
broadly elliptic to oblong-lanceolate, usually mucronate, narrowing
gradually into a long slender stalk. Fl.-stems 8–30(–40) cm., angled,
corymbosely branched usually (but not always) well above the middle;
spikes short, dense, spreading, ± recurved; spikelets closely set in 2 rows
on the upper side of the spike. Corolla 8 mm. diam., blue-purple.
Anthers yellow. Fl. 7–10. Muddy salt-marshes. Great Britain, northwards to Fife and Dumfries.

2. L. humile Miller   Lax-flowered Sea Lavender.

Resembling *L. vulgare* but with oblong-lanceolate and obscurely veined
lvs; fl.-stems branched from below the middle; infl. not corymbose;
spikes long, lax, erect or somewhat incurved; spikelets distant; anthers
reddish. Fl. 7–8. Muddy salt-marshes. Great Britain northwards to
Dumfries and Northumberland; all round the coast of Ireland but rare
in the north.

   L. vulgare and *L. humile* can be distinguished most certainly by microscopic
characters of stigma and pollen, the former species being dimorphic, the latter
monomorphic with the pollen of one and the stigma of the other form of
L. vulgare (see larger *Flora*).

3. L. bellidifolium (Gouan) Dumort.   Matted Sea Lavender.

Perennial, with deep tap-root and much-branched woody stock. Lvs
1·5–4 cm., few in each rosette, obovate- or lanceolate-spathulate, blunt,
narrowed into a slender stalk, dying before flowering is over. Fl.-stems
7–30 cm., rough, decumbent, spreading in a circle, much branched from
near the base, with numerous repeatedly forked barren branches below;
fertile spikes dense, spreading, recurved; spikelets close-set in 2 rows on
the upper side of the spike; bracts with a very broad scarious margin.
Corolla 5 mm. diam., pale lilac. Fl. 7–8. Drier parts of sandy salt-
marshes in Norfolk and Suffolk.

4. L. binervosum (G. E. Sm.) C. E. Salmon, *sensu lato*
                                              Rock Sea Lavender.

Perennial, with ascending branched woody stock. Lvs 2–12·5 cm.,
numerous, very variable in shape from obovate-spathulate to linear-
oblong, usually 3-veined below and narrowed into an obscurely 3-veined
(rarely 1-veined) stalk. Fl.-stems 5–30(–50) cm., slender, wavy, branched
from near the base, barren branches few or 0; spikes varying in length

and straightness; spikelets usually in 2 rows on the upper side of the spike and usually not close-set, but sometimes crowded and sometimes fls in subglobose heads; outermost bract scarious only at the margin. Corolla 4–8 mm. diam. Fl. 7–9. Maritime cliffs, rocks and stabilized shingle. Great Britain northwards to Wigtown and Lincoln; local in Ireland and mainly in the east. Very variable. For descriptions of local forms see the larger *Flora*.

2. ARMERIA Willd.

Perennial herbs with branched woody stocks and basal rosettes of long ± narrow entire lvs. Fls stalked in a solitary terminal hemispherical head consisting of a close aggregate of cymose spikelets with a scarious involucre; top of fl.-stem enclosed by the downward prolongation of the bracts to form a tubular sheath. Calyx funnel-shaped, 5–10-ribbed below, scarious and pleated above; petals very nearly free, persistent; stamens inserted on the base of the petals; styles hairy below. Fr. enclosed in the persistent corolla, 5-ribbed above.

Lvs linear, 1–3-veined; calyx-teeth acute or shortly awned.
1. maritima
Lvs linear-lanceolate, 3–5-veined; calyx-teeth with awns half their length. Robust plant, 20–60 cm., with glabrous scape and involucral sheath 20–40 mm. Sand-dunes in Jersey.
A. arenaria (Pers.) Schultes (*A. plantaginea* Willd.)

1. A. maritima (Miller) Willd.   Thrift, Sea Pink.

Lvs 2–15 cm., linear, 1 (rarely 3)-veined, somewhat fleshy, gland-dotted, glabrous to pubescent. Fl.-stems 5–30 cm., lfless, usually downy (but 20–55 cm. and glabrous in ssp. **elongata** (Hoffm.) Bonnier, very local inland in Lincs.); involucral sheath 8–14 mm.; heads 1·5–2·5 cm. diam. Calyx-tube with hairy ribs; calyx-teeth acute or with awns less than half their length. Corolla 8 mm. diam., rose-pink or white. Fl. 4–10. Coastal salt-marshes, pastures, rocks and cliffs, and also to 4200 ft. on mountains inland; in suitable localities throughout the British Is.

96. PRIMULACEAE

Herbs, usually perennial, rarely undershrubs. Fls actinomorphic, 5 (4–9)-merous. Corolla present (except *Glaux*). Stamens inserted on the corolla-tube and opposite its lobes, sometimes alternating with staminodes. Ovary superior (½-inferior in *Samolus*), 1-celled; style 1, with capitate stigma. Fr. a capsule opening by valves or splitting transversely.

1   Lvs all in a rosette on a very short stem; fl. stems lfless.   **2**
    Stems long, lfy.   **3**

2   Corolla-lobes incurved or spreading; stock not a corm. 1. PRIMULA
    Corolla-lobes strongly reflexed; stock a corm. Very rarely
       naturalized.   *Cyclamen hederifolium* Aiton

3   Water plant; lvs submerged, pinnate; fls lilac.   2. HOTTONIA
    Land plants; lvs not pinnate; fls not lilac.   **4**

4   Fls yellow.   **5**
    Fls not yellow.   **6**

5   Fls not in dense racemes in axils of lvs; corolla 10 mm. or more.
                   3. LYSIMACHIA
    Fls in dense racemes in axils of lvs; corolla 4–5 mm. (very local).
          Naumburgia thyrsiflora (L.) Reichenb.

6   Fls white; stems erect.   **7**
    Stems prostrate or ascending; fls not white (except rarely in *Glaux*,
       and then with no corolla).   **8**

7   Lvs mostly in one whorl; fls solitary or few; ovary superior.
                   4. TRIENTALIS
    Lvs not whorled; fls numerous; ovary ½-inferior.   7. SAMOLUS

8   Capsule dehiscing transversely; corolla present.   5. ANAGALLIS
    Capsule dehiscing by 5 valves; corolla 0 but calyx petal-like.
                   6. GLAUX

1. PRIMULA L.

Perennials. Fl.- or infl.-stalks lfless. Fls in umbels or whorls, rarely apparently solitary, white, yellow, pink or purple. Calyx 5-toothed. Corolla funnel- or salver-shaped, 5-lobed. Stamens shorter than corolla-tube. Capsule opening by valves.

1   Lvs with white meal beneath; fls lilac or purple.   **2**
    Lvs without white meal; fls yellow.   **3**

2   Lvs finely crenate; fls lilac; corolla-lobes not touching one another;
       calyx-teeth ±acute; capsule cylindrical, much longer than calyx.
       Damp, base-rich soils; N. England and S. Scotland, local. Bird's-
       eye Primrose.   *P. farinosa* L.
    Lf-margins entire; fls blue-purple; corolla-lobes touching one
       another; calyx-teeth blunt; capsule little longer than calyx. N.
       Scotland and Orkney; very local. Scottish Bird's-eye Primrose.
                 P. scotica Hooker

3 Fls in an umbel on a long stalk; fl.-stalks shortly hairy; limb of
corolla concave, rarely more than 20 mm. diam.   4
   Fls apparently solitary on long stalks with shaggy hairs; limb of
corolla flat, usually more than 30 mm. diam.   **2. vulgaris**

4 Calyx uniformly pale green; corolla with folds in the throat; fr.
ovoid, enclosed in calyx.   **1. veris**
   Calyx with midribs conspicuously darker green than the remainder;
corolla without folds in the throat; fr. oblong-ovoid, longer than
calyx. Woods on chalky boulder clay in E. England; local. Oxlip.
                                     P. elatior (L.) Hill

1. P. veris L.   Cowslip, Paigle.

Hairy and ± glandular. Rhizome short, stout, covered with swollen lf-
bases. Lvs 5–15(–20) cm., ovate-oblong, blunt, finely crenate or
± toothed, finely hairy on both sides, abruptly narrowed at base; lf-
stalks about as long as blade, winged. Infl.-stalk 10–30 cm., 1–30-fld.
Fl.-stalks 1 cm. or more. Fls 10–15 mm. diam., deep yellow or buff with
distinct orange spots at base of lobes, nodding. Calyx 12–15 × 6–8 mm.,
finely hairy; teeth 2–3 mm., ovate, blunt with small point. Corolla-
tube c. 15 mm., lobes strongly concave, notched. Fr. c. 10 mm.; fr.-
stalks erect. Fl. 4–5. Meadows and pastures, especially on calcareous
soils, locally abundant. Forms hybrids with *P. vulgaris*.

2. P. vulgaris Hudson   Primrose.

Lvs 8–15(–20) cm., obovate-spathulate, blunt, irregularly toothed, hairy
beneath, glabrous above except on the veins, narrowed gradually at base;
stalk short or 0. Fls c. 30 mm. diam., yellow (rarely pink), ± erect,
apparently solitary on stalks 5–10 cm. Calyx 15–17 × 4 mm., nearly
cylindrical, with shaggy hairs; teeth 4–6 mm., narrowly triangular,
acuminate. Corolla-tube c. 15 mm., mouth with thickened folds and
greenish markings; lobes flat, shallowly notched. Fr. ovoid, shorter
than calyx; fr.-stalks curved downwards. Fl. 12–5. Woods and hedge-
banks and, in the west, in open grassy places; common, but now scarce
or extinct near many large towns, owing to depredations of 'flower-
lovers'.

2. HOTTONIA L.

Floating herbs with submerged lvs. Fls whorled. Calyx 5-partite;
corolla salver-shaped, lobes 5, fringed at base. Capsule opening by
valves which remain attached at the top.

1. H. palustris L.   Water Violet.

Nearly glabrous. Stems floating and rooting. Lvs up to c. 10 cm., apparently whorled, 1–2-pinnate, lobes very narrow. Infl. stem 40 cm. or more, lfless, projecting from the water. Fls 20–25 mm. diam., lilac with a yellow throat, 3–8 in a whorl, sometimes cleistogamous. Fl.-stalks 1–2 cm., finely glandular-hairy, ascending in fl., deflexed in fr. Calyx divided almost to base into narrow teeth as long as corolla-tube. Fr. c. 5 mm., globose, 5-valved. Ponds and ditches, widely distributed, but local except in the east.

3. Lysimachia L.

Lvs opposite or whorled, rarely alternate, quite entire. Fls in lf-axils or in terminal panicles, solitary or in racemes, yellow (British spp.). Corolla wheel-shaped. Capsule subglobose, 5-valved.

1	Plant creeping or trailing.	2
	Plant erect.	3
2	Lvs acute; calyx-teeth awl-shaped.	**1. nemorum**
	Lvs obtuse; calyx-teeth ovate.	**2. nummularia**
3	Plant with numerous elongate bulbils in axils of lvs; corolla yellow with dark dots or streaks; lvs narrow-lanceolate. Naturalized in a few localities.	

L. terrestris (L.) Britton, E. E. Sterns & Poggenb.

	Plant without bulbils; corolla plain yellow; lvs lanceolate to ovate.	4
4	Corolla-lobes glandular-ciliate. Naturalized in a few localities.	

L. punctata L.

	Corolla-lobes not ciliate.	5
5	Lvs dotted with orange or black glands; fl. stalks 1 cm. or less; margins of calyx-teeth orange; corolla-lobes not glandular.	

3. vulgaris

Lvs not gland-dotted; fl.-stalks 2–4 cm.; margins of calyx-teeth green; corolla-lobes densely glandular towards base. Naturalized in a few localities.   *L. ciliata* L.

1. L. nemorum L.   Yellow Pimpernel.

Up to 40 cm., glabrous, slender. Lvs 2–4 cm., ovate, rounded at base, stalks short. Fls c. 12 mm. diam., solitary on slender stalks which are longer than the subtending lf. Calyx-teeth c. 5 mm. Corolla-lobes spreading, not ciliate. Fr. c. 3 mm. diam., globose. Fl. 5–9. Woods and shady hedge-banks, common except in the driest parts of the country.

2. L. nummularia L.   Creeping Jenny.

Up to 60 cm., glabrous. Lvs 1·5–3 cm., broadly ovate or suborbicular, rounded or almost truncate at base, gland-dotted, stalks short. Fls 15–25 mm. diam., solitary on rather stout stalks which are usually shorter than the subtending lvs. Calyx-teeth 8–10 mm. Corolla-lobes somewhat concave, gland-dotted and minutely fringed. Fr. very rare in Britain. Fl. 6–8. Hedge-banks and grassy places, local.

3. L. vulgaris L.   Yellow Loosestrife.

60–150 cm., hairy, rhizomatous. Lvs 5–12 cm., often in whorls, lanceolate to ovate-lanceolate, acute, subsessile. Fls c. 15 mm. diam., in terminal panicles; fl.-stalks 1 cm. or less, slender. Calyx-teeth narrow-triangular, ciliate. Corolla concave, lobes not ciliate. Fr. globose. Fl. 7–8. In fens and besides rivers and lakes, locally common.

4. TRIENTALIS L.

Erect, unbranched, glabrous herbs with slender rhizomes. Lvs in one whorl of 5–6 at top of stem, with a few small alternate lvs below. Fls white, solitary. Calyx 5–9-partite. Corolla wheel-shaped, 5–9-partite. Capsule globose, 5-valved.

1. T. europaea L.   Chickweed Wintergreen.

10–25 cm., slender. Lvs 1–8 cm., ± obovate, stiff and shiny, acute or obtuse, entire or finely toothed in upper part, base cuneate; stalk short or almost 0. Fls 15–18 mm. diam., erect, usually solitary; stalk 2–7 cm., very slender. Calyx-teeth 4–5 mm., very narrow, acuminate. Corolla-lobes usually 7, ovate, acute. Fr. c. 6 mm., valves falling off. Fl. 6–7. Local in pine-woods and among moss in grassy and boggy places in the north.

5. ANAGALLIS L.

Annual or perennial herbs. Lvs opposite, entire. Fls solitary on slender stalks in the axils of lvs. Corolla wheel- or funnel-shaped, 5-lobed. Stamens inserted at base of corolla-tube. Capsule globose, opening by a transverse split.

1 Plant 1–4(–7) cm.; corolla not more than 1 mm. diam., exceeded by
       calyx-teeth.   **3. minima**
  Plant 5–30 cm.; corolla at least 6 mm. diam., not exceeded by calyx-
       teeth.   *2*

Stems subterete, rooting at nodes; lvs broadly ovate, obtuse; corolla
    2–3 times as long as calyx.   **1. tenella**
Stems quadrangular, not rooting at nodes; lvs lanceolate to ovate,
    acute; corolla less than twice as long as calyx.   **2. arvensis**

1. A. tenella (L.) L.   Bog Pimpernel.

Slender glabrous perennial, 5–15 cm. Lvs ovate to suborbicular,
short-stalked. Fl.-stalks much longer than subtending lvs. Corolla up
to 14 mm. diam., funnel-shaped, pink. Capsule c. 3 mm. diam. Fl. 6–8.
In damp peaty and grassy places and in bogs, rather local and commonest in the west.

2. A. arvensis L.   Scarlet Pimpernel, Shepherd's Weatherglass.

Glabrous annual or perennial 6–30 cm. Stems gland-dotted. Lvs ovate
to lanceolate, sessile, dotted with black glands beneath. Corolla wheel-
shaped, up to 14 mm. diam., red, pink, blue or lilac. Capsule c. 5 mm.
diam. Fl. 6–8. Cultivated land, etc., widespread. Two sspp. occur in
Britain, ssp. **arvensis** (usually red fls, rarely blue, with densely gland-
fringed petals) commonly and ssp. **foemina** (Miller) Schinz & Thell.
(blue fls, with non-overlapping sparsely fringed petals) rarely.

3. A. minima (L.) E. H. L. Krause

                           (*Centunculus minimus* L.) Chaffweed.

Small erect glabrous annual. Lvs 3–5 mm., ovate, subsessile. Fls subsessile. Calyx divided nearly to base. Corolla pink or white. Capsule c.
1·5 mm. diam. Fl. 6–7. Damp sandy places on heaths and cart-tracks,
especially near the sea. Local but widely distributed north to Lewis.

6. GLAUX L.

A small glabrous succulent herb. Stem creeping and rooting. Lvs entire,
4-ranked. Fls axillary, sessile. Calyx white or pink. Corolla 0. Stamens
5, alternating with the calyx-lobes. Capsule 5-valved, few-seeded.

1. G. maritima L.   Sea Milkwort, Black Saltwort.

Procumbent to suberect perennial 10–30 cm. Lvs 4–12 mm., elliptic-
oblong to obovate, subsessile. Fls 5 mm. diam., usually pink. Fr. c.
3 mm. Fl. 6–8. Grassy salt-marshes, rock-crevices, etc., by the sea and in
inland saline districts, locally common.

7. SAMOLUS L.

Herbs. Lvs alternate or mostly basal. Fls white, bracteate. Corolla ± bell-shaped. Stamens 5, alternating with staminodes, filaments very short. Ovary ½-inferior. Capsule 5-valved.

1. S. valerandi L.   Brookweed.

Glabrous perennial 5–45 cm. Stem lfy. Lvs 1–8 cm., obovate to spathulate, entire. Infl. simple or branched. Pedicels with bracts adnate to about the middle. Corolla 2–3 mm. diam. Stamens not projecting. Fl. 6–8. Wet places, especially near the sea, locally common.

97. BUDDLEJACEAE

Shrubs or trees, rarely herbs, with simple opposite lvs and glandular hairs. Fls regular, hermaphrodite, 4-merous. Corolla tubular below. Stamens equalling in number and alternate with corolla-lobes, inserted on the tube, hypogynous. Ovary superior, 2-celled, with numerous ovules on axile placentae; style 1. Fr. a capsule or berry; seeds endospermic.

1. BUDDLEJA L.

Lvs with interpetiolar stipules, often reduced to a line. Twigs and lvs with stellate and glandular hairs. Fls 4-merous. Fr. a capsule with numerous very small seeds.

*1. B. davidii Franchet

Shrub 1–5 m., with pubescent, pithy, somewhat angled twigs. Lvs 10–25 cm., ovate-lanceolate or lanceolate, acuminate, serrate, dark green above, white-tomentose beneath. Fls with a very heavy pungent smell, in dense cymes forming an interrupted narrow terminal panicle, 10–30 cm.; infl.-axes, fl.-stalks and calyx ± tomentose. Corolla lilac or violet with an orange ring at the mouth; tube cylindric, c. 1 cm.; lobes 1–2 mm. Fl. 6–10. Introduced and increasingly naturalized in waste places in S. England. Much visited by butterflies.

98. OLEACEAE

Trees or shrubs. Lvs usually opposite. Fls hermaphrodite or unisexual, actinomorphic, usually 4-merous. Calyx small. Corolla usually gamopetalous. Stamens 2, very rarely more, usually attached to the corollatube. Ovary superior, 2-celled.

1 Tree; lvs nearly always pinnate; bark grey; buds black; fr. winged.
                                                                   1. FRAXINUS

    Shrubs; lvs simple; bark and buds brown or greenish; fr. not
    winged.   2

2 Lvs 5–12 cm.; fls usually lilac; fr. a capsule. Planted and ±natura-
    lized. Lilac.   *Syringa vulgaris* L.

    Lvs 3–6 cm.; fls white; fr. a berry.   2. LIGUSTRUM

1. FRAXINUS L.

Fls polygamous or dioecious; sepals and petals 0 (British sp.). Fr.
flattened, dry, winged at tip.

1. F. excelsior L.   Ash.

15–25 m., deciduous. Bark smooth, becoming fissured on old branches.
Buds black, large (terminal 5–10 mm.). Lvs to 30 cm., imparipinnate
(rarely simple); lflets c. 7 cm., 7–13, lanceolate to ovate, serrate. Fls in
panicles, purplish, appearing before the lvs. Fr. c. 3 cm., pale brown.
Fl. 4–5. Common on calcareous soils, particularly in the wetter parts of
Britain where it forms woods; less frequent on acid soils and in drier
regions.

2. LIGUSTRUM L.

Shrubs or small trees. Lvs often evergreen. Fls hermaphrodite, in
terminal panicles, strong-scented. Corolla funnel-shaped.

1. L. vulgare L.   Privet.

Up to 5 m., becoming ± lfless late in winter. Bark smooth. Young twigs
shortly hairy. Lvs lanceolate. Panicle 3–6 cm. Fls 4–5 mm. diam.;
corolla-tube as long as lobes; anthers projecting from tube but shorter
than lobes. Fr. 6–8 mm., black and shiny when ripe. Fl. 6–7. Common
in hedges and scrub on calcareous soils north to Durham, widely
naturalized.

**L. ovalifolium* Hassk., with elliptic-oval to elliptic-oblong evergreen lvs,
young twigs glabrous, and corolla-tube 2–3 times as long as lobes, is widely
used for hedges. Native of Japan.

99. APOCYNACEAE

Woody plants, often climbing, rarely herbs, with milky latex and internal
phloem. Lvs usually opposite, entire; stipules 0. Fls hermaphrodite,
actinomorphic, hypogynous, solitary or in cymes. Corolla gamo-

petalous, contorted in bud. Stamens as many as corolla-lobes and alternating with them, epipetalous, with short filaments; anthers convergent on the stylar head. Ovary of 2 carpels, free below but with a common style; ovules numerous. Fr. various; seeds often winged or plumed, usually endospermic.

1. Vinca L.

Creeping shrubs or perennial herbs with evergreen lvs in opposite pairs. Fls solitary, axillary. Corolla blue or white, salver-shaped, with 5 broad asymmetric lobes and an obconic tube fluted and hairy within; stamens 5, with short sharply kneed filaments and introrse anthers ending upwards in broadly triangular convergent hairy flaps; styles united to a column with an enlarged head bearing the stigmatic surface as a band round its broadest part and surmounted by a plume of white hairs. Fr. of 2 follicles each with several long narrow seeds.

Lf-stalks very short; fls 2·5–3 cm. diam.; calyx-lobes glabrous.

1. minor

Lf-stalks c. 1 cm.; fls 4–5 cm. diam.; calyx-lobes ciliate. Trailing shrub with ovate somewhat cordate lvs. Introduced. Copses and hedgerows in S. England; occasional in Ireland. Greater Periwinkle.

*V. major L.

1. V. minor L.   Lesser Periwinkle.

Procumbent plant with trailing stems, 30–60 cm., rooting at intervals, and short erect flowering stems. Lvs 25–40 mm., very short-stalked, lanceolate-elliptic, glabrous. Fl.-stems each with 1 (rarely 2) axillary fl., 25–30 mm. diam. Calyx-lobes lanceolate, glabrous. Corolla blue, mauve or white. Ripe follicles (rarely seen in Britain) 25 mm., each with 1–4 blackish seeds. Fl. 3–5. Doubtfully native. Local in woods, copses and hedge-banks throughout Great Britain.

100. GENTIANACEAE

Herbs, usually glabrous. Lvs opposite, entire, exstipulate, usually sessile. Infl. usually a dichasial cyme. Fls hermaphrodite, regular. Petals joined below into a tube which persists round the capsule; the free lobes contorted in bud. Stamens borne on the corolla-tube, as many as the petals and alternating with them. Ovary superior, 1-celled with 2 parietal placentae, sometimes 2-celled; style 1 or lacking. Fr. usually a septicidal capsule.

1 Fls with 6–8 yellow petals; stem-lvs broadly joined in pairs.
                                              2. BLACKSTONIA
   Fls with 4–5 petals; stem-lvs not joined.   2

2 Corolla pink or yellow, rarely white; style distinct, very slender,
     soon falling.   3
   Corolla blue or purple, rarely white; style 0 or ovary gradually
     tapering into style; stigmas persisting in fr.   5

3 Calyx-lobes triangular, less than half as long as the calyx-tube;
     corolla yellow; stigma peltate. A slender narrow-leaved annual
     of damp, sandy and peaty places near the sea, chiefly in S.W.
     England and S.W. Ireland.   *Cicendia filiformis* (L.) Delarbre
   Calyx divided into very narrow lobes extending almost to the base;
     corolla pink; stigma 2-cleft or stigmas 2.   4

4 Anthers ovate, not twisted; calyx-lobes flat; petals 4. A slender
     narrow-leaved annual of sandy commons in two spots in
     Guernsey.   *Exaculum pusillum* (Lam.) Caruel
   Anthers linear, twisting after fl.; calyx-lobes keeled; petals usually 5.
                                              1. CENTAURIUM

5 Corolla blue, with small lobes between the large ones, but without a
     fringe at the top of the tube.   3. GENTIANA
   Corolla purple, rarely white, without small lobes but with a fringe at
     the top of the tube.   4. GENTIANELLA

1. CENTAURIUM Hill   Centaury.

Annual, rarely perennial, herbs. Fls usually with 5 corolla-lobes. Calyx
deeply divided into linear keeled lobes. Corolla ± funnel-shaped, pink,
rarely white. Anthers slender, twisting spirally after dehiscence. Style
thread-like, soon falling.

1 Erect annuals without decumbent non-flowering shoots; corolla-
     lobes 7 mm. or less.   2
   Perennial, with decumbent non-flowering shoots bearing short-
     stalked roundish lvs; corolla-lobes 8–9 mm. Confined to grassy
     cliffs and dunes in Pembroke and Cornwall.
                                              C. portense (Brot.) Druce

2 Individual fls stalked, not clustered; corolla-lobes 3–4 mm.; no
     basal rosette of lvs.   3
   Fls sessile or nearly so, ± clustered; corolla-lobes 5–7 mm.; basal
     rosette of lvs present.   4

3 Branches spreading at a wide angle; infl. lax; internodes 2–4.
                                              1. pulchellum

Branches ascending at a narrow angle; infl. rather dense; internodes
5–9. A rare annual of damp grassy places near the sea in the Isle
of Wight, Dorset and the Channel Is.

C. tenuiflorum (Hoffmanns. & Link) Fritsch

4 Stem-lvs narrowly oblong, the basal spoon-shaped, 5 mm. broad or
     less, all minutely toothed on the margin.   **3. littorale**
   Lvs ovate to oblong, the basal usually more than 5 mm. broad.   5

5 Stamens inserted at the top of the corolla-tube; infl. various.
                                                            2. erythraea
   Stamens inserted at the base of the corolla-tube; infl. always very
     dense. A small annual of dry, usually calcareous, grassland near
     the sea. Very local from the south coast of England northwards to
     Lancashire and Northumberland; Channel Is.

C. capitatum (Willd.) Borbás

1. C. pulchellum (Swartz) Druce

A glabrous annual with no basal lf-rosette; stems erect, slender, 2–
15 cm., unbranched or with branches spreading at a wide angle; stem
internodes 2–4. Lvs 2–15 × 1–10 mm., ovate or ovate-lanceolate, acute,
the upper usually longer than the lower. Fls in a lax infl.; pedicels c.
2 mm. Corolla-tube usually longer than the calyx; lobes 2–4 mm. Capsule about equalling calyx. Fl. 6–9. Common in damp grassy places
near the sea, more local inland, in S. and C. England, rarer northwards but reaching S. Scotland; in Ireland chiefly on the S. and E. coast
from Cork to Dublin.

2. C. erythraea Rafn (C. minus Moench; Erythraea centaurium (L.) Pers.)

A glabrous annual 2–50 cm., with a basal rosette of obovate or elliptic
lvs 1–5 cm. × (4–)8–20 mm., usually blunt, prominently 3–7 veined;
stem-lvs shorter, sometimes narrower but never parallel-sided. Fls
sessile or nearly so, often clustered in a ± dense flat-topped infl. Corolla-
tube and capsule longer than the calyx; corolla-lobes 5–6 mm. Fl. 6–10.
Common in dry grassland, dunes, wood-margins, etc., throughout
England, Wales and Ireland, less so in Scotland but reaching Ross and
the Outer Hebrides.

3. C. littorale (D. Turner) Gilmour

An annual 2–25 cm., somewhat rough to the touch, with a basal rosette
of narrowly spathulate lvs 1–2 cm. × 3–5 mm., blunt, indistinctly
3-veined; stem-lvs shorter, parallel-sided; all minutely toothed at the
margin. Fls sessile, clustered in dense flat-topped infls. Corolla-tube

not longer than the calyx; lobes 6–7 mm. Capsule much longer than the
calyx. Fl. 7–8. On dunes and in sandy places by the sea; in Great Britain
almost confined to coasts north of a line from S. Wales to Northumberland; in Ireland only in Derry.

2. BLACKSTONIA Hudson

Annuals. Corolla yellow, with a short tube and 6–8 lobes. Anthers
narrow, sometimes slightly twisted after dehiscence. Style thread-like,
with 2 deeply 2-cleft stigmas, soon falling.

1. B. perfoliata (L.) Hudson   Yellow-wort.
An erect glaucous annual 15–45 cm., with a basal rosette of obovate lvs
1–2 cm., blunt, free; stem-lvs ovate-triangular, acute, the members of an
opposite pair broadly joined by their bases. Fls 10–15 mm. diam. Capsule ovoid. Fl. 6–10. Calcareous grassland and dunes: common in
S. England and reaching Durham, Lancashire and Kirkcudbright;
S. Ireland to Meath and Sligo; Jersey.

3. GENTIANA L.   Gentian.

Perennial, rarely annual, glabrous herbs. Calyx-teeth 5, joined by a
membrane continuous with the membranous top of the calyx-tube.
Corolla usually blue, with small lobes between the 5 large 3-veined lobes
but without a fringe at the top of the tube. Anthers neither twisted nor
versatile. Ovary gradually tapering into the style, or style 0; stigmas 2,
persistent on the fr. Nectaries at the base of the ovary.

1 Corolla sky-blue with 5 green lines outside, its tube obconical;
    lvs 1·5–4 cm., very narrow. Very local on wet heaths from Dorset
    and Kent to Anglesey, Cumberland and Yorks. Bog Gentian.
                                G. pneumonanthe L.
  Corolla brilliant deep blue, its tube cylindrical; lvs 1·5 cm. or less,
    ovate or ovate-oblong.   2

2 Perennial, with numerous lf-rosettes; corolla c. 15 mm. diam.
    A very local plant of grassy places on limestone in Upper Teesdale
    and in Clare, Galway and Mayo in W. Ireland. Spring Gentian.
                                    G. verna L.
  Erect annual; corolla c. 8 mm. diam. A rare plant of rock-ledges on
    mountains in Perth and Angus. Small Gentian.   *G. nivalis* L.

4. GENTIANELLA Moench   Gentian.

Annual or biennial glabrous herbs. Calyx tubular, nowhere mem-branous. Corolla purple or whitish, with no small lobes between the 4 or 5 large 5–9-veined lobes but with a fringe at the top of the tube. Anthers versatile, not twisted. Ovary gradually tapering into the style, or style 0. Stigmas 2, persistent on the fr. Nectaries on the corolla.

1   Calyx-lobes 4, the two outer much larger than the inner and over-
       lapping them.   **1. campestris**
     Calyx-lobes 4 or 5, usually almost equal, never overlapping.   2

2   Corolla 25–30 mm., bluish-purple; internodes 9–15. A robust
       biennial with ovate lvs and the corolla-tube much longer than
       the calyx. Local, in calcareous grassland, in S. and C. England
       and Wales.   *G. germanica* (Willd.) Börner
     Corolla 13–20(–23) mm.; internodes 0–9.   3

3   Uppermost internode and terminal fl.-stalk together forming less
       than half the total height of the plant.   **2. amarella**
     Uppermost internode and terminal fl.-stalk together forming more
       than half the total height of the plant.   4

4   Upper stem-lvs lanceolate; calyx-teeth ±appressed to corolla. A
       small biennial of calcareous grassland in S. England. Fl. 4–6.
                                   G. anglica (Pugsley) E. F. Warburg
     Upper stem-lvs ovate or ovate-lanceolate; calyx-teeth ±spreading.
       A very local annual or biennial of dune-slacks in S. Wales. Fl.
       8–11.   *G. uliginosa* (Willd.) Börner

1. G. campestris (L.) Börner   Field Gentian.

Biennial (or annual) 3–25 cm. Basal lvs 1–2·5 cm., ovate or spathulate, blunt; upper stem-lvs 2–3 cm., oblong or lanceolate. Calyx divided nearly to the base into 4 lobes; 2 outer lobes ovate-elliptic, acute or acuminate, overlapping and almost hiding the 2 lanceolate inner lobes. Corolla 15–30 mm., bluish-lilac, rarely white, its tube equalling or exceeding the calyx. Fl. 7–10. Common in N. England and Scotland in grassland and dune-slacks, and widespread in Ireland but rare and local in C. and S. England.

2. G. amarella (L.) Börner   Felwort.

Biennial 8–25 cm., the oblong rosette-lvs of the 1st year dying in autumn; basal lvs of 2nd year 5–20 mm., spathulate, blunt, often dead at fl.; stem-lvs ovate-lanceolate to lanceolate, subacute. Calyx-lobes usually 5, almost equal. Corolla dull purple, rarely dull red or whitish, its tube

longer than the calyx. Fl. 8–10. Rather common over most of the British
Is in dry pastures and dunes, usually calcareous.

Variable; 4 sspp. occur in Britain. See larger *Flora*.

101. MENYANTHACEAE

Like Gentianaceae but aquatic or bog plants with spirally arranged
lvs (except sometimes on fl.-stems) and having the corolla falling after
fl., its lobes valvate in bud.

Lvs ternate; fls pink or white.	1. MENYANTHES
Lvs simple, orbicular; fls yellow.	2. NYMPHOIDES

1. MENYANTHES L.

Aquatic or bog plant. Lvs ternate, all spirally arranged. Fls in a raceme
on a lfless scape, heterostylous, 5-merous. Capsule subglobose, opening
by 2 valves.

1. M. trifoliata L.   Buckbean, Bogbean.

A glabrous aquatic or bog plant with the lvs and fls raised above the
water-surface; rhizome creeping. Lflets 3·5–7 cm., obovate or elliptic,
entire; their stalks 7–20 cm., with long sheathing base. Scape 12–30 cm.,
c. 10–20-fld; fl.-stalks 5–10 mm. Calyx-lobes ovate. Corolla pink outside, paler or white within, c. 15 mm. across, much fimbriate. Fl. 5–7.
Ponds, edges of lakes, and in wetter parts of bogs and fens, to 3000 ft.
Rather common in suitable habitats throughout the British Is.

2. NYMPHOIDES Séguier

Aquatic. Lvs simple, orbicular, deeply cordate, those of the fl.-stems
opposite. Fls on long stalks in clusters in the lf-axils, 5-merous. Capsule
ovoid, beaked, opening irregularly.

1. N. peltata (S. G. Gmelin) Kuntze   Fringed Water-lily.

A glabrous aquatic with floating lvs and fls; rhizome creeping; lvs
alternate. Fl.-stems long, floating, their lvs opposite. Lvs 3–10 cm.,
orbicular, deeply cordate at base, entire or sinuate, purplish below and
purple-spotted above, long-stalked. Fls in 2–5-fld axillary clusters, their
stalks 3–7 cm. Calyx-lobes oblong-lanceolate. Corolla c. 1 cm. across,
yellow, lobes fimbriate-ciliate. Fl. 7–8. Ponds and slow rivers in E. and
C. England, north to Shropshire and S. Yorks.

102. POLEMONIACEAE

Annual or perennial herbs and a few shrubs. Lvs spirally arranged or opposite; stipules 0. Fls actinomorphic, hermaphrodite. Sepals 5. Corolla with basal tube and 5 free lobes above. Stamens 5, inserted on the corolla and alternating with its lobes. Ovary superior, 3-celled, with a single style and usually 3 stigmas. Fr. a capsule; seeds endospermic.

1. POLEMONIUM L.

Usually perennial herbs with spirally arranged pinnate lvs and showy fls. Stamens inserted all at the same height on the corolla-tube, the broadened downwardly-curved hairy bases of the filaments almost closing the throat of the corolla.

1. P. caeruleum L.   Jacob's Ladder.

Perennial, with short creeping rhizome and erect lfy stem, 30–90 cm., hollow, angled, somewhat glandular above. Lvs 10–40 cm., with terminal and 6–12 pairs of lateral lflets; lower lvs stalked, upper sub-sessile; lflets 2–4 cm., ovate-lanceolate or oblong, acuminate, entire, glabrous. Infl. corymbose. Fls 2–3 cm. diam., drooping. Calyx bell-shaped with acute teeth. Corolla rich blue or white, with short tube (2 mm.). Capsule erect, included in the calyx-tube; seeds shortly winged. Fl. 6–7. Local on grassy slopes, screes and rock-ledges, mainly on lime-stone and on slopes facing north; England from Stafford and Derby northwards to the Cheviots. The common garden form has paler fls and broader lflets and is apparently not the British plant.

103. BORAGINACEAE

Usually rough or stiffly hairy herbs. Stems usually terete. Lvs nearly always alternate, entire. Fls hermaphrodite, often in scorpioidal cymes, usually actinomorphic. Calyx 5-toothed. Corolla 5-lobed, wheel-, funnel- or bell-shaped, often pink in bud, becoming bright blue; tube often closed at top with scales or hairs. Stamens 5, joined to corolla-tube and alternating with its lobes. Ovary superior, in British spp. deeply 4-lobed. Fr. of 4 nutlets.

  1 Lvs all stalked, rounded or cordate at base. Naturalized in woods
       in a few places. Blue-eyed Mary.
                                              Omphalodes verna Moench
     At least the upper lvs sessile, though sometimes narrowed towards
       base.   *2*

2   Calyx-lobes toothed, enlarging greatly in fr. and forming a
     flattened 2-lipped covering round fr.; corolla-lobes purple when
     first opened, 3 mm. diam.; plant procumbent. Naturalized,
     rare. Madwort.   *Asperugo procumbens* L.
     Not as above.   3

3   Nutlets covered with hooked or barbed bristles; calyx-teeth
     spreading nearly horizontally, not concealing fr.
                                              1. CYNOGLOSSUM
     Nutlets without bristles; calyx ±concealing fr.   4

4   Some or all stamens projecting far beyond the corolla; plant
     stiffly hairy.   5
     Stamens all shorter than corolla-tube or, if slightly projecting, then
     plant glabrous.   7

5   All stamens projecting; corolla regular.   6
     Some stamens shorter than corolla-tube; corolla ±zygomorphic.
                                                8. ECHIUM

6   Stamens glabrous; anthers 8–10 times as long as broad, prolonged
     into a point at tip; annual. Garden escape. Borage.
                                     Borago officinalis L.
     Stamens hairy; anthers 2–3 times as long as broad, not prolonged
     into a point; perennial. Naturalized in woods.
                               Trachystemon orientalis (L.) D. Don

7   Plant glabrous, very glaucous; lvs covered with rough dots. Sea-
     shores in the north, local. Oyster-plant.
                          Mertensia maritima (L.) S. F. Gray
     Plant ±hairy, not glaucous; lvs without rough dots.   8

8   Fls nodding.   2. SYMPHYTUM
     Fls erect.   9

9   Corolla with hairy or papillose oblong scales or folds in the throat.   10
     Corolla with glabrous rounded scales in the throat.   6. MYOSOTIS

10 Calyx divided for ¼–½ its length; fl. spring.   5. PULMONARIA
     Calyx divided almost to base; fl. summer.   11

11 Corolla with long hairy folds in the throat.   7. LITHOSPERMUM
     Corolla with conspicuous scales in the throat.   12

12 Perennial; lvs ovate; corolla-tube straight.   3. PENTAGLOTTIS
     Annual; lvs linear-lanceolate to oblong; corolla-tube curved.
                                             4. ANCHUSA

1. CYNOGLOSSUM L.

Annual or biennial. Calyx deeply 5-lobed. Corolla funnel-shaped with
prominent scales at top of tube, lobes blunt. Nutlets flat or convex,
attached to the conical receptacle by a narrow outgrowth.

Lvs grey with silky ±appressed hairs; fr. with a thickened border.
                                                        1. officinale
Lvs green, with a few stiff hairs, upper surface nearly glabrous; fr. with-
    out a thickened border. S. and C. England. Rare. *C. germanicum* Jacq.

1. C. officinale L.   Hound's-tongue.

30–90 cm., smelling of mice. Lower lvs up to 30 cm., lanceolate to
ovate, stalked; upper lvs sessile; all hairy on both surfaces. Cymes
usually branched, eventually 10–25 cm. Fls c. 1 cm. diam., dull red-
purple, rarely whitish; their stalks c. 1 cm., recurved in fr. Nutlets
5–6 mm., flattened, spines all about the same length. Fl. 6–8. In grassy
places and at borders of woods on light rather dry soils, particularly
near the sea; widespread but local.

2. SYMPHYTUM L.

Perennial. Lower lvs stalked, upper usually sessile or decurrent. Fls in
terminal forked scorpioidal cymes, nodding. Corolla funnel-shaped or
subcylindrical, with 5 short, broad lobes. Scales 5, narrow, usually
shorter than corolla-tube. Nutlets ovoid, smooth or granulate, base
ring-shaped, toothed.

1 Rhizome slender, far-creeping; fl. stems rarely more than 20 cm.,
        unbranched. Locally naturalized.   **S. grandiflorum* DC.
     Rhizome stout, plant tufted; fl. stems taller and nearly always
        branched.   2
2 Rhizome swollen and tuberous; roots fibrous; stems up to 50 cm.,
        unbranched or with 1–2 short branches near top; middle stem-lvs
        considerably larger than lower ones. Damp woods and hedges;
        local. Tuberous Comfrey.   *S. tuberosum* L.
     Roots thick and tuberous; stems much-branched; lower lvs largest.   3
3 Calyx-teeth at least as long as tube; lvs lanceolate.   4
     Calyx-teeth c. ½ as long as tube; lvs ovate. Locally naturalized.
                                                         *S. orientale L.
4 Upper lvs shortly stalked; calyx-teeth blunt. Rarely naturalized.
        Rough Comfrey.   **S. asperum* Lepech.
     Upper lvs sessile and often decurrent; calyx-teeth acute.   5
5 Stem-lvs strongly decurrent, so that the stem is broadly winged; fls
        yellowish-white, purplish or pinkish.   **1. officinale**
     Stem-lvs slightly decurrent or stem narrowly winged; fls blue or
        purplish-blue.   *asperum × officinale*

1. S. officinale L.   Comfrey.

30–120 cm., erect, stiffly hairy. Lower lvs 15–25 cm., ovate-lanceolate,
stalked; upper narrower, broadly decurrent, the wing usually extending
to the next lf below. Calyx-teeth very narrowly lanceolate, long-
acuminate. Corolla 15–17 mm. Fl. 5–6(–9). Beside rivers, streams and
wet ditches; widely distributed but probably introduced in the north.

*S. asperum Lepech. was formerly cultivated but is now rather uncommon by roads and at edges of fields. Stiffly hairy with stout
bulbous-based bristles. Upper lvs shortly stalked. Fls blue; calyx
enlarging in fr., teeth blunt. Hybrids between this and *S. officinale*
(*S. × uplandicum* Nyman) are common and show many different combinations of the characters of the parents; they occur in hedge-banks, open
woods, etc., but seldom near water. Certain types have been described
as spp. (e.g. *S. peregrinum* Ledeb.).

3. PENTAGLOTTIS Tausch

Perennial. Fls blue, in bracteate cymes. Corolla wheel-shaped, tube
closed by scales. Nutlets concave at base, with a stalk-like attachment
at the inner side.

1. P. sempervirens (L.) Tausch   Alkanet.

30–100 cm., stiffly hairy. Lvs ovate, acute or acuminate, lower up to
30 cm., stalked. Cymes on long stalks arising from the lf-axils, the fls
apparently in heads with a pair of small lvs close below them. Calyx-
teeth very narrow; corolla c. 10 mm. diam., bright blue with white scales.
Fl. 5–6. Hedgerows, etc.; common and perhaps native in S.W., widespread but local elsewhere.

4. ANCHUSA L.

Annual or perennial. Fls usually blue or purple; corolla often ± irregular, tube closed by scales or hairs. Nutlets strongly concave at base with
no stalk-like projection.

1. A. arvensis (L.) Bieb. (*Lycopsis arvensis* L.)   Bugloss.

Annual, 15–50 cm. covered with hairs with swollen bulbous bases. Lvs
up to c. 15 cm., obovate-lanceolate to narrowly oblong, margins undulate and with a few teeth; lower narrowed into a stalk, upper sessile
and ½-clasping the stem. Cymes elongating during fl. Fls subsessile,

bracts lf-like. Calyx-teeth narrow; corolla c. 5 mm. diam., bright blue; scales white. Fl. 6–9. On light soils; widely distributed and locally common.

5. PULMONARIA L. Lung-wort.

Perennials with creeping rhizomes. Fl.-stems unbranched. Fls purple or blue, often pink in bud. Corolla funnel-shaped with 5 tufts of hairs alternating with the stamens. Nutlets with a raised ring round the base.
The two following are uncommon:

Basal lvs lanceolate, gradually narrowed at base; corolla 5–6 mm. diam.; nutlets flattened, crested. Woods in Dorset, Hants and Isle of Wight. *P. longifolia* (Bast.) Boreau
Basal lvs ovate, abruptly narrowed at base; corolla c. 10 mm. diam.; nutlets ovoid, acute. Naturalized in hedge-banks, mainly in the south. **P. officinalis* L.

6. MYOSOTIS L.

Annual or perennial. Cymes terminal, scorpioidal. Calyx 5-toothed, sometimes divided nearly to base. Corolla wheel-shaped, usually pink in bud, becoming blue; tube closed by 5 notched scales. Nutlets shiny, biconvex and bluntly 3-angled, narrow at base.

1	Hairs on calyx-tube appressed or almost 0.	*2*
	At least some hairs on calyx-tube spreading, short, stiff, hooked or curled; calyx never subglabrous.	*6*
2	Most lvs 3–5 times as long as broad; fls bright blue, rarely white.	*3*
	Lvs scarcely more than twice as long as broad; fls very pale blue. Wet places in mountain districts in N. England and Scotland; rare. *M. Stolonifera* Gay (*M. brevifolia* C. E. Salmon)	
3	Lower part of stem with spreading hairs; corolla (3–)4–10 mm. diam.; perennial.	*4*
	Lower part of stem glabrous or with appressed hairs; corolla 2–4(–5) mm. diam.; annual or biennial.	*5*
4	Stalk of fr. 1–2 times as long as calyx; calyx-tube at least twice as long as teeth; bracts 0. **1. scorpioides**	
	Stalk of fr. 3–5 times as long as calyx; calyx-tube as long as teeth; bracts present at base of cyme. **2. secunda**	
5	Calyx hairy; corolla-lobes flat; style rarely more than ½ as long as calyx-tube. **3. caespitosa**	
	Calyx almost glabrous; corolla-lobes concave; style as long as calyx-tube. Damp places behind dunes; Jersey. *M. sicula* Guss.	
6	Stalk of fr. as long as or longer than calyx.	*7*
	Stalk of fr. shorter than calyx.	*9*

7   Corolla 4–10 mm. diam., lobes flat.   8
    Corolla usually smaller, lobes concave.   **5. arvensis**

8   Stalk of fr. about as long as calyx (slightly longer in lowest frs); nut-
    lets black. Alpine rocks in N. England and Scotland; rare.
                                          M. alpestris F. W. Schmidt
    Stalk of fr. 1½–2 as long as calyx; nutlets dark brown (woods).
                                                              4. sylvatica

9   Corolla at first yellow or white, tube about twice as long as calyx in
    mature fl.; calyx-teeth nearly erect in fr.   **6. discolor**
    Corolla never yellow, tube shorter than calyx; calyx-teeth spreading.
                                                          7. ramosissima

1. M. scorpioides L.   Water Forget-me-not.

Perennial 15–45 cm., ± hairy. Lower lvs up to c. 7 cm., oblong-lanceolate to obovate-lanceolate, narrowed but scarcely stalked at base; upper
lvs narrower. Fr.-stalks spreading or reflexed. Calyx-teeth triangular,
¼–⅓ length of calyx. Corolla 3–10 mm. diam., sky-blue, rarely white;
lobes flat, shallowly notched. Style at least as long as calyx-tube. Nutlets c. 1·6 × 1 mm., narrowly ovoid, black. Fl. 5–9. Common and
generally distributed in wet places.

2. M. secunda A. Murray   Water Forget-me-not.

± hairy perennial 20–60 cm. Main stem with ± prostrate laterals arising
from its base, the sterile ones rooting at the nodes. Lower lvs ovate-
spathulate, upper oblong-lanceolate. Fr.-stalks reflexed. Calyx-teeth
lanceolate, ½ length of calyx. Corolla 4–6 mm. diam., blue; lobes very
slightly notched. Style at least as long as calyx-tube. Nutlets c.
1 × 0·7 mm., broadly ovoid, dark brown. Fl. 5–8. Fairly generally
distributed on wet non-calcareous soils.

3. M. caespitosa C. F. Schultz   Water Forget-me-not.

± hairy annual or biennial 20–40 cm. Lvs lanceolate, with appressed
hairs above and, at least the lower, subglabrous beneath. Fr.-stalks
spreading. Calyx-teeth triangular-ovate, ½–¾ length of calyx. Corolla
2–5 mm. diam., sky-blue, rarely white; lobes rounded. Style c. ½ length
of calyx-tube, rarely longer. Nutlets c. 1·3 × 1 mm., broadly ovoid,
truncate at base, dark brown. Fl. 5–8. Common in damp places.

4. M. sylvatica Hoffm.   Wood Forget-me-not.

Hairy perennial 15–45 cm. Lower lvs ovate-spathulate, upper lanceolate to oblong, all with ± spreading hairs on both surfaces. Bracts 0.

Fr.-stalks spreading. Calyx-teeth oblong-lanceolate, $\frac{2}{3}-\frac{3}{4}$ length of calyx, spreading in fr. Corolla 6–10 mm. diam., bright blue, rarely white; lobes flat, rounded: tube at least as long as calyx. Style longer than calyx-tube. Nutlets c. 2×1 mm., ovoid, dark brown. Fl. 5–6(–9). Locally abundant in damp woods.

5. M. arvensis (L.) Hill   Common Forget-me-not.

Hairy annual 15–30 cm. Lower lvs broadly ovate, stalked, upper oblong-lanceolate, sessile, all with ± spreading hairs on both surfaces. Bracts 0. Fr.-stalks up to twice as long as calyx, spreading. Calyx-teeth narrow-triangular, $\frac{2}{3}-\frac{3}{4}$ length of calyx, spreading in fr. Corolla usually less than 5 mm. diam., bright blue; lobes concave; tube shorter than calyx. Nutlets c. $1\cdot5 \times 0\cdot7$ mm., ovoid, dark brown. Fl. 4–9. Common in ± open but not wet habitats.

6. M. discolor Pers.   Yellow-and-blue Forget-me-not.

Hairy annual 8–25 cm. Lvs oblong-lanceolate, lower narrowed at base, upper sessile, all hairy. Bracts 0. Cymes in fr. not much longer than lfy part of stem. Fr.-stalks ascending, shorter than calyx. Calyx-teeth oblong-lanceolate, c. $\frac{1}{2}$ length of calyx, erect in fr. Corolla c. 2 mm. diam., lobes concave; tube at length c. twice as long as calyx. Nutlets c. $1\cdot2 \times 0\cdot5$ mm., ovoid, almost black. Fl. 5–9. Locally common in open communities on light soils.

7. M. ramosissima Rochel (*M. hispida* Schlecht.) Early Forget-me-not.

Hairy annual 2–25 cm. Lower lvs ovate-spathulate, forming a rosette; upper oblong, sessile; all hairy. Bracts 0. Cymes in fr. much longer than lfy part of stem. Fr.-stalks ascending, shorter to slightly longer than calyx. Calyx-teeth narrowly triangular, $\frac{1}{3}-\frac{1}{2}$ length of calyx, spreading in fr. Corolla c. 2 mm. diam., lobes concave; tube shorter than calyx. Nutlets c. $0\cdot7 \times 0\cdot5$ mm., ovoid, pale brown. Fl. 4–6. Locally common on dry shallow soils.

7. LITHOSPERMUM L.

Annual or perennial herbs or small shrubs. Fls subsessile (British spp.) in bracteate cymes. Calyx 5-partite. Corolla wheel- or funnel-shaped, hairy, with folds or small scales at top of tube. Nutlets very hard, smooth or warty, base truncate.

1 Fls 12–15 mm. diam., reddish-purple, then blue; non-flowering
    stems creeping, flowering erect. On limestone and chalk in the
    south; very local. Blue Gromwell.   *L. purpurocaeruleum* L.
  Fls 3–4 mm. diam., white, rarely blue; all stems erect.   *2*

2 Lvs with prominent lateral nerves; nutlets smooth, shiny, white;
   perennial.   **1. officinale**
   Lateral nerves not apparent; nutlets warty, grey-brown; annual.
                                                                 2. arvense

1. L. officinale L.   Gromwell (Grummel).

Hairy ± rough much-branched perennial 30–80 cm. Lvs up to c. 7 cm.,
usually lanceolate, sessile. Corolla yellowish- or greenish-white. Nutlets
ovoid, blunt. Fl. 6–7. Hedges, etc., mainly on basic soils; not uncommon in England and Wales, rare in Scotland.

2. L. arvense L.   Corn Gromwell, Bastard Alkanet.

Hairy ± rough simple or little-branched annual 10–50(–90) cm. Lvs
up to 3(–5) cm., lower obovate, narrowed into a stalk, upper narrowly-
lanceolate, sessile. Corolla white, sometimes blue, tube violet or blue.
Nutlets trigonous-conical. Fl. 5–7. Arable fields, rarely elsewhere;
commonest on light soils in England.

8. ECHIUM L.

Usually stout, stiffly hairy or rough herbs or shrubs. Fls in simple or
branched unilateral cymes. Calyx 5-partite. Corolla funnel-shaped with
a straight tube but unequal lobes. Stamens unequal, at least some projecting. Nutlets with flat, nearly triangular bases.

Basal lvs with no apparent lateral veins; upper stem-lvs rounded at
   base; fls blue, with 4 projecting stamens.   **1. vulgare**
Basal lvs with prominent lateral veins; upper stem-lvs cordate at base;
   fls red, becoming purple-blue, with 2 projecting stamens. S.W.
   England, Channel Is; very local.   *E. lycopsis* L.

1. E. vulgare L.   Viper's Bugloss.

Very rough biennial 30–90 cm. Lvs up to 15 cm., lanceolate or oblong,
rarely ovate, lower stalked, upper sessile. Cymes in a large terminal
panicle. Fls 15–18 mm., subsessile. Calyx-teeth narrow, shorter than
corolla-tube. Corolla pinkish-purple in bud, becoming bright blue,
rarely white. Fl. 6–9. Locally abundant on light soils and mainly in the
south.

104. CONVOLVULACEAE

Herbs, shrubs or rarely trees, often climbing, juice usually milky. Lvs
spirally arranged. Fls actinomorphic, often large and showy. Sepals 5,
usually free. Corolla usually funnel-shaped and shallowly 5-lobed or

2-angled. Stamens 5, inserted towards the base of the corolla-tube and alternating with the lobes. Ovary superior. Fr. a capsule.

1 Green plants with large lvs; fls more than 1 cm. diam.   2
  Reddish or yellowish parasites with scale-like lvs; fls less than 5 mm.
    diam.   3. CUSCUTA
2 Bracteoles small, distant from calyx.   1. CONVOLVULUS
  Bracteoles large, overlapping calyx.   2. CALYSTEGIA

1. CONVOLVULUS L.

Herbs, often climbing. Fls in 1–few-fld axillary or terminal groups. Stigmas very narrow.

1. C. arvensis L.   Bindweed, Cornbine.

20–75 cm., scrambling or climbing. Roots penetrating the earth to depths of 3 m. or more. Stems climbing by twisting in a counter-clockwise direction round the stems of other plants. Lvs 2–5 cm., oblong or ovate, hastate or sagittate at base; stalk shorter than blade. Fls c. 2 cm., solitary or 2–3 together; stalks longer than subtending lvs. Bracteoles c. 2 mm., awl-shaped. Corolla up to 3 cm. diam., white or pink. Fl. 6–9. Widespread and common except on the poorer soils.

2. CALYSTEGIA R.Br.

Like *Convolvulus* except for the large bracteoles enclosing the calyx and the broad stigmas.

Stems climbing; lvs ovate-cordate or sagittate; bracteoles longer than
    calyx.   **1. sepium**
Stems not climbing; lvs reniform; bracteoles shorter than calyx.
                                                              2. soldanella

1. C. sepium (L.) R.Br.   Bellbine, Larger Bindweed.

1–3 m., climbing and stoloniferous. Rhizomes seldom more than 30 cm. deep. Lvs up to 15 cm. Fls 5–7 cm., solitary, white or pink; bracteoles nearly flat to strongly inflated. Corolla funnel-shaped. Fl. 7–9.

Ssp. **sepium**. ± Glabrous. Fls usually white; bracteoles not or scarcely inflated, their margins not overlapping. Hedges, fens, etc., and as a garden weed. Common in S. England and Wales, becoming rarer northwards.

*Ssp. **pulchra** (Brummitt & Heywood) Tutin. Hairy at least on fl.-stalks and lf-stalks and sinuses. Fls bright pink; bracteoles slightly inflated, longer than broad, with overlapping margins. Rather uncommon, in the neighbourhood of gardens.

*Ssp. **silvatica** (Kit.) Maire. ± Glabrous. Fls white, rarely with pale pink stripes on the outside; bracteoles strongly inflated, as broad as long, with overlapping margins. Hedges etc., common in S.E. England and the Midlands but extending to Scotland.

2. C. soldanella (L.) R.Br.   Sea Bindweed.

10–60 cm., trailing. Lf-blade 1–4 cm., reniform, rather fleshy; stalk usually longer than blade. Fls c. 5 cm., solitary, stalks longer than the subtending lvs; bracteoles 10–15 mm. Corolla 2·5–4 cm. diam., pink or pale purple. Fl. 6–8. Locally common on sandy or shingly sea-shores.

3. CUSCUTA L.   Dodder.

Twining, pink, yellow or white, rootless parasites, attached to the host by suckers. Fls in lateral, often dense, heads or short spikes; bracts often present. Corolla urceolate or bell-shaped usually with small petaloid scales below the insertion of the stamens.

Styles shorter than ovary; stamens not projecting; scales small, not
   closing corolla-tube. Usually on *Urtica*; local.   *C. europaea* L.
Styles longer than ovary; stamens somewhat projecting; scales large,
   ± closing corolla-tube.   **1. epithymum**

1. C. epithymum (L.) L.   Common Dodder.

Annual or sometimes perennating. Stems reddish, c. 1 mm. diam. Fls pinkish, scented, sessile in dense heads 5–10 mm. diam. Calyx divided for ¾ of its length, lobes acute. Corolla-lobes somewhat spreading, acuminate. Fl. 7–9. On *Ulex*, *Calluna* and various other plants; locally common in England and Wales.

105. SOLANACEAE

Herbs or shrubs. Lvs spirally arranged or sometimes apparently opposite. Fls usually actinomorphic. Calyx (4–)5(–10)-lobed. Corolla usually 5-lobed. Ovary superior. Fr. a capsule or berry.

1 Stamens conspicuous; anthers much longer than filaments, opening
     by apical pores and forming a cone.   3. SOLANUM
    Stamens not particularly conspicuous; anthers shorter than fila-
     ments, opening by slits and never forming a cone.   2

2 Shrubs, often spiny.   1. LYCIUM
    Herbs.   3

3 Plant viscid-hairy; fls in scorpioidal cymes.   2. HYOSCYAMUS
   Plant not viscid; fls not in scorpioidal cymes.   4

4 Fls up to 3 cm., drooping; lvs entire; tall perennial with black
      berries 1·5–2 cm. diam. On calcareous soils; local. Deadly
      Nightshade.   *Atropa belladonna* L.
   Fls c. 7 cm., erect, lvs ± toothed; coarse annual with usually prickly
      capsule 4–5 cm. Rare casual; occasionally ± naturalized. Thorn-
      Apple.   **Datura stramonium* L.

1. LYCIUM L.

Shrubs. Lvs entire, often crowded on short shoots. Corolla funnel-
shaped. Stamens inserted at the mouth of the calyx-tube, projecting;
anthers short. Fr. a berry.

Usually spiny; lvs lanceolate, grey-green; corolla-lobes shorter than or
   rarely as long as tube.   **1. barbarum**
Usually unarmed; lvs ovate or rhombic, bright green; corolla-lobes as
   long as or longer than tube.   **2. chinense**

*1. L. barbarum L. (*L. halimifolium* Miller)
                                        Duke of Argyll's Tea-plant.
Stems up to 2·5 m., greyish-white, often spiny. Lvs up to c. 6 cm.,
lanceolate, tapering into a short stalk. Fls solitary or few on short
axillary shoots. Corolla c. 1 cm., rose-purple turning pale brown. Fr.
1–2 cm., scarlet. Fl. 6–9. Widely naturalized in hedges, on old walls, etc.

*2. L. chinense Miller
Like 1 but less spiny; lvs broader and bright green, and fr. 1·5–2·5 cm.
Fl. 6–9. Naturalized, particularly near the sea.

2. HYOSCYAMUS L.

Fls in axils of bracts, often arranged in scorpioidal cymes. Calyx urceolate. Corolla somewhat zygomorphic, bell- or funnel-shaped; lobes
blunt. Fr. a capsule enclosed in the calyx, constricted in the middle and
circumscissile near the top.

1. H. niger L.   Henbane.
Annual or biennial up to 80 cm. Hairs soft, glandular. Stem stout.
Lvs up to 15–20 cm., with few large teeth or nearly entire, lower stalked,
upper sessile and amplexicaul. Fls c. 2 cm., in 2 rows; bracts lf-like,
Corolla 2–3 cm. diam., lurid yellow usually veined with purple. Calyx-
tube in fr. 15–20 mm. diam., strongly veined. Fl. 6–8. On light soils,
especially near the sea, and in farmyards, etc.

3. Solanum L.

Herbs, shrubs or small trees. Fls solitary or in cymes, white, purple or blue. Corolla wheel-shaped. Stamens inserted near the top of the usually short corolla-tube. Fr. a berry.

1   Calyx not enlarging as the fr. develops; ripe fr. red or black.   2
    Calyx enlarging greatly as the fr. develops; ripe fr. green.   3

2   Woody perennial; fls c. 1 cm., purple (very rarely white); fr. ovoid, red.   **1. dulcamara**
    Herbaceous annual; fls c. 0·5 cm., white; fr. globose, black.
                                           2. nigrum

3   Lvs deeply pinnately divided into narrow lobes; fr. 10 mm. or more in diam. Naturalized in E. Anglia.   **S. triflorum* Nutt.
    Lvs entire or weakly toothed; fr. c. 5 mm. diam. A weed in E. Anglia and on rubbish-dumps.   **S. sarrachoides* Sendt.

1. S. dulcamara L.   Bittersweet, Woody Nightshade.

30–200 cm., scrambling, glabrous or ± hairy. Lvs up to c. 8 cm., ovate, entire or with 1–4 deep lobes or stalked lflets at base. Cymes ± opposite the lvs, stalked, umbellate; fl.-stalks erect in fl., recurved in fr. Fls c. 1 cm.; corolla-lobes 3–4 times as long as calyx. Anthers cohering in a cone. Fr. c. 1 cm. Fl. 6–9. Common in hedges, woods, on waste ground and shingle-beaches.

2. S. nigrum L.   Black Nightshade.

Up to c. 60 cm., glabrous or hairy. Lvs ovate or rhombic, entire or sinuate-toothed; blade decurrent on petiole. Cymes usually below and on the opposite side of the stem to the lvs, umbellate; fl.-stalks erect in fl., deflexed in fr. Fls c. 0·5 cm., corolla-lobes c. twice as long as calyx. Anthers not or scarcely cohering in a cone. Fr. c. 8 mm. Fl. 7–9. A weed of cultivation; common in England, rare or local elsewhere.

106. SCROPHULARIACEAE

Mostly herbs, sometimes hemiparasitic, with exstipulate lvs. Fls hermaphrodite, zygomorphic. Calyx 5- or 4-lobed. Corolla gamopetalous, imbricate in bud, from regularly 5-lobed to strongly 2-lipped with the lobes obscure. Stamens borne on the corolla, 5, alternating with the corolla-lobes, or more frequently 4 (the upper being absent or represented by a staminode), sometimes only 2. Ovary superior, 2-celled with numerous ovules on axile placentae; style terminal, simple or 2-lobed. Fr. a capsule; seeds numerous, endospermic.

1   Stamens 5; fls many in a large erect terminal raceme or panicle;
      yellow or white.   1. VERBASCUM
      Stamens 4 or 2.   2

2   Stamens 2; fls blue or more rarely white or pinkish.   11. VERONICA
      Stamens 4; fls never a true blue.   3

3   Corolla-tube saccate or spurred at base.   4
      Corolla-tube not saccate or spurred.   8

4   Lvs lanceolate to linear or oblong, cuneate or tapered at base, not
      or shortly stalked.   5
      Lvs ovate, orbicular, reniform or hastate, rounded to cordate at
      base, stalked.   7

5   Corolla saccate at base.   2. ANTIRRHINUM
      Corolla spurred.   6

6   Fls in a terminal raceme, the bracts much shorter than the lvs;
      mouth of corolla closed; capsule opening by valves.
                                       3. LINARIA
      Fls axillary, the bracts scarcely differing from the lvs; mouth of
      corolla slightly open; capsule opening by pores.
                                4. CHAENORRHINUM

7   Lvs entire or hastate, pinnately veined; decumbent annuals; fls
      yellow with upper lip purple.   5. KICKXIA
      Lvs lobed, palmately veined; creeping perennials (chiefly on walls);
      fls lilac with an orange spot on the palate.   6. CYMBALARIA

8   Stems creeping and rooting at the nodes, or lvs all basal; corolla
      flat, 5 mm. diam. or less, with very short tube; fls solitary.   9
      Stems not creeping and rooting, lfy; fls in infls; corolla with dis-
      tinct tube, usually strongly zygomorphic (except *Erinus*).   10

9   Lvs oblong to awl-shaped, entire.   9. LIMOSELLA
      Lvs reniform, long-stalked, crenately 5–7-lobed, in clusters at the
      nodes; fls 1–2 mm. diam., 5-lobed, the 2 upper lobes yellowish,
      the 3 lower pink. Very local in moist shady places in S.W.
      England and S. Wales; very rare in S.W. Ireland; Channel Is.
      Cornish Moneywort.   *Sibthorpia europaea* L.

10   Calyx 5-lobed, not inflated.   11
      Calyx 4-lobed, or if 5-lobed strongly inflated after flowering and
      lobes usually lf-like.   14

11   Lvs opposite.   12
      Lvs not opposite.   13

12   Corolla with nearly globular tube and small lobes.
                                    7. SCROPHULARIA
      Corolla with broad straight tube and large lobes.   8. MIMULUS

13 Corolla-tube narrow, scarcely longer than calyx, lobes 5, nearly
      equal, purple; a low tufted perennial with ±spathulate lvs.
      Introduced and naturalized in a few places.   *Erinus alpinus L.
    Corolla-tube bell-shaped, tube several times longer than calyx,
      lobes small; tall erect plant.   10. DIGITALIS

14 Calyx inflated, at least after flowering.   15
    Calyx not inflated.   16

15 Lvs spirally arranged, pinnately cut; fls pink.   12. PEDICULARIS
    Lvs opposite, crenate or toothed; fls yellow or brown.
                                                    13. RHINANTHUS

16 Upper lip of corolla laterally flattened, mouth nearly closed; fls
      ±yellow.   14. MELAMPYRUM
    Upper lip of corolla not flattened, mouth open.   17

17 Upper lip of corolla 2-lobed, only slightly concave; corolla white or
      lilac with violet lines and usually a yellow spot on lower lip,
      rarely purple or yellow.   15. EUPHRASIA
    Upper lip of corolla entire or nearly so, arched.   18

18 Fls 4–8 mm., in 1-sided spikes, pink.   16. ODONTITES
    Fls more than 1 cm.; spikes not 1-sided.   19

19 Plant perennial, with opposite sessile ovate lvs; fls c. 2 cm., dull
      purple; seeds few, large, winged or ribbed. Very local in moun-
      tain meadows and on rock-ledges on basic soils in N. England
      and Scotland.   Bartsia alpina L.
    Plant annual, sticky, with opposite lanceolate lvs; fls yellow; seeds
      minute.   17. PARENTUCELLIA

1. VERBASCUM L.

Mostly biennial herbs with rosettes of basal lvs and tall erect stems
with spirally arranged lvs. Fls in terminal racemes or panicles. Calyx
deeply 5-lobed, nearly regular. Corolla usually yellow, flat, with
5 nearly equal lobes and very short tube. Stamens 5, the filaments of all
or at least the 3 upper hairy, those of the 2 lower longer than the rest.
Capsule septicidal; seeds small, numerous.

1 Hairs on the filaments white.   2
   Hairs on the filaments purple.   4

2 Lower filaments glabrous or much less hairy than the upper; fls 15–
     30 mm. diam., in a dense spike-like infl.   1. thapsus
   Filaments all equally hairy; fls less than 20 mm., in a laxer raceme or
     panicle.   3

3 Stem angled; lvs nearly glabrous above, mealy below; fls 15–20 mm.
     diam., white, rarely yellow. Very local in waste places and on
     calcareous banks in S. England and Wales.   V. lychnitis L.

Stem terete; lvs thickly clothed on both sides with a mealy white
wool, easily rubbed off; fls yellow. Roadsides in E. Anglia, very
local.   *V. pulverulentum* Vill.

4   Lvs and stems hairy; anthers all equal and all transversely attached.
                                                       2. nigrum
Lvs and stems glabrous or nearly so; lower anthers larger than the
upper, obliquely attached.   5

5   Fl.-stalks longer than calyx; fls always solitary, yellow or whitish.
Rare and usually casual, chiefly in S. England. Moth Mullein.
                                                       V. blattaria L.
Fl.-stalks shorter than calyx; fls 1–5 together, yellow. S.W. England;
casual elsewhere.   *V. virgatum* Stokes

1. V. thapsus L.   Mullein, Aaron's Rod.

Biennial, rarely annual 30–200 cm., erect, densely clothed with soft
whitish wool. Lvs obovate-lanceolate to oblong, ± acute, crenate; the
basal 15–45 cm., narrowed into a winged stalk; stem-lvs smaller with
the base decurrent nearly to the next lf below. Fls in a dense spike-like
infl. Fls with stalk very short or 0. Corolla 1·5–3 cm. diam., yellow.
Three upper filaments clothed with whitish hairs, the two lower glabrous
or nearly so, their anthers obliquely attached. Capsule ovoid. Fl. 6–8.
Rather common on sunny banks and waste places, especially on dry
soils. Throughout most of the British Is.

2. V. nigrum L.   Dark Mullein.

Erect biennial 50–120 cm. Stem angled, stellate-pubescent. Lvs dark
green above, thinly hairy, pale and more conspicuously stellate-
pubescent beneath, crenate; the basal 10–30 cm., ovate to lanceolate,
cordate at base, long-stalked; stem-lvs smaller, broadly cuneate at base,
upper almost sessile. Fls 5–10 in the axil of each bract of the terminal
raceme, their stalks of varying length. Corolla 12–22 mm. diam., usually
yellow with small purple spots at the base of each lobe. Filaments all
clothed with purple hairs; anthers all equal and transversely attached.
Capsule ovoid. Fl. 6–10. Waysides and open banks, etc., chiefly on
calcareous soil. Rather common in S. England and reaching Nottingham and N. Wales.

2. ANTIRRHINUM L.

Herbs with lower lvs opposite, upper spirally arranged. Fls solitary and
axillary or forming terminal racemes. Calyx deeply 5-lobed. Corolla
strongly 2-lipped, the lower 3-lobed with a projecting 'palate' closing
the mouth of the fl., the upper 2-lobed; tube broad, saccate at base.
Stamens 4. Capsule with 2 unequal cells, opening by 2 or 3 pores.

Erect annual 20–50 cm., with narrow entire lvs 3–5 cm.; fls subsessile
in axils of upper lvs; calyx-lobes linear; corolla 10–15 mm., pinkish-
purple; capsule shorter than calyx. Local in cultivated ground in
England and Wales and S. Ireland.   *A. orontium* L.

Erect perennial 30–80 cm., with narrow entire lvs 3–5 cm.; fls in a
terminal raceme; calyx-lobes ovate; corolla 3–4 cm., reddish-purple
in wild continental form; capsule longer than calyx. Introduced and
naturalized on old walls in many parts of England and Ireland.
Snapdragon.   *A. majus* L.

3. LINARIA Miller

Herbs. Lvs all opposite or whorled or more commonly the upper spirally
arranged; pinnately veined. Fls in terminal racemes, the bracts small.
Fls as in *Antirrhinum* but corolla spurred. Capsule opening by 4–10
apical valves of varying length.

1 Plant glabrous or only infl. glandular; corolla 8 mm. or more
(excluding spur).   *2*
Whole plant sticky with glandular hairs, annual 5–15 cm., with
lanceolate lvs 3–10 mm.; corolla 4–6 mm., yellow, often with
violet spur. Introduced on dunes at Braunton Burrows (N.
Devon).   *L. arenaria* DC.

2 Corolla either violet or whitish striped with purple.   *3*
Corolla yellow.   *5*

3 Annual 15–30 cm.; upper lvs 1–3 cm., linear; fls 10 or fewer; corolla
10–15 mm. violet, with straight spur almost as long. Jersey only.
                                            L. pelisseriana (L.) Miller
Perennial 30–90 cm.; fls numerous; spur short or strongly curved. *4*

4 Infl. 15–40-fld, dense; corolla c. 8 mm., violet, with incurved spur
more than half as long. Introduced and sometimes naturalized.
Purple Toadflax.   *L. purpurea* (L.) Miller
Infl. 10–30-fld, lax; corolla 7–14 mm., white or pale lilac, violet-
striped; spur straight, about a quarter as long.   **1. repens**

5 Erect perennial; sepals ovate to lanceolate, acute.   **2. vulgaris**
Decumbent glaucous annual 5–15 cm.; lvs 1–3 cm.; sepals linear,
blunt; corolla 10–15 mm., yellow, with spur ± straight, almost as
long. Doubtfully native on sand in Cornwall.
                                            L. supina (L.) Chazelles

1. L. repens (L.) Miller   Pale Toadflax.

Glabrous glaucous perennial 30–80 cm., with creeping rhizome and
numerous erect stems. Lvs 1–4 cm., linear, whorled below, spirally
arranged above. Fls in long lax racemes, 10–30-fld. Sepals linear-
lanceolate, acute. Corolla 7–14 mm., white or pale lilac, striped with

violet veins, palate with an orange spot; spur straight, about a quarter
as long as corolla. Capsule subglobose, longer than calyx. Fl. 6–9.
Local in dry stony fields and waste places, usually calcareous, from
Yorks and Lancashire southwards and in Ireland.

2. L. vulgaris Miller   Yellow Toadflax.

Glaucous perennial 30–80 cm., glabrous or sometimes with glandular
infl.; stems numerous, erect, mostly root-borne. Lvs 3–8 cm., linear-
lanceolate. Fls in a long dense raceme, c. 20-fld. Sepals ovate or lanceolate, acute. Corolla 15–25 mm., yellow with orange palate; spur
± straight, nearly as long as corolla. Capsule ovoid, more than twice as
long as calyx. Fl. 7–10. Common in grassy and cultivated fields,
hedge-banks and waste places northwards to Ross; Ireland.

4. CHAENORRHINUM (DC.) Reichenb.

Differs from *Linaria* in the axillary fls with somewhat open-mouthed
corolla, and the unequal cells of the capsule which open by pores.

1. Ch. minus (L.) Lange   Small Toadflax.

Erect annual 8–25 cm., usually glandular-pubescent; branches ascending. Lvs 1–2·5 cm., spirally arranged, linear-lanceolate to oblong,
entire, blunt, narrowed at base to a short stalk. Fls solitary, stalked,
in the axils of upper lvs. Sepals ± linear, blunt. Corolla 6–8 mm.,
purple outside, paler inside; spur short, blunt. Capsule ovoid, shorter
than calyx. Fl. 5–10. In arable land and waste places; common in S.
England but extending to Kincardine and Argyll; Ireland.

5. KICKXIA Dumort.   Fluellen.

Like *Linaria* but lvs spirally arranged except for the few lowest, fls
axillary, and the cells of the capsule opening by the detachment of
circular lids.

Lvs ovate or orbicular; fl.-stalks shaggy.   **1. spuria**
Lvs hastate; fl.-stalks glabrous.   **2. elatine**

1. K. spuria (L.) Dumort.

Hairy and glandular decumbent annual 20–50 cm., branching from the
base. Lvs short-stalked, ovate or orbicular, rounded or subcordate at
base, the lowest to 6 cm., smaller upwards. Fl.-stalks shaggy. Sepals
ovate. Corolla 8–11 mm., yellow with deep purple upper lip; spur

curved, about equalling corolla. Capsule globose, glabrous, shorter than calyx. Fl. 7–10. Rather local in arable land, usually in cornfields on calcareous or base-rich loams; S. England and Wales northwards to Cardigan and Lincoln; S. Ireland.

2. K. elatine (L.) Dumort.

Like *K. spuria* but more slender and less hairy and glandular; upper and middle lvs hastate; fl.-stalks glabrous; corolla 7–9 mm., with upper lip paler purple and spur straight. Fl.7–10. Same habitat as *K. spuria* but extending northwards to Cumberland and Yorks; S. Ireland.

6. CYMBALARIA Hill

Like *Linaria* but with palmately-veined stalked lvs, spirally arranged above, opposite below; fls axillary; capsule opening by 2 lateral pores, each pore with 3 valves.

*1. C. muralis P. Gaertner, B. Meyer & Scherb. Ivy-leaved Toadflax.

Glabrous perennial with trailing or drooping often purplish stems 10–80 cm., rooting at intervals. Lvs 2·5 cm., mostly spirally arranged, usually 5-lobed, thick, sometimes purplish beneath, long-stalked. Fl.-stalks c. 2 cm., recurved. Sepals linear-lanceolate. Corolla 8–10 mm., lilac (rarely white) with a whitish palate, a yellow spot at the mouth and darker lines on the upper lip; spur curved, about ⅓ as long as the corolla. Capsule globose. Fl. 5–9. Introduced and now common on old walls and rarely on rocks almost throughout the British Is.

7. SCROPHULARIA L.

Herbs with square stems and opposite lvs. Fls in axillary cymes, usually forming a terminal panicle. Calyx 5-lobed. Corolla usually dingy in colour, with a nearly globular tube and 5 small lobes, the 2 upper united at base. Fertile stamens 4, bent downwards; the fifth usually represented by a staminode at the base of the upper lip, sometimes 0. Stigma capitate. Capsule septicidal.

1 Sepals blunt, with a scarious border; fls brownish or purplish; staminode present.   *2*
  Sepals acute, without a scarious border; fls yellowish; staminode 0.
    A softly glandular-hairy biennial or perennial herb 30–80 cm., with broadly ovate-cordate lvs. Introduced locally in plantations and waste places over much of Great Britain.   **S. vernalis* L

2   Lvs and stems glabrous or nearly so.   3
    Lvs (on both surfaces) and stems downy; lvs ovate-cordate, rugose;
      fls dull purple. Hedge-banks in S.W. England and Jersey.
                                                         S. scorodonia L.

3   Stems 4-angled, not winged; sepals with narrow scarious border.
                                                 1. nodosa
    Stems 4-winged; sepals with broad (0·5–1 mm.) scarious border.   4

4   Lvs crenate; staminode orbicular or reniform, entire.   **2. auriculata**
    Lvs serrate; staminode with 2 spreading lobes. Like *S. aquatica* but
      stem more broadly winged and lvs never cordate. Rare in damp
      shady places in Great Britain; very rare in Ireland.
                                                    S. umbrosa Dumort.

1. S. nodosa L.   Figwort.

Perennial 40–80 cm., glabrous except for the glandular infl.; rhizome short, irregularly tuberous. Stem sharply 4-angled but not winged. Lvs 6–13 cm., ± ovate, acute, coarsely serrate, ± truncate at base and often unequally decurrent down the lf-stalk, which is not winged. Infl. a panicle of cymes in the axils of bracts, the lowest bracts lf-like. Fl.-stalk 2–3 times as long as fl. Calyx-lobes ovate, blunt, with very narrow scarious border. Corolla to c. 1 cm., with greenish tube and usually reddish-brown upper lip. Staminode broader than long, retuse. Capsule 6–10 mm., ovoid, acuminate. Fl. 6–9. Common in damp and wet woods and hedge-banks throughout most of the British Is.

2. S. auriculata L. (*S. aquatica* auct.)   Water Betony.

Perennial 50–100 cm., ± glabrous except for the glandular infl.; rhizome not tuberous. Stem 4-winged. Lvs 6–12 cm., ± ovate, blunt, crenate, often with 1–2 small pinnae at the base, lf-stalk winged. Infl. a panicle of cymes in the axils of bracts unlike the lvs. Fl.-stalks about equalling the fls. Calyx-lobes rounded, with broad (0·5–1 mm.) scarious border. Corolla to 1 cm., brownish-purple above, greenish on the underside. Staminode suborbicular, entire. Capsule 4–6 mm., sub-globose, apiculate. Fl. 6–9. Common on the edges of ponds and streams and in wet woods and meadows northwards to S. Scotland; Ireland.

8. MIMULUS L.

Herbs with opposite lvs. Fls from the axils of lvs or bracts. Calyx tubular, 5-angled and 5-toothed. Corolla with a long tube, hairy in the throat, 2-lipped; the upper lip 2-lobed, the lower longer, 3-lobed, the lobes all flat. Stamens 4. Stigma with 2 flat lobes. Capsule included in the calyx, loculicidal; seeds small, numerous.

1 Plant glabrous or pubescent only above; lower lvs stalked, upper
    sessile; corolla 2·5–4·5 cm., red-spotted.   *2*
    Plant viscid-hairy all over; all lvs short-stalked; corolla 1–2 cm.,
    yellow, not red-spotted. Introduced and occasionally naturalized
    in wet places. Musk.   *M. moschatus* Douglas ex Lindley

2 Calyx and fl.-stalks (1·5–3 cm.) pubescent; corolla with small red
    spots only, markedly 2-lipped, throat nearly closed.   **1. guttatus**
    Calyx and fl.-stalks (3·5 cm. or more) glabrous; corolla with large
    red blotches or variegated with purple, only slightly 2-lipped,
    throat wide open. In similar places to *M. guttatus* but much more
    local and chiefly in the north.   *M. luteus* L.

*1. M. guttatus DC.   Monkey-flower.

Perennial 20–50 cm. with ascending or decumbent fl.-stems, glabrous
below, ± pubescent above. Lvs 1–7 cm., irregularly toothed; lower
ovate or oblong, upper broader. Infl. with lowest bracts lf-like, upper
smaller. Calyx inflated in fr.; teeth ± deltate, the upper much longer.
Corolla 2·5–4·5 cm., yellow with small red spots, markedly 2-lipped
with the lower lip much longer and with a prominent palate nearly
closing the throat. Capsule oblong, blunt. Fl. 7–9. Introduced and
now rather common on banks of streams, etc., through almost the whole
of the British Is.

9. LIMOSELLA L.

Annual herbs growing on wet mud at the edges of pools or where water
has stood, creeping by runners; the lvs all basal. Fls small, axillary,
solitary. Calyx 5-toothed. Corolla flat, nearly regular, 5-lobed with
short tube. Stamens 4. Capsule septicidal.

Lvs of mature plants 5–15 mm., elliptic, long-stalked; calyx longer
    than corolla-tube; corolla 2–5 mm. diam., white or lavender, its tube
    1·5 mm. Very local. Mudwort.   *L. aquatica* L.
Lvs 10–25 mm., all awl-shaped; calyx shorter than corolla-tube;
    corolla up to 4 mm. diam., white, with orange tube c. 3 mm. Wales;
    very rare.   *L. subulata* Ives

10. DIGITALIS L.

Tall biennial or perennial herbs with spirally arranged lvs, the lowest
forming a rosette. Fls nodding, in terminal 1-sided racemes. Calyx
deeply 5-lobed. Corolla with a long bell-shaped tube, shortly 5-lobed,
the 2 upper forming an upper lip, the lowest usually longer. Stamens 4.
Capsule septicidal with many small seeds.

1. D. purpurea L.   Foxglove.

Erect biennial, rarely perennial, 50–150 cm. Lvs ovate to lanceolate, crenate, green and softly pubescent above, grey-tomentose beneath, narrowed into a winged stalk. Raceme 20–80-fld, with lanceolate bracts diminishing upwards. Lower sepals ovate, upper lanceolate, all acute. Corolla 4–5 cm., much longer than calyx, usually pinkish-purple with deeper spots on a white ground inside the lower part of the tube, ciliate and with a few long hairs within. Stamens and style included. Capsule ovoid, rather longer than calyx. Fl. 6–9. Common in open places in woods and on heaths and mountain-rocks on acid substrata, to 2900 ft. Throughout most of the British Is.

11. VERONICA L.   Speedwell.

Annual or perennial herbs or low shrubs, with opposite lvs. Fls usually blue. Calyx usually with 4 lobes, the upper being absent. Corolla flat, 4-lobed, with a very short tube; upper lobe longest, lower smallest. Stamens 2. Capsule ± laterally flattened; seeds few.

1	Fls in axillary racemes.	2
	Fls in terminal racemes, or solitary.	8
2	Plant glabrous or, if hairy, lvs linear-lanceolate; growing in wet places.	3
	Plant ± hairy, lvs ovate or oblong; growing in drier places.	6
3	Racemes from one only of a pair of lvs.	**4. scutellata**
	Racemes in opposite pairs.	4
4	Lvs stalked, blunt.	**1. beccabunga**
	Lvs sessile, acute.	5
5	Corolla pale blue; fl.-stalks ascending after flowering.	**2. anagallis-aquatica**
	Corolla pinkish; fl-stalks widely spreading after flowering.	**3. catenata**
6	Fl.-stalks 2 mm. or less, shorter than bract and calyx; racemes rather dense, pyramidal.	**5. officinalis**
	Fl.-stalks 4 mm. or more, as long as or longer than bracts and calyx; raceme lax.	7
7	Stem hairy all round; capsule longer than calyx; lvs with stalk 5–15 mm.	**6. montana**
	Stem hairy on two opposite sides only; capsule shorter than calyx; lvs sessile or with stalk less than 5 mm.	**7. chamaedrys**
8	Fls in dense many-fld terminal spike-like racemes; corolla-tube longer than broad. Pubescent perennial 8–60 cm., with upper	

lvs lanceolate to linear, ±crenate. Very local on limestone in
W. England and Wales and on dry basic sand in E. Anglia.
                                                              V. spicata L.

Fls in lax racemes or few-fld head-like racemes, or solitary; corolla-
tube very short, much broader than long.   9

9 Fls in racemes, with at least the upper bracts very different from
the lvs.   10
Fls solitary in the axils of lvs (the upper lvs sometimes rather
smaller).   18

10 Lvs glabrous (or very finely and shortly hairy) entire or nearly so;
perennial (except *peregrina*).   11
Lvs conspicuously pubescent, often glandular, toothed or lobed;
annual.   15

11 Perennial; bracts (at least the upper) shorter or scarcely longer
than fls.   12
Glabrous annual; bracts all much longer than fls, the upper lanceolate, entire. Introduced and naturalized, especially in N.W.
Ireland.   **V. peregrina* L.

12 Shrubby at base; corolla c. 1 cm. diam., bright blue. Very local on
alpine rocks in Scotland. Rock Speedwell.   *V. fruticans* Jacq.
Herbs; corolla white, pale or dull blue or pink, smaller.   13

13 Fls pink; fl.-stalks longer than bracts. Plant with slender
creeping and rooting stems and short lateral flowering branches.
Introduced and naturalized in a few places in N. England and
S. Scotland.   **V. repens* Clarion ex DC.
Fls not pink; fl.-stalks shorter than bracts.   14

14 Fl.-stalks longer than calyx; fr. broader than long, scarcely exceeding calyx, style about as long; fls white or pale blue.

                                                      8. serpyllifolia

Fl.-stalks shorter than calyx; fr. longer than broad, much exceeding calyx, style very short; fls dull blue in a dense head-like
raceme. Damp alpine rocks in Scotland. Alpine Speedwell.
                                                              V. alpina L.

15 Fl.-stalks much shorter than calyx.   16
Fl.-stalks longer than calyx.   17

16 Lvs toothed.   **1. arvensis**
Lvs pinnately 3–7-lobed to more than half-way. Erect annual,
local in dry grassland in the E. Anglian breckland. Spring
Speedwell.   *V. verna* L.

17 Lvs ovate, deeply toothed; fr. longer than calyx; erect annual, very
local in dry grassland in the E. Anglian breckland; ? introduced.
                                                              V. praecox All.
Lvs digitately 3–7-lobed; fr. shorter than calyx. Suberect annual,
local in sandy arable fields in E. England.   *V. triphyllos* L.

18 Lvs with 5–7 large teeth in the lower half; fls pale lilac; sepals
    cordate at base.   **10. hederifolia**
    Lvs regularly crenate-serrate; fls usually blue or blue and white;
    sepals narrowed at base.   *19*

19 Decumbent annuals; lvs not reniform; fl.-stalks not twice as long
    as lvs.   *20*
    Creeping pubescent perennial often forming mats; lvs reniform, c.
    5 mm., crenate; fl.-stalks thread-like, several times as long as lvs.
    Introduced and naturalized in many places.
                                  *V. filiformis Sm.

20 Lobes of fr. divergent; fls 8–12 mm. diam.   **11. persica**
    Lobes of fr. not divergent; fls 4–8 mm. diam.   *21*

21 Sepals ovate, ±acute; fr. with short crisped glandless hairs and a
    few longer glandular hairs; corolla usually uniformly blue.
                                    12. polita
    Sepals oblong, ±blunt; fr. without short crisped hairs; lower lobe
    of corolla usually white or pale.   **13. agrestis**

1. V. beccabunga L.   Brooklime.

Glabrous perennial 20–60 cm.; stems creeping and rooting at base,
then ascending, fleshy. Lvs 3–6 cm., rather thick and fleshy, oval or
oblong, blunt, rounded at base, shallowly crenate-serrate, short-stalked.
Racemes in opposite pairs, lax, 10–30-fld; bracts linear-lanceolate.
Calyx-lobes narrowly ovate, acute. Corolla 7–8 mm. diam., blue.
Capsule ± orbicular, shorter than calyx. Fl. 5–9. In streams, ponds,
marshes and wet places in meadows; common throughout the British Is.

2. V. anagallis-aquatica L.   Water Speedwell.

Perennial or sometimes annual, glabrous except for the sometimes
glandular infl. Stems 20–30 cm., fleshy, green, shortly creeping and rooting below, then ascending. Lvs 5–12 cm., ovate-lanceolate or lanceolate,
acute, half-clasping, remotely toothed. Racemes in opposite pairs,
rather lax, 10–50-fld., ascending. Bracts linear, acute, shorter than or
equalling the fl.-stalks at flowering. Calyx-lobes ovate-lanceolate, acute.
Corolla c. 5–6 mm. diam., pale blue. Fl.-stalks ascending after flowering. Capsule ± orbicular, usually slightly longer than broad. Fl. 6–8. In
ponds, streams, wet meadows and on wet mud. Common throughout
much of the British Is.

3. V. catenata Pennell

Like *V. anagallis-aquatica* but stems usually purple-tinged; racemes
more spreading; bracts lanceolate, longer than the fl.-stalks at flowering;

calyx-lobes elliptic or oblong; corolla pink with darker lines; fl.-stalks spreading ± at right angles after flowering; capsule usually broader than long. Fl. 6–8. In similar places to *V. anagallis-aquatica* and often with it (and forming sterile hybrids).

4. V. scutellata L.   Marsh Speedwell.

Perennial, usually glabrous but sometimes glandular or pubescent. Stems 10–50 cm., creeping below, then ascending. Lvs 2–4 cm., linear-lanceolate or linear, acute, half-clasping, distantly toothed, yellowish-green, often tinged with purple. Racemes single and alternating at successive nodes, lax, up to 10-fld. Fl.-stalks twice as long as the linear bracts, spreading in fr. Corolla c. 6–7 mm. diam., white or pale blue with purple lines. Capsule flat, broader than long, deeply emarginate. Fl. 6–8. In ponds, bogs, wet meadows, etc., often on acid soils. Rather common throughout the British Is.

5. V. officinalis L.   Common Speedwell.

Perennial with stems 10–40 cm., hairy all round, creeping below, rooting and often forming large mats, then ascending. Lvs 2–3 cm., ± elliptic-oblong, crenate, subsessile, hairy on both sides. Racemes 1(–2) at each upper node, long-stalked, dense and pyramidal, 15–25-fld. Fl.-stalks half as long as the linear bracts. Corolla c. 6 mm. diam., lilac (like the stamens and style). Capsule broader above, longer than calyx. Fl. 5–8. Grassland, heaths and open woods, often on dry soils. Common throughout the British Is.

6. V. montana L.   Wood Speedwell.

Perennial; stems 20–40 cm., hairy all round, creeping and rooting below, then ascending. Lvs 2–3 cm., ± broadly ovate, truncate or broadly cuneate at base, coarsely crenate-serrate, light green, hairy on both sides; stalk 5–15 mm. Racemes 1–2 at each upper node, long-stalked, lax, 2–5-fld. Fl.-stalks much longer than the small linear bracts. Corolla c. 7 mm. diam., lilac-blue. Capsule ± orbicular, longer than calyx. Fl. 4–7. Damp woods. Rather local throughout England, Wales and Ireland and northwards to Inverness.

7. V. chamaedrys L.   Germander Speedwell.

Perennial; stems 20–40 cm.; prostrate and rooting below, then ascending, with long white hairs in two lines on opposite sides. Lvs 1–2·5 cm., triangular-ovate, sessile or short-stalked (to 5 mm.), ± cordate at base, coarsely crenate-serrate, dull green, hairy. Racemes 1(–2) at each upper node, long-stalked, lax, 10–20 fld. Fl.-stalks 4–6 mm., not shorter than the lanceolate bracts. Corolla c. 1 cm. diam., deep bright blue with white eye. Capsule obcordate, shorter than calyx. Fl. 3–7. In grassland, woods, hedges, etc. Very common throughout the British Is.

8. V. serpyllifolia L. Thyme-leaved Speedwell.

Perennial; stems 10–30 cm., finely downy, creeping and rooting below, then usually ascending. Lvs 1–2 cm., oval or oblong, almost entire, rounded at both ends, subsessile or short-stalked, glabrous, light green. Racemes terminal, up to 30-fld., lax. Fl.-stalks shorter than bracts. Corolla white or pale blue with darker lines; filaments and style white, anthers slaty-violet. Capsule obcordate, about equalling calyx and style. Fl. 3–10. Grassland, heaths, waste places, etc., especially where moist; common throughout the British Is. (Also, locally, in damp places on mountains, with stems rooting almost throughout their length and larger bluer fls: ssp. **humifusa** (Dicks.) Syme)

9. V. arvensis L. Wall Speedwell.

Erect annual 3–25 cm., pubescent and sometimes glandular. Lvs to 1·5 cm., triangular-ovate, coarsely crenate-serrate, only the lowest stalked. Racemes terminal, long. Upper bracts lanceolate, entire, longer than the fls.; lower passing into the lvs. Fl.-stalks less than 1 mm. Corolla blue, shorter than calyx. Capsule about as long as broad, obcordate, shorter than calyx; seeds flat. Fl. 3–10. Cultivated ground and in ± open habitats in grassland and heaths and on rock-ledges; common throughout the British Is.

10. V. hederifolia L. Ivy-leaved Speedwell.

Annual, branched at base, with decumbent branches 10–60 cm.; stem hairy. Lvs to 1·5 cm., ± reniform, with 2–3 large teeth or small lobes on each side in the lower half, rather thick, light green, stalked, blunt at apex, ± truncate at base, 3-veined, ciliate. Fls solitary, axillary; fl.-stalks usually shorter than lvs. Sepals ovate-cordate. Corolla shorter than calyx, pale lilac. Capsule glabrous, hardly flattened, broader than long. Fl. 3–5(–8). Common in cultivated ground throughout the British Is.

11. V. persica Poiret Large Field Speedwell.

Annual, branched at base, with decumbent branches 10–40 cm.; stem hairy. Lvs 1–3 cm., triangular-ovate, short-stalked, coarsely crenate-serrate, light green, hairy on veins beneath. Fls solitary, axillary; fl.-stalks longer than lvs but not twice as long, decurved in fr. Sepals ovate, enlarging and spreading in fr. Corolla 8–12 mm. diam., bright blue with the lower lobe often paler or white. Capsule 2-lobed, the lobes sharply keeled and divergent, so that the fr. is nearly twice as broad as long. Fl. 1–12. Introduced but now very common in cultivated land throughout the British Is.

12. V. polita Fries   Grey Speedwell.

Pubescent annual, branched at base, with decumbent branches. Lvs 5–15 mm., ovate, short-stalked, dull green, blunt, ± truncate at base, coarsely crenate-serrate. Fls solitary, axillary; fl.-stalks equalling or shorter than lvs, decurved in fr. Sepals ovate, ± acute, enlarging in fr. Corolla 4–8 mm. diam., usually uniformly bright blue. Capsule 2-lobed, the lobes erect, not keeled, clothed with short crisped glandless hairs and some longer glandular hairs. Fl. 1–12. Common in cultivated ground throughout the British Is.

13. V. agrestis L.   Field Speedwell.

Like *V. polita* but sepals oblong or ovate-oblong, ± blunt; corolla pale blue or pinkish with the lower lobe or 3 lobes white or very pale; capsule-lobes obscurely keeled, with long glandular hairs, often also with glandless hairs but these never short and crisped. Fl. 1–12. Cultivated ground, probably throughout the British Is, but much less common in the south than *V. polita*.

12. PEDICULARIS L.

Hemiparasitic herbs with spirally arranged pinnately-cut lvs. Fls in terminal lfy spikes or racemes. Calyx with 2–5 lf-like lobes, becoming inflated in fr. Upper lip of corolla laterally flattened, entire or with 2 or 4 small teeth; lower lip 3-lobed. Capsule flattened, with a few large seeds in the lower part.

Annual: calyx pubescent outside; upper lip of corolla with 4 teeth.
                                                                     1. palustris
Perennial; calyx usually glabrous outside; upper lip of corolla with
   2 teeth.   **2. sylvatica**

1. P. palustris L.   Marsh Lousewort.

Annual 8–60 cm., often purplish, nearly glabrous. Stem single, branching from near the base. Lvs 2–4 cm., pinnately cut, the lobes toothed. Bracts similar but smaller. Calyx pubescent, often reddish, with 2 broad irregularly-cut lobes. Corolla 2–2·5 cm., purplish-pink; upper lip with a tooth at each side near the tip and another lower down. Capsule curved, longer than calyx. Fl. 5–9. Wet meadows, marshes, fens and wet heaths, to 2800 ft.; rather common throughout the British Is.

2. P. sylvatica L. Lousewort.

Perennial 8–25 cm., nearly glabrous, with many decumbent branches from the base. Lvs to 2 cm., pinnately cut, with toothed lobes. Racemes 3–10-fld; bracts lf-like. Calyx usually glabrous outside, 5-angled, with 4 small lf-like 2–3-lobed teeth and a small linear 5th (upper) tooth at a lower level. Corolla 2–2·5 cm., pink; upper lip with a tooth on each side near the tip only. Capsule obliquely truncate, about equalling the calyx. Fl. 4–7. Damp heaths, moors and bogs to 3000 ft.; rather common throughout the British Is.

13. RHINANTHUS L.

Annual hemiparasitic herbs with opposite toothed lvs. Fls in terminal lfy spikes. Calyx flattened and inflated in fl., enlarging in fr., with 4 entire teeth. Corolla yellow or brown, its upper lip laterally flattened with 2 teeth at the end; lower lip 3-lobed. Stamens 4, included in the upper lip. Capsule flattened, with a few large usually winged seeds.

For a fuller treatment of this difficult genus the larger *Flora*, ed. 2, should be consulted.

Lvs 3·5–7 cm. × 8–18 mm.; bracts yellow-green, their lower teeth usually long; corolla-tube curved upwards; teeth of upper lip of corolla long (c. 2 mm.), twice as long as broad; mouth of corolla closed. In cornfields, etc., local, chiefly in N. England and Scotland.
　　　　　　　　　　　　　　　　　R. serotinus (Schönh.) Oborny
Lvs 2–3(–4) cm. × 2–8 mm.; bracts green, their lowest teeth not reaching half-way to midrib; corolla-tube straight; teeth rounded, short, not longer than broad; mouth of corolla somewhat open.
　　　　　　　　　　　　　　　　　　　　　　　　　　1. minor

1. R. minor L., *sensu lato* Yellow-rattle.

Stem to 50 cm., 4-angled, often dark-spotted. Lvs oblong to linear, 2–3 cm. × (2–)5–8 mm., sessile, subcordate at base, crenate-toothed, rough at least above, hairy or not. Fls solitary, subsessile, in the axils of lf-like green triangular bracts which are longer than the calyx; lower teeth of bracts not much longer than the upper. Calyx shortly hairy on the margin (rarely also on the surface). Corolla (13–)15 mm., usually yellow with violet teeth, rarely brownish; tube straight; mouth somewhat open; teeth rounded, not longer than broad. Capsule shorter than calyx; seeds all winged. Fl. 5–9. In pastures, damp meadows, fens and grassy places on mountains throughout the British Is.

14. MELAMPYRUM L.

Annual hemiparasitic herbs with opposite, mostly entire, ± sessile, narrow lvs. Fls in terminal racemes or spikes, the bracts often toothed. Calyx tubular, 4-toothed. Corolla 2-lipped; upper lip laterally flattened; lower lip 3-lobed, shorter, with prominent palate nearly closing the mouth. Stamens 4, included in the upper lip. Capsule flattened, with 1–4 ovoid seeds.

1   Fls in ± dense spikes.   2
      Fls in pairs in the axils of lf-like bracts forming a very lax, 1-sided,
      interrupted raceme.   3

2   Spikes 4-sided, very dense; bracts ovate-acuminate, cordate, the
         lower part bright rosy-purple and pectinate, the upper green,
         entire, recurved; corolla pale yellow, variegated with purple.
         Very local at wood-edges in E. and C. England. Crested Cow-
         wheat.   *M. cristatum* L.
      Spikes conical or cylindrical, laxer; bracts lanceolate, erect, pink at
         first, with long slender teeth to 8 mm.; corolla with pink tube,
         yellow throat and deep pink lips. Very local and rare in arable
         fields in S.E. England. Field Cow-wheat.   *M. arvense* L.

3   Corolla 12–22 mm., its tube much longer than calyx; lower lip
         straight; mouth closed or somewhat open.   **1. pratense**
      Corolla 8–10 mm., deep yellow, its tube equalling or shorter than
         calyx; lower lip deflexed; mouth widely open. Very local in moun-
         tain woods. Wood Cow-wheat.   *M. sylvaticum* L.

1. M. pratense L.   Common Cow-wheat.

Very variable. Annual 8–60 cm., glabrous or somewhat hispid, with spreading branches. Lvs 1·5–10 cm., ± sessile, ovate-lanceolate to linear-lanceolate, entire. Fls in the axils of each of distant opposite green lf-like bracts, the two at a node turned to the same side of the stem; bracts entire to pectinate. Calyx with very narrow lobes, rather longer than the tube. Corolla 12–22 mm., varying from deep yellow to whitish, sometimes variegated with red or purple; tube about twice as long as calyx; mouth closed or somewhat open; lower lip directed forward. Capsule usually 4-seeded. Fl. 5–10. Common in woods, heaths, etc., on acid humus; to 3000 ft. Throughout the British Is.

15. EUPHRASIA L.   Eyebright.

Annual hemiparasitic herbs. Lvs opposite, or the upper spirally arranged, rather small. Fls solitary and sessile in the axils of bracts which usually differ somewhat from the other lvs, thus forming a terminal spike. Calyx bell-shaped, 4-toothed. Upper lip of corolla slightly concave with 2 lobes, the lower with 3 emarginate lobes. Stamens 4. Seeds small, numerous.

The numerous spp. included in *E. officinalis* L. are difficult to determine and the larger *Flora* should be consulted.

Bracts generally more than ½ as broad as long, with contiguous teeth; capsule ciliate with long straight hairs.   **1. officinalis**
Bracts (like the lvs) generally less than ½ as broad as long, with more distant narrow teeth; capsule glabrous or rarely with a few weak marginal bristles. On limestone and dunes in W. Ireland.
                                                    E. salisburgensis Funck

1. E. officinalis L., *sensu lato*

Stem 2–30(–40) cm., erect or ascending, wiry, often much branched below, sometimes glandular-hairy. Lvs ovate to lanceolate, crenate-serrate, with or without glandular or eglandular hairs. Bracts lf-like, but usually with more acute teeth than the lvs. Corolla with the lobes of the upper lip spreading, commonly white, often with purplish upper lip, sometimes wholly purplish. Capsule ciliate, variable in shape and in length in relation to the calyx. Fl. 6–9. Meadows, pastures, fens, bogs, etc., throughout the British Is. Very variable, and divisible into numerous species which are largely interfertile.

16. ODONTITES Ludwig

Annual hemiparasitic herbs with opposite lvs. Fls in a 1-sided terminal spike-like raceme; bracts like the lvs but smaller. Calyx 4-toothed. Corolla 2-lipped; upper lip concave, entire or emarginate; lower with 3 entire lobes. Stamens 4. Seeds small, rather few, furrowed.

1. O. verna (Bellardi) Dumort.   Red Rattle.

Erect branching pubescent annual, often purple-tinted, up to 50 cm. Lvs sessile, lanceolate to linear-lanceolate, distantly toothed. Racemes terminal on main stem and upper branches. Calyx bell-shaped. Corolla c. 9 mm., purplish-pink. Anthers slightly exserted. Capsule ± equalling calyx. Fl. 6–8. Common in cultivated fields and waste places throughout the British Is.

17. PARENTUCELLIA Viv.

Annual hemiparasitic herbs with opposite lvs. Fls in terminal spike-like racemes; bracts like the lvs. Calyx tubular, 4-toothed. Corolla 2-lipped; upper lip entire or emarginate, forming a hood; lower with 3 entire lobes. Stamens 4. Capsule narrow; seeds many, minute.

1. P. viscosa (L.) Caruel   Yellow Bartsia.

Erect annual 10–50 cm., usually unbranched, sticky with glandular hairs. Lvs 1·5–4 cm., lanceolate, ± acute, sessile, toothed. Corolla c. 2 cm., yellow, the lower lip much longer than the upper. Stamens included in the upper lip. Fl. 6–10. Damp grassy places, usually near the south and west coasts northwards to Argyll, but occasionally in inland counties; Ireland, especially in the south-west.

107. OROBANCHACEAE

Annual to perennial herbaceous root-parasites lacking chlorophyll, usually with erect scaly aerial shoots bearing fls in dense terminal racemes or spikes. Fls zygomorphic, hermaphrodite. Calyx tubular below; corolla with a somewhat curved tube below, 2-lipped above; stamens 4, borne on the corolla-tube in 2 pairs, their anthers close together; ovary superior, 1-celled, with numerous ovules on 4 (2–6) parietal placentae; style 1, with a 2-lobed stigma. Fr. an incompletely 2-valved capsule; seeds very small and numerous, endospermic.

Plants with scaly rhizome; calyx equally 4-lobed; lower lip of corolla
    almost parallel to upper, not spreading.   1. LATHRAEA
Plants not rhizomatous; calyx laterally 2-lipped; lower lip of corolla
    divergent from upper, spreading.   2. OROBANCHE

1. LATHRAEA L.

Rhizome branched, creeping, covered with broad fleshy whitish scales and bearing rootlets which are swollen at points of attachment to the host-root. Fls borne singly in the axils of scales. Calyx bell-shaped, equally 4-lobed; corolla 2-lipped, the upper lip strongly concave, entire, the lower smaller and 3-lobed. Fr. opening elastically from the tip.

Aerial stem 8–30 cm.; fls white or dull purple, short-stalked.
                                             1. squamaria
No aerial stem; fls bright purple, long-stalked, arising singly from the
    rhizome and grouped in corymbose clusters. Introduced. Natural-
    ized locally on roots of poplars and willows in damp shady places.
    Purple Toothwort.   *L. clandestina L.

1. L. squamaria L.   Toothwort.

Perennial, with stout erect fl.-stem, 8–30 cm., white or pale pink, slightly downy above and with a few whitish scales below. Infl. a 1-sided raceme with ovate scaly bracts each with a single short-stalked fl. in its axil. Calyx glandular, tubular below and with 4 broadly triangular teeth above. Corolla white tinged with dull purple, slightly longer than calyx.

Fr. ovoid, acuminate. Fl. 4–5. On roots of various woody plants, especially elm and hazel, in moist woods and hedgerows on good soils, and locally common in some limestone districts; Great Britain northwards to Perth and Inverness; throughout Ireland except the north-west.

2. OROBANCHE L.

'Broomrapes', with underground tubers attached to host-roots. From these tubers arise erect scaly fl.-stems with terminal spikes of almost sessile fls. Calyx laterally 2-lipped; corolla 2-lipped, the upper lip erect and ± 2-lobed, the lower spreading and 3-lobed; stigma 2-lobed. Valves of fr. often remaining attached at their tips.

1 Fls each with 1 bract at the front and 2 lateral bracteoles (apart
    from the calyx-lobes); valves of fr. free above.   *2*
    Fls each with 1 bract only (apart from the calyx-lobes); valves of
    fr. united above.   *3*

2 Fls 10–16 mm., yellowish-white tinged marginally with blue-
    purple; stem usually branched; chiefly on hemp and tobacco.
    Probably native in Channel Is, but introduced in E. and S.
    England. Branched Broomrape.   *O. ramosa* L.
    Fls 18–30 mm., dull blue-purple; stem usually simple, bluish;
    chiefly on *Achillea millefolium*. Rare in S. England and S. Wales;
    common on walls in Channel Is. Purple Broomrape.
                                          O. purpurea Jacq.

3 Stigma-lobes yellow, at least at first.   *4*
    Stigma-lobes purple, red or brown throughout.   *7*

4 Corolla 18–25 mm.   *5*
    Corolla 10–20 mm.   *6*

5 Upper lip of corolla almost entire; stamens inserted close to base
    of corolla-tube, their filaments glabrous below; parasitic on
    shrubby *Papilionaceae*, especially gorse and broom.
                              1. rapum-genistae
    Upper lip of corolla finely toothed; stamens inserted well above
    base of corolla-tube, their filaments hairy below; parasitic chiefly
    on *Centaurea scabiosa*.   **2. elatior**

6 Stem reddish or purple; parasitic on ivy.   **4. hederae**
    Stem yellow (Channel Is only).   **3. minor var. flava**

7 Stem and corolla both purplish-red; parasitic on thyme and per-
    haps other Labiatae. Local on rocky slopes, screes, etc., chiefly
    in W. Cornwall, W. Scotland, the Hebrides and N. and W.
    Ireland, but scattered elsewhere. Red Broomrape.
                              O. alba Stephan
    Stem and corolla variously coloured but not both purplish-red; not
    parasitic on thyme or other Labiatae.   *8*

8   Corolla 20–30 mm.; parasitic on *Galium mollugo*. Only in Kent and
    Argyll. Clove-scented Broomrape.   *O. caryophyllacea* Sm.
    Corolla 10–20(–22) mm.   9

9   Fl.-stem purple; calyx usually shorter than corolla-tube; lower lip
    of corolla unequally 3-lobed, the reniform middle lobe the
    largest. A rare coastal plant of S. England and the Channel Is,
    usually parasitic on *Daucus* Carrot Broomrape.

                                   O. maritima Pugsley
    Fl.-stem yellowish, tinged with red or purple; calyx usually about
    equalling the corolla-tube; lower lip of corolla about equally
    3-lobed.   10

10 Corolla 10–16(–18) mm., its back regularly curved throughout.

                                         3. minor
    Corolla 15–22 mm., its back almost straight from near the base,
    then abruptly curved into the upper lip.   11

11 Corolla with pale glands; filaments densely hairy below. On
    Picris and *Crepis* spp.; rare and local in S. England and S.
    Wales. Picris Broomrape.   *O. picridis* F. W. Schultz ex Koch
    Corolla with dark glands; filaments glabrous or sparsely hairy
    below. On thistles; very local in Yorks. Thistle Broomrape.
                                      O. reticulata Wallr.

1. O. rapum-genistae Thuill.   Greater Broomrape.

Fl.-stem 20–80 cm., yellowish, stout, with numerous scales near the
base, glandular-hairy throughout. Fls in a long compact spike; bracts
exceeding the fls. Calyx-lips usually equally bifid and nearly equalling
the corolla-tube. Corolla 20–25 mm., yellowish tinged with purple,
glandular; back of tube curved throughout; upper lip of corolla almost
entire, lower with middle lobe much larger than the 2 laterals; both lips
waved. Stamens inserted at base of corolla-tube; filaments glabrous at
base. Stigma-lobes pale yellow, distant. Fl. 5–7. Parasitic on shrubby
Papilionaceae, chiefly on *Ulex* and *Sarothamnus*, occasionally on
Genista tinctoria, etc. Throughout England and Wales but only in S.W.
Scotland and S. and S.E. Ireland.

2. O. elatior Sutton   Tall Broomrape.

Fl.-stem 15–70 cm., yellowish or reddish, stout, with numerous scales
below, glandular-hairy throughout. Fls in a dense spike; bracts about
as long as fls. Calyx-lips somewhat unequally bifid, usually shorter
than corolla-tube. Corolla 18–25 mm., pale yellow, usually purple-
tinged, glandular; back of tube curved throughout; upper lip of corolla
2-lobed, lower lip nearly equally 3-lobed; all lobes finely toothed and
waved. Stamens inserted well above (4–6 mm.) base of corolla-tube;

filaments hairy below. Stigma-lobes yellow. Fl. 6–7. Parasitic chiefly on *Centaurea scabiosa* on dry calcareous soils; locally common and widely distributed in England and Wales.

3. O. minor Sm. Lesser Broomrape.

Fl.-stem 10–50 cm., usually yellowish tinged with red, ± glandular-hairy, sparsely scaly. Fls in a spike lax below; bracts at least equalling fls. Calyx-lips unequally bifid or entire, about equalling corolla-tube. Corolla 10–16 mm., yellowish, ± tinged and veined with purple, sparsely glandular; back of tube regularly curved throughout; upper lip of corolla notched, lower ± equally 3-lobed, the middle lobe broadly reniform; all lobes finely toothed and waved. Stamens inserted just above base of corolla-tube, their filaments usually nearly glabrous throughout. Stigma-lobes distant, purple (but yellow in var. *flava* E. Regel: Channel Is only). Fl. 6–9. Parasitic chiefly on *Trifolium* spp. and other herbaceous Papilionaceae; var. *compositarum* Pugsl., on *Crepis capillaris*, *Hypochaeris radicata* and other Compositae, has somewhat larger corolla (12–18 mm.) and filaments densely hairy below. Throughout England and Wales and common in clover-fields in the south; very rare in Scotland.

4. O. hederae Duby Ivy Broomrape.

Fl.-stem 10–60 cm., purplish, glandular-hairy, with a few scales below. Fls in a long lax spike; bracts serrate near the tip, at least equalling fls. Calyx-lips entire or unequally bifid, about equalling the corolla-tube. Corolla 12–20 mm., cream, strongly purple-veined, sparsely glandular; back of tube straight except at the inflated base; upper lip of corolla entire to notched, lobes of lower lip waved and finely toothed. Stamens inserted above base of corolla-tube, their filaments almost glabrous below. Stigma-lobes partially united, yellow. Fl. 6–7. On *Hedera helix*. Local in S. England and Wales northwards to Anglesey and Leicester; rather common in Ireland but absent from the north-east.

108. LENTIBULARIACEAE

Insectivorous herbs with lvs spirally arranged, sometimes all basal. Calyx 5-lobed or 2-lipped. Corolla gamopetalous, 2-lipped, spurred. Stamens 2, inserted on base of corolla-tube. Ovary superior, 1-celled with numerous ovules on a free-central placenta; stigma often sessile, unequally 2-lobed. Fr. a capsule opening irregularly by 2 or 4 valves; seeds small, numerous, non-endospermic.

Lvs entire, in a basal rosette; insectivorous by sticky glands covering
the whole plant.   1. PINGUICULA
Lvs divided into thread-like segments, spirally arranged on the long
stems; insectivorous by bladders borne on the lvs.   2. UTRICULARIA

1. PINGUICULA L.

Perennial herbs with lvs confined to a basal rosette, clothed all over
(except corolla) with sticky glands which catch insects. Lvs entire,
sessile, with involute margins. Fls solitary on naked scapes. Calyx
5-lobed, the lobes unequal. Corolla 2-lipped, upper lip 2-, lower 3-lobed,
spurred, open at mouth. Capsule opening by 2 valves.

1   Fls pale lilac or white, 10 mm. or less (excluding spur); spur
     2–4 mm.   *2*
    Fls bright violet, 10 mm. or more; spur more than 4 mm.   *3*

2   Corolla pale lilac, 6–7 mm.; spur cylindric.   **1. lusitanica**
    Corolla white, 8–10 mm.; spur conic. Formerly in Ross, now
    believed extinct. Alpine Butterwort.   *P. alpina* L.

3   Corolla 10–15 mm.; lobes of lower lip widely separated; spur
     4–7 mm.   **2. vulgaris**
    Corolla 15–20 mm.; lobes of lower lip contiguous or overlapping;
     spur 10 mm. Bogs and wet rocks in Kerry and Cork.
                                     P. grandiflora Lam.

1. P. lusitanica L.   Pale Butterwort.

Overwintering as a rosette. Lvs 1–2 cm., oblong, yellowish, tinged
purple. Scapes 3–15 cm., very slender. Corolla 6–7 mm., pale lilac,
yellow in the throat; lobes of upper lip roundish, of lower emarginate;
spur 2–4 mm., bent downwards, ± cylindric, blunt. Capsule globose.
Fl. 6–10. Local in bogs and on wet heaths in the west from Cornwall
to Hants and northwards to Orkney and Outer Hebrides; throughout
Ireland but rare in the centre.

2. P. vulgaris L.   Common Butterwort.

Overwintering as a rootless bud. Lvs 2–8 cm., ovate-oblong, bright
yellow-green. Scapes 5–15 cm. Corolla 10–15 × c. 12 mm., violet with a
short broad white patch at the mouth; lobes of lower lip deep, much
longer than broad, flat, entire, divergent; spur 4–7 mm., straight or
somewhat curved, slender, acute. Capsule ovoid. Fl. 5–7. Bogs, wet
heaths and among wet rocks. Common throughout the British Is,
except S. England and S. Ireland.

2. UTRICULARIA L.

Perennial rootless herbs overwintering by turions. Stems long, lfy. Lvs divided into thread-like segments, some or all bearing small bladders which trap animals. Fls in short racemes on lfless scapes. Calyx 2-lipped, divided nearly to base; lips entire or obscurely toothed. Corolla-lips entire, lower with a projecting palate, upper smaller. Capsule globose, opening irregularly.

1 Stems of one kind, all bearing green floating lvs furnished with
       numerous bladders.   **1 and 2. vulgaris** agg.
     Stems of two kinds: (*a*) bearing green floating lvs or without
       bladders, and (*b*) colourless, bearing bladders on much reduced
       lvs, often beneath the surface of the substratum.   **2**

2 Lf-segments finely toothed, with bristles on the teeth; bladders con-
       fined (or nearly so) to colourless stems.   **3. intermedia**
     Lf-segments entire, without bristles; bladders present on green lvs.
                                                                4. minor

1 and 2. U. vulgaris agg.   Greater Bladderwort.

Free-floating. Stems 15–45 cm., all alike, bearing green lvs with numerous bladders. Lvs 2–2·5 cm., broadly ovate in outline, pinnately divided; segments finely toothed, with small bristles on the teeth; bladders c. 3 mm. Scape 10–20 cm., 2–8-fld. Corolla 12–18 mm., bright yellow; spur conic. Fl. 7–8. Lakes, ponds, ditches and small pools, often in relatively deep water. Throughout the British Is except Channel Is. Not flowering freely in many parts of its range, so that the following constituent spp. cannot always be distinguished:

1. U. vulgaris L.

Lf-segments with groups of bristles. Fl.-stalks 6–17 mm., rather stout. Upper lip of corolla about as long as palate; lower with margins deflexed almost at right angles. Fr. freely-produced (if the plants fl.). Commonly in base-rich water. Great Britain northwards to Dunbarton and Angus; rare in the west and in Ireland.

2. U. neglecta Lehm.

Like *U. vulgaris* L. but lf-segments with solitary, rarely grouped, bristles; fl.-stalks 11–26 mm., rather slender; upper lip of corolla about twice as long as palate; lower ± flat, somewhat wavy; fls rare; fr. very rare. Ditches, clay-pools, etc., but often in base-poor acid water; northwards to Perth but rare in the east; Ireland.

3. U. intermedia Hayne   Intermediate Bladderwort.

Stems slender, of two kinds: (a) bearing green lvs without (or occasionally with very few) bladders, and (b) colourless, often buried in the substratum, bearing bladders on much reduced lvs. Lvs in 2 ranks, 4–12 mm., circular in outline, palmately divided; segments finely toothed, with 1–2 bristles on each tooth. Bladders c. 3 mm. Fls very rarely produced, 2–4, on a scape 9–16 cm. Corolla 8–12 mm., bright yellow with reddish-brown lines; spur 5–6·5 mm., conic. Fl. 7–9. Very local in lakes, pools and ditches, usually in shallow acid peaty water. Commonest in N. Scotland and W. Ireland, but widely scattered.

4. U. minor L.   Lesser Bladderwort.

Like U. intermedia but with bladders on the green stems as well as the colourless; lvs only 3–6 mm., with entire segments lacking bristles; bladders 2 mm.; scape 4–15 cm.; corolla 6–8 mm., pale yellow; spur very short, blunt. Fl. 6–9. Ponds, ditches, bog- and fen-pools; throughout the British Is, but rather local.

109. ACANTHACEAE (p. xxv)

110. VERBENACEAE

Herbs, shrubs, trees and woody climbers usually with opposite or whorled lvs; stipules 0. Fls usually in many-fld infls, hermaphrodite, ± zygomorphic, hypogynous. Calyx of 5(–4) sepals, sometimes 2-lipped; corolla gamopetalous, commonly 2-lipped; stamens 4, rarely 5 or 2; ovary superior, syncarpous, initially 1-celled but becoming 2-, 4- or 5-celled by ingrowth of the placentae, and most commonly 4-(8- or 10-) celled through growth of additional septa between the placentae; style terminal; ovules usually 1 per cell. Fr. usually a drupe with 1, 2 or 4 stones, sometimes a capsule, rarely of 4 nutlets; seeds usually non-endospermic.

1. VERBENA L.

Herbs or dwarf shrubs with spikes, corymbs or panicles of smallish fls. Calyx tubular below, unequally 5-toothed above; corolla somewhat 2-lipped with 5 spreading lobes, its tube downy within; stamens usually 4, in pairs, not exserted; ovary 4-celled with 1 ovule in each cell. Fr. of 4 nutlets separating when ripe.

1. V. officinalis L.   Vervain.

Perennial, with woody stock and stiffly erect tough stems, 30–60 cm., roughly hairy. Lvs 2–7·5 cm., oblanceolate or rhombic in outline, pinnatifid, with ovate to oblong lobes, lowest sometimes much the largest; upper lvs narrower, less divided; all dull green, roughly hairy. Fls in slender terminal spikes, elongating in fr. Fls 4 mm. diam. Calyx ribbed, shortly hairy. Corolla pale lilac, its tube almost twice as long as calyx. Nutlets reddish-brown. Fl. 7–9. Waysides and waste places. Local throughout England and Wales; Fife; naturalized in Ireland, mainly in southern half.

111. LABIATAE

Usually herbs with ± 4-angled stems and opposite simple lvs; stipules 0. Fls in cymes which are usually contracted so as to form a whorl-like infl., the whorls often close together and forming spikes or heads. Calyx usually 5-toothed, often 2-lipped. Corolla zygomorphic, 4- or 5-lobed, usually 2-lipped. Stamens 4 in 2 pairs, or 2. Ovary superior; style usually gynobasic, branched into 2 above. Fr. of 4 nutlets.

1	Corolla with 4 nearly equal lobes, small.		2
	Corolla 2-lipped or 1-lipped.		3
2	Stamens 4; lvs entire or toothed.	1. MENTHA	
	Stamens 2; lvs pinnately lobed.	2. LYCOPUS	
3	Corolla clearly 2-lipped.		4
	Corolla 1-lipped or the upper lip represented by 2 short teeth.		23
4	Calyx 2-lipped, lips entire, the upper with a scale on the back.	19. SCUTELLARIA	
	Calyx ± equally 5-(rarely10-) toothed or 2-lipped with toothed lips; scale 0.		5
5	Stamens included in the corolla-tube.	18. MARRUBIUM	
	Stamens exserted.		6
6	Stamens 2, each with a unilocular anther.	8. SALVIA	
	Stamens 4, with bilocular anthers.		7
7	Stamens, at least the longer pair, longer than the upper lip of the corolla, diverging.		8
	Anthers placed under the upper lip of the corolla, stamens converging or parallel.		9
8	Fls in a loose terminal panicle; bracteoles ovate, conspicuous.	3. ORIGANUM	
	Fls in a dense head or spike which may be interrupted below; bracteoles linear.	4. THYMUS	
9	Stamens curved, connivent; upper lip of corolla flat (or weakly concave in *Melissa*).		10
	Stamens straight; upper lip of corolla concave (strongly so except in *Melittis*).		13

10 Corolla-tube straight.   *11*
     Corolla-tube curved upwards. Sweet-scented perennial 30–60 cm.
       high with white or pinkish fls, often escaping and sometimes
       naturalized. Balm.   **Melissa officinalis* L.

11 Calyx-tube straight; fls in opposite axillary stalked cymes.
                          5. CALAMINTHA
     Calyx-tube curved; fls in opposite axillary whorls without a com-
       mon stalk.   *12*

12 Whorls many-fld, dense; calyx-tube not gibbous at base; fls rosy-
       purple.   7. CLINOPODIUM
     Whorls 3–8-fld; calyx-tube gibbous at base; fls violet. 6. ACINOS

13 Outer pair of stamens longer than the inner (sometimes curving
       downwards after dehiscence and then appearing shorter).   *14*
     Outer pair of stamens shorter than the inner.   *22*

14 Calyx 2-lipped.   *15*
     Calyx ± equally 5-toothed.   *16*

15 Calyx closed after flowering, lobes of lower lip acute; corolla
       1·5 cm. or less.   9. PRUNELLA
     Calyx broadly bell-shaped, open after flowering, lobes of lower lip
       obtuse; corolla 2·5–4 cm., pink or white and pink. Very local in
       woods and hedge-banks in S. England and Wales. Bastard
       Balm.   *Melittis melissophyllum* L.

16 Lower lvs deeply lobed. Perennial 60–120 cm. with white or pink
       purple-spotted fls c. 12 mm. Sometimes naturalized. Mother-
       wort.   **Leonurus cardiaca* L.
     Lvs not lobed.   *17*

17 Calyx funnel-shaped (i.e. with a spreading rim below the base of
       the teeth), teeth short and broad.   12. BALLOTA
     Calyx tubular or bell-shaped, teeth usually longer.   *18*

18 Lateral lobes of lower lip of corolla short, obscure, with one or
       more small teeth.   14. LAMIUM
     Lateral lobes of lower lip of corolla well-developed.   *19*

19 Corolla deep yellow; stoloniferous perennial.   13. LAMIASTRUM
     Perennials with purple fls or annuals with purple or pale yellow fls.   *20*

20 Corolla with 2 bosses at base of lower lip; upper lip of corolla
       laterally compressed, helmet-shaped; annual.   15 GALEOPSIS
     Corolla without bosses; upper lip concave but not laterally com-
       pressed; annual or perennial.   *21*

21 Perennials or biennials with numerous stem-lvs, or annuals; corolla-
       tube with a ring of hairs within.   10 STACHYS
     Perennial with lvs mostly in a basal rosette and only 2–3 pairs of
       stem-lvs; corolla-tube with scattered hairs within. 11. BETONICA

22   Fls many in each whorl, forming a terminal infl., white.
                                      16. NEPETA
    Fls 2–6 in each whorl in the lf-axils, violet.   17. GLECHOMA

23   Corolla of a single 5-lobed lip; tube glabrous inside.
                                      20. TEUCRIUM
    Upper lip of corolla consisting of 2 short teeth; lower lip con-
    spicuous, 3-lobed; tube with a ring of hairs inside   21. AJUGA

1. MENTHA L.

Perennials with a characteristic pleasing smell and creeping rhizome.
Fls in whorls which often form a spike or head, small. Calyx nearly
equally 5-toothed, 10–13-nerved. Corolla with 4 nearly equal lobes.
Stamens 4, diverging, ± equal.

    Many hybrids or intermediates between the following species (except
M. pulegium) occur. See larger *Flora*.

1   Lvs less than 1 cm. broad, entire or obscurely toothed; throat of
    calyx hairy; the two lower calyx-teeth narrower than the three
    upper.   2
    Lvs more than 1·5 cm. broad; throat of calyx naked; calyx-teeth
    equal.   3

2   Lvs 3–5 mm., ovate-orbicular; stems creeping, filiform; whorls 2–6-
    fld; corolla-tube not gibbous; occasionally naturalized.
                                        *M. requienii Bentham
    Lvs 5–15 mm., oval or oblong; stems not filiform, whorls many-fld;
    corolla-tube somewhat gibbous below the mouth.   **1. pulegium**

3   Whorls all axillary, the axis terminated by lvs or with very few fls in
    the axils of the uppermost pair.   4
    Whorls forming terminal spikes or heads.   5

4   Calyx-teeth scarcely longer than broad; calyx campanulate, hairy;
    stamens normally exserted.   **2. arvensis**
    Calyx-teeth much longer than broad; calyx either ± tubular or
    glabrous; stamens often included.   Hybrids

5   Fls in a head often with axillary whorls below (if calyx-tube glabrous,
    lvs thinly hairy and stamens included, a hybrid).   **3. aquatica**
    Fls in a spike.   6

6   Lvs stalked.   Hybrids
    Lvs sessile or subsessile.   7

7   Pedicels and calyx-tube glabrous; lvs glabrous or near so.
                                              4. spicata
    Pedicels and calyx-tube hairy; lvs ± densely hairy, at least beneath.   8

8 Lvs 3–8 cm., lanceolate, not rugose.   **5. longifolia**
  Lvs 2–4 cm., oblong or suborbicular, rugose.   **6. rotundifolia**
  (Hybrids between **5** and **6** are rather frequent).

1. M. pulegium L.   Penny-royal.

Stems prostrate or less often erect, then up to 30 cm. Lvs 0·8–2 cm., oblong or oval, obtuse, minutely puberulous, obscurely crenate-serrate, shortly stalked. Fls in distant whorls, lilac. Calyx with the 2 lower teeth narrower than the 3 upper, hairy in the throat within and outside. Corolla-tube gibbous. Fl. 8–10. Wet places on sandy soil. Very local from Ayr and Berwick southwards and in Ireland.

2. M. arvensis L.   Corn Mint.

Stems 10–60 cm., erect or ascending, ± hairy. Lvs 2–6·5 cm., suborbicular to lanceolate, usually obtuse, shallowly toothed, stalked, hairy. Fls in distant whorls, lilac. Calyx hairy outside, the teeth scarcely longer than broad. Fl. 8–10. Variable. Arable fields and damp open habitats throughout the British Is; common.

3. M. aquatica L.   Water Mint.

Stems 15–90 cm., ± erect, ± hairy. Lvs 2–6 cm., ± ovate, ± hairy, serrate, stalked. Fls in a terminal head c. 2 cm. across, usually with axillary whorls below, lilac. Pedicels and calyx hairy. Fl. 7–10. Variable. Swamps, marshes, etc., throughout the British Is; common.

*4. M. spicata L.   Spearmint.

Perennial 30–90 cm., with aerial stolons and glabrous erect stems, smelling pungent. Lvs 4–9 cm., lanceolate, acute or acuminate, serrate, glabrous or nearly so, ± sessile. Fls in a terminal cylindrical spike, 3–6 cm., lilac; spikes often clustered. Pedicels and calyx glabrous. Fl. 8–9. Commonly grown as a pot-herb, naturalized in many places by roadsides and in waste places.

5. M. longifolia (L.) Hudson   Horse-mint.

Perennial 60–90 cm., with creeping underground stems and hairy erect aerial ones, smell not pungent. Lvs 3–8 cm., lanceolate, acute or acuminate, serrate, green and hairy above, grey and felted or velvety beneath, not or scarcely wrinkled, sessile. Fls in a terminal cylindrical spike 3–10 cm., lilac, hairy outside. Pedicels and calyx hairy. Fl. 8–9. Scattered over the British Is in damp roadsides and waste places but rather local and doubtfully native.

6. M. rotundifolia (L.) Hudson   Apple-scented Mint.

Stoloniferous perennial 60–90 cm., with hairy erect usually branched stems, densely white-hairy, strongly fragrant. Lvs 2–4 cm., oblong to round, rounded or with a very small point at the apex, hairy above, grey or white velvety beneath, subcordate at base, wrinkled, sessile. Fls in dense terminal spikes 3–5 cm., often forming a panicle, pinkish-lilac, hairy outside. Pedicels and calyx hairy. Fl. 8–9. Ditches, roadsides and waste places, doubtfully native; local in England and Wales, still more so in Scotland and Ireland.

2. LYCOPUS L.

Differs from *Mentha* in the absence of scent; stamens 2.

1. L. europaeus L.   Gipsy-wort.

Perennial with creeping rhizome and stiffly erect stems 30–100 cm. Lvs to 10 cm., ovate-lanceolate to elliptic, shortly stalked, acute, pinnately lobed with numerous triangular lobes. Fls c. 3 mm., in many-fld distant axillary whorls, white with a few small purple dots on the lower lip. Fl. 6–9. River-banks, marshes, etc., common in England, Wales and Ireland, less so in Scotland.

3. ORIGANUM L.

Fls in dense cymes forming a terminal panicle, Bracteoles ovate, imbricate, conspicuous. Calyx 13-nerved with 5 nearly equal teeth. Corolla 2-lipped, Stamens 4, straight, diverging, longer than the corolla.

1. O. vulgare L.   Wild Marjoram.

Hairy aromatic perennial herb of somewhat bushy appearance. Lvs 1·5–4·5 cm., ovate, stalked, entire or obscure toothed. Bracteoles purple. Gynodioecious. Fls 6–8 mm., rose-purple. Fl. 7–9. Dry pastures, hedge-banks, etc., usually calcareous. Common in the south, local in the north.

4. THYMUS L.

Low aromatic shrubs. Fls in few-fld whorls forming a spike or head. Bracteoles minute. Calyx 2-lipped. Corolla and stamens as in *Origanum*.

1   Fl.-stem sharply angled, hairy only on the angles; plant tufted or with the branches ascending; lateral veins of lvs not prominent below when dry; infl. elongated on well-developed plants.

                                               1. pulegioides

Fl.-stem obscurely angled, hairy on at least two sides; plant with
long creeping branches forming a mat; lateral veins of lvs prominent below when dry; infl. usually capitate.   2

2   Fl.-stem equally hairy all round. Breckland of E. Anglia, very
local.   *T. serpyllum* L.
Fl.-stem hairy on 2 sides only or more hairy on 2 sides than on the
other two.   **2. drucei**

1. T. pulegioides L.   Larger Wild Thyme.

Tufted undershrub up to 25 cm. with ascending fl.-branches and short
creeping branches. Fl.-stems sharply 4-sided with long hairs only at the
angles, two opposite faces narrow and shortly pubescent, the other two
broader and glabrous. Lvs 6–10 × 3–6 mm., strongly aromatic, ovate to
elliptic, obtuse, ciliate at base otherwise glabrous, slightly folded upwards about the midrib, shortly stalked. Infl. elongated when normally
developed (short when grazed). Corolla rose-purple. Fl. 7–8. In dry,
usually calcareous grassland, rather common in S.E. England, rare
elsewhere.

2. T. drucei Ronn.   Wild Thyme.

Mat-like undershrub, rarely more than 7 cm., with long creeping
branches, the fl.-stems in rows on the branches. Fl.-stems below infl.
4-angled with two opposite sides densely hairy with hairs of varying
length, the other two sides less hairy or glabrous. Lvs 4–8 × 1·5–4 mm.,
usually only faintly aromatic, nearly round to elliptic or oblanceolate,
obtuse, ciliate otherwise glabrous or hairy, flat, shortly stalked. Infl.
usually capitate, strongly aromatic. Fl. 5–8, earlier than the other spp.
Common in dry grassland, heaths, dunes, screes and among rocks
throughout the British Isles.

5. CALAMINTHA Miller

Perennial herbs. Fls in opposite axillary stalked cymes, bracts like the
cauline lvs but smaller. Calyx 13-nerved, hairy within, tube straight.
Corolla 2-lipped, tube straight, naked within. Stamens 4, shorter than
corolla, curved and convergent.

1   Corolla more than 15 mm.; lvs 3·5–6 cm. Very rare, on chalky
banks in the Isle of Wight. Wood Calamint.   *C. sylvatica* Bromf.
Corolla not more than 15 mm.; lvs rarely more than 4 cm.   2

2   Hairs in throat of calyx included; calyx-teeth long-ciliate; partial
peduncles 0 or very short; lvs on main axis mostly 2–4 cm.
1. ascendens

Hairs in throat of calyx protruding after fl.; calyx-teeth shortly and sparsely ciliate; partial peduncles obvious; lvs on main axis mostly 1–2 cm. **2. nepeta**

1. C. ascendens Jordan   Common Calamint.

Hairy perennial with shortly creeping rhizome and ± numerous erect little-branched stems 30–60 cm. Lf-blades 2–4 cm., ovate, obtuse, with 5–8 shallow teeth on each side, green, with long ± appressed hairs on both surfaces; stalk c. 1 cm. Calyx 6–8 mm., with long hairs and shining sessile glands; upper teeth spreading widely, lower teeth much longer. Corolla lilac with darker spots. Fl. 7–9. Dry banks, usually calcareous, from Yorks and Cumberland southwards, rather local; Ireland except N.E.

2. C. nepeta (L.) Savi   Lesser Calamint.

Similar to *C. ascendens* but rhizome long-creeping and whole plant greyer and more branched so as to appear bushy. Lf-blades 1–2 cm., with 5 or fewer teeth with short scurfy and few long hairs above, more numerous long hairs beneath, stalk usually 5 mm. or less. Calyx 4–6 mm., with short hairs; upper teeth nearly straight, lower teeth little longer. Corolla lilac, scarcely spotted. Fl. 7–9. Dry banks, usually calcareous from Yorks to Pembroke and Kent, local.

6. ACINOS Miller

Allied to *Calamintha* but fls in c. 6-fld axillary whorls, bracts like the lvs. Calyx-tube curved, gibbous at base.

1. A. arvensis (Lam.) Dandy   Basil-thyme.

A branching, rather wiry, somewhat hairy, usually annual herb 10–20 cm. Lvs 5–15 mm., ovate, subacute, stalked, obscurely toothed, glabrescent. Fls 7–10 mm., forming a loose terminal infl., violet with white markings on the lower lip. Fl. 5–9. Arable fields and bare ground in grassland, etc., on dry and usually calcareous soils, throughout Great Britain, rather local especially in the north; S.E. and central Ireland.

7. CLINOPODIUM L.

Allied to *Calamintha* but fls in remote many-fld terminal and axillary whorls, bracts like the lvs. Calyx-tube curved, not gibbous, glabrous or nearly so within.

1. C. vulgare L.   Wild Basil.

Hairy ±odourless perennial with shortly creeping rhizome and numerous erect, nearly simple stems 30–80 cm. Lf-blades 2–5 cm., ovate, stalked, subobtuse, shallowly toothed. Fls 15–20 mm., rose-purple. Fl. 7–9. Hedges, etc., less often in grassland, on dry, usually calcareous, soils. From Inverness southwards, local in the north, common in the south; a rare alien in Ireland.

8. Salvia L.

Fls in axillary whorls forming a ± interrupted terminal spike, the bracts differentiated. Calyx 2-lipped. Corolla 2-lipped. Stamens 2, the filaments short, the connective very long with one anther cell aborted.

1   Upper lip of calyx with conspicuous ±equal teeth; corolla-tube
      with ring of hairs within (rare alien of waste places).
                                                            *S. verticillata L.
      Upper lip of calyx with very short teeth, the two lateral connivent
      over the middle one; corolla-tube without a ring of hairs.   *2*

2   Calyx pubescent and glandular but without long white hairs; corolla
      10–25 mm., the smaller fls female only, upper lip glandular out-
      side. Calcareous grassland in S. and E. England, rare.
                                                            S. pratensis L.
      Calyx glandular and pilose with long white hairs; corolla 6–15 mm.,
      fls all hermaphrodite, the smaller cleistogamous; upper lip not
      glandular outside.   *3*

3   Basal lvs usually less than twice as long as broad; corolla with
      2 white spots at the base of the lower lip.   **1. horminoides**
      Basal lvs usually more than twice as long as broad; corolla with-
      out white spots. Dunes, Guernsey.   *S. verbenaca* L.

1. S. horminoides Pourret   Wild Clary.

Pubescent perennial with large basal rosette of lvs and little-branched stems 30–80 cm., with smaller lvs, glandular and purple-tinged above. Basal lvs 4–12 cm., oblong or ovate, obtuse, rugose, variously lobed or toothed, stalked. Fls violet-blue, usually small, cleistogamous and with the corolla shorter than the calyx but sometimes open and up to 15 mm. Dry pastures and roadsides from Ross southwards, local; S.E. Ireland.

9. Prunella L.

Fls in few-fld whorls forming a dense terminal oblong infl. Bracts orbicular, very different from lvs. Calyx strongly 2-lipped. Upper lip of corolla very concave. Stamens 4, straight, the outer pair the longer.

Upper lvs entire or shallowly toothed; fls normally violet; sinus between
upper calyx-teeth gradually rounded. **1. vulgaris**
Upper lvs lyrate or pinnatifid; fls normally cream; sinus between upper
calyx-teeth parallel-sided. Dry calcareous grassland in S. England,
very local and probably introduced. *P. laciniata* L.

1. P. vulgaris L.
Self-heal.

Sparingly pubescent, rather untidy perennial herb with short rhizome
and flowering stems 5–30 cm. Lf-blades 2–5 cm., ovate, entire or
shallowly toothed, stalked. Bracts and calyx with long white hairs,
usually purplish. Fls 10–14 mm., violet, rarely pink or pure white. Very
common in grassland, clearings in woods and waste places throughout
the British Isles.

10. STACHYS L.

Herbs without a well-marked basal rosette and with numerous stem-lvs.
Fls in whorls, the upper forming a terminal spike. Calyx with 5 narrow
± equal teeth. Upper lip of corolla concave, tube with a ring of hairs
within. Stamens 4, straight, the outer pair the longer and diverging
laterally from the corolla after flowering, anther cells divaricate.

1 Fls yellowish-white; annual 10–30 cm.; lvs oblong. Fls in 3–6-fld
whorls 11–13 mm. long. Usually a casual, occasionally naturalized. **S. annua* (L.) L.
Fls purple. *2*

2 Annual; corolla 7 mm. or less. **1. arvensis**
Perennial or biennial; corolla 12 mm. or more *3*

3 Stem and lvs densely clothed with long white silky hairs. A robust
usually biennial plant of hedge-banks and rough ground now
confined to Oxfordshire. *S. germanica* L.
Lvs green. *4*

4 Bracteoles nearly as long as calyx-tube; calyx-teeth broad, less than
half as long as tube. A robust perennial of open woods in Glos.
and Denbigh. *S. alpina* L.
Bracteoles very short, scarcely longer than pedicel; calyx-teeth
narrow, more than half as long as tube. *5*

5 Lvs lanceolate, with short petioles or subsessile; fls dull purple.
2. palustris
Lvs ovate, with long petioles; fls claret-coloured. **3. sylvatica**

1. S. arvensis (L.) L.
Field Woundwort.

A hairy annual 10–25 cm., with slender ascending stems usually
branched at the base. Lvs 1·5–3 cm., ovate, obtuse, truncate or cordate
at base, crenate-serrate, stalked. Fls pale purple in 2–6-fld whorls in

the axils of lf-like bracts which become smaller upwards forming a lax spike. Corolla 6–7 mm. Fl. 4–11. Arable fields on non-calcareous soils throughout the British Isles, rather common in the west, local in the east.

2. S. palustris L.   Marsh Woundwort.

Green, stiffly hairy, odourless perennial 40–100 cm., with long creeping rhizome, producing small tubers at apex in autumn. Stems hollow, not or slightly branched. Lvs 5–12 cm., oblong-lanceolate, toothed, very shortly (5 mm. or less) or not stalked. Fls dull purple in c. 6-fld whorls forming a spike dense above, interrupted below. Calyx not or sparingly glandular, hairy. Corolla 12–15 mm. Fl. 7–9. Common by streams and ditches and in swamps and fens throughout the British Is.

3. S. sylvatica L.   Hedge Woundwort.

Green, stiffly hairy, perennial 30–100 cm., foetid when bruised, tubers 0. Stems solid, often branched. Lvs 4–9 cm., ovate, toothed, long-stalked. Fls claret-coloured in c. 6-fld whorls forming an interrupted spike. Calyx glandular, hairy. Corolla 13–15 mm. Fl. 7–8. Common in shady places on the richer soils throughout the British Is.

11. BETONICA L.

Differs from *Stachys* in having a well-marked basal rosette and few stem-lvs, corolla-tube usually without a ring of hairs, outer stamens not diverging from the corolla after flowering and anther-cells nearly parallel.

1. B. officinalis L.   Betony.

A green somewhat hairy perennial 15–60 cm. with a short woody rhizome and stiff, usually simple stems with 2–3 pairs of lvs. Lvs 3–7 cm., mostly in a basal rosette, ± oblong, cordate at base, obtuse, crenate, on very long petioles which become much shorter up the stem. Bracts ovate or lanceolate, only the lowest pair somewhat lf-like. Fls bright reddish-purple in a spike, dense above, often interrupted below. Corolla c. 15 mm., hairy outside, less so within. Fl. 6–9. Open woods, hedgebanks, grassland and heaths, usually on the lighter soils. Common in England and Wales, local in Scotland (absent from the north) and Ireland.

12. BALLOTA L.

Perennial herbs. Fls in many-fld axillary whorls, the bracts lf-like. Calyx funnel-shaped, 10-nerved with 5 broadly ovate, acuminate,

± equal teeth. Upper lip of corolla somewhat concave; tube shorter than calyx with a ring of hairs within. Stamens 4, parallel, the outer pair the longer; anther-cells diverging.

1. B. nigra L. Black Horehound.

Hairy perennial 40–100 cm. with a short stout rhizome and branched erect stems, smelling unpleasant. Lvs 2–5 cm., ovate, coarsely crenate, stalked. Fls purple, in dense many-fld whorls. Corolla 12–18 mm., hairy. Fl. 6–10. Common on roadsides and hedge-banks in England and Wales, local in Scotland (absent from the north) and Ireland.

13. LAMIASTRUM Heister ex Fabr.

Stoloniferous perennial herbs. Fls in dense axillary whorls, the bracts lf-like. Calyx tubular, campanulate, with ± equal teeth. Corolla 2-lipped; upper lip laterally compressed, helmet-shaped; lower lip 3-lobed, the middle lobe only slightly larger than the lateral; tube longer than calyx, dilated above, with a ring of hairs within. Anther-cells divaricate, glabrous.

1. L. galeobdolon (L.) Ehrend. & Polatsch.
    (*Galeobdolon luteum* Hudson) Yellow Archangel.

Sparingly hairy perennial herb 20–60 cm., with long lfy runners produced at or after flowering. Lvs of fl-stems 4–7 cm., ovate, acute or acuminate, truncate or rounded at base, irregularly toothed, stalked; lvs of runners shorter and relatively broader. Corolla c. 2 cm., yellow with brownish markings. Fl. 5–6. Common in woods, usually on the heavier soils, in England and Wales; S. Scotland, doubtfully native; S.E. Ireland.

14. LAMIUM L.

Annual or perennial herbs. Fls in dense axillary whorls, the bracts lf-like. Calyx tubular or tubular-campanulate with 5 nearly equal mucronate teeth. Corolla 2-lipped, the upper lip laterally compressed, forming a hood; lower lip 3-lobed, the lateral lobes very small, each with a small tooth (thus differing from all the other genera); tube dilated above. Anther-cells divaricate, hairy.

1 Annuals; corolla 15 mm. or less; tube not suddenly contracted near
    base.   2
  Perennials; corolla 2 cm. or more; tube suddenly contracted near
    base.   5
2 Bracts, at least the upper, sessile, differing somewhat from the lvs.   *3*
  Bracts stalked, resembling the lvs.   *4*

3 Calyx in fl. 7 mm. or less, tube densely clothed with somewhat
     spreading white hairs, teeth connivent in fr.   **1. amplexicaule**
   Calyx in fl. 8 mm. or more, tube with a moderate number of stiff
     appressed hairs, teeth spreading in fr.   **2. molucellifolium**
4 Lvs irregularly cut; corolla-tube without or with a faint ring of hairs
     within.   **3. hybridum**
   Lvs crenate-serrate; corolla-tube with a conspicuous ring of hairs
     towards the base within.   **4. purpureum**
5 Corolla white; tube with an oblique ring of hairs within towards the
     base.   **5. album**
   Corolla purple; tube with a transverse ring of hairs towards the base;
     lvs usually with large whitish blotch. Garden escape, sometimes
     ±naturalized. Spotted Dead-nettle.   *L. maculatum* L.

1. L. amplexicaule L.   Henbit.

Finely pubescent annual 5–25 cm., with ascending branches from base.
Lvs 1–2·5 cm., ±orbicular, obtuse, truncate, rounded or subcordate at
base, shallowly crenately lobed; stalk long (3–5 cm. on lower lvs).
Bracts like the lvs but sessile and clasping the stem, often larger.
Whorls few, rather distant. Calyx 5–6 mm., tubular, clothed with
white, ±spreading hairs; teeth usually rather shorter than tube, connivent in fr. Corolla when well-developed c. 15 mm., pinkish-purple,
long-exserted but more frequently small, included and cleistogamous;
tube glabrous within. Fl. 4–8. Cultivated ground, usually on light dry
soils through nearly the whole British Is, but local in Ireland.

2. L. molucellifolium Fries   Intermediate Dead-nettle.

Like *L. amplexicaule* but usually more robust; lvs more triangular;
bracts not clasping the stem; infl. denser; calyx 8–12 mm. in fl., with
appressed hairs, teeth longer than tube, spreading in fr.; corolla-tube
scarcely longer than calyx with a faint ring of hairs within. Fl. 5–9.
Cultivated ground from Huntingdon and Derby northwards; N. and
W. Ireland.

3. L. hybridum Vill.   Cut-leaved Dead-nettle.

Like *L. purpureum* but more slender and less hairy; lvs often smaller,
the upper truncate at base and ±decurrent down the stalk, all irregularly
cut; corolla-tube less exserted, without or with a faint ring of hairs
within towards the base. Fl. 3–10. Cultivated ground; spread over the
whole British Is, but local.

4. L. purpureum L. Red Dead-nettle.

Hairy annual 10–45 cm., branched from the base, often purple-tinged. Lvs 1–5 cm., ovate, obtuse, cordate at base, regularly crenate-serrate, stalked. Bracts similar but rounded or cordate at base, stalked or the upper subsessile. Infl. rather dense. Calyx 5–6 mm., tubular-campanulate, hairy, teeth about as long as tube, spreading in fr. Corolla 10–15 mm., pinkish-purple, tube longer than calyx, with a ring of hairs near the base within. Fl. 3–10. Very common in cultivated ground and waste places throughout the British Is.

5. L. album L. White Dead-nettle.

Hairy perennial 20–60 cm., with creeping rhizome and erect stems. Lvs 3–7 cm., ovate, acuminate, cordate at base, coarsely toothed, stalked. Bracts similar. Fls white, in mostly distant whorls. Corolla c. 2 cm., tube with an oblique ring of hairs near the base within. Fl. 5–12. Common in hedge-banks and waste places in England, becoming very rare in the west and in N. Scotland; introduced in E. Ireland.

15. GALEOPSIS L.

Annual herbs. Fls in dense terminal and axillary whorls, bracts like the lvs but often smaller. Calyx with 5 somewhat unequal spiny-pointed teeth. Corolla 2-lipped; upper laterally compressed, helmet-shaped; lower lip 3-lobed with 2 conical projections at base; tube longer than calyx, straight, dilated above, with a ring of hairs within. Anther-cells parallel, superposed, opening separately, ciliate.

1 Stem softly hairy or nearly glabrous, not swollen at nodes. 2
  Stem bristly, swollen at nodes. 4

2 Lvs and calyx hairy but not silky; fls rosy purple. 3
  Lvs and calyx velvet-silky; fls pale yellow. Arable land in a few
    places in England and Wales. *G. segetum* Necker

3 Lvs linear-lanceolate to oblong-lanceolate with 1–4 teeth on each
    side. **1. angustifolia**
  Lvs ovate or ovate-lanceolate with 3–8 teeth on each side. Rarely
    naturalized in cultivated or waste ground. **G. ladanum* L.

4 Corolla 13–20 mm., pink, purple or white, very rarely pale yellow
    with a violet spot, tube nearly always scarcely exceeding calyx.
    2. tetrahit
  Corolla c. 30 mm., pale yellow, usually with a violet spot on lower
    lip, tube much exceeding calyx. **3. speciosa**

1. G. angustifolia Hoffm.   Narrow-leaved Hemp-nettle.

Stem 10–80 cm., hairy or nearly glabrous, not swollen at nodes. Lvs
1·5–8 cm., linear-lanceolate to oblong-lanceolate, acute, tapered at base
into a short stalk, often abundantly appressed-hairy with 1–4 small teeth
on each side. Calyx tubular, often with appressed whitish hairs, without or with few glands. Corolla 1·5–2·5 cm., bright rosy purple. Rather
common in arable land in England and Wales, absent from N. Scotland;
E. Ireland.

2. G. tetrahit L., *sensu lato*   Common Hemp-nettle.

Stem 10–100 cm., bristly and with red- or yellow-tipped glandular hairs.
Lvs 2·5–10 cm., ± ovate, acuminate, cuneate at base, toothed, hairy,
stalked. Calyx bristly with rather prominent veins. Corolla 13–20 mm.,
usually pink, purple or white with network of darker markings on lower
lip, rarely yellow with violet lower lip; tube scarcely longer than calyx
(rarely much longer in plants with the calyx-teeth much shorter than
usual). Fl. 7–9. Common in arable land, less common in woods, fens
and wet heaths throughout the British Is.

3. G. speciosa Miller   Large-flowered Hemp-nettle.

Like *G. tetrahit* but usually more robust; glandular hairs always yellow-
tipped; veins of calyx less prominent; corolla 22–34 mm., pale yellow
with lower lip mostly violet (rarely yellow throughout), tube about
twice as long as calyx. Fl. 7–9. Arable land, often on black peaty
soil, local but scattered over the whole British Is.

16. NEPETA L.

Perennial herbs. Fls in many-fld axillary whorls forming a terminal infl.
Calyx tubular, 15-nerved, 5-toothed. Upper lip of corolla flat; tube
rather suddenly curved and dilated at the middle, glabrous within.
Stamens 4, outer pair shorter than inner; anther cells divergent, opening
by a common slit.

1. N. cataria L.   Cat-mint.

A strongly scented grey-pubescent perennial 40–100 cm. with branched
stems. Lvs 3–7 cm., ovate, cordate at base, coarsely serrate, white-
tomentose beneath; bracts similar but much smaller. Upper whorls
crowded, lower more widely spaced. Calyx-teeth narrow, the upper the
longest, the two lower the shortest. Corolla c. 12 mm., white with small
purple spots. Fl. 7–9. Hedge-banks and roadsides, usually on calcareous
soil, rather local in England and Wales; very rare in S. Scotland;
scattered over Ireland except the S.W., but introduced.

17. GLECHOMA L.

Allied to *Nepeta* but fls in 2–4-fld secund axillary whorls, bracts not differing from lvs; corolla-tube straight, hairy within at base of lower lip; anther-cells at right-angles, each opening by a separate slit.

1. G. hederacea L.   Ground Ivy.

Hairy perennial 10–30 cm., with long creeping and rooting stems and ascending flowering branches. Lvs 1–3 cm. diam., kidney-shaped or ovate-cordate, obtuse, crenate, long-stalked. Gynodioecious, the hermaphrodite fls larger than the female. Corolla 15–20 mm., violet, with purple spots on the lower lip. Fl. 3–5. Woods, grassland and waste places, mainly on the damper and heavier soils, common throughout nearly the whole British Isles (rare in N. Scotland).

18. MARRUBIUM L.

Perennial herbs. Fls in many-fld axillary whorls, bracts like the lvs. Calyx 10-nerved, 5- or 10-toothed. Upper lip of corolla nearly flat; tube shorter than calyx, naked or nearly so within. Stamens and style included in corolla-tube. Anther-cells diverging.

1. M. vulgare L.   White Horehound.

White-tomentose perennial 30–60 cm., with short stout rhizome and branched stems. Lvs 1·5–4 cm., ± orbicular, crenate, obtuse, rugose, green above, white beneath, stalked. Whorls broader than high. Calyx with 10 small, hooked teeth. Corolla 1·5 cm., whitish. Fl. 6–11. Local on downs, waste places and roadsides in England, S. Scotland and S. Ireland.

19. SCUTELLARIA L.

Perennial herbs. Fls in axillary pairs sometimes forming a terminal raceme. Calyx 2-lipped, the lips entire, the upper with a small scale on the back. Corolla 2-lipped, the lateral lobes small. Stamens 4, the outer pair the longer, anthers of outer pair 2-celled, of inner 1-celled.

1 Fls 10–20 mm., blue-violet.   *2*
    Fls 6–10 mm., pale pinkish-purple.   **2. minor**

2 Lvs cordate at base, toothed; fls not forming a well-marked raceme;
    corolla-tube slightly curved.   **1. galericulata**
    Lvs hastate at base, otherwise entire; fls forming a short terminal
    raceme; corolla-tube strongly curved; introduced in a wood in
    Norfolk.   **S. hastifolia* L.

1. S. galericulata L.   Common Skull-cap.

Perennial 15–50 cm., with creeping rhizome and erect stems. Lvs 2–5 cm., ovate-lanceolate or oblong-lanceolate, obtuse or subacute, cordate at base, somewhat toothed, shortly stalked. Bracts like the lvs but gradually decreasing in size upwards. Corolla 10–20 mm., blue-violet, several times as long as calyx, tube slightly curved below. Fl. 6–9. Common on edges of streams and in fens and water-meadows throughout the British Isles.

2. S. minor L.   Lesser Skull-cap.

Smaller in all its parts than *S. galericulata*. Stems 10–15 cm. Lvs 1–3 cm., toothed only near the base. Bracts entire, often truncate at base. Corolla 6–10 mm., pale pinkish-purple with darker spots, 2–4 times as long as calyx, tube nearly straight. Fl. 7–10. Wet heaths, etc., throughout most of the British Is, but absent in N.E. Scotland and N.E. Ireland.

20. Teucrium L.

Herbs or undershrubs. Calyx 5-toothed, 10-nerved. Corolla of one 5-lobed lip (the lower), the 4 lateral lobes short, the middle lobe large. Stamens 4, exserted, the outer pair the longer.

1 Fls in terminal racemes, the bracts very different from the lvs; upper tooth of calyx much larger than others; corolla pale yellow-ish green.   **1. scorodonia**
   Fls in axillary whorls, sometimes forming a terminal spike, bracts like the lvs though sometimes smaller; calyx-teeth equal or nearly so; corolla purple.   2

2 Lvs pinnatifid; a rare annual or biennial of bare chalky ground in S.E. England. Cut-leaved Germander.   *T. botrys* L.
   Lvs toothed; perennials.   3

3 Whorls forming a terminal spike, upper bracts shorter than fls; a low tufted almost woody plant sometimes naturalized on walls. Wall Germander.   **T. chamaedrys* L.
   Whorls distant, bracts all longer than fls. A creeping plant with ascending fl.-stems. Calcareous river banks in E. England and W. Ireland and dune slacks in Devon. Water Germander.
                                                    T. scordium L.

1. T. scorodonia L.   Wood Sage.

Hairy perennial 15–30 cm., with creeping rhizome and erect branched stems. Lvs 3–7 cm., ovate, cordate at base, subacute, crenate, rugose, stalked. Bracts ovate-lanceolate, entire, shorter than fls. Calyx c.

5 mm., the upper tooth much the largest. Corolla pale yellowish-green; tube c. 8 mm., lip 5–6 mm., deflexed. Fl. 7–9. Woods, grassland, heaths and dunes usually on acid or slightly basic soils. Common in Great Britain, less common in Ireland.

21. AJUGA L.

Herbs. Calyx nearly equally 5-toothed, 10-nerved. Corolla with a very short upper lip and conspicuous 3-lobed lower lip; tube ±exserted, with a ring of hairs within. Stamens exserted.

1   Lvs divided into 3 linear lobes; corolla yellow. A very local annual
      of bare chalky ground in S.E. England. Ground Pine.

A. chamaepitys L.

    Lvs toothed or entire; corolla blue or violet.   *2*

2   With aerial stolons; stem hairy on two opposite sides.   **1. reptans**
    Without aerial stolons; stem hairy all round.   *3*

3   Upper bracts shorter than fls; basal lvs withering before fl.-time.
      Rarely naturalized in England.   ***A. genevensis* L.
    Bracts all longer than fls; basal lvs persistent. Rock-crevices in
      Westmorland, Scotland and West Ireland, rare. *A. pyramidalis* L.

1. A. reptans L.   Bugle.

Perennial 10–30 cm., with short rhizome and long lfy rooting aerial stolons. Stems simple, hairy on two opposite sides. Basal lvs 4–7 cm., forming a rosette, glabrescent, obovate or oblong, entire or obscurely toothed, obtuse, tapered into a long stalk; stem-lvs few, shorter, subsessile. Upper bracts shorter than fls, ±blue-tinged. Corolla blue, rarely pink or white. Fl. 5–7. Common in damp woods and fields throughout the British Is.

112. PLANTAGINACEAE

Herbs. Lvs usually all basal and spirally arranged. Scapes arising in lf-axils. Fls small, actinomorphic, usually hermaphrodite and 4-merous, arising in the axils of bracts and usually arranged in racemose heads or spikes. Perianth ±scarious; corolla tubular; stamens generally inserted on corolla. Ovary superior. Fr. a circumscissile capsule or 1-seeded and indehiscent.

Terrestrial; stolons 0; fls hermaphrodite, many together; fr. dehiscent.

1. PLANTAGO

Aquatic; stolons present; fls unisexual, male solitary, female few
    together; fr. indehiscent.   2. LITTORELLA

1. Plantago L.

Fls in heads or spikes, mostly hermaphrodite. Stamens inserted on the corolla-tube. Fr. dehiscent.

1	Stem long; lvs opposite; lower bracts with lf-like tips. Rare alien.
	*P. indica L.
	Stems short; lvs spirally arranged, forming a basal rosette. 2
2	Lvs not very narrow or pinnatifid; corolla-tube glabrous. 3
	Lvs very narrow or pinnatifid; corolla-tube hairy. 5
3	Scape deeply furrowed; corolla-lobes with brown midrib.
	3. lanceolata
	Scape not furrowed; corolla-lobes without midrib. 4
4	Lvs abruptly narrowed at base; stalk usually as long as blade; scape scarcely longer than lvs. **1. major**
	Lvs gradually narrowed at base, stalk much shorter than blade; scape much longer than lvs. **2. media**
5	Lvs 3–5(–7)-nerved, never pinnatifid; corolla-lobes with a brown midrib. **4. maritima**
	Lvs 1-nerved, usually pinnatifid; corolla-lobes without a midrib.
	5. coronopus

1. P. major L.   Rat-tail Plantain.

Lvs 10–15(–30) cm., ovate or elliptic, entire or irregularly toothed. Infl. (1–)10–15(–50) cm. Bracts acute, brownish with a green keel. Fls c. 3 mm., corolla yellowish-white, lobes triangular. Anthers at first lilac then dirty yellow. Fr. 8–16-seeded. Fl. 5–9. Farmyards, cultivated ground, etc., widely distributed. Variable.

2. P. media L.   Hoary Plantain.

Finely hairy. Lvs 4–6(–30) cm., elliptic to ovate, weakly and irregularly toothed, 5–9-ribbed. Infl. 2–6(–8) cm.; scape often 30 cm. Bracts acute, membranous at margins. Fls c. 2 mm., scented; corolla whitish, lobes lanceolate, blunt. Anthers lilac or white. Fr. usually 4-seeded. Fl. 5–8. Widely distributed on base-rich soils; rare in Scotland.

3. P. lanceolata L.   Ribwort.

Lvs (2–)10–15(–30) cm., lanceolate to ovate-lanceolate, entire or slightly toothed, 3–5-ribbed; stalk usually c. ½ as long as blade. Infl. 1–2(–5) cm.; scape much longer than lvs, deeply furrowed. Bracts ovate-acuminate. Fls c. 4 mm., brownish. Stamens white. Fr. 2-seeded. Fl. 4–8. Common except on the poorest soils.

4. P. maritima L. Sea Plantain.

Usually glabrous. Lvs 5–15(–30) cm., narrow, fleshy, entire or slightly toothed, faintly 3–5-nerved, rarely up to 15 mm. broad and 7-nerved. Infl. 2–6(–12) cm.; scape as long as or longer than lvs. Bracts ovate, ± blunt. Fls c. 3 mm.; 2 lower sepals united, not winged; corolla-lobes with a rather broad indistinct midrib. Stamens pale yellow. Fr. 2-seeded. Fl. 6–8. In salt-marshes and short maritime turf and on mountains, widely distributed.

5. P. coronopus L. Buck's-horn Plantain.

Usually hairy. Lvs 2–6 cm., narrow, very variable but most often 1–2-pinnatifid, 1-nerved. Infl. 0·5–4 cm.; scape somewhat longer than lvs. Bracts ovate, often long-acuminate, sometimes obtuse. Fls c. 3 mm.; 2 lower sepals united and winged; corolla-lobes without a midrib. Stamens pale yellow. Fr. 3–4-seeded. Fl. 5–7. On light soils and in rock-crevices, common near the sea.

2. LITTORELLA Bergius

Aquatic, flowering only when exposed. Fls monoecious. Stamens with long filaments. Fr. indehiscent.

1. L. uniflora (L.) Ascherson Shore-weed.

Stoloniferous, often forming a turf in shallow water. Lvs 2–10(–25) cm., usually ½-cylindrical and narrow, with a sheathing base, borne in rosettes. Scape usually shorter than lvs, slender. Male fl. 5–6 mm.; stamens 1–2 cm. Female fls 4–5 mm., subsessile; style c. 1 cm. Fl. 6–8. In shallow water of non-calcareous lakes and ponds.

113. CAMPANULACEAE

Mostly herbs with latex. Lvs usually spirally arranged, simple, exstipulate. Fls hermaphrodite, actinomorphic. Corolla gamopetalous, usually bell-shaped. Stamens alternate with corolla-lobes and inserted low down. Ovary inferior, rarely superior. Fr. a capsule or fleshy.

1 Fls solitary or in racemes or panicles; corolla-lobes usually shorter
    than tube, ovate or broadly triangular.   2
  Fls very numerous, in heads or spikes; corolla-lobes much longer
    than tube, narrow.   4

2 Stem creeping; lvs all similar, long-stalked.   1. WAHLENBERGIA
  Stem erect or ascending; upper lvs much smaller or narrower than
    basal, sessile or nearly so.   3

3   Ovary and capsule not or little longer than wide.   2. CAMPANULA
    Ovary and capsule at least 3 times as long as wide.   3. LEGOUSIA

4   Plant nearly or quite glabrous; fl. buds curved (local, on chalk in
        south England).   4. PHYTEUMA
    Plant hairy; fl. buds straight (on lime-free soils).   5. JASIONE

1. WAHLENBERGIA Schrader

Calyx-tube hemispherical or oblong-obconic. Corolla bell-shaped or
almost wheel-shaped. Capsule dehiscing by 2–5 loculicidal valves.

1. W. hederacea (L.) Reichenb.   Ivy Campanula.

Creeping, glabrous perennial up to 30 cm. Lf-blade 5–10 mm., suborbicular, angled or obscurely lobed. Fls solitary, ± nodding. Corolla
6–10 mm., pale blue. Fl. 7–8. Damp and peaty places in the south and
west, local but extending to Argyll.

2. CAMPANULA L.

Calyx-tube ovoid or subglobose. Corolla bell- or wheel-shaped. Capsule dehiscing by lateral pores or valves.

1   Calyx with broadly cordate, reflexed appendages between the teeth;
        stigmas 5; stiffly hairy. Naturalized on railway banks. Canter-
        bury Bell.   *C. medium L.
    Calyx without appendages; stigmas 3.   2

2   Fls sessile or nearly so.   **4. glomerata**
    Fls distinctly stalked.   3

3   Middle stem-lvs ovate (2–4 times as long as broad).   4
    Middle stem-lvs narrower (5–10 times as long as broad).   6

4   Calyx-teeth awl-shaped to narrowly lanceolate, spreading; corolla
        not more than 30 mm.   **3. rapunculoides**
    Calyx-teeth ovate-lanceolate, erect; corolla (25–)30–55 mm.   5

5   Stem sharply angled, bristly; lower lvs strongly cordate at base;
        corolla 25–35 mm.   **2. trachelium**
    Stem bluntly angled, glabrous or with short soft hairs; lower lvs
        rounded or decurrent at base; corolla 40–55 mm.   **1. latifolia**

6   Fls nodding; basal lvs orbicular, cordate; lower stem-lvs stalked.
                                                       5. rotundifolia
    Fls erect or suberect; basal lvs neither orbicular nor cordate; lower
        stem-lvs sessile.   7

7   Glabrous; corolla 25–35 mm. Locally naturalized.
                                                       *C. persicifolia L.
    Hairy or scabrid; corolla not more than 25 mm.   8

8 Root fibrous; basal lvs narrowed gradually at base; bracteoles
about the middle of the fl. stalks. Very local.   *C. patula* L.
Root fleshy; basal lvs abruptly narrowed at base; bracteoles at
base of fl. stalks. Locally naturalized. Rampion.

C. rapunculus L.

1. C. latifolia L.   Large Campanula.

Erect perennial 50–120 cm. Lvs ovate to ovate-oblong, toothed,
acuminate. Infl. simple or with short branches. Fls suberect; stalks c.
2 cm. Calyx nearly or quite glabrous, teeth narrowly triangular. Corolla
blue-purple or white, lobes suberect, ciliate. Capsule 12–15 mm., ovoid,
nodding. Fl. 7–8. Woods and hedge-banks, local and much commoner
in the north than the south.

2. C. trachelium L.   Bats-in-the-Belfry.

Erect perennial 50–100 cm. Lower lvs broadly ovate, upper narrower,
coarsely and irregularly toothed, acuminate. Infl. a lfy panicle with
short branches. Fls suberect; stalks c. 1 cm. Calyx usually stiffly hairy,
teeth triangular. Corolla blue-purple, lobes suberect, ciliate. Capsule c.
7 mm., hemispherical, nodding. Fl. 7–9. Woods and hedge-banks,
rather local.

*3. C. rapunculoides L.   Creeping Campanula.

Slightly hairy erect perennial 30–60 cm., with slender underground
stolons. Lower lvs stalked, ovate, cordate or rounded at base, upper
sessile, narrower. Infl. a raceme or panicle; bracts small. Fls nodding;
stalks c. 5 mm. Calyx with appressed hairs, teeth narrow. Corolla blue-
purple, lobes spreading, ciliate. Capsule c. 7 mm., hemispherical,
nodding. Fl. 7–9. Naturalized in waste ground, etc., and a bad weed in
gardens.

4. C. glomerata L.   Clustered Bellflower.

Erect downy perennial 3–20(–60) cm. Lower lvs stalked, ovate, ± cordate and toothed, upper narrower and sessile. Infl. subcapitate,
often with a few short branches. Fls erect, sessile. Corolla blue-purple,
rarely white, lobes suberect, later spreading. Capsule erect. Fl. 5–9.
Usually in grassland on calcareous soils, locally common.

5. C. rotundifolia L.   Harebell, Bluebell (in Scotland).

Slender perennial 15–40(–60) cm., with underground stolons. Basal
lvs c. as broad as long, lower stem-lvs lanceolate, upper very narrow.
Infl. ± branched or sometimes of a single terminal fl. Buds erect, fls

nodding, stalks very slender. Calyx-teeth very narrow, spreading. Corolla blue, rarely white. Capsule subglobose, nodding. Fl. 7–9. Dry grassy places, often on poor shallow soils.

3. LEGOUSIA A. DC.

Like *Campanula* but ovary and capsule much elongated and corolla ± wheel-shaped.

1. L. hybrida (L.) Delarbre   Venus' Looking-glass.

Annual 5–30 cm., stiffly hairy. Lvs sessile, oblong or oblong-ovate. Fls erect, mostly in terminal few-fld cymes. Corolla 8–15 mm. diam., scarcely ½ as long as calyx, red-purple or lilac. Capsule 15–30 mm., subcylindrical. Fl. 5–8. Arable fields.

4. PHYTEUMA L.

Two very local spp. may be distinguished as follows:

Infl. capitate; fls violet. Locally abundant on chalk in S. England.
                                                P. tenerum R. Schulz
Infl. spicate; fls yellowish. Sussex only.   *P. spicatum* L.

5. JASIONE L.

Fls small, numerous, sessile or subsessile in a terminal head. Corolla blue, rarely white, divided nearly to base. Anthers joined for a short distance at their base. Capsule dehiscing loculicidally by 2 short valves within the persistent calyx-teeth.

1. J. montana L.   Sheep's-bit.

5–50 cm., hairy, spreading or ascending. Lvs narrowly oblong to lanceolate, ciliate, all on the lower part of the stem. Infl. 5–35 mm. diam., depressed globose, surrounded by an involucre of numerous bracts. Calyx-teeth awl-shaped. Corolla c. 5 mm., persistent. Fl. 5–8. In rough grassy places on light lime-free soils, locally abundant.

114. LOBELIACEAE

Mostly herbs with latex. Lvs alternate, simple, exstipulate. Fls zygomorphic. Corolla gamopetalous, 1–2-lipped. Anthers cohering in a tube round the style.

1. LOBELIA L.

Fls in bracteate racemes or solitary in lf-axils. Corolla-tube curved, split to base along back, 2-lipped.

Terrestrial; stems lfy; lvs toothed; fls erect or spreading, blue. Very
    local in S. England.   *L. urens* L.
Aquatic; stems lfless except for a few small scales; lvs entire; fls nodding,
    pale lilac.   **1. dortmanna**

1. L. dortmanna L.   Water Lobelia.

20–60 cm., glabrous, stoloniferous. Stems slender, hollow. Lvs 2–4(–8) cm., narrow, recurved. Racemes few-fld, rising above the water. Calyx-teeth ovate, blunt, erect. Corolla 15–20 mm. Fl. 7–8. Stony lakes and tarns with acid water; locally common in mountain districts.

115. RUBIACEAE

Woody plants or herbs with opposite and decussate or whorled simple entire lvs whose stipules may stand between the lvs (interpetiolar) or between lf and stem. Fls usually small, in terminal and axillary cymes; usually actinomorphic, hermaphrodite. Sepals usually free, often very small or represented only by a ring-like ridge; corolla tubular below; stamens equalling the corolla-lobes in number and alternating with them, inserted on the corolla-tube; ovary inferior, usually 2-celled, with 1 to many ovules on an axile placenta in each cell. Fr. a capsule, berry or drupe, or dry and splitting into 1-seeded halves; seeds endospermic.

The British spp. all belong to tribe Galieae with 2 or more lf-like interpetiolar stipules, so that there appear to be 4 or more lvs in a whorl, of which, however, only 2 have axillary buds. The fr. is usually of 2 separating indehiscent 1-seeded parts.

1   Fls in small heads with a basal involucre of lf-like bracts; corolla-
      tube at least twice as long as the free lobes.   *2*
    Fls not in involucrate heads; corolla-tube not twice as long as the
      free lobes, often much shorter.   *3*

2   Calyx of 4(–6) distinct teeth, persistent in fr.; corolla 4–5 mm., pale
      lilac; annual.   **1. SHERARDIA**
    Calyx an inconspicuous ridge; corolla 10 mm., whitish; perennial.
                                2. *Asperula taurina*

3   Corolla pinkish with tube at least as long as the free lobes.
                                  2. ASPERULA

Corolla white, yellow or greenish, with tube usually shorter than
the free lobes, often very short.   *4*

4   Corolla with 4(–3) free lobes; fr. dry.   3. GALIUM
    Corolla with 5 free lobes; fr. a berry.   4. RUBIA

1. SHERARDIA L.

Lvs 4–6 in a whorl. Fls few in small terminal head-like clusters with a
basal involucre of lfy bracts. Calyx with 4–6 distinct sepals; corolla
funnel-shaped with 4 free lobes; stamens 4; style bifid. Fr. dry.

1. S. arvensis L.   Field Madder.

Annual with numerous prostrate or decumbent spreading stems,
5–40 cm., ±glabrous, with rough angles. Lower lvs 4 in a whorl,
obovate-cuspidate, soon withering; upper 5–18 mm., 5–6 in a whorl,
elliptic-acute, the margins and underside of the midrib rough with
forwardly directed bristles. Fls 3 mm. diam., 4–8 in terminal heads.
Involucre of 8–10 bracts longer than the fls. Sepals 4–6, enlarging in fr.
Corolla 4–5 mm., pale lilac, funnel-shaped, with a long slender tube
about twice as long as the 4 free lobes. Fr. 4 mm., crowned by the sepals,
of 2 obovoid halves rough with short appressed bristles. Fl. 5–10.
Arable fields and waste places. Common throughout the British Is.

2. ASPERULA L.

Usually perennial herbs. Fls in heads or panicles. Calyx an indistinct
ridge or of 4–5 small teeth not persisting in fr.; corolla funnel-shaped, its
tube longer than the 4(–3) free lobes; styles ±joined. Fr. of two dry
1-seeded separating parts.

Upper lvs linear, 4–6 in a whorl; fls pink, long-stalked.   **1. cynanchica**
Upper lvs ovate-lanceolate, 4 in a whorl; fls white tinged with yellowish-
  pink, in involucrate head-like infls. Introduced. Naturalized locally.
  Pink Woodruff.   **A. taurina* L.

1. A. cynanchica L.   Squinancy Wort.

Perennial, with a woody non-creeping stock producing numerous slender
prostrate or ascending shoots, 8–40 cm., much branched, 4-angled,
glabrous. Lower lvs elliptical to obovate, 4 in a whorl, withering early;
middle and upper lvs 6–25 mm., linear, 4(–6) in a whorl, often very
unequal; all mucronate, glabrous, firm, recurved. Fls 3–4 mm. diam., in
lax, long-stalked, few-fld terminal and axillary corymbs. Corolla 6 mm.,

funnel-shaped, the 4(–3) acute lobes almost as long as the tube, white within, pale pink-lilac and rough on the outside. Fr. 3 mm., densely tubercled and somewhat wrinkled. Fl. 6–7. Gynodioecious. Dry calcareous pastures and calcareous sand-dunes. Locally abundant in S. England and S. Wales and extending northwards to Yorks and Westmorland; S.W. Ireland, to Waterford and Galway.

3. GALIUM L.

Annual to perennial, with whorls of 4–10 lvs. Fls small, hermaphrodite or polygamous. Calyx a minute ridge; corolla with tube shorter than the 4(3–5) free lobes and usually very short; stamens 4; styles 2, joined below. Fr. of two dry 1-seeded separating parts, with the seed adherent to the ovary-wall.

1	Lvs 3-veined, in whorls of 4.	*2*
	Lvs 1-veined, or with strong midrib and weaker diverging laterals.	*3*
2	Fls yellow; fr. smooth, glabrous.	**2. cruciata**
	Fls white; fr. rough with hooked bristles.	**3. boreale**
3	Fls funnel-shaped, white, with tube almost equalling the free lobes.	
		1. odoratum
	Fls rotate, i.e. with free lobes spreading horizontally from the top of the very short tube, yellow or white.	*4*
4	Lvs blunt or acute but never mucronate.	*5*
	Lvs cuspidate or mucronate.	*6*
5	Lvs linear-lanceolate to broadly oblanceolate; submerged autumn lvs similar to lvs on aerial shoots; infl. pyramidal or oblong, its branches soon spreading.	**8. palustre**
	Lvs narrowly linear; submerged lvs very narrow, limp, up to 2 cm.; infl. obconical; branches erect-ascending throughout. Very local on pond-margins in Devon, Hants and Channel Is. Slender Marsh Bedstraw.	*G. debile* Desv.
6	Stems markedly rough with recurved prickles on the angles.	*7*
	Stems smooth or slightly rough on the angles.	*11*
7	A perennial plant of wet peaty places; fls white.	**9. uliginosum**
	Annual plants not usually in wet places; fls green, cream or reddish, not pure white.	*8*
8	Fls in axillary 3-fld cymes shorter than the subtending lvs; fr.-stalks strongly recurved.	**11. tricornutum**
	Fls in axillary cymes longer than the subtending lvs; fr.-stalks straight.	*9*
9	Fls 2 mm. diam.; fr. 4–6 mm., covered with hooked hairs with swollen bases.	**11. aparine**

Fls not more than 1 mm. diam.; fr. 1–3 mm., smooth or with hairs
   with non-swollen bases.   10

10   Lvs with forwardly directed marginal prickles; fls 0·5 mm. diam.,
      reddish outside; fr. 1 mm., glabrous, granulate. Plant with
      decumbent or ascending weak slender glabrous stems and lvs
      becoming reflexed. Rare and local in S.E. England. Wall
      Bedstraw.   *G. parisiense* L.
   Lvs with backwardly directed marginal prickles; fls about 1 mm.
      diam., greenish-white; fr. 1·5–3 mm., with white hooked hairs,
      rarely glabrous. Doubtfully native. A rare and local weed of
      arable land in S. England to Stafford and Leicester. False
      Cleavers.   *G. spurium* L.

11   Fls yellow.   12
   Fls white.   13

12   Fls golden yellow.   **5. verum**
   Fls very pale yellow.   *G.* × *pomeranicum* Retz. (*verum* × *mollugo*)

13   Robust decumbent to erect plants with fls in large terminal pyra-
      midal panicles; corolla-lobes long-cuspidate; fr. wrinkled.
                                                                    4. mollugo
   Slender ± decumbent plants with fls in small axillary and terminal
      corymbs; corolla-lobes acute; fr. with small acute tubercles or
      low dome-shaped papillae.   14

14   Lvs on fl.-shoots obovate to oblanceolate, their margins with
      forwardly directed prickles.   **6. saxatile**
   Lvs on fl.-shoots oblanceolate to linear, their margins with few to
      many backwardly directed prickles.   15

15   Fr. covered with acute tubercles.   **7. sterneri**
   Fr. covered with low dome-shaped papillae. A local plant of chalk
      and limestone grassland in S. England and northwards to
      Lincoln. Slender Bedstraw.   *G. pumilum* Murray

1. G. odoratum (L.) Scop. (*Asperula odorata* L.)   Sweet Woodruff.
Perennial, hay-scented when dried, with slender branched far-creeping
rhizomes and erect simple 4-angled stems, 15–45 cm., hairy beneath the
nodes, otherwise glabrous. Lvs firm, in distant whorls, lanceolate or
elliptical, ± cuspidate, glabrous but with forwardly directed marginal
prickles; lowest lvs small, 6 in a whorl; middle lvs 2·5–4 × 0·6–1·5 cm.,
6–8(9) in a whorl. Fls about 6 mm. diam., in long-stalked terminal and
lateral cymes forming a ± umbellate infl. Corolla 4–6 mm., 4-lobed to
about half-way, pure white. Fr. 2–3 mm., rough with hooked black-
tipped bristles. Fl. 5–6. Locally abundant in woods on damp well-
drained calcareous or base-rich soils. Throughout most of the British Is.

2. G. cruciata (L.) Scop. Crosswort, Mugwort.

Perennial, with slender creeping stock and slender 4-angled hairy decumbent or ascending stems, 15–70 cm., much branched near the base. Lvs up to 2·5 cm., ovate-elliptical, 3-veined, hairy on both sides, 4 in a whorl, yellowish-green. Fls 2–2·5 mm. diam. in axillary c. 8-fld cymes usually shorter than the subtending lvs; terminal fls hermaphrodite, lateral male. Corolla pale yellow. Fr. 1·5 mm., almost globose, glabrous, smooth, ultimately blackish, on recurved stalks. Fl. 5–6. Open woodland and scrub, hedges, waysides and pastures, especially on calcareous soils. Throughout Great Britain northwards to Moray and Inverness; Inner Hebrides; very local in Ireland and probably introduced.

3. G. boreale L. Northern Bedstraw.

Perennial, with creeping stock and erect rigid 4-angled stems, 20–45 cm., glabrous or pubescent. Lvs 1–4 cm., 4 in a whorl, lanceolate or elliptical, 3-veined, rough on the margins and the underside of the midrib, turning black when dried. Fls 3 mm. diam., in a terminal pyramidal lfy panicle. Corolla white, 4-lobed. Fr. 2·5 mm., olive-brown, densely hispid with hooked bristles. Fl. 7–8. Rocky slopes, streamsides, moraine, scree and shingle, stable dunes, etc. Locally common in Wales and N. Britain northwards from Lancashire and Yorks, and in W. and N. Ireland; Inner Hebrides; doubtfully native in Orkney and Shetland.

4. G. mollugo L. (incl. *G. erectum* Hudson) Great Hedge Bedstraw.

Perennial, with stout stock and decumbent to erect, glabrous or pubescent, 4-angled stems, 25–120 cm., not blackening when dried. Lvs 8–25 mm., 6–8 in a whorl, linear to obovate, mucronate to cuspidate, 1-veined, rough on the margins with stout forwardly directed ± appressed prickles. Fls 3–4 mm. diam., in rather lax cymes forming a lax terminal panicle. Corolla white, its lobes cuspidate. Fr. 1–2 mm., glabrous, finely wrinkled, blackening when dry; fr.-stalks spreading. Fl. 6–7 and 9. Very variable. Hedge-banks, open woodland, scrub, grassy slopes, pastures, etc.; common in the south and reaching Sutherland and Orkney; scattered throughout Ireland. The forms on dry calcareous soils often have narrower lvs, a more erect habit and smaller flowers.

5. G. verum L. Lady's Bedstraw.

Perennial, stoloniferous, with slender stock and erect to decumbent, glabrous or sparsely pubescent, bluntly 4-angled stems, 15–100 cm., blackening when dried. Lvs 6–25 × 0·5–2 mm., 8–12 in a whorl, linear, mucronate, 1-veined, dark green and rough above, pale and pubescent beneath, with revolute margins. Fls 2–4 mm. diam., in a terminal lfy panicle. Corolla bright yellow, its lobes apiculate. Fr. 1·5 mm., smooth, glabrous, blackening when dry. Fl. 7–8. In grassland on all but the most acid soils, hedge-banks, stable dunes, etc. Abundant throughout the British Is.

6. G. saxatile L. (*G. harcynicum* Weigel)   Heath Bedstraw.

Perennial mat-forming herb with numerous prostrate non-flowering and decumbent or ascending flowering shoots, 10–20(–30) cm., 4-angled, glabrous, smooth, turning black when dried. Lvs 7–10 mm., 6–8 in a whorl, obovate on non-flowering and often narrower on flowering shoots, mucronate, with small straight forwardly-directed marginal prickles. Fls 3 mm. diam., in few-fld cymes shorter than the stem-internodes and forming a cylindrical panicle. Corolla pure white, its 4 free lobes acute, not cuspidate. Fr. c. 2 mm., glabrous, covered with high acute tubercles. Fl. 6–8. Heaths, moors, grasslands and woods on acid soils. Common throughout the British Is.

7. G. sterneri Ehrendorfer (*G. pumilum* Murray, p.p.)

Sterner's Bedstraw.

Perennial mat-forming herb with some prostrate non-flowering and many erect or ascending flowering shoots, 15–25 cm., glabrous or hairy, much branched at base giving a tangled growth; lf-whorls crowded below. Plant turning dark green when dried. Lvs on flowering shoots 10–14 mm., 6–8 in a whorl, oblanceolate to linear, mucronate, with many curved backwardly-directed marginal prickles. Fls 3–4 mm. diam., in compact cymes longer than the stem-internodes and forming a pyramidal panicle. Corolla creamy white. Fr. 1·25 mm., glabrous, covered with acute tubercles. Fl. 6–7. On grassy calcareous slopes and on calcareous or basic rocks; common on Carboniferous Limestone in Brecon, N. England and W. Ireland, occasional on other rocks from N. Wales and Northumberland northwards to Sutherland and Orkney and in N.E. Ireland.

8. G. palustre L.   Marsh Bedstraw.

Perennial, with slender creeping stock and decumbent to ascending shoots, 15–120 cm., 4-angled, glabrous, smooth or more usually rough-ish on the angles, weak, turning black when dried. Lvs 0·5–2(–3·5) cm., 4–6 in a whorl, usually broadly oblanceolate, blunt or subacute but never mucronate, usually rough on the margins with small backwardly directed prickles. Fls 3–4·5 mm. diam., in spreading cymes forming a loose pyramidal panicle. Corolla white, with 4 acute lobes. Fr. 1·2–

1·7 mm., glabrous, wrinkled, turning black. Fl. 6–7. Very variable in size. Marshes, fens, flushes, ditches, stream-banks, etc. Common throughout the British Is.

9. G. uliginosum L. Fen Bedstraw.

Perennial, with slender creeping stock and weak decumbent or ascending glabrous 4-angled stems, 10–60 cm., very rough on the angles with downwardly-directed prickles, not blackening when dried. Lvs 0·5–1(–1·5) cm., (4–)6–8 in a whorl, linear-oblanceolate, mucronate, glabrous, the margins rough with backwardly-directed prickles. Fls 2·5–3 mm. diam., in small corymbs forming a narrow panicle. Corolla white, with 4 acute lobes. Fr. 1 mm., glabrous, finely wrinkled, turning dark brown; fr.-stalks deflexed. Fl. 7–8. Fens. Locally frequent throughout Great Britain northwards to Ross; in Ireland northwards to Mayo and Meath, and mostly near the coast.

10. G. tricornutum Dandy (G. tricorne auct.)

Rough Corn Bedstraw.

Annual, with decumbent or ascending-scrambling stems, 10–60 cm., glabrous, sharply 4-angled, the angles very rough with downwardly-directed prickles. Lvs 2–3 cm., 6–8 in a whorl, linear-lanceolate, mucronate, glabrous, with strong backwardly-directed hooked marginal prickles. Fls 1–1·5 mm. diam., in 1–3-fld stalked axillary cymes. Corolla cream-coloured, with 4 acute lobes. Fr. 3–4 mm., wider than the corolla, pale, often of a single spherical half-fr., granulate with large papillae, not bristly, their stalks strongly recurved. Fl. 6–9. Doubtfully native. Cornfields, chiefly on calcareous soils. Local in England and Wales; perhaps introduced more recently further northwards to Moray.

11. G. aparine L. Goosegrass, Cleavers, Hairif.

Annual, with prostrate or scrambling-ascending diffusely branched stems, 15–120 cm., glabrous or hairy above the nodes, 4-angled, the angles very rough with downwardly-directed prickles. Lvs 12–50 mm., 6–8 in a whorl, linear-oblanceolate or narrowly elliptical, the margin with prickles backwardly-directed except those near the tip. Fls 2 mm. diam., in stalked 2–5-fld axillary cymes, the stalks bearing a whorl of lf-like bracts. Corolla greenish-white, with 4 acute lobes. Fr. 4–6 mm., wider than the corolla, olive or purplish, covered with white hooked bristles with swollen bases, their stalks spreading. Fl. 6–8. Hedges, waste places, drained fen-peat, limestone scree, maritime shingle, etc. Abundant throughout the British Is.

4. RUBIA L.

Perennials, often woody below. Fls yellowish, in terminal and axillary cymes. Calyx a narrow ridge; corolla with very short tube and 5 horizontally spreading free lobes; stamens 5; styles 2, joined below. Fr. succulent, usually globose and 1-seeded, derived from only 1 half of the ovary.

1. R. peregrina L.   Wild Madder.

Evergreen plant with long creeping stock and trailing or scrambling-ascending glabrous stems, 30–120 cm., woody below, sharply 4-angled above, the angles rough with downwardly-directed prickles. Lvs 1·5–6 cm., 4–6 in a whorl, ovate- to elliptic-lanceolate, rigid, leathery, shining above, the margins and midrib beneath rough with curved bristles. Fls 5 mm. diam. Corolla pale yellowish-green, with 5 long-cuspidate lobes. Fr. 4–6 mm. diam., subglobose, black. Fl. 6–8. Hedges, thickets, scrub and on stony ground. Only in S. and S.W. England and coastal counties of Wales northwards to Caernarvon; Ireland.

116. CAPRIFOLIACEAE

Shrubs, rarely herbs, with opposite lvs; stipules 0 or small. Fls hermaphrodite. Calyx often small. Corolla gamopetalous, lobes imbricate. Stamens inserted on the corolla-tube, equal in number and alternate with the corolla-lobes, or rarely 1 stamen missing. Ovary inferior, 2–5-celled, with 1–many axile ovules. Fr. various. Seeds endospermic.

1	Lvs pinnate.	**1. SAMBUCUS**
	Lvs simple.	*2*
2	Fls in compound cymes.	**2. VIBURNUM**
	Fls in short, spike-like racemes.	**3. SYMPHORICARPOS**
	Fls in pairs, heads or whorls.	*3*
3	Erect or climbing shrub; fls sessile.	**4. LONICERA**

Prostrate creeping undershrub with roundish lvs, 5–15 mm. and small pink bell-shaped fls in stalked pairs. A rare plant of woods or shaded rocky places in Scotland.   *Linnaea borealis* L.

1. SAMBUCUS L.

Deciduous shrubs or small trees, rarely herbs. Lvs pinnate; stipules present or 0. Fls in compound umbel-like or panicle-like cymes, regular, 5-merous. Calyx very small. Corolla white with a short tube and flat spreading lobes. Ovary 3–5-celled with 1 ovule per cell. Style short with 3–5 stigmas. Fr. a drupe.

1 Cymes flat-topped; fr. black.   *2*
    Cymes ovoid, panicle-like; fr. scarlet. A shrub to 4 m.; stipules represented by large glands. Introduced and sometimes naturalized, especially in Scotland.   *S. racemosa* L.
2 Herb; stipules conspicuous.   **1. ebulus**
    Shrub; stipules very small or 0.   **2. nigra**

1. S. ebulus L.   Danewort.

Perennial foetid glabrous herb, 60–120 cm., with creeping rhizome and numerous stout erect grooved stems. Lflets 5–15 cm., 7–13, oblong, acuminate, sharply serrate; stipules conspicuous, ovate. Infl. flat-topped, 7–10 cm. diam., with 3 primary rays. Anthers purple. Fr. black. Fl. 7–8. Local by roadsides and in waste or damp grassy places northwards to Orkney; Ireland.

2. S. nigra L.   Elder, Bourtree.

Shrub or small tree to 10 m., often with vigorous erect shoots from the base; branches often arching. Bark brownish-grey, deeply furrowed, corky. Twigs stout, greyish, with prominent lenticels. Lflets 3–9 cm., usually 5–7, ± ovate-elliptical, acuminate, sparsely hairy on the veins beneath, serrate; stipules 0 or very small and narrow. Infl. flat-topped, 10–20 cm. diam., with 5 primary rays. Fls c. 5 mm. diam. Anthers cream. Fr. 6–8 mm., black. Fl. 6–7. Woods, scrub, roadsides and waste places; especially characteristic of disturbed, base-rich and nitrogen-rich soils; rabbit-resistant. Throughout the British Is.

2. VIBURNUM L.

Shrubs or small trees. Lvs simple. Stipules 0 or small. Fls in compound umbel-like cymes, regular, 5-merous. Calyx-teeth very small. Corolla funnel-shaped or bell-shaped with flat spreading lobes. Ovary 1-celled with 1 ovule. Stigmas 3, sessile. Fr. a drupe.

Fls all alike; lvs finely toothed; buds naked.   **1. lantana**
Outer fls sterile and much larger than inner; lvs lobed; buds scaly.
                                       2. opulus

1. V. lantana L.   Wayfaring Tree, Mealy Guelder Rose.

Deciduous shrub, 2–6 m. Twigs and naked buds greyish, scurfily stellate-pubescent. Lvs 5–10 cm., ovate, ± acute, cordate at base, serrulate, rugose and sparingly stellate-pubescent above, densely stellate-tomentose beneath; stipules 0. Infl. 6–10 cm. diam., umbel-like, dense, short-stalked. Fls all alike and fertile. Corolla c. 6 mm., cream-white. Fr. c. 8 mm., ovoid, at first red, changing quickly to black. Fl. 5–6. Scrub, woods and hedges on calcareous soils; common in S. England but probably not native north of Yorks.

2. V. opulus L.   Guelder Rose.

Deciduous shrub, 2–4 m. Twigs greyish, glabrous, slightly angled. Buds scaly. Lvs 5–8 cm., with 3(–5) acuminate irregularly toothed lobes, glabrous above, sparingly pubescent beneath; stalk with discoid glands close below the blade; stipules awl-shaped. Infl. 5–10 cm. diam., rather loose, its stalk 1–4 cm. Inner fls fertile, c. 6 mm. diam.; outer sterile, 15–20 mm., white. Fr. c. 8 mm., subglobose, red. Fl. 6–7. Woods, scrub and hedges, especially on damp soils; common in England, Wales and Ireland, less so in Scotland but reaching Caithness.

3. SYMPHORICARPOS Duhamel

Deciduous shrubs. Buds scaly. Stipules 0. Fls regular, 4–5-merous. Calyx small. Ovary with 2 fertile cells, each with 1 ovule, and 2 sterile cells with numerous ovules. Style slender. Fr. a berry.

*1. S. rivularis Suksdorf   Snowberry.

Shrub 1–3 m., spreading underground, with many slender erect stems. Twigs slender, yellowish-brown, glabrous. Lvs of twigs 2–4 cm., ovate, blunt, entire or a few lobed, dull green beneath; lvs of sucker shoots often conspicuously lobed. Fls 3–7 in terminal spike-like racemes. Corolla 5–6 mm., bell-shaped, pink, hairy at the throat within. Fr. 1–1·5 cm., globose, white. Fl. 6–9. Introduced. Commonly planted and naturalized in many places. Native of western N. America.

4. LONICERA L.

Deciduous, rarely evergreen, shrubs or woody climbers. Buds scaly. Stipules 0. Fls sessile in axillary long-stalked pairs or in heads or whorls. Calyx with 5 small teeth. Corolla in British spp. strongly 2-lipped with 4-lobed upper and entire lower lip. Stamens 5. Ovary 2-celled with numerous axile ovules. Style slender. Fr. a few-seeded berry.

1   Upright shrub; fls in pairs, c. 1 cm., yellowish often tinged with red;
    corolla-tube short. Very rare in Sussex; introduced elsewhere.
    Fly Honeysuckle.   *L. xylosteum* L.
  Woody climbers; fls in heads or whorls; corolla 4–5 cm., with long
    tube and spreading limb.   *2*

2 Lvs all free; bracts small; corolla glandular.   **1. periclymenum**
   Upper lvs and bracts connate in pairs; bracts large; corolla not
   glandular. Introduced in several places.   **L. caprifolium* L.

1. L. periclymenum L.   Honeysuckle.

Twining shrub to 6 m., but often low and trailing or scrambling, glabrous or somewhat pubescent. Lvs 3–7 cm., ovate, dark green above,
glaucous beneath, usually acute, short-stalked. Fls in terminal heads.
Bracts small, not exceeding ovary. Corolla cream-white within, turning
darker after pollination, purplish or yellowish and glandular outside.
Fr. red, globose. Fl. 6–9. Woods, hedges, scrub and shady rocks, to
2000 ft.; common throughout the British Is.

117. ADOXACEAE

A single sp. of obscure relationships. Fls with 2–3-lobed calyx (or
bracts) and a 4–5-lobed corolla (or perianth). Stamens epipetalous, 4–5,
but each divided to the base with a single anther-lobe to each half.
Ovary 3–5-celled, half-inferior, tapered above into the styles; 1 pendulous ovule per cell. Fr. a drupe; seed endospermic.

1. Adoxa L.

1. A. moschatellina L.   Moschatel, Townhall Clock.

A glabrous perennial herb 5–10 cm. Rhizome far-creeping, with fleshy
whitish scales at its tip. Basal lvs ternate, long-stalked, light green,
dull above, glossy beneath; lflets 1–3 cm., long-stalked, ternate or
deeply 3-lobed with the divisions often further lobed. Fl.-stems erect,
unbranched; stem-lvs 2, opposite, 8–15 mm., short-stalked, ternate or
3-lobed. Infl. of 5 light green fls in a close terminal head, the top fl. with
2-lobed calyx, 4-lobed corolla and 4 stamens, the lateral fls with 3-lobed
calyx, 5-lobed corolla and 5 stamens. Fr. green, rarely produced. Fl.
4–5. Woods, hedge-banks and mountain-rocks on fairly rich substrata;
ascending to 3600 ft. Throughout Great Britain; Antrim only in Ireland.

118. VALERIANACEAE

Herbs, often with strong-smelling rhizomes. Lvs opposite or basal,
entire or pinnatifid, exstipulate. Infl. cymose. Fls generally small, often
zygomorphic. Calyx often forming a feathery pappus in fr. Corolla
funnel-shaped. Stamens 1–4, inserted towards base of corolla-tube.
Ovary inferior, with 1 pendulous ovule.

1   Stamens 3; corolla without a long spur at base, lilac or pale pink.   2
    Stamen 1; corolla with a long spur at base, bright red or, less often,
       white.   3. CENTRANTHUS

2   Annual; corolla not saccate at base, lilac; calyx not forming a
       feathery pappus in fr.   1. VALERIANELLA
    Perennial; corolla saccate at base, pale pink; calyx forming a
       feathery pappus in fr.   2. VALERIANA

1. VALERIANELLA Miller

Small herbs with apparently dichotomous branching. Fls solitary in the
forks of the branches and in terminal cymose heads. Calyx a toothed or
funnel-shaped rim, sometimes almost 0. Fr. of one 1-seeded cell and
two, sometimes small, sterile cells.

Ripe fr. is essential for the determination of the spp.

1   Fr. flattened or nearly quadrangular in section; calyx in fr. very
       small and inconspicuous.   2
    Fr. ovoid or oblong, flat on one face, convex on the other; calyx in
       fr. distinct.   3

2   Fr. flattened, nearly as broad as long inside view; fertile cell corky on
       back.   1. locusta
    Fr. oblong, nearly quadrangular in section, more than twice as long
       as broad; fertile cell not corky on back. Rare.   V. carinata Loisel.

3   Fr. oblong; calyx deeply 5–6 toothed, as broad as fr., strongly
       veined. Rare.   V. eriocarpa Desv.
    Fr. ovoid; calyx nearly entire or with 1 tooth much larger than
       others, not more than ⅓ as broad as fr., scarcely veined.   4

4   Calyx minutely toothed, c. ⅓ as broad as fr.; sterile cells together
       larger than fertile. Local.   V. rimosa Bast.
    One tooth of calyx much larger than others, calyx ½ as broad as fr.;
       sterile cells much smaller than fertile.   2. dentata

1. V. locusta (L.) Betcke   Lamb's Lettuce, Corn Salad.

Slender erect nearly glabrous annual 7–40 cm. Stems much-branched,
weakly angled, rather brittle. Lvs 2–7 cm., lower spathulate, upper
oblong. Calyx very small. Corolla small. Fr. c. 2.5×2 mm. Fl. 4–6.
Arable land, hedge-banks and dunes, widely distributed.

2. V. dentata (L.) Poll.

Like 1 but calyx with one prominent tooth and fr. c. 2×1.5 mm. Fl. 6–7.
Locally common in cornfields.

2. VALERIANA L.

Herbs, mostly with bitter taste and peculiar smell. Infl. terminal, sub-capitate in fl., branches lengthening in fr. Calyx-teeth inrolled in fl., enlarging and forming a pappus in fr. Corolla-tube with a slight bulge at base on one side, lobes unequal. Ovary apparently of one carpel.

1   Basal lvs pinnate with a terminal lflet.   **1. officinalis**
    Basal lvs entire.   2

2   Blades of lower lvs 2–5 cm., cuneate to rounded at base; stolons
      present; fls unisexual.   **2. dioica**
    Blades of lower lvs 10–15 cm., cordate at base; stolons 0; fls
      hermaphrodite. Locally naturalized.   **V. pyrenaica* L.

1. V. officinalis L.   Valerian.

20–150 cm., nearly glabrous. Lvs up to 20 cm., all pinnate; lflets lanceolate, entire or irregularly toothed. Corolla c. 5 mm. diam., pale pink. Very variable. Fl. 6–8. Rough grassy and bushy places, widely distributed.

2. V. dioica L.   Marsh Valerian.

15–30(–50) cm., nearly glabrous. Basal lvs elliptic to ovate, blunt and quite entire, long-stalked; stem-lvs pinnatifid, sessile or nearly so. Plant dioecious. Male fls c. 5 mm. diam., female half the size. Fl. 5–6. In wet places, widely distributed but commoner in the north.

3. CENTRANTHUS DC.

Fls in terminal panicled cymes. Calyx inrolled in fl., enlarging and forming a pappus in fr. Corolla-tube with a spur or conical projection at base. Ovary apparently of one carpel.

*1. C. ruber (L.) DC.   Red Valerian.

30–80 cm., somewhat glaucous. Lvs c. 10 cm., ovate or ovate-lanceolate, entire or upper toothed. Fls c. 5 mm., red or white; corolla-tube 8–10 mm., slender; spur 1–2 times length of ovary. Fl. 6–8. Abundantly naturalized on old walls, cliffs, etc., particularly in the south and west.

119. DIPSACACEAE

Usually herbs with opposite or whorled lvs; stipules 0. Infl. usually a head with a calyx-like involucre of bracts at its base. Fls zygomorphic, hermaphrodite or ± male-sterile, each surrounded at its base by an epicalyx or *involucel* of united bracteoles and often subtended by a bract borne on the receptacle. Calyx small, cup-shaped or ± deeply cut into 4–5 segments or into numerous teeth or hairs; corolla with a basal tube and 4–5 lobes which are ± equal or form 2 lips; stamens 4 or 2, borne on the corolla-tube and alternating with the corolla-lobes, exserted, with free anthers; ovary inferior, 1-celled with 1 ovule; style slender; stigma simple or 2-lobed. Fr. dry, indehiscent, 1-seeded, enclosed in the involucel and often surmounted by the persistent calyx; seeds endospermic.

1   Stem spiny or prickly; involucral and receptacular bracts spine-
        tipped.   1. DIPSACUS
    Stem not spiny or prickly; involucral and receptacular bracts not
        spine-tipped.   2

2   Receptacle hemispherical, hairy; receptacular bracts 0; calyx of
        8(–16) teeth or bristles.   2. KNAUTIA
    Receptacle elongated, not hairy; receptacular bracts present; calyx
        of 5 teeth or bristles.   3

3   Stem-lvs pinnatifid or pinnate; marginal fls larger than the central;
        corolla 5-lobed.   3. SCABIOSA
    Stem-lvs entire or faintly toothed; fls equal; corolla 4-lobed.
                                                          4. SUCCISA

1. DIPSACUS

Biennial or perennial herbs, often large, with spiny or prickly stems, lvs in opposite pairs, often connate, and fls in cylindrical or conical heads. Involucre of 1–2 rows of narrow erect or spreading spine-tipped bracts, usually much longer than the ± spine-tipped receptacular bracts. Involucel 4-angled below, a very short ± 4-lobed cup above. Calyx a 4-angled cup, ciliate above. Corolla with long tube and 4 unequal lobes. Fr. crowned by the persistent calyx.

Basal lvs oblong, short-stalked; stem-lvs with connate bases; head
    conical, usually pale lilac.   **1. fullonum**
Basal lvs ovate, long-stalked; stem-lvs not connate; head spherical,
    white.   **2. pilosus**

1. D. fullonum L.   Teasel.

Biennial, with stout yellowish tap-root and erect glabrous angled stems, 50–200 cm., prickly on the angles. Basal lvs in a rosette, dying early in 2nd season, oblong, short-stalked, entire, glabrous but with scattered prickles; stem-lvs narrowly lanceolate, connate at the base into a water-collecting cup, prickly only on the midrib beneath, entire, crenate or distantly toothed. Heads 3–8 cm., conical, blunt, always erect; fls rose-purple, rarely white. Fl. 7–8. The wild teasel is ssp. **sylvestris** (Huds.) Clapham, with involucral bracts curving upwards, the longest at least equalling the head, receptacular bracts exceeding the fls and ending in a long straight spine, corolla-tube 9–11 mm. Copses, stream-banks, road-sides, rough pasture, etc., especially on clay soils; locally common in the south and extending northwards to Perth; in Ireland local and mainly in the south and east.

The fullers' teasel, ssp. **fullonum**, has the involucral bracts spreading hori-zontally, receptacular bracts almost equalling the fls and ending in a stiff recurved spine, and the corolla-tube 13 mm. It is still cultivated and occa-sionally escapes.

2. D. pilosus L.   Small Teasel.

Biennial with erect angled and furrowed stem, 30–120 cm., with sparse weak prickles on the angles. Basal lvs in a rosette, ovate, long-stalked, crenate-toothed, hairy; stem-lvs short-stalked, ovate to narrowly ellipti-cal, simple or the upper ones with a basal pair of small lflets. Heads 2–2·5 cm. diam., spherical, long-stalked, at first drooping, then erect. Involucral bracts spreading or slightly reflexed, narrowly triangular, with sparse long silky hairs. Receptacular bracts equalling the fls, abruptly contracted into the straight spiny point, long-ciliate. Corolla 6–9 mm., whitish. Anthers dark violet. Fr. 5 mm., hairy. Fl. 8. Damp woods, hedge-banks, ditch-sides, etc., especially on chalk or limestone. Local in England and Wales northwards to S. Lancs and Yorks.

2. Knautia L.

Herbs with lvs in opposite pairs. Heads long-stalked, flat, with an involucre of numerous herbaceous bracts. Receptacle hemispherical, hairy; receptacular bracts 0. Involucel 4-angled, ending above in a very short obscurely toothed cup. Calyx a short cup below, prolonged up-wards as 8(–16) teeth or bristles, falling in fr.

1. K. arvensis (L.) Coulter   Field Scabious.

Perennial, with branched erect stock and tap-root. Stem 25–100 cm., erect, rough below with downwardly-directed bristles, glabrous above. Basal lvs in an overwintering rosette, oblanceolate, short-stalked, com-monly simple but sometimes lyrate-pinnatifid, entire or crenate-toothed; stem-lvs usually deeply pinnatifid, but some or all of the uppermost simple; all dull green, hairy. Heads commonly 3–4 cm. diam., their stalks hairy. Involucral bracts ovate-lanceolate, hairy, in 2 rows. Fls bluish-lilac, marginal larger than central. Calyx with 8 teeth. Corolla

unequally 4-lobed. Fr. 5–6 mm., densely hairy. Fl. 7–9. Dry grassy
fields, dry pastures, banks, etc. Throughout the British Is, except Outer
Hebrides and Shetland; less common in the north.

3. SCABIOSA L.

Herbs with lvs in opposite pairs. Heads long-stalked, convex, with an
involucre of numerous herbaceous bracts in 1–3 rows. Receptacle elongated, with non-spinous bracts. Involucel cylindrical with 8 furrows (at
least above), ending upwards in a ± scarious pleated funnel-shaped cup.
Calyx cup-shaped below, usually prolonged upwards as 5 narrow teeth.
Corolla with short tube and 5 unequal lobes, marginal fls usually larger
than central. Fr. crowned by the persistent calyx.

1. S. columbaria L.   Small Scabious.

Perennial, with long tap-root and erect branching stock. Stems 15–
70 cm., erect, slender, sparsely hairy. Basal lvs long-stalked, obovate to
oblanceolate, simple and crenate or ± lyrate-pinnatifid; stem-lvs successively shorter-stalked and more deeply and narrowly pinnatifid, the
uppermost with linear segments. Heads 1·5–3·5 cm. diam., with long
pubescent stalks. Involucral bracts c. 10 in 1 row, linear-lanceolate.
Corolla bluish-lilac, rarely pink or white; outer fls much larger than
central. Fr. head ovoid; fr. 3 mm., downy, surmounted by the hairy
8-furrowed involucel and blackish calyx-teeth. Fl. 7–8. Dry calcareous
pastures, banks, rocks and cliffs. Great Britain northwards to Stirling
and Angus; Channel Is.

4. SUCCISA Haller

Perennial herb with lvs in opposite pairs. Heads long-stalked, hemispherical, with an involucre of numerous herbaceous bracts. Receptacle
elongated, not hairy, with non-spinous bracts. Involucel 4-angled with
2 furrows in each face, ending above in 4 erect triangular herbaceous
lobes. Calyx a short cup prolonged upwards as 5(–4) narrow teeth.
Corolla almost equally 4-lobed, marginal not appreciably larger than
central. Calyx persistent in fr.

1. S. pratensis Moench   Devil's-bit Scabious.

Root-stock short, erect, ending abruptly below as though bitten off;
roots long, stout. Stem 15–100 cm., erect or ascending, glabrous or
appressed-hairy above. Basal lvs 5–30 cm., in a rosette, obovate-lanceolate to narrowly elliptical, narrowed into a short stalk, firm, net-veined
beneath, usually entire, sparsely hairy; stem-lvs few, narrower, some-

times all bract-like. Heads 1·5–2·5 cm. diam. Involucral bracts broadly lanceolate, pubescent, ciliate. Receptacular bracts elliptical, purple-tipped. Corolla 4–7 mm., mauve to dark blue-purple, rarely white. Fr. 5 mm., downy, surmounted by the involucel and 5(–4) reddish-black calyx-teeth. Fl. 6–10. Some plants have ± abortive anthers in all the fls of somewhat larger heads. Marshes, fens, meadows and pastures, damp woods. Common throughout the British Is.

120. COMPOSITAE

Herbaceous or sometimes woody plants of very diverse habit, often with latex. Lvs exstipulate. Fls small (*florets*), aggregated into heads (*capitula*) simulating single larger fls and surrounded by calyx-like *involucres* of one or more rows of bracts. Receptacle of the head expanded, with or without *receptacular scales* each subtending a floret. Florets all similar (head *homogamous*) or central and marginal florets differing (head *heterogamous*), and then the central florets usually hermaphrodite or rarely male, the outer female or rarely neuter. Calyx never typically herbaceous but represented by a *pappus* of numerous simple or feathery hairs, or a smaller number of membranous scales, teeth or bristles, or by a continuous membranous ring; sometimes 0. Corolla gamopetalous, of two main types in British spp.: (*a*) *tubular*, the corolla-tube being surmounted by 5 equal teeth; (*b*) *ligulate*, the corolla-tube being prolonged only along one side as a strap-shaped ligule, usually 3- or 5-toothed at its tip. In homogamous heads the florets may be all of either type, but in heterogamous heads the central or *disk-florets* are usually tubular and regular and the marginal or *ray-florets* usually ligulate or, if tubular, distinctly larger and more conspicuous than the disk-florets. Stamens 5, inserted on the corolla-tube, their anthers usually united laterally so that they form a closed cylinder round the style and dehisce into this cylinder. Ovary inferior, 1-celled, with 1 basal ovule; style single below but branching above into 2 stigmatic arms. Pollination by insects, rarely by wind (*Artemisia*, etc.); some genera have apomictic spp. (*Hieracium*, *Taraxacum*, etc.). Fr. an achene crowned by the pappus, sometimes with a slender *beak* interposed between them.

1 A small very rare plant of Scottish mountains, 3–6 cm. high, with
   silky-white subpalmate lvs and usually solitary nodding yellow
   heads c. 12 mm. diam.; all florets tubular.
                                    25. *Artemisia norvegica*
  Not as above.   2

2   Flowering heads appearing early in the year, before or with the
       large basal lvs.   3
    Flowering heads appearing after the lvs.   4

3   Heads in stout racemes; florets all tubular, white or purplish.
                                      7. PETASITES
    Heads solitary, with yellow ligulate ray-florets.   6. TUSSILAGO

4   Plant with milky latex; florets all ligulate.   48
    Plant without milky latex; at least the central florets tubular.   5

5   Stem-lvs mostly opposite.   6
    Lvs all basal or stem-lvs spirally arranged.   8

6   Florets all tubular, pale pink; stem-lvs 3–5-partite.
                                  18. EUPATORIUM
    Disk-florets tubular, yellow; ray-florets ligulate or 0.   7

7   Heads exceeding 1 cm. diam.; ray-florets yellow or 0.   1. BIDENS
    Heads less than 1 cm. diam.; ray-florets usually 5, very small, white.
                                    2. GALINSOGA

8   Heads unisexual; the fruiting heads, which enclose 2 achenes,
       covered with ± hooked spines and ending in 2 horn-like
       processes.   3. XANTHIUM
    Achenes not permanently enclosed within a spinous and horned
       fruiting head.   9

9   Involucral bracts slender, spreading, hook-tipped, the fruiting
       head forming an adhesive bur containing several achenes.
                                  27. ARCTIUM
    Not as above.   10

10   Heads small, inconspicuous, wind-pollinated, in long racemose
       panicles; pappus 0; lvs very aromatic.   25 ARTEMISIA
    Not as above.   11

11   Heads with ligulate ray-florets.   12
    Florets all tubular.   28

12   Ray-florets yellow.   13
    Ray-florets not yellow.   20

13   Pappus of hairs.   14
    Pappus not of hairs, or 0.   18

14   Involucral bracts in many unequal imbricating rows.   15
    Involucral bracts equal, in 1 or 2 rows, sometimes with a few
       much shorter bracts at the base.   17

15   Heads in elongated racemes or racemose panicles.
                                14. SOLIDAGO
    Heads solitary or in cymose panicles.   16

16 Pappus of an inner row of hairs and an outer row of scales; plants
    of moist or wet places. 9. PULICARIA
    Pappus of 1 row of hairs; not plants of wet places   8. INULA

17 Heads 4 cm. or more in diam.; involucral bracts in 2 equal rows.
    5. DORONICUM
    Heads not exceeding 3·5 cm. diam.; involucral bracts equal, in
    1 row, or with a few much shorter outer bracts.   4. SENECIO

18 Heads 7–12 cm. diam. Tall perennial herb, 50–250 cm., with
    pinnate or pinnatifid lvs; disk-florets brownish; receptacle
    conical. Introduced and locally naturalized.
    *Rudbeckia laciniata L.
    Heads less than 7 cm. diam.   *19*

19 Lvs green, deeply pinnatifid with toothed lobes, densely hairy
    beneath; receptacular scales present.   19. *Anthemis tinctoria*
    Lvs glaucous, toothed or pinnately lobed, glabrous; receptacular
    scales 0.   24. *Chrysanthemum segetum*

20 Pappus of hairs.   *21*
    Pappus not of hairs, or 0.   *22*

21 Ray-florets in 1 row.   15. ASTER
    Ray-florets in 2 or more rows, numerous and very narrow.
    16. ERIGERON

22 Lvs all basal; heads solitary.   17. BELLIS
    Lvs not all basal.   *23*

23 Receptacular scales present.   *24*
    Receptacular scales 0.   *26*

24 Heads 4–6 mm. diam. or, if larger, lvs simple, narrow, finely
    toothed; ray-florets with short broad ligules; achenes strongly
    compressed.   21. ACHILLEA
    Heads exceeding 1 cm. diam.; lvs finely divided into narrow
    segments; ray-florets with long narrow ligules; achenes not or
    little compressed.   *25*

25 Annual; tube of disk-florets flattened and somewhat winged
    below; achenes ribbed on both faces, truncate above.
    19. ANTHEMIS
    Perennial; tube of disk-florets not flattened or winged; achenes
    slightly ribbed on one face, rounded above.   20. CHAMAEMELUM

26 Lvs finely divided into linear segments; involucral bracts equal.   *27*
    Lvs not divided into linear segments; outer involucral bracts
    shorter than inner.   24. CHRYSANTHEMUM

27 Ray-florets soon reflexed; receptacle conical from the first, hollow;
    achene with 5 slender ribs on one face.   23. MATRICARIA

Ray-florets spreading; receptacle becoming conical in fr., solid; achenes with 3 broad corky ribs on one face.

22. TRIPLEUROSPERMUM

28 Florets red, purplish, blue or rarely white.   29
Florets not red, purplish or blue; commonly yellow.   37

29 Basal lvs long-stalked, reniform, crenate-toothed; stem 10–30 cm. with 2 distant ± scale-like lvs and a solitary pale violet head. A very rare Scottish alpine, probably introduced.

Homogyne alpina (L.) Cass.

Not as above.   30

30 Involucral bracts with a scarious or spinous terminal appendage, differing in colour and texture from the basal part; pappus 0, or of scales, or of hairs not exceeding the length of the achenes.

32. CENTAUREA

Involucral bracts spinous or not, but not appendaged; pappus of hairs exceeding the length of the achenes.   31

31 Pappus-hairs feathery, i.e. with slender lateral hairs readily visible to the naked eye.   32
Pappus-hairs simple or toothed but not feathery to the naked eye.   34

32 Lvs spinous at least at the margins; involucral bracts ± distinctly spine-tipped.   29. CIRSIUM
Neither lvs nor involucral bracts spinous.   33

33 Heads usually 4–10 in a close corymb; lvs entire or distantly toothed; a rare alpine.   *Saussurea alpina* (L.) DC.
Heads usually solitary, rarely 2–3; lvs with fine prickles.

29. CIRSIUM

34 Neither lvs nor involucral bracts spinous.   33. SERRATULA
Lvs and involucral bracts spinous; thistles.   35

35 Lvs glabrous, conspicuously white-veined above; stem not spinous-winged.   30. SILYBUM
Lvs not conspicuously white-veined above, decurrent down the stem in spinous wings.   36

36 Lvs white with dense cottony hairs above and beneath; stem cottony-white; achenes 4-angled.   31. ONOPORDUM
Lvs not white on both sides; achenes not 4-angled.   28. CARDUUS

37 Involucre spinous.   38
Involucre not spinous.   39

38 Lvs spinous; inner involucral bracts yellow, scarious, spreading in fr.   26. CARLINA
Lvs not spinous; stem with wavy non-spinous wings; involucral bracts with a terminal spinous appendage; a rare weed or casual.

32. *Centaurea solstitialis* L.

39 Lvs simple, entire or toothed. 40
    Lvs pinnately lobed or divided. 46

40 Lvs glabrous. **15. ASTER**
    Lvs not glabrous. 41

41 Pappus 0; a very rare white-woolly maritime plant. Cottonweed.
                        Otanthus maritimus (L.) Hoffmanns. & Link
    Pappus of hairs. 42

42 Lvs ovate to ovate-oblong, pubescent but not woolly or cottony
        beneath; heads c. 1 cm. diam.; involucral bracts neither scarious
        nor woolly. **8. *Inula conyza***
    Lvs linear to broadly lanceolate, cottony or woolly at least beneath;
        heads usually small but if as much as 1 cm. diam. then with
        scarious or woolly involucral bracts. 43

43 Annual woolly herbs whose small stem-lvs are usually held erect
        and whose stems branch cymosely, the main stem and the suc-
        cessively overtopping branches each ending in a small roundish
        cluster of heads; outer involucral bracts herbaceous, ±woolly;
        receptacle with marginal scales. **10. FILAGO**
    Perennial or rarely annual herbs; heads variously arranged, but if
        as in *Filago* then with spreading stem-lvs; all involucral bracts
        scarious; receptacular scales 0. 44

44 Perennial herbs, 5–25 cm., with basal rosettes of obovate-
        spathulate lvs and long lfy stolons; stem-lvs erect, 1–2 cm.; heads
        in a close terminal cluster; involucral bracts white or pink;
        dioecious. **13. ANTENNARIA**
    Not as above. 45

45 Perennial herbs, 25–100 cm., with short stolons; stem-lvs spread-
        ing, narrowly elliptical, 5–10 cm.; heads in a cymose panicle;
        involucral bracts shining white; subdioecious. **12. ANAPHALIS**
    Perennial herbs with heads in spike-like racemes, or annuals with
        heads in roundish clusters; involucral bracts straw-coloured or
        brownish; not dioecious. **11. GNAPHALIUM**

46 Pappus of hairs. **4. *Senecio vulgaris***
    Pappus not of hairs. 47

47 Lvs repeatedly divided into linear segments.
                        23. *Matricaria matricarioides*
    Lvs not divided into linear segments; heads numerous, in a flat-
        topped corymb. **24. *Chrysanthemum vulgare***

48 Pappus of scales, or 0. 49
    Pappus, at least of the central achenes, of hairs. 51

49 Heads blue; pappus of scales. **34. CICHORIUM**
    Heads yellow; pappus not of scales. 50

50   Stem lfy.   35. LAPSANA
     Lvs all basal.   36. ARNOSERIS

51   All pappus-hairs, or the inner of 2 rows, feathery, i.e. with lateral
         hairs visible to the naked eye.   52
     Pappus-hairs simple or shortly toothed but not feathery to the
         naked eye.   56

52   Involucral bracts in 1 row.   40. TRAGOPOGON
     Involucral bracts in more than 1 row.   53

53   Outer involucral bracts either narrow and lax or broadly cordate;
         stems rough with hooked bristles.   39. PICRIS
     Outer involucral bracts appressed, not broadly cordate.   54

54   Stem-lvs linear-lanceolate to linear, entire; lateral hairs of pappus
         interlocking; a very rare marsh plant.   *Scorzonera humilis* L.
     Stem-lvs minute or 0.   55

55   Receptacular scales present.   37. HYPOCHAERIS
     Receptacular scales 0.   38. LEONTODON

56   Achenes strongly compressed.   57
     Achenes not strongly compressed.   60

57   Achenes beaked or at least markedly narrowed upwards.   58
     Achenes neither beaked nor markedly narrowed upwards.   59

58   Involucre of many unequal rows of bracts; pappus-hairs in 2 equal
         rows.   41. LACTUCA
     Involucre of an inner row of equal long bracts and an outer row of
         much shorter bracts; pappus-hairs in 2 rows, the inner long,
         outer short.   42. MYCELIS

59   Heads yellow.   43. SONCHUS
     Heads blue; a very rare alpine plant. Blue Sowthistle.
                                                  Cicerbita alpina (L.) Wallr.

60   Achenes beaked.   61
     Achenes not beaked.   62

61   Achenes strongly muricate at the base of the beak; lvs all basal.
                                                            46. TARAXACUM
     Achenes not muricate; stem lfy.   45. CREPIS

62   Involucre usually of many unequal rows of bracts; achene not or
         hardly narrowed upwards; pappus brownish, rough and brittle
         (but see also *Crepis paludosa*).   44. HIERACIUM
     Involucre usually of bracts of equal length except for some shorter
         ones near the base; achene usually distinctly narrowed upwards;
         pappus usually white and soft (but *C. paludosa*, with glabrous
         stem and lvs, has outer bracts c. $\frac{1}{2}$ as long as inner, achenes
         hardly narrowed and pappus dirty white).   45. CREPIS

Subfamily CARDUOIDEAE: plants without milky latex; corolla of disk-florets never ligulate.

1. BIDENS L.   Bur-Marigold.

Annual to perennial herbs with opposite lvs. Heads heterogamous or homogamous; involucral bracts in 2 rows, the outer usually long and lfy; receptacle with scales. Ray-florets ligulate and neuter or commonly 0; disk-florets tubular, hermaphrodite, yellow. Achenes 4-angled, ± flattened; pappus represented by 2–4 stiff barbed bristles.

1 Lvs simple, sessile; heads drooping.   **1. cernua**
  Most or all lvs deeply lobed or compound; heads ± erect.   2

2 Most lvs deeply cut into 3(–5) lobes; achenes greenish-brown, with downwardly-directed barbs on the angles and with few or no tubercles on the faces.   **2. tripartita**
  Most lvs pinnate with 3–5 separate lflets; achenes blackish, their angles not barbed but the faces strongly and densely tubercled. Introduced and locally naturalized.   **B. frondosa* L.

1. B. cernua L.

Annual, with erect glabrous or sparsely hairy stems, 8–60 cm. Lvs 4–15 cm., simple, lanceolate-acuminate, sessile, coarsely toothed, ± glabrous, pale green. Heads 15–25 mm. diam. (without disk-florets), long-stalked, drooping, solitary and terminal on the main stem and branches. Outer involucral bracts 5–8, lanceolate, spreading, much longer than the inner broadly ovate dark-streaked bracts. Ray-florets usually 0; when present c. 12 mm., yellow. Achenes 5–6 mm., broadening upwards, with 4 barbed angles prolonged into 4 (rarely 3) barbed bristles. Fl. 7–9. Locally common by ponds and streams and especially where water stands only in winter. Throughout the British Is northwards to Moray and Argyll.

2. B. tripartita L.

Annual, with erect usually much branched glabrous or somewhat downy stems, 15–60 cm. Lvs 5–15 cm., usually 3(–5)-partite but sometimes simple, narrowed below into a short winged stalk; segments lanceolate, acute, coarsely toothed, the end segment sometimes broader and 3-lobed. Heads 15–25 mm. diam., solitary, suberect. Outer involucral bracts 5–8, oblong, spreading; inner broadly ovate, brownish. Ray-florets usually 0. Achenes 7–8 mm., obovoid-oblong, much flattened, with barbed angles, 2 of which are prolonged upwards into barbed bristles with or without 1–2 shorter ones. Fl. 7–9. Locally common in ditches and by ponds, lakes and streams throughout the British Is northwards to Moray and Kintyre.

2. GALINSOGA Ruiz & Pavon

Annual herbs with simple 3-veined opposite lvs and small heads in dichasial cymes. Involucre of a few ovate bracts. Heads with few yellow hermaphrodite tubular disk-florets and 4–8 white female ligulate ray-florets. Pappus of several distinct scarious scales.

Stem ±glabrous; stalks of heads with short ascending hairs (less than 0·5 mm.) and some spreading glandular hairs; receptacular scales usually trifid; pappus-scales not awned.   **1. parviflora**

Stem hairy, its upper part and the stalks of heads ±densely clothed with spreading flexuous simple hairs (0·5 mm. or more long) and glandular hairs; receptacular scales not trifid; at least the longest pappus-scales narrowed into an awn. An infrequent casual in S. England and S. Wales.   **G. ciliata* (Rafin.) Blake

*1. G. parviflora Cav.   Gallant Soldier.

Stem 10–75 cm., erect, much-branched, ±glabrous. Lvs stalked, ovate-acuminate, waved and distantly toothed at the margins. Heads 3–5 mm. diam. Receptacle shortly conical, with trifid scales. Ray-florets usually 5, the white ligule c. 1 mm., about as broad as long, 3-lobed. Central achenes c. 1 mm., ovoid, black, covered with short white bristles, with pappus of 8–20 silvery fimbriate unawned scales which about equal the achenes; marginal achenes flattened, 3-angled, somewhat curved, bristly only above and with pappus-scales short or abortive. Fl. 5–10. Introduced and well-established as a weed of arable land and waste places in S. England.

3. XANTHIUM L.

Annual monoecious herbs with spirally arranged triangular or palmately lobed lvs, often with spines at the base of the lf-stalk. Heads solitary or in axillary clusters, male in the upper part of the plant, female below. Male heads with subglobose involucre, its bracts free, and numerous tubular florets with rudimentary ovaries. Female heads with ovoid involucre ending above in 2 horn-like processes, its bracts united and covered with recurved hooks; florets 2, tubular, very slender, enclosed in the involucre except for the styles protruding through lateral holes in the terminal horns; stamens 0. Achenes remaining enclosed in pairs in the hardened involucres which are animal-dispersed.

The following are frequent casuals and may establish themselves for a time:

Plant with 1–2 strong trifid yellow spines at the base of each lf-stalk; lvs short-stalked, narrowly rhombic, wedge-shaped at base, 3–5-lobed, white-felted beneath. Spiny Cocklebur.   **X. spinosum* L

Plant not spiny; lvs long-stalked, ± triangular, cordate at base, coarsely
toothed to palmately lobed, roughly hairy. Cocklebur.

X. strumarium L.

4. SENECIO L.

Herbs or woody plants with spirally arranged lvs. Heads usually heterogamous; involucral bracts herbaceous, mostly in 1 row with a few short
outer ones; receptacular scales 0. Ray-florets ligulate, female, yellow, or
0; disk-florets tubular, hermaphrodite, yellow. Achenes cylindrical,
ribbed; pappus of simple hairs.

1	Lvs pinnately lobed or cut.	2
	Lvs simple, entire or toothed.	10
2	Lvs densely white-felted beneath.	**10. cineraria**
	Lvs glabrous, pubescent or cottony beneath.	3
3	Heads with ligule of ray-florets short (5 mm. or less) and soon becoming revolute, or 0.	4
	Heads with ray-florets conspicuous and spreading.	7
4	Ray-florets usually 0; heads at first subsessile in dense clusters, later stalked.	**7. vulgaris**
	Ray-florets present; heads stalked from the first, in loose corymbs.	5
5	Involucre conical; ligule of ray-florets very short and strongly revolute from the first.	6
	Involucre broadly cylindrical; ligule of ray-florets 4–5 mm., at first erect or spreading, then revolute; achenes 3–3·5 mm. A very rare plant, intermediate between *S. vulgaris* and *S. squalidus* and known only from one roadside in Flint. Welsh Ragwort.	
		S. cambrensis Rosser
6	Stem and lvs very sticky with glandular hairs; involucral bracts c. 20, the outermost 2 or 3 more than one-third as long as the rest; achenes glabrous.	**6. viscosus**
	Stem and lvs not or slightly glandular; involucral bracts c. 13, the outermost very small; achenes pubescent.	**5. sylvaticus**
7	Heads in dense flat-topped corymbs.	8
	Heads in irregular loose corymbs or cymes.	9
8	Plant with short creeping stolons; lower stem-lvs with small narrow acute terminal lobe; small outer involucral bracts about half as long as the rest.	**3. erucifolius**
	Stolons 0; lower stem-lvs with blunt terminal lobe; small outer involucral bracts less than half as long as the rest.	**1. jacobaea**
9	Weed of waste places, banks, walls, etc.; 5–13 small outer invo-lucral bracts; all bracts conspicuously black-tipped.	**4. squalidus**

Biennial marsh plant; 2–5 small outer involucral bracts; bracts not
   black-tipped.   **2. aquaticus**

10 Tall plants (to 150 cm.) with numerous heads in large compound
   corymbs; involucre with a few short outer bracts.   *11*
   Small plants, 7–30 cm. (very rarely to 90 cm.) with 1–6(–12) heads;
   involucre with no short outer bracts.   **9. integrifolius**

11 Stem-lvs glaucous, thick, fleshy, decurrent; ray-florets 4–6; heads
   1·5–2 cm. diam. Introduced. Established in wet meadows and
   by streams in a few places.   ***S. doria** L.*
   Stem-lvs green, thin, not decurrent; ray-florets 6–8; heads 3 cm.
   diam.   **8. fluviatilis**

1. S. jacobaea L.   Ragwort.

Biennial to perennial, with short erect stock and erect grooved glabrous
or cottony stems, 30–150 cm.; stolons 0. Basal lvs (usually dead before
flowering) in a rosette, stalked, lyrate-pinnatifid with large ovate blunt
terminal lobe and 0–6 pairs of much smaller oblong lateral lobes, all
further lobed or toothed; stem-lvs pinnatifid with blunt terminal lobe,
lower stalked, upper half-clasping; all lvs glabrous or somewhat cottony
beneath, firm, waved. Heads 1·5–2(–2·5) cm. diam., in a large flat-
topped compound corymb. Inner involucral bracts c. 13, often brown-
tipped; outer usually 3–6, less than half as long. Ray florets c. 13, rarely
0. Achenes 2 mm. Fl. 6–10. Abundant on sand-dunes and a weed of
waste land, waysides and neglected or overgrazed pastures throughout
the British Is.

2. S. aquaticus Hill   Marsh Ragwort.

Usually biennial with short erect stock and erect stems, 25–80 cm.,
glabrous or cottony above, often reddish; stolons 0. Basal lvs long-
stalked, simple, elliptical to ovate, or lyrate-pinnatifid; lower stem-lvs
stalked, ± lyrate-pinnatifid, upper half-clasping, pinnatifid; all crenate
to coarsely serrate, ± glabrous, firm, slightly waved, often purplish
beneath. Heads (2–)2·5–3 cm. diam. in an irregular loose corymb or
cyme. Involucral bracts green with white margins, the outermost 2–5
much less than half as long as the rest. Ray-florets 13–20. Achenes
2·5–3 mm. Fl. 7–8. Marshes, wet meadows and ditches. Common
throughout the British Is.

3. S. erucifolius L.   Hoary Ragwort.

Perennial, stoloniferous, with short creeping stock and erect furrowed
stems, 30–120 cm., sparsely cottony. Basal lvs stalked, obovate-lanceolate in outline, ± pinnatifid; stem-lvs deeply pinnatifid (lower stalked,
upper clasping), with small narrow terminal and long linear-oblong

lateral lobes, all ± acute and entire or with a few teeth; all lvs somewhat hairy and often sparsely cottony beneath. Heads c. 2 cm. diam., in dense flat-topped compound corymbs. Inner involucral bracts c. 13, broadly lanceolate; outer somewhat spreading, usually 5–7, of which some are at least half as long as the inner. Ray-florets c. 13, broader than in *S. jacobaea*. Achenes 2 mm. Fl. 7–8. Roadsides, field-borders, shingle-banks, grassy slopes, etc., chiefly on lowland calcareous and heavy soils; locally common in England and Wales but rare in Scotland and extending northwards only to Fife; only in E. Ireland.

*4. S. squalidus L. Oxford Ragwort.

Overwintering annual, rarely longer-lived, with ± glabrous, tough, diffusely branched, ascending stems, 20–40 cm. Lower lvs narrowed into a winged stalk, upper half-clasping; all usually deeply pinnatifid with oblong entire or toothed lobes, but very variable in depth of lobing and sometimes simple and coarsely toothed. Heads 1·5–2·5 cm. diam., in an irregular loose corymb. Involucre bell-shaped, broadening below but remaining constricted above in older heads; inner bracts usually 21; outer 5–13, very short; all conspicuously black-tipped. Ray-florets usually 13, broad. Achenes 1·5–2 mm. Fl. 6–12. Introduced. Spreading rapidly on old walls, waste ground, railway banks, bombed sites, etc., throughout England to Lancs and Yorks and locally in S. Ireland; very rare in S. Scotland.

5. S. sylvaticus L. Wood Groundsel.

Annual, with erect slender grooved stem, 30–70 cm., somewhat cottony or pubescent, not or slightly glandular. Lvs yellow-green, deeply and irregularly pinnatifid, the lobes unequal, ± cut or toothed; lower lvs oblanceolate to obovate in outline, short-stalked, upper oblong, sessile or clasping; all cottony at first but becoming ± glabrous. Heads 5 mm. diam. in a large flat-topped corymb. Involucre conical, glandular-hairy; outer bracts less than one-quarter as long as inner. Ray-florets 8–14, very short and revolute. Achenes 2·5 mm. Fl. 7–9. In open vegetation on sandy non-calcareous substrata. Locally common throughout most of the British Is.

6. S. viscosus L. Stinking Groundsel.

Annual, foetid, with erect very viscid glandular-hairy stems, 10–60 cm. Lvs dark green, glandular and very viscid, deeply pinnatifid with nearly equal toothed or pinnatifid lobes; lower obovate in outline, short-stalked, upper oblong, sessile. Heads 8 mm. diam., long-stalked, in a large irregular rounded compound corymb. Involucre ovoid-conical, densely glandular, its outer bracts almost half as long as the inner. Ray-florets c. 13, short and revolute. Achenes 3 mm. Fl. 7–9. Probably native. Waste ground, railway banks and tracks, sea-shores, etc. Locally common throughout lowland Great Britain; introduced in Ireland.

7. S. vulgaris L.   Groundsel.

Annual or overwintering, with erect or ascending rather weak and succulent stems, 8–45 cm., glabrous or with non-glandular hairs. Lvs glabrous or cottony, pinnatifid, with distant, oblong, blunt, irregularly toothed lobes; lower lvs short-stalked, upper oblong, half-clasping. Heads 4 mm. diam., at first almost sessile, later stalked, in dense corymbose clusters. Involucre cylindrical, usually glabrous, the outer bracts black-tipped and about one-quarter as long as the inner. Ray-florets usually 0, rarely up to 11, shortly revolute. Achenes 1·5–2 mm. Fl. 1–12. Cultivated ground and waste places. Abundant throughout the British Is.

*8. S. fluviatilis Wallr.   Broad-leaved Ragwort.

Perennial, with creeping stock and long stolons. Stems 80–120 cm., erect, very lfy, glabrous below, slightly downy and sometimes glandular above. Lvs 10–20 × 2–5 cm., elliptical, acute, sessile, ± glabrous, the margins with small irregular cartilaginous teeth. Heads 3 cm. diam., numerous, in large compound corymbs. Involucre broadly ovoid, the 12–15 inner bracts about twice as long as the 5 outer, all pubescent. Ray-florets 6–8, spreading. Achenes 3–4 mm. Fl. 7–9. Introduced. Naturalized in fens and fen-woods and by streams in many places in Great Britain northwards to Ross and in Ireland.

9. S. integrifolius (L.) Clairv.   Field Fleawort.

Perennial, with short erect stock and erect ± cottony stems, usually 7–30 cm. Basal lvs 3–5 cm., in a rosette, broadly ovate, rounded at base, narrowing into a winged stalk, entire or with small distant teeth; stem-lvs few, oblong-lanceolate, entire, sessile, sometimes slightly clasping; all pubescent on both sides, cottony when young. Heads 1·5–2·5 cm. diam., short-stalked, 1–6 in a subumbellate cluster. Involucral bracts all equal, cottony below. Ray-florets c. 13. Achenes 3–4 mm. Fl. 6–7. Chalk and limestone grassland and calcareous banks. Local in England from Dorset and Kent northwards to Gloucester and N. Lincoln; larger forms (to 90 cm., with up to 12 heads) in Anglesey and Westmorland.

*10. S. cineraria DC.   Cineraria.

Perennial, woody below, with erect white-woolly stems, 30–60 cm. Lvs ovate or ovate-oblong, lowest coarsely toothed or shallowly lobed, upper deeply pinnatifid, densely white-felted beneath, cottony but green above. Heads 8–12 mm. diam., in large flattish dense corymbs. Involucre white-woolly. Ray-florets 10–12. Fl. 6–8. Introduced. Naturalized on maritime cliffs in S. and S.W. England, S. Wales, E. Ireland and Channel Is.

5. DORONICUM L.   Leopard's-bane.

Perennial herbs with locally tuberized stolons, spirally arranged simple lvs and large long-stalked yellow heads. Bracts herbaceous in 2–3 almost equal rows. Heads heterogamous: disk-florets hermaphrodite, tubular, yellow; ray-florets in 1 row, female, ligulate, yellow. Achenes ± cylindrical, ribbed; pappus of 1–2 rows of simple hairs (or 0 in marginal florets).

Basal lvs broadly ovate-cordate; uppermost lvs ovate, clasping; heads
  usually several, 4–6 cm. diam.   **1. pardalianches**
Basal lvs ovate-elliptical, narrowed gradually into the stalk; uppermost lvs elliptical to lanceolate, sessile, somewhat decurrent; all lvs
  with curving main veins; head usually solitary, 5–8 cm. diam. Introduced and rarely naturalized in woods, hedgerows, etc.
                                                             * D. plantagineum L.

*1. D. pardalianches L.

Perennating by stout underground stolons with tuberous tips. Basal lvs long-stalked, broadly ovate-cordate, ± entire, ciliate and hairy on both sides. Flowering stems 30–90 cm., erect, ± woolly. Lower stem-lvs with a long stalk winged above and clasping below; middle stem-lvs constricted above the clasping base; upper stem-lvs ovate, clasping; all pale green, thin, hairy, entire to distantly toothed. Heads 4–6 cm. diam., bright yellow, usually several. Involucral bracts narrowly triangular, glandular-ciliate. Achenes black, those of the disk downy and with pappus, of the ray glabrous and with no pappus. Fl. 5–7. Introduced and naturalized locally in woods and plantations throughout Great Britain northwards to Ross and Moray.

6. TUSSILAGO L.

Heads solitary, heterogamous; involucral bracts mostly in 1 row. Ray-florets ligulate, numerous (up to 300), in many rows, female; disk-florets few, male. Pappus of many rows of long, simple hairs.

1. T. farfara L.   Coltsfoot.

Perennial herb with long stoutish white scaly stolons. Lvs 10–20(–30) cm. across, all basal, roundish-polygonal with 5–12 very shallow, acute, distantly blackish-toothed lobes, cordate at base, at first white-felted above and below, later only below; lf-stalk furrowed above. Heads 15–35 mm. diam., solitary on purplish scaly and woolly stems 5–15 cm., opening long before the lvs appear, erect in bud. Involucral bracts numerous, linear, somewhat hairy, with a few broader basal scales grading into those of the flowering stem. Florets bright pale yellow. Fl. 3–4. Abundant throughout the British Is, especially on stiff soils, in arable fields, waste places, banks, land-slides, boulder-clay, cliffs, etc.; also in dunes, screes and stream-side shingle and in calcareous spring-flushes; reaching 3500 ft.

7. PETASITES Miller

Perennial rhizomatous ± dioecious herbs with large basal lvs. Heads in spike-like racemes or panicles; bracts irregularly 2–3-rowed. 'Male' heads of tubular staminate florets, with short non-stigmatic style-arms, and 0–5 marginal fertile florets; 'female' of threadlike tubular or shortly ligulate fertile florets and 1–7 central staminate florets. Achenes cylindrical, glabrous; pappus of long slender simple hairs.

1 Fls lilac, vanilla-scented, appearing at the same time as the new lvs; corolla of marginal florets shortly but distinctly ligulate.

2. fragrans

Fls not vanilla-scented, appearing before the lvs; corolla of marginal florets tubular, not at all ligulate.   2

2 Fls pale pinkish-violet; lvs up to 60(–90) cm. diam., greyish beneath at maturity, with the lower part of each basal lobe bounded by a lateral vein.   **1. hybridus**

Fls whitish; lvs up to 30 cm. diam., white-woolly beneath at maturity, with no part of their basal lobes bounded by lateral veins. Introduced and naturalized locally from Leicester and Merioneth northwards to Aberdeen; rare in N. Ireland.

***P. albus** (L.) Gaertner

1. P. hybridus (L.) P. Gaertner, B. Meyer & Scherb.   Butterbur.

Rhizome stout, far-creeping. Lvs 10–90 cm. across, mostly basal, long-stalked, roundish, deeply cordate, at first downy on both sides but later green above and greyish beneath; stalks stout, hollow, channelled above; blade with large distant teeth and smaller ones intervening; lower part of each basal lobe bordered by a lateral vein. Flowering stems 10–40 cm. (–80 cm. in fr.), appearing before the lvs, stout, purplish below, covered with lanceolate scales. Heads 1–3 in the axils of linear bracts, pale pinkish-violet; 'male' heads 7–12 mm., very short-stalked; 'female' heads 3–6 mm., lengthening in fr., longer-stalked.

Involucral bracts narrow, blunt, glabrous, purplish. Achenes 2–3 mm., yellowish-brown; pappus whitish. Fl. 3–5. In wet meadows and copses and by streams, the 'male' plant locally common throughout the British Is, the 'female' not uncommon in Lancs, Yorks, Cheshire and Derby but rare or absent elsewhere.

*2. P. fragrans (Vill.) C. Presl   Winter Heliotrope.
Rhizome far-creeping. Lvs 10–20 cm. across, long-stalked, roundish, deeply cordate, equally fine-toothed, slightly pubescent below, persistent through the winter. Flowering stems 10–25 cm., with a few scales. Infl. short, lax, of about 10 lilac vanilla-scented heads; involucral bracts narrow, acute. 'Male' heads chiefly of tubular florets, 'female' of slender florets with a short broad ligule. Fl. 1–3. Introduced and locally naturalized on the sides of streams and roads, etc., throughout Great Britain and Ireland.

8. INULA L.

Annual to perennial herbs with spirally arranged usually simple lvs. Heads heterogamous; involucral bracts herbaceous, in many rows; receptacle-scales 0. Ray-florets in 1 row, female, ligulate or rarely almost tubular; disk-florets hermaphrodite, tubular; all yellow. Achenes with pappus-hairs in 1 row.

1 Heads 6–8 cm. diam.; outer involucral bracts broadly ovate; stem-
    lvs large, ovate, clasping. A large perennial, 60–150 cm., with
    basal lvs 25–40 cm., elliptical. Introduced and widely scattered
    but uncommon in fields, waysides, copses, etc. Elecampane.
                                                            *I. helenium L.
    Heads less than 5 cm. diam.; outer involucral bracts narrow.   *2*

2 Marginal florets inconspicuous, not ligulate; lvs pubescent, resem-
    bling those of *Digitalis*.   **1. conyza**
    Ligulate ray-florets present.   *3*

3 A maritime plant with ± linear succulent lvs often 3-toothed at the
    apex; heads up to 2·5 cm. diam.   **2. crithmoides**
    Not maritime; lvs elliptical, ciliate and usually hairy on the veins
    beneath, otherwise glabrous; heads 2·5–3 cm. diam. A very rare
    plant of one lake-shore in Ireland.   *I. salicina* L.

1. I. conyza DC.   Ploughman's Spikenard.
Biennial to perennial with an oblique stock and erect or ascending stems, 20–130 cm., often reddish, softly downy. Basal lvs ovate-oblong, narrowed into a flattened stalk; stem-lvs elliptical to lanceolate, subsessile, acute; all downy, especially beneath, and with small irregular teeth. Heads c. 1 cm. diam., numerous, in a terminal corymb. Outer

involucral bracts green, downy, with spreading or recurved tips; inner longer, narrow, ± scarious, often purple. Marginal florets tubular or nearly so, shorter than the inner bracts; all yellowish. Achenes c. 2 mm., dark brown; pappus reddish-white. Fl. 7–9. Dry or rocky slopes, cliffs and open scrub-woodland on chalk or limestone; locally common in England and Wales.

2. I. crithmoides L.   Golden Samphire.

Perennial glabrous maritime herb with a branched woody stock and ascending very fleshy stems, 15–90 cm. Lvs 2·5–6 cm., glabrous, fleshy, linear or oblanceolate, narrowed below to the sessile base, those on the main stem 3-toothed at the tip. Heads few, c. 2·5 cm. diam., in a terminal corymb, the lvs below them bract-like. Involucral bracts linear-lanceolate, glabrous, not spreading. Ray-florets numerous, narrow, golden-yellow, almost twice as long as the inner bracts; disk-florets orange-yellow. Fl. 7–8. Salt-marshes, shingle-banks and maritime cliffs and rocks of the south and west coasts from Essex round to Wigtown and Kirkcudbright; S. and E. Ireland from Kerry to Dublin.

9. PULICARIA Gaertner

Annual to perennial herbs differing from *Inula* in the 2-rowed pappus whose inner row is of hairs and outer of short scales.

Perennial; stem-lvs 3–8 cm., distinctly cordate; heads 1·5–3 cm. diam.;
  ray-florets much longer than disk-florets; outer pappus-scales united
  in a toothed cup.   **1. dysenterica**
Annual; stem-lvs 2–4 cm., not or hardly cordate; heads c. 1 cm. diam.;
  ray-florets hardly longer than disk-florets; outer pappus-scales free to
  the base. A rare and decreasing plant of places where water stands
  only in winter; S. England and Wales to Norfolk and Merioneth.
                                                    P. vulgaris Gaertner

1. P. dysenterica (L.) Bernh.   Fleabane.

Stoloniferous. Stems 20–60 cm., erect, sparsely hairy. Basal lvs oblong, narrowed to the base; middle and upper stem-lvs oblong-lanceolate to lanceolate, clasping and distinctly cordate at base; all 3–8 cm., entire or distantly toothed, densely and softly hairy. Heads 1·5–3 cm. diam., in a loose corymb; involucral bracts linear, scarious-tipped, hairy, glandular. Ray-florets numerous, ligulate, almost twice as long as the disk-florets; all florets golden-yellow. Achenes 1·5 mm., hairy; outer pappus a toothed cup, inner of long hairs. Fl. 8–9. Common in marshes, wet meadows, ditches, etc., throughout the British Is, northwards to Stirling and Kintyre.

10. FILAGO L.

Mostly annual herbs with the stems and spirally arranged lvs covered with woolly hairs. Heads heterogamous, small, in roundish terminal and axillary clusters; involucre 5-angled, of numerous bracts, the outer herbaceous, woolly, the inner scarious; receptacle conical, with scales only round the margin. Florets all tubular; outer very slender, female, in the axils of receptacular scales; central stouter, hermaphrodite. Achenes somewhat compressed; pappus of central florets of several rows of simple hairs, of the outer florets of 1 row or 0.

1   8–40 heads in each cluster; involucral bracts cuspidate, not spread-
    ing in fr.   2
    Fewer than 8 heads in each cluster; involucral bracts not cuspidate,
    spreading like a star in fr.   4
2   Lvs not apiculate, often waved at the margin; 20–40 heads in
    each cluster; clusters not overtopped by lvs at their base.
                                        1. vulgaris
    Lvs apiculate, hardly waved at the margin; 8–20 heads in each
    cluster; clusters overtopped by 1–5 lvs at their base.   3
3   Each cluster of heads with 3–5 lvs overtopping it and resembling
    an involucre; outer involucral bracts with recurved points;
    plants white-woolly, branching from the base. A rare plant of
    sandy fields and waysides in S. England northwards to Gloucester
    and Norfolk.   *F. pyramidata* L. (*F. spathulata* C. Presl)
    Clusters with 1–2 overtopping lvs; bracts with straight points;
    plant yellowish-woolly, branching above the middle. A rare plant
    of sandy fields in S. and C. England northwards to Worcester and
    S. Yorks.   *F. lutescens* Jordan (*F. apiculata* G. E. Sm. ex Bab.)
4   Lvs linear, 8–20 × 1 mm.; clusters of heads much overtopped by the
    lvs at their base; all achenes with pappus-hairs. Introduced and
    naturalized in a few places.   **F. gallica* L.
    Lvs 5–10 mm., linear-lanceolate; clusters of heads not overtopped
    by lvs at their base; outer achenes with no pappus.   **2. minima**

1. F. vulgaris Lam. (*F. Germanica* L. non Hudson) Common Cudweed.

Annual, with erect or ascending densely woolly stems, 5–30(–45) cm., simple or branched at the base and with further branches immediately beneath the terminal clusters of heads. Lvs 1–2(–3) cm., erect, lanceo-late, blunt or tapering to an acute tip, entire and usually waved at the margin, covered with white woolly hairs. Heads in ± sessile clusters of 20–40 terminating main stem and branches and half sunk in white woolly hairs; each cluster c. 12 mm. diam. Involucral bracts linear, erect, folded lengthwise; the outer short, woolly; the inner longer, scarious, yellowish, with a yellow awn-like point. Florets small, yellow. Achenes 0·6 mm.; outer with no pappus. Fairly common on heaths, dry pastures, fields and waysides, usually on acid sandy soils, throughout England, Wales, S. Scotland and Ireland; rare in N. Scotland.

2. F. minima (Sm.) Pers.   Slender Cudweed.

Annual, with slender erect or ascending stems, 5–15(–30) cm., branched above the middle, the branches ± erect. Lvs 5–10 mm., linear-lanceolate, acute, entire. Stem and lvs covered with greyish silky hairs. Heads in small clusters of 3–6 which are not overtopped by the lvs at their base. Involucral bracts lanceolate, blunt, woolly, with glabrous scarious yellowish tips, spreading like a star in fr. Achenes 0·5 mm.; outer with no pappus. Fl. 6–9. Locally common on acid sandy heaths and fields throughout Great Britain and Ireland, but rare in the extreme north.

11. GNAPHALIUM L.

Herbs or dwarf shrubs with spirally arranged simple lvs usually covered with white woolly hairs. Heads small, like those of *Filago* but with the involucral bracts all scarious and spreading in fr., receptacular scales 0 and pappus of 1 row of slender brittle hairs.

1	Perennials with heads in spikes or racemes, sometimes solitary; bracts with a dark marginal band.	2
	Annuals with heads in dense clusters; bracts almost uniform in colour.	4
2	A densely tufted plant usually only 2–12 cm. high; heads 1–7 in a short compact spike. An alpine plant of Scottish mountains. Dwarf Cudweed.	*G. supinum* L.
	Erect not densely tufted plants usually more than 12 cm. high; heads many, in long spikes.	3
3	Lvs narrow, 1-veined, steadily diminishing in size upwards; heads forming a long interrupted spike more than half the length of the stem.	**1. sylvaticum**
	Lvs lanceolate, 3-veined, not diminishing in size until well above the middle of the stem; spike short, compact at least above, about ¼ the length of the stem. A rare plant of high Scottish mountains. Highland Cudweed.	*G. norvegicum* Gunnerus
4	Lvs narrow, acute; clusters of heads overtopped by lvs at their base; bracts brownish.	**2. uliginosum**
	Lvs oblong, lower blunt; clusters of heads not overtopped by lvs; bracts straw-coloured. A rare plant of sandy fields and waste places in the Channel Is; probably introduced in S. and E. England. Jersey Cudweed.	*G. luteo-album* L.

1. G. sylvaticum L. Wood Cudweed.

Perennial, with a short woody stock producing short non-flowering and erect flowering shoots, the latter 8–60 cm., covered with whitish woolly hairs. Lvs of basal rosettes and lower stem-lvs 2–8 cm., lanceolate-acute, 1-veined, narrowed to a stalk; others very narrow, acute, sessile, diminishing steadily up the stem; all glabrous above, woolly beneath. Heads c. 6 mm., 1–8 in the axils of the upper stem-lvs and forming an interrupted lfy spike more than half the length of the stem. Bracts with a central green stripe and broad scarious margins, brownish towards the tip, the outer woolly below. Florets pale brown, only the 3–4 central ones hermaphrodite. Fl. 7–9. Locally common in dry open woods, heaths and dry pastures on acid soils throughout the British Is.

2. G. uliginosum L. Marsh Cudweed.

Annual with decumbent or ascending stems, 4–20 cm., much and diffusely branched near the base, densely covered with woolly hairs. Lvs 1–5 cm., narrowly oblong, narrowed below, woolly on both sides. Heads 3–4 mm., in dense ovoid sessile terminal clusters of 3–10, over-topped by the lvs at their base. Bracts pale brown and woolly below, darker and glabrous towards the tip. Florets yellowish, mostly female with a few central ones hermaphrodite. Pappus deciduous. Fl. 7–8. Common throughout the British Is in damp places in sandy fields, heaths, waysides, etc., on acid soils.

12. ANAPHALIS DC.

Perennial woolly herbs closely resembling *Antennaria* but more robust and less completely dioecious.

*1. A. margaritacea (L.) Bentham, Pearly Everlasting, has erect robust stems, 30–100 cm., elliptical acute ± entire lvs, 6–10 × 1–1·5 cm., woolly at least beneath, and corymbs of numerous heads, c. 10 mm. diam., whose brown involucral bracts have white shining scarious rounded tips. Introduced. Established in moist meadows, by rivers, on walls and in sandy and waste places in many localities but especially in parts of Wales and E. Scotland.

13. ANTENNARIA Gaertner

Perennial herbs or dwarf shrubs with spirally arranged narrow simple lvs, stems and lvs covered with whitish woolly or silky hairs. Heads clustered, ± dioecious; involucral bracts scarious, white or coloured, not spreading in fr.; heads of female plants with very slender tubular florets each with many rows of pappus-hairs; of male plants with broader hermaphrodite but usually sterile tubular florets each with a pappus of a few hairs thickened above like the antennae of a butterfly.

1. A. dioica (L.) Gaertner   Cat's-foot.

Perennial, with an above-ground creeping stock producing stolons which root at the nodes and erect unbranched woolly flowering shoots, 5–20 cm. Lvs 1–4 cm., mostly in rosettes, obovate-spathulate; upper stem-lvs erect, narrow, acute; all green and glabrous (or sparsely hairy) above, white-woolly beneath. Heads 2–8 in a close umbel; the female c. 12 mm., male c. 6 mm. diam. Outer involucral bracts woolly below, glabrous at the tip; of male heads broad, usually white, spreading like ray-florets; of female heads narrow, usually rose-pink, erect. Fl. 6–7. On heaths, dry pastures and dry mountain slopes throughout the British Is, but rare in the south.

14. SOLIDAGO L.

Perennial herbs with simple spirally arranged lvs. Heads small, yellow, panicled; involucral bracts herbaceous, in many rows; receptacle flat, naked. Ray-florets in 1 row, ligulate, female or neuter; disk-florets tubular, hermaphrodite. Achenes many-ribbed; pappus-hairs shortly ciliate, usually in 1 row.

1. S. virgaurea L.   Golden-rod.

Stock stout, obliquely ascending; stem erect, 5–75 cm., glabrous or downy. Basal lvs 2–10 cm., obovate or oblanceolate, narrowing to a short stalk, usually toothed; stem-lvs elliptical, ± acute, entire or obscurely toothed. Heads 6–10 mm., short-stalked, in a panicle with straight erect branches or in a simple raceme. Bracts greenish-yellow, very narrow, scarious-margined, glabrous or nearly so. Florets all yellow; ray-florets 6–12, spreading. Fl. 7–9. Common in dry woods and grassland, on rocks, cliffs, hedge-banks, dunes, etc., throughout the British Is, but rare in the south-east, where it avoids calcareous soils.

15. ASTER L.

Usually perennial herbs with spirally arranged simple lvs. Bracts lfy, in many rows; receptacle flat, naked, pitted; ray-florets in 1 row, ligulate, female or neuter, sometimes 0; disk-florets tubular, hermaphrodite. Achenes compressed, not ribbed; pappus of 2 to several rows of shortly ciliate hairs.

1 A maritime plant with markedly fleshy lvs; heads with or without
   bluish ray-florets. **1. tripolium**
   Lvs not markedly fleshy. 2

2 Lvs linear to linear-lanceolate, entire, rough on the margins; ray-
   florets 0. A very rare late-flowering plant of limestone rocks near
   the sea. Goldilocks.
                       A. linosyris (L.) Bernh. (*Linosyris vulgaris* Less.)
   Ray-florets bluish, reddish or whitish. Garden escapes, locally well
   established. Michaelmas Daisies.
               **A. novi-belgii* L., **A. novae-angliae* L., etc. (see larger *Flora*).

1. A. tripolium L. Sea Aster.

A biennial or short-lived perennial maritime herb with a short swollen
suberect stock and stout erect stems, 15–100 cm., glabrous. Lvs 7–
12 cm., fleshy, glabrous, faintly 3-veined, entire or almost so; basal lvs
oblanceolate to obovate, narrowing into a long stalk; stem-lvs narrow.
Heads 8–20 mm. diam. in flat-topped infls. Bracts few, narrow, blunt,
the outer scarious-tipped. Ray-florets spreading, blue-purple or whitish,
or 0; disk-florets yellow. Fl. 7–10. A common salt-marsh plant, found
also on maritime cliffs, rocks and sea-walls, all round the coasts of the
British Is, but almost confined to estuaries in N. England and Scotland;
the form with no ray-florets is very rare in N. England and Scotland.
   The China Aster of gardens is *Callistephus chinensis* Nees.

16. ERIGERON L. Fleabane.

Herbs with spirally arranged simple lvs, closely resembling *Aster* but
with the ligulate ray-florets narrower, more numerous and in 2 or more
rows and with the pappus usually of only 1 row of hairs.

1 Heads solitary, rarely 2–3; ray-florets purple. A rare alpine plant of
   C. Scotland. *E. borealis* (Vierh.) Simmons
   Heads usually several, in a panicle or corymb. 2

2 Heads 3–5 mm. diam., greenish-white, in a long panicle.
                                                       2. canadensis
   Heads at least 1 cm. diam., in a corymb. 3

3 Annual or biennial; lvs all entire, softly hairy; ray-florets pale
   purple. **1. acer**
   Perennial; some of the lower lvs usually 3-lobed or 3(–5)-toothed at
   the tip; lvs sparsely hairy or glabrous; ray-florets white above,
   purple beneath. Introduced and naturalized in S.W. England and
   Channel Is. **E. mucronatus* DC.

1. E. acer L.   Blue Fleabane.

Annual or biennial with an erect slender reddish stem, 8–10 cm., rough
with long hairs. Basal lvs 3–7·5 cm., obovate-lanceolate, stalked; stem-
lvs numerous, linear-lanceolate, half-clasping; all entire, hairy. Heads
1–2 cm. diam., 1–several in a corymbose panicle; involucral bracts
linear, red-tipped. Ray-florets pale purple, erect, little longer than the
yellow disk-florets; outer disk-florets very slender, female, inner broader,
hermaphrodite. Achenes 2–3 mm., yellowish; pappus reddish-white.
Fl. 5–9. Locally common in dry grassland, sand-dunes, banks and walls,
especially on calcareous substrata, throughout England and Wales;
local in Ireland, very rare in Scotland.

*2. E. canadensis L. (*Conyza canadensis* (L.) Cronq.)
                                                       Canadian Fleabane.

Annual with stiffly erect very lfy stems, 8–100 cm., much branched,
sparsely hairy or subglabrous. Basal lvs obovate-lanceolate, stalked,
± toothed, soon dying; stem-lvs 1–4 cm., narrow, acute, ± entire, hairy
and ciliate. Heads 3–5 mm. diam., numerous, in long panicles; involucral bracts linear-oblong, acute, with broad scarious margins, glabrous.
Ray-florets whitish, erect, female; disk-florets pale yellow, hermaphrodite. Achenes 1·5 mm., pale yellow, downy; pappus yellowish. Fl. 8–9.
Introduced and well established as a weed of waste ground, waysides,
cultivated land on light soils, dunes, walls, etc., throughout England and
Wales but rare in the north and only in a few places in S. Scotland.

17. BELLIS L.

Herbs with spirally arranged lvs often confined to a basal rosette. Heads
solitary; involucre of many herbaceous bracts in (1–)2 rows; receptacle-
scales 0. Ray-florets in 1 row, white or pink, female; disk-florets hermaphrodite. Achenes obovate, compressed, bordered, not ribbed; pappus 0.

1. B. perennis L.   Daisy.

Perennial, with short erect stock and stout roots. Lvs 2–4(–8) cm., confined to a basal rosette, obovate-spathulate, rounded at the end, crenate-
toothed, narrowed abruptly into a short broad stalk, sparsely hairy.
Scapes 3–12(–20) cm., lfless, hairy. Head 1·5–2·5 cm. diam.; involucral
bracts oblong, blunt, green or black-tipped, hairy. Ray-florets
numerous, narrow, spreading, white or pink; disk bright yellow.
Achenes 1·5–2 mm., pale, strongly compressed, ± downy. Fl. 1–10.
Abundant in short grassland (to 3000 ft.) throughout the British Is.

18. EUPATORIUM L.

Perennial herbs or shrubs with usually opposite lvs. Heads few-fld, white, pink or purplish, in terminal corymbs or panicles. Involucral bracts few, in 2–3 loose rows; receptacle-scales 0. Florets all tubular, hermaphrodite. Achenes 5-angled; pappus of 1 row of hairs.

1. E. cannabinum L.   Hemp Agrimony.

Perennial herb with a woody stock and erect downy striate stems, 30–120 cm. Basal lvs oblanceolate, stalked; stem-lvs subsessile, 3(–5)-partite, with elliptical-acuminate toothed segments, 5–10 cm.; branch-lvs simple, ovate or lanceolate; all shortly hairy and gland-dotted. Heads in dense corymbs, each head with 5–6 reddish-mauve or whitish florets and c. 10 oblong purple-tipped ± scarious involucral bracts. Styles white, long. Achenes blackish, gland-dotted; pappus whitish. Fl. 7–9. A common gregarious plant of marshes and fens, stream-banks and moist woods throughout most of the British Is.

19. ANTHEMIS L.

Usually strongly scented herbs with spirally arranged 1–3 times pinnately-divided lvs whose ultimate segments are linear. Heads solitary, usually heterogamous; involucral bracts blunt, scarious-margined; receptacle with narrow scarious scales. Ray-florets ligulate, long and narrow, female or neuter, sometimes 0; disk-florets hermaphrodite, yellow, tubular, the tube flattened and somewhat winged below. Achenes not or little flattened, ribbed on both faces, truncate above with a flattened apical disk as broad as the achene; pappus represented by a small membranous rim.

1 Ray-florets yellow; heads 2·5–4 cm. diam.; lvs white-woolly beneath.
      Introduced and rarely naturalized. Yellow Chamomile.
                                                                  *A. tinctoria L.
   Ray-florets white; heads not exceeding 3 cm.   2
2 Plant glabrous or slightly hairy, with a strong unpleasant smell;
      receptacle-scales linear-acute; ray-florets usually without styles;
      achenes tubercled.   **1. cotula**
   Plant pubescent or woolly, scarcely scented; receptacle-scales
      lanceolate-cuspidate; ray-florets with styles; achenes not
      tubercled.   **2. arvensis**

1. A. cotula L.   Stinking Mayweed.

Annual foetid herb with erect sparsely hairy stems, 20–60 cm. Lvs 1·5–5 cm., the ultimate segments narrowly linear, acute, ± glabrous. Heads

12–25 mm. diam., rather short-stalked. Involucral bracts oblong, blunt, ± glabrous, with green central stripe and broad scarious margins. Receptacle long-conical with linear-acute scales only near the apex. Ray-florets usually neuter (without styles), white, at first spreading, then reflexed. Achenes 2 mm., yellowish-white, 10-ribbed, tubercled on the back; pappus-rim inconspicuous, crenate. Fl. 7–9. A locally common weed of arable land and waste places, especially on heavy soils, in S. and C. England and Ireland, rarer elsewhere.

2. A. arvensis L.   Corn Chamomile.

Annual scarcely scented herb with decumbent or ascending stems, 12–50 cm., much branched below. Lvs 1·5–5 cm., the ultimate segments short, oblong, acute, hairy or even ± woolly beneath, especially when young. Heads 2–3 cm. diam., long-stalked. Involucral bracts oblong, scarious-tipped, pale, downy. Receptacle conical, with lanceolate cuspidate scales just exceeding the disk-florets. Ray-florets female (with style), white, spreading. Achenes 2–3 mm., whitish, ribbed all round, glabrous, not tubercled, wrinkled on the top; pappus-rim crenate. Fl. 6–7. Locally common in arable land and waste places on calcareous soils throughout Great Britain; introduced in E. Ireland.

20. CHAMAEMELUM Miller

Like *Anthemis* but the tube of the disk-florets cylindrical with an enlarged and persistent base covering the top of the fr., and the small ± unribbed achenes slightly flattened, rounded above, the apical disk being much less than the full width of the achene; pappus-rim inconspicuous.

The only British sp. is a perennial herb with white rays, and so cannot be confused with any of the three *Anthemis* spp. on p. 399; the receptacle-scales separate it from *Tripleurospermum* and *Matricaria*.

1. C. nobile (L.) All. (*Anthemis nobilis* L.)   Chamomile.

Perennial pleasantly scented herb with a short much-branched creeping stock and decumbent or ascending hairy stems, 10–30 cm. Lvs 1·5–5 cm., the ultimate segments short, linear-acute, sparsely hairy. Heads 1·8–2·5 cm. diam., long-stalked. Involucral bracts oblong, downy, with broadly scarious and laciniate white margins. Receptacle conical, its scales oblong, blunt, concave, often torn at the tip. Ray-florets female (with styles), broad, white, spreading, rarely 0. Achenes 1–1·5 mm. Fl. 6–7. Local on sandy commons and pastures and grassy roadsides throughout England, Wales and Ireland; casual elsewhere.

21. ACHILLEA L.

Usually perennial herbs with spirally arranged pinnately dissected or rarely simple lvs. Heads in corymbs, heterogamous; involucral bracts many-rowed, scarious-margined; receptacle with narrow scarious scales. Ray-florets female, ligulate, usually short and broad; disk-florets herm-aphrodite, tubular. Achenes strongly compressed, truncate above, not ribbed; pappus 0.

Lvs finely pinnatisect; heads 4–6 mm. diam., numerous, in a dense
   corymb.   **1. millefolium**
Lvs simple, broadly linear, finely toothed; heads 12–18 mm. diam.,
   few, in a lax corymb.   **2. ptarmica**

1. A. millefolium L.   Yarrow, Milfoil.

Perennial strongly scented far-creeping stoloniferous herb with erect, furrowed, usually simple, ± woolly stems, 8–45(–60) cm. Lvs 5–15 cm., lanceolate in outline, 2–3 times pinnate, the ultimate segments finely linear; basal lvs long, stalked; upper shorter, sessile, often with 2–3 small axillary lvs. Heads 4–6 mm. diam., numerous, in dense terminal corymbs. Involucre ovoid, its bracts rigid, oblong, blunt, keeled, ± glabrous, with a broad dark scarious margin. Ray-florets usually 5, as broad as long, 3-toothed at the tip, white, less often pink or reddish; disk-florets whitish. Achenes 2 mm., shining greyish, somewhat winged. Fl. 6–8. Common in meadows and pastures, on grassy banks, hedgerows and waysides on all but the poorest soils throughout the British Is.

2. A. ptarmica L.   Sneezewort.

Perennial herb with a creeping woody stock and erect angular stems, 20–60 cm., glabrous below but hairy above. Lvs 1·5–8 cm., linear-lanceolate, sessile, acute, ± glabrous, sharply and finely toothed. Heads 12–18 mm. diam., few, in a rather lax corymb. Involucre hemispherical, its bracts lanceolate to oblong, blunt, green-centred, ± woolly, with reddish-brown scarious margins. Ray-florets 8–13, ovate, white; disk-florets greenish-white. Achenes 1·5 mm., pale grey. Fl. 7–8. Common throughout the British Is in damp meadows and marshes and by streams.

22. TRIPLEUROSPERMUM Schultz Bip.

Annual to perennial herbs with spirally arranged repeatedly pinnately-cut lvs, the ultimate segments narrowly linear. Heads solitary, hetero-gamous; involucre of usually 2 rows of equal bracts with scarious margins; receptacle becoming conical in fr., solid; receptacular scales 0.

Ray-florets ligulate, white; disk-florets tubular, hermaphrodite. Achenes with 3 broad corky ribs on the inner face and 2 dark brown oil-glands at the top of the outer face; pappus represented by a membranous border crowning the achene.

1. T. maritimum (L.) Koch (*Matricaria maritima* L.)
Scentless Mayweed.

Annual to perennial almost scentless herbs with prostrate to erect glabrous stems, 10–60 cm. Lvs oblong in outline, glabrous. Heads 1·5–4(–5) cm. diam., long-stalked; involucral bracts oblong, blunt, with a narrow brown scarious margin. Ray-florets spreading. Achenes 2–3 mm. Fl. 7–9. Abundant throughout the British Is as an erect annual weed of arable and waste land; longer-lived prostrate or decumbent forms with fleshy lf-segments are locally common by the sea on drift-lines, shingle-beaches, cliffs, rocks, walls, etc.

23. MATRICARIA L.

Like *Tripleurospermum* but with hollow receptacle, conical from the first, and achenes with 5 slender ribs on the inner face; no oil-glands on the outer face of the achene. Ray-florets white or 0.

Ray-florets 0.   **2. matricarioides**
Ray-florets white, soon reflexed.   **1. recutita**

1. M. recutita L. (*M. chamomilla* L.)   Wild Chamomile.

Annual pleasantly aromatic herb resembling *Tripleurospermum maritimum*. Heads 12–22 mm. diam.; involucral bracts linear, blunt, yellowish-green with the narrow scarious margins much the same colour. Receptacle hollow, markedly conical from the first. Ray-florets soon reflexed, sometimes 0. Achenes 1–2 mm., pale grey, slender, obliquely truncate above and unrimmed, with 4–5 ribs on the inner face but no oil-glands on the outer. Fl. 6–7. Locally abundant as a weed of sandy or loamy arable soil and waste places throughout England and Wales; probably introduced elsewhere.

*2. M. matricarioides (Less.) Porter
Pineapple-weed, Rayless Mayweed.

Annual strongly aromatic herb with erect glabrous stems, 5–30 cm., much branched above, the branches rigid. Lf-segments bristle-pointed. Heads 5–8(–12) mm. diam., short-stalked; involucral bracts oblong, blunt, with broad scarious margins. Receptacle hollow, conical. Ray-florets 0; disk-florets dull greenish-yellow. Achenes 1·5 mm., with 4 inconspicuous ribs on the inner face and an obscure rim above (rarely a well-developed toothed or lobed crown). Fl. 6–7. Introduced. An abundant and increasing weed of waysides and waste places and especially of trampled places throughout the British Is.

24. CHRYSANTHEMUM L.

Herbs or shrubs with spirally arranged toothed or variously dissected lvs. Heads solitary or in corymbs, usually heterogamous; involucre hemispherical, its bracts with scarious margins; receptacle-scales 0. Ray-florets ligulate, female, yellow or white, or 0; disk-florets tubular, hermaphrodite, yellow. Pappus 0 or represented by a minute raised rim.

1 Heads with conspicuous ray-florets.   2
  Heads apparently without ray-florets.   **4. vulgare**
2 Ray-florets yellow.   **1. segetum**
  Ray-florets white.   3
3 Lvs simple, toothed; heads 3–5 cm. diam., solitary.
     2. leucanthemum
  Lvs pinnate; heads 12–22 mm. diam., in corymbs.   **3. parthenium**

1. C. segetum L.   Corn Marigold.

Annual herb with erect glabrous stems, 20–50 cm. Lvs 2–8 cm., glabrous, glaucous, somewhat fleshy; lower oblong-cuneate, narrowed into a winged stalk, coarsely toothed or pinnately lobed; upper oblong, sessile, clasping, toothed or not. Heads 3·5–6·5 cm. diam., solitary, with a long stalk thickened upwards; involucral bracts broadly ovate, glaucous, with very broad pale brown scarious margins. Ray-florets golden yellow, rarely 0. Achenes 2·5 mm., pale, strongly ribbed, those of ray with 2 narrow wings, of disk unwinged; pappus 0. Fl. 6–8. Perhaps introduced. Locally common as a weed of acid arable soils throughout the British Is.

2. C. leucanthemum L.   Moon-Daisy, Ox-eye Daisy, Marguerite.

Perennial herb with woody stock producing non-flowering lf-rosettes and erect flowering stems, 20–70 cm., sparsely hairy or subglabrous. Lower lvs roundish to obovate-spathulate, crenate or toothed, long-stalked; upper stem-lvs oblong, blunt, toothed to pinnate-lobed, sessile, half-clasping; all dark green, very sparsely hairy. Heads 3–6 cm. diam., solitary, long-stalked; involucral bracts green with narrow dark purplish scarious margins and tips. Ray-florets white, rarely 0. Achenes 2–3 mm., cylindrical, pale grey, 5–10-ribbed; only those of the ray with a pappus-rim. Fl. 6–8. Common in grassland on all the better types of soil throughout the British Is.

3. C. parthenium (L.) Bernh.   Feverfew.

Perennial strongly aromatic herb with ± vertical stock and an erect somewhat downy stem, 25–60 cm. Lvs 2·5–8 cm., yellowish-green, downy to subglabrous; lower lvs long-stalked, ovate in outline, pinnate, the lflets narrowly ovate, pinnate-lobed; upper lvs shorter stalked and less divided. Heads 12–22 mm. diam., long-stalked, in ± lax corymbs; involucral bracts bluntly keeled, downy, with narrow pale scarious margins and laciniate tips. Ray-florets white, short and broad. Achenes 1·5 mm., 8–10-ribbed, all with a short pappus-rim. Fl. 7–8. Probably introduced. Frequent on walls, waste places, hedge-banks, etc., throughout Great Britain and in several places in Ireland.

4. C. vulgare (L.) Bernh. (*Tanacetum vulgare* L.)   Tansy.

Perennial strongly aromatic herb with creeping stoloniferous stock and stiffly erect lfy stems, 30–100 cm., tough, angled, usually reddish, subglabrous. Lvs 15–25 cm., pinnate; the lower stalked, oblong in outline, with c. 12 pairs of deeply pinnate-lobed lflets whose segments are lanceolate, acute, sharp-toothed; the uppermost sessile, half-clasping, with long narrow acute sharp-toothed simple lflets; all dark-green and gland-dotted. Heads 7–12 mm. diam., numerous, in a dense corymb; outer involucral bracts ovate-lanceolate, inner narrower, all blunt with scarious margins. Ray-florets yellow, with so short a ligule as to appear tubular and like the disk-florets. Achenes 1·5 mm., greenish-white, 5-ribbed; pappus a short unevenly toothed membranous cup. Fl. 7–9. Roadsides, hedgerows, waste places, etc., throughout the British Is.

25. ARTEMISIA L.

Perennial herbs or small shrubs, rarely annuals, with spirally arranged usually pinnately dissected lvs. Heads small, 1–many, in racemes or racemose panicles, homogamous or heterogamous; involucre cylindrical or globular, of many bracts with narrow scarious margins; receptacle without scales. Marginal florets tubular and female or 0; disk-florets tubular, hermaphrodite. Achenes ± cylindrical, not strongly ribbed; pappus 0.

1 Plant only a few cm. high, with a single head (rarely 2) c. 12 mm. diam.; basal lvs silky white, cuneate or palmately lobed, the lobes

deeply 3–5-toothed. Known only on two mountains in W.
Scotland. *A. norvegica* Fries
Not as above. 2

2 Ultimate lf-segments at least 2 mm. wide. 3
Ultimate lf-segments not more than 1 mm. wide. 5

3 Lf-segments ± blunt, whitish, with hairs on both sides; heads
broadly bell-shaped. 4
Lf-segments usually acute, glabrous above, whitish beneath; heads
ovoid. **1. vulgaris**

4 Stems, lvs and heads densely white-felted; heads 5–7 mm. diam.,
longer than wide. Introduced and established in a few places in
S. England and near Dublin. Old Woman. **A. stellerana* Besser
Stem, lvs and heads covered with silky white hairs, not densely
felted; heads 3–5 mm. diam., wider than long. **2. absinthium**

5 Plant strongly aromatic; lvs white-woolly beneath; heads 2 mm.
diam.; maritime. **3. maritima**
Plant hardly scented lvs, ± glabrous or sparsely silky; heads 3–4 mm.
diam. A very local plant of the 'breckland' heaths of E. Anglia;
casual elsewhere. Field Southernwood. *A. campestris* L.

1. A. vulgaris L. Mugwort.

Perennial aromatic herb with erect, sparsely hairy, reddish, grooved
and angled stems, 60–120 cm. Lvs 5–8 cm., dark green and ± glabrous
above, white-woolly beneath; basal lvs short-stalked, lyrate-pinnate-
lobed; stem-lvs sessile, clasping, bipinnate, the uppermost
simply pinnate; ultimate lf-segments 3–6 mm. wide, acute. Heads 2–3 mm.
diam., ovoid, ± erect, in dense lfy racemose panicles; involucral bracts
narrow, woolly. Florets yellow or purplish, all fertile. Achenes c. 1 mm.
Fl. 7–9. Common in waste places, waysides, hedgerows, etc., throughout
the British Is.

2. A. absinthium L. Wormwood.

Perennial aromatic herb with non-flowering rosettes and erect silky-
hairy grooved and angled stems, 30–90 cm., ± woody below. Lvs
2·5–5(–10) cm., those of the barren rosettes and the lower stem-lvs tri-
pinnate, middle stem-lvs bipinnate and the uppermost simply pinnate or
undivided; ultimate lf-segments c. 2–3 mm. wide, usually blunt, gland-
dotted, whitish with silky hairs on both sides. Heads 3–4 mm. diam.,
rather wider than long, drooping, in a much-branched racemose panicle;
involucral bracts silky-hairy, the outer linear and inner ovate, all blunt
and broadly scarious-margined. All florets fertile, yellow. Achenes
1·5 mm. Fl. 7–8. Not infrequent in waste places throughout much of
the British Is.

3. A. maritima L.   Sea Wormwood.

Perennial strongly aromatic herb with non-flowering rosettes and decumbent then erect flowering shoots, 20–50 cm., usually downy. Lvs 2–5 cm., the ultimate segments c. 1 mm. wide, blunt; lower lvs bipinnate, stalked; upper simply pinnate, sessile; uppermost pinnate-lobed or entire; all ± woolly on both sides, not gland-dotted. Heads 1–2 mm. diam., longer than wide, in lfy racemose short-branched panicles; outer involucral bracts herbaceous, downy, inner with broad scarious margins. Florets yellowish or reddish, the central ones sometimes sterile. Achenes rarely ripened. Fl. 8–9. Locally common on the drier parts of salt and brackish marshes, also on drift-lines of the foreshore, sea-walls and sea-cliffs. Great Britain northwards to N. Aberdeen and Dunbarton; local in Ireland.

26. CARLINA L.

Thistle-like herbs or low shrubs with pinnate-lobed spinous lvs. Heads homogamous; outer involucral bracts lf-like, inner longer, scarious, coloured, shining, spreading in dry weather. Florets all tubular, hermaphrodite. Achenes cylindrical, covered with appressed forked hairs; pappus of feathery hairs in 1 row, united basally in groups of 2–4.

1. C. vulgaris L.   Carline Thistle.

Biennial with taproot and stiffly erect flowering stems, 10–60 cm., usually purplish, somewhat cottony. Rosette-lvs of the first season 7–13 cm., dying before flowering, oblanceolate-oblong, acute, narrowed into the sessile base, cottony especially beneath; stem-lvs shorter, half-clasping, subglabrous; all with waved and lobed weakly spiny margins. Heads 2–4 cm. diam., 2–5 or more in a corymb, rarely 1. Outer involucral bracts spiny, shorter than the many long linear straw-yellow inner ones which resemble ray-florets when dry. Achenes 2–4 mm., rusty-hairy. Fl. 7–10. Chalk and limestone grassland; locally common throughout most of lowland Great Britain and Ireland.

27. ARCTIUM L.

Tall biennial herbs with long stout tap-roots and large spirally arranged ovate non-spinous lvs. Heads in racemes or corymbs, homogamous; involucre ovoid-conical to globose, its bracts numerous, with appressed bases and long stiff slender spreading hooked tips; receptacle flat, with rigid awl-shaped scales. Florets all tubular, hermaphrodite, purple or white. Achenes obovoid-oblong, somewhat flattened; pappus of rough hairs in several rows, free to the base.

Lf-stalks solid; heads few, sub-corymbose; achenes pale.   **1. lappa**
Lf-stalks hollow; heads few to many, ±racemose; achenes
  brownish.   **2. minus**

1. A. lappa L.   Great Burdock.

Stems 90–130 cm., stout, furrowed, ±woolly, with many spreading-ascending branches. Lvs with ovate-cordate, entire or distantly toothed, usually blunt blades up to 40 cm. long, green and sparsely cottony above, grey-cottony beneath; lf-stalk up to 30 cm., solid, furrowed above. Heads 3–4 cm. overall diam., long-stalked, few, in subcorymbose terminal clusters, globose in bud, widely open above in fr. Involucral bracts green, almost glabrous, the basal reflexed in fr. Florets reddish-purple, the wider upper part of the corolla distinctly shorter than the slender lower part. Achenes 5–7 mm., pale. Fl. 7–9. Waste places and waysides; rarely in woods. Scattered throughout lowland Great Britain northwards to Argyll and Fife; in Ireland only in the north midlands.

2. A. minus Bernh., *sensu lato*   Lesser Burdock.

Stems 60–130 cm., furrowed, often reddish, somewhat woolly, with many spreading or down-curving branches. Basal lvs as in *A. lappa* but narrower (ovate-oblong) and acute, nearly glabrous above, sparsely cottony but green beneath; lf-stalk hollow. Heads in racemes or the top 2–4 subcorymbose. Florets red-purple, the wide upper part of the corolla about equalling the slender lower part. Achenes brownish. Fl. 7–9.

Ssp. minus has heads 1·5–2·2 cm. overall diam. (including spines), ovoid in bud and contracted at the top in fr., short-stalked or subsessile, in racemes.

Ssp. pubens (Bab.) J. Arènes differs from ssp. *minus* in having heads 2–3 cm. overall diam., open at the top in fr., with stalks to 15 cm.

Ssp. nemorosum (Lejeune) Syme (*A. vulgare* sensu A. H. Evans) is a more robust plant with heads 3–4 cm. overall diam., globular in bud and open at the top in fr., short-stalked or subsessile, in racemes but with the top 2–4 in a subcorymbose cluster.
Waste places, waysides, scrub and open woodland throughout the British Is.

28. Carduus L.

Thistles, whose spirally arranged simple or pinnate lvs have spiny margins. Heads solitary or in terminal clusters, homogamous; involucral bracts narrow, usually spine-tipped, in many rows; receptacle densely bristly. Florets all tubular, hermaphrodite. Achenes obovate-oblong, glabrous; pappus of many rows of simple but rough (not feathery) hairs, united below.

1 Heads oblong-cylindrical, falling when the achenes are ripe; corolla equally 5-lobed. 2

   Heads ovoid or hemispherical, not falling; corolla distincly 2-lipped with one entire and one 4-lobed lip. 3

2 Stem lfy up to close beneath fl.-heads and with continuous broad spinous wings; lvs somewhat cottony beneath; heads 2–10(–20) in a cluster; inner bracts equalling or exceeding florets.
                                                1. tenuiflorus

   Stem naked close beneath fl.-heads and with interrupted narrow spinous wings; lvs densely cottony beneath; heads solitary or 2–4 in a cluster; inner bracts falling short of florets. Introduced. Naturalized at Plymouth and a rare casual elsewhere.
                                     ****C. pycnocephalus* L.**

3 Heads 3–5 cm. diam., usually solitary, drooping; bracts lanceolate above then contracted abruptly into an oblong base, the middle and outer strongly reflexed. **2. nutans**

   Heads 1–3 cm. diam., solitary or clustered, erect; bracts linear, straight or the outermost recurved at the tip, not contracted above the base. **3. crispus**

1. C. tenuiflorus Curtis   Slender Thistle.

Annual to biennial with stout tap-root and erect broadly and continuously spinous-winged stems, 15–120 cm., ±cottony. Basal lvs oblanceolate in outline, blunt, narrowed to the base; stem-lvs decurrent, acute; all sinuate-pinnatifid, spinous-margined, ±cottony beneath. Heads about 15 × 8 mm., cylindrical, sessile, in dense terminal clusters of 3–10(–20), with the stem winged close beneath them. Bracts glabrous, ovate-lanceolate, acuminate, broadly scarious-margined, ending in a ±outwardly curved spine; inner bracts scarious, at least as long as the pale purple-red, rarely white, florets. Achenes 3·5 mm., transversely wrinkled, tubercled above. Fl. 6–8. Waysides and waste places, especially by the sea. Locally common throughout lowland Great Britain northwards to the Clyde Is and Moray; throughout Ireland but rare in the north midlands.

2. C. nutans L.   Musk Thistle.

Usually biennial with erect interruptedly spinous-winged stem, 20–100 cm., naked for some distance beneath the heads, ±cottony throughout. Basal lvs elliptical, narrowing into a stalk-like base, sinuate; stem-lvs oblong-lanceolate, decurrent, deeply pinnatifid with triangular

2–5-lobed spine-tipped segments; all with undulate and spinous margins, sparsely hairy on both sides, woolly on the veins beneath. Heads 3–5 cm. diam., solitary or 2–4 in a loose corymb, hemispherical, drooping. Involucre cottony, often purplish, the outer and middle bracts strongly reflexed, all spine-tipped, lanceolate-acuminate, contracted abruptly into an oblong base. Florets red-purple, distinctly 2-lipped. Achenes 3–4 mm., with fine transverse wrinkles. Fl. 5–8. Pastures, waysides, arable fields and waste places on calcareous soils. Locally common throughout Great Britain northwards to Ross; Inner Hebrides; S. and W. Ireland, very local.

3. C. crispus L.   Welted Thistle.

Biennial, with slender tap-root and erect cottony stem, 30–120 cm., usually naked just beneath the fl.-heads but otherwise with a continuous narrow wavy spinous-margined wing. Basal lvs elliptical in outline, sinuate-pinnatifid, narrowed into a stalk-like base; stem-lvs decurrent, deeply pinnatifid with 3-lobed ovate segments; all dull green, cottony beneath, with weakly spinous margins. Heads 20 × 10–20 mm., erect, almost globose, usually in dense clusters of 3–5, rarely solitary. Involucral bracts slightly cottony, linear, not contracted above the base, ending in a weak slender spine, the outermost spreading, green, the inner erect, purplish. Florets red-purple or white, 2-lipped. Achenes 3 mm., with fine transverse wrinkles and a prominent terminal tubercle. Fl. 6–8. Damp grass verges and stream-sides, hedgerows, waste places. A lowland plant common in the south but becoming rare in the north though reaching Ross and Inner Hebrides; throughout E. Ireland, very rare in the west.

29. CIRSIUM Miller

Thistles with spirally arranged simple or pinnatifid lvs, usually with spinous margins. Plants often gynodioecious; heads solitary or in corymbs or dense clusters, homogamous; florets all tubular, hermaphrodite or female. Involucre, receptacle and florets as in *Carduus* but achenes with a pappus of many rows of feathery hairs united at the base.

1   Plant usually stemless, with a single sessile head at the centre of the
    lf-rosette.   **5. acaule**
    Plant with elongated stem.   2
2   Lvs bristly on the upper surface, and dull, not shining.   3
    Lvs not bristly (though sometimes pubescent) on the upper surface
    and often more or less glossy.   4

3   Stem not winged; involucre very cottony.   **1. eriophorum**
      Stem interruptedly spiny-winged; involucre not or scarcely cottony.
                                                 2. vulgare

4   Stem continuously spiny-winged.   **3. palustre**
      Stem not winged, or, if so, only for very short distances.   5

5   Heads yellow, overtopped by large pale ovate bract-like uppermost
      lvs; most stem-lvs ovate-acuminate, sessile and clasping, hardly
      spinous on the margins. Introduced. Naturalized in a few places.
                                      C. oleraceum (L.) Scop.
      Heads purple-red, rarely white.   6

6   Heads 1·5–2·5 cm.; florets with the broad upper part of corolla
      shorter than the slender basal tube, 5-cleft almost to its base.
                                             4. arvense
      Heads 2·5–5 cm.; florets with the broad upper part of the corolla
      about equalling the slender basal tube, 5-cleft to about the middle.   7

7   Lvs densely white-felted beneath with softly prickly-ciliate margins;
      head 3·5–5 cm.   **6. heterophyllum**
      Lvs at most whitish-cottony beneath, not densely felted; head
      2·5–3 cm.   8

8   Lvs deeply pinnatifid, green beneath; heads solitary or 1–4 in a
      cluster; roots swollen, spindle-shaped. A very local plant of cal-
      careous pastures in England and Wales. Tuberous Thistle.
                                      C. tuberosum (L.) All.
      Lvs sinuate-toothed, sometimes ±lobed, whitish-cottony beneath;
      heads usually solitary; roots not swollen.   **7. dissectum**

1. C. eriophorum (L.) Scop.   Woolly Thistle.

Biennial, with thick tap-root and stout erect unwinged furrowed stem,
60–150 cm., cottony, not prickly. Basal lvs up to 60 cm., ovate-oblong
in outline, narrowing into a short stalk, deeply pinnatifid and strongly
undulate, the narrowly lanceolate distant spine-tipped segments usually
2-lobed with 1 lobe directed upwards and 1 downwards; stem-lvs similar
but sessile, half-clasping the stem with basal auricles; all bristly but
green above, white-cottony beneath, spiny and ciliate on the margins.
Heads 3–3·5 × 4–7 cm., usually solitary, ± erect. Involucre very cottony,
its bracts lanceolate-acuminate, ending in a long narrow spreading
ciliate point slightly broadened just beneath its tip; outermost spine-
tipped. Florets pale red-purple. Anthers blue-purple. Achenes 6 mm.,
smooth, shining. Fl. 7–9. Grassland, open scrub and roadsides on
calcareous soil. Local in England and Wales northwards to Westmorland and Durham; introduced further north.

2. C. vulgare (Savi) Ten. Spear Thistle.

Biennial, with long tap-root and erect interruptedly spiny-winged cottony furrowed stems, 30–150 cm. Basal lvs 15–30 cm., obovate-lanceolate in outline, narrowed into a short stalk-like base, ± deeply pinnatifid and waved, with the segments usually 2-lobed, the upper lobe toothed, the lower entire, lobes and teeth tipped with long stout spines; stem-lvs sessile; all hairy and prickly above (and not glossy), rough or cottony beneath. Heads 3–5 × 2–4 cm., solitary or 2–3 in a cluster, short-stalked. Involucre at most slightly cottony, its bracts narrow, green, with the long point recurved and spine-tipped in the outer, erect and scarious in the inner. Florets pale red-purple. Achenes 3·5 mm., yellow streaked with black. Fl. 7–10. Fields, waysides, gardens, waste places. Common throughout the British Is.

3. C. palustre (L.) Scop. Marsh Thistle.

Biennial, with short erect stock and erect, furrowed, narrowly but con-tinuously spiny-winged, hairy and cottony stem, 30–150 cm. Basal lvs narrowly oblanceolate in outline, stalked, pinnatifid, the lobes shallow with spinous margins; stem-lvs sessile, deeply pinnatifid and waved, the segments each with 2–3 spine-tipped and spiny-ciliate lobes; all pubescent above, slightly cottony beneath. Heads 1·5–2 × 1–1·5 cm., short-stalked in crowded lfy clusters. Involucre slightly cottony, its bracts narrow, purplish, appressed. Florets dark red-purple, rarely white. Achenes 3 mm., pale fawn. Fl. 7–9. Marshes, moist grassland, hedgerows, woods. Abundant throughout the British Is.

4. C. arvense (L.) Scop. Creeping Thistle, Field Thistle.

Perennial, initially with slender tap-root from which arise far-creeping whitish lateral roots bearing numerous adventitious non-flowering and flowering shoots, the latter 30–90(–150) cm., erect, furrowed, unwinged, glabrous or cottony above. Basal lvs not in a compact rosette, oblong-lanceolate in outline, narrowed to a short stalk-like base, usually ± pinnatifid and waved with triangular toothed and spiny-ciliate lobes ending in strong spines; upper lvs sessile, half-clasping, more deeply pinnatifid; all ± glabrous, or cottony beneath. Plants ± dioecious; heads 1·5–2·5 cm., short-stalked, solitary or in terminal subcorymbose clusters of 2–4; involucre glabrous or somewhat cottony, that of male heads globose, of female ovoid; involucral bracts purplish, mostly appressed, outer short, ovate-mucronate, spine-tipped, inner longer, lanceolate, scarious-tipped. Florets dull pale purple or whitish, the broad upper part of the corolla shorter than the slender basal tube and

5-cleft to its base. Achenes 4 mm., dark brown. Male heads sometimes ripen a few fr.; female heads have abortive anthers. Fl. 7–9. Fields, waysides, waste places, etc. Abundant throughout the British Is. Very variable.

5. C. acaule Scop.   Stemless Thistle.

Perennial, with short rhizome and a rosette of lvs, 10–15 × 2–3 cm., oblong-lanceolate in outline, stalked, ± deeply pinnatifid, strongly and stiffly waved, the segments with 3–4 triangular teeth or lobes, stoutly spine-tipped and spiny-ciliate, ± glabrous and shining above, hairy on the veins beneath. Plants gynodioecious; heads 3–4 × 2·5–5 cm., ovoid, 1(–3), usually sessile on the rosette, rarely on an elongated stem. Involucre glabrous, purplish, the outer bracts ovate with short spiny point, inner narrow, blunt; all appressed. Florets bright red-purple. Achenes 3–4 mm. Fl. 7–9. In closely grazed pastures, especially on chalk or limestone. Locally common in England northwards to Yorks; Glamorgan and Denbigh.

6. C. heterophyllum (L.) Hill   Melancholy Thistle.

Perennial, stoloniferous, with obliquely ascending stock and erect grooved cottony unwinged stem, 45–120 cm. Basal lvs 20–40 × 4–8 cm., elliptic-lanceolate, long-stalked, finely toothed; all but the lowest stem-lvs lanceolate-acuminate, unlobed, broad-based, clasping the stem with rounded auricles, entire or somewhat toothed; all flat and soft, green and glabrous above, white-felted beneath, with softly prickly-ciliate margins. Heads 3·5–5 × 3–5 cm., usually solitary. Involucre broadly ovoid, its bracts ovate-lanceolate, appressed, purplish-tipped. Florets red-purple, rarely white. Achenes 4–5 mm., fawn. Fl. 7–8. Hilly pastures and stream-sides, upland scrub and open woods. Locally common in Great Britain from Merioneth, Stafford and Derby northwards; Hebrides; Fermanagh.

7. C. dissectum (L.) Hill   Marsh Plume Thistle.

Perennial, shortly stoloniferous, with obliquely ascending stock, non-swollen roots and erect terete cottony unwinged stem, 15–80 cm., usually with a few small bract-like lvs above the middle. Basal lvs 12–25 × 1·5–3 cm., elliptic-lanceolate, long-stalked, sinuate-toothed or pinnatifid; stem-lvs usually only 3–5, half-clasping; all green and hairy above, whitish-cottony beneath, the margins with soft prickles. Heads 2·5–3 × 2–2·5 cm., usually solitary. Involucre purplish, cottony, its bracts lanceolate, appressed, outer spine-tipped. Florets dark red-purple.

Achenes 2·5 mm., pale fawn. Fl. 6–8. Fens and bog-margins, always on wet peat. Local in England and Wales northwards to Yorks; throughout Ireland and common in the west.

30. SILYBUM Adanson

Annual or biennial thistles with white-veined or otherwise variegated lvs. Heads homogamous; involucre of many rows of spiny bracts; receptacle hairy. Florets all tubular and hermaphrodite, red-purple; stamen-filaments united at their base into a tube. Achenes obovoid, flattened, crowned by a membranous rim; pappus of many rows of rough hairs united below into a basal ring.

*1. S. marianum (L.) Gaertner   Milk-Thistle.

Annual to biennial with erect grooved slightly cottony unwinged stem, 40–120 cm. Lvs oblong, sinuate-lobed or pinnatifid, with strongly spinous margins, glabrous, pale shining green variegated with white along the veins above; basal lvs narrowed into a sessile base, stem-lvs clasping, with rounded spiny-ciliate auricles. Heads 4–5 × 1–2 cm., solitary, erect or ± drooping. Involucre glabrous, the outer bracts with an ovate-oblong base surmounted by a triangular spiny-ciliate lfy appendage which, in all but the basal bracts, ends in a stout yellowish spine, long and spreading or recurved in the middle bracts, shorter and erect in the innermost. Achenes 6–7 mm., blackish, grey-flecked and with a yellow ring near the tip, transversely wrinkled. Fl. 6–8. Introduced. Naturalized locally in waste places and a common casual throughout lowland Great Britain and Ireland northwards to Aberdeen.

31. ONOPORDUM L.

Biennial thistles. Heads homogamous; involucre of many rows of leathery spiny bracts; receptacle not hairy, deeply pitted. Florets all tubular, hermaphrodite, usually red-purple. Achenes obovoid, flattened or 4-angled; pappus of many rows of rough hairs united into a basal ring.

1. O. acanthium L.   Scotch Thistle, Cotton Thistle.

A large biennial thistle with stout tap-root and stiffly erect continuously and broadly spinous-winged white-woolly stem, 45–150 cm. Lvs sessile, elliptic-oblong, with sinuate teeth or triangular lobes ending in strong spines, cottony on both sides. Heads 3–5 × 3–5 cm., usually solitary.

Involucre subglobose, cottony, its bracts narrow, green, tipped with yellowish spreading or reflexed spines. Florets pale purple, rarely white. Achenes 4–5 mm., dark-mottled, transversely wrinkled; pappus pale reddish. Fl. 7–9. Doubtfully native. Fields, roadsides and waste places. Scattered throughout Great Britain northwards to Ross, but rare in Scotland.

32. CENTAUREA L.

Usually herbs with spirally arranged non-spinous lvs. Heads homogamous or heterogamous; involucral bracts numerous, each with a membranous or scarious terminal appendage which is pectinate, laciniate, toothed, ciliate or spiny, rarely entire; receptacle bristly. Florets tubular, either all similar and hermaphrodite or, more commonly, the marginal larger and neuter. Achenes flattened, smooth, with an oblique attachment-scar; pappus of several rows of rough hairs free to the base or of scaly bristles or 0.

1 Involucral bracts with a non-spiny terminal appendage which is at
      most minutely prickly.   2
   Bracts with a distinctly spiny appendage which is simple, palmate or
      pinnate.   6

2 Appendages of involucral bracts decurrent, i.e. extending some way
      down the sides of the basal part of the bract.   3
   Appendages not or very slightly decurrent.   4

3 Florets red-purple; perennial.   **1. scabiosa**
   Florets blue; annual herb with erect wiry stem, 20–90 cm., narrow
      grey-cottony lvs, outer bracts with silvery appendages, and heads
      with large bright blue marginal florets. Weed of cornfields and
      waste places, formerly common, now rare. Cornflower, Blue-
      bottle.   *C. cyanus* L.

4 Involucre 5–6 mm. diam.; appendages of bracts slightly decurrent,
      triangular, pectinate-ciliate, pale brown. Introduced. Established
      in Jersey, casual elsewhere. Panicled Knapweed.
                                                                   *C. paniculata L.
   Involucre at least 10 mm. diam.; appendages not decurrent.   5

5 Appendages of involucral bracts dark brown or blackish, dull, flat,
      those of outer bracts deeply pectinate; pappus usually present.
                                                                               2. nigra
   Appendages pale brown, scarious, shining, concave, their margins
      often laciniate but not pectinate; pappus 0. Resembling *C. nigra*
      and hybridizing with it. Introduced. Established but rare in
      S. England, casual elsewhere. Brown-rayed Knapweed.
                                                                     *C. jacea L.

6   Appendages of involucral bracts spreading or reflexed, with short
     equal palmately arranged spines; bracts yellowish, appendages
     red-brown; florets pale red-purple. A rare plant of Channel Is;
     naturalized in S. Wales, casual elsewhere. Rough Star-Thistle.
                                                 C. aspera L.
     Appendages with a long terminal and much shorter lateral spines.   7
7   Heads red-purple; terminal bract-spine 2–2·5 cm., yellow. Rare in
     waysides and waste places in S. England and S. Wales. Star
     Thistle.   *C. calcitrapa* L.
     Heads yellow; terminal bract-spine not more than 2 cm., yellow.
     Introduced. Rare in cultivated land in S. and E. England, casual
     elsewhere. St. Barnaby's Thistle.   *C. solstitialis* L.

1. C. scabiosa L.   Greater Knapweed.

Perennial, with stout oblique woody stock enclosed above in fibrous
scales, and erect grooved ± pubescent stems, 30–90 cm. Basal lvs
10–25 cm., oblanceolate in outline, stalked, usually deeply pinnatifid
with entire or pinnatifid lobes, sometimes merely toothed or entire;
stem-lvs sessile, deeply pinnatifid; all firm, roughly but sparsely hairy on
both sides, often shining above. Heads 3–5 cm. diam., solitary, long-
stalked. Involucre ovoid-globose, its bracts with blackish-brown horseshoe-shaped pectinate decurrent appendages. Florets red-purple, with
or without a larger marginal row. Achenes 4–5 mm., greyish; pappus
present. Fl. 7–9. Dry grassland, hedge-banks, roadsides, cliffs, etc.,
especially on calcareous substrata. Throughout lowland Great Britain
but rare in Scotland; Ireland.

2. C. nigra L.   Lesser Knapweed, Hardheads.

Perennial, with stout oblique stock and erect tough rigid stems, 15–
60(–90) cm., grooved, usually ± roughly hairy. Basal lvs stalked, entire
or sinuate-toothed or somewhat pinnatifid; stem-lvs sessile, entire or
with a few teeth towards the base; all rather softly hairy, often cottony
beneath when young. Heads 2–4 cm. diam., subsessile, solitary, the
stem often conspicuously swollen just beneath. Involucre ovoid-globose,
its bracts with brown or blackish, non-shining, flat, triangular, non-
decurrent, pectinate appendages. Florets red-purple, usually without a
larger marginal row. Achenes 3 mm., pale brown; pappus of short
bristly hairs. Fl. 6–9. Grassland, waysides, cliffs, etc. Throughout the
British Is. except Shetland. Variable. Forms with pale but pectinate
involucral bracts, with a marginal row of larger florets, or with no
pappus, may result from hybridization with *C. jacea*.

33. SERRATULA L.

Perennial herbs with spirally arranged non-spinous lvs. Heads homogamous; involucre with imbricating acute but not spinous bracts in many rows; receptacle with dense chaffy scales. Florets all tubular. Achenes slightly flattened, glabrous; pappus of many rows of stiff rough simple hairs, all free to the base.

1. S. tinctoria L.   Saw-wort.

Stock short, stout, ± erect; stem 30–90 cm., erect, slender, wiry, grooved. Lvs 12–25 cm., ovate-lanceolate to lanceolate in outline, ± glabrous, undivided to almost pinnate; basal and lower stem-lvs stalked, upper sessile, all with fine bristle-tipped marginal teeth. Heads 1·5–2 cm., stalked or subsessile, dioecious, the female larger than the male with its florets exceeding the ovoid-cylindrical involucre, those of the male equalling the narrower involucre; all involucral bracts appressed, purplish where exposed. Florets reddish-purple, rarely white; anthers of male dark blue, of female white, abortive. Fl. 7–9. Wood margins, clearings and rides and open grassland on moist basic soils over limestone or chalk; local throughout England and Wales and S.W. Scotland; 1 station in S. Ireland.

Subfamily CICHORIOIDEAE: plants with milky latex; all florets ligulate, the ligules 5-toothed at their tips.

34. CICHORIUM L.

Usually perennial herbs with spirally arranged often runcinate basal lvs. Heads terminal and axillary; bracts of the cylindrical involucre in 2 rows; receptacle naked. Florets usually blue. Achenes obovoid, angled, flat-topped, crowned with 1–2 rows of short blunt scales.

1. C. intybus L.   Chicory, Wild Succory.

Perennial, with long stout tap-root, short erect stock and stiffly erect, tough, grooved stems, 30–120 cm., glabrous to roughly hairy. Basal lvs short-stalked, oblanceolate in outline, runcinate-pinnatifid or toothed; stem-lvs less divided to entire, sessile, clasping the stem with basal auricles; all glandular-ciliate. Heads 2·5–4 cm. diam., one terminating the stem and others in subsessile axillary clusters of 2–3. Outer involucral bracts spreading above; inner twice as long, erect; all green. Florets bright blue. Achenes with a pappus of fimbriate scales. Fl. 7–10.

Probably native. Roadsides and pastures; locally common especially on calcareous lowland soils in England and Wales; rare and probably introduced in Scotland and Ireland.

35. LAPSANA L.

Annual herbs with small yellow heads in loose panicles or corymbs. Involucral bracts erect, in 1 row with a few very small basal scales; receptacle naked. Achenes curved, about 20-ribbed, with neither pappus nor border.

Heads 1·5-2 cm. diam.; involucre 6-8 mm.   **1. communis**
Heads 2·5-3 cm. diam.; involucre 8-9 mm. Introduced. Known only
   on a railway bank in Beds.   **L. intermedia* Bieb.

1. L. communis L.   **Nipplewort.**

Stem 20-90 cm., erect, usually with spreading hairs below. Lower lvs long-stalked, lyrate-pinnatifid, with a very large ovate terminal lobe and usually a few small lateral lobes; upper lvs ovate to lanceolate, acute, short-stalked, entire or nearly so; all thin, usually hairy. Heads 1·5-2 cm. diam., many, in a corymbose panicle. Involucre 6-8 mm., of 8-10 linear abruptly blunt-tipped bracts, erect and strongly keeled in fr. Fl. 7-9. Waysides, hedges, wood-margins, walls and waste places; common throughout the British Is.

36. ARNOSERIS Gaertner

Small annual herb with lvs confined to a basal rosette and yellow heads. Involucral bracts in 1 row apart from a few tiny basal scales; receptacular scales 0. Achenes broadly obovoid, strongly ribbed, crowned by a very short membranous rim.

1. A. minima (L.) Schweigg. & Koerte   Lamb's or Swine's Succory.

Lvs 5-10 cm., obovate-oblong, narrowed into a short stalk, distantly and rather coarsely toothed, ciliate, glabrous or somewhat hairy. Scapes 7-30 cm., many, simple or sparingly branched above, the branches (in the axils of minute bracts) curving upwards and overtopping the main stem; all hollow and markedly club-shaped beneath the heads. Heads 7-11 mm. diam., solitary, terminal. Involucral bracts ± glabrous with a prominent pale keel, connivent in fr. Achenes 1·5 mm., with 5 strong and 5 weaker intermediate ribs. Fl. 6-8. Arable fields on sandy soils. Local in S. and E. Great Britain from Dorset and Kent northwards to Moray.

37. HYPOCHAERIS L.

Herbs, with lvs confined to a basal rosette or with a few small lvs on the flowering stems (scapes) which are usually thickened beneath the heads. Involucre of many imbricating rows of narrow bracts. Receptacle with numerous narrow chaffy scales. Florets yellow. Achenes cylindrical, ribbed, at least the inner beaked; pappus of 1 row of feathery hairs often with an outer row of shorter simple hairs.

1   Annual; lvs ± glabrous; florets about equalling the involucre; heads
    opening only in full sun. **2. glabra**
   Perennial; lvs roughly hairy; florets exceeding the involucre; heads
    opening in dull weather *2*

2   Heads usually several; scape thickened for some distance beneath
    the heads; lvs not spotted; pappus-hairs in 2 rows, the inner
    feathery, outer simple. **1. radicata**
   Heads 1(–4); scape thickened only just beneath the heads or not at
    all; lvs usually purple-spotted; pappus of 1 row of feathery hairs.
    A rare plant of calcareous pastures and grassy cliffs, with large
    lemon-yellow heads. Spotted Cat's Ear. *H. maculata* L.

1. H. radicata L.
Cat's Ear.

Perennial, with short erect branching stock. Lvs 7–25 cm., broadly oblong-lanceolate, narrowed into a broad stalk-like base, somewhat toothed or lobed, rough with simple hairs, dull green above. Scapes 20–60 cm., often many from each rosette, lfless or with 1–2 small lvs below, forking, enlarged beneath the heads and bearing many small scale-like bracts. Heads 2·5–4 cm. diam. Florets bright yellow, exceeding the involucre, the outer green or grey-violet beneath. Achenes orange, strongly muricate. Fl. 6–9. Meadows, pastures, grassy dunes, waysides, etc.; common throughout the British Is.

2. H. glabra L.
Smooth Cat's Ear.

Annual, with a rosette of oblanceolate pale green glabrous lvs, somewhat toothed or lobed and narrowed to a stalk-like base. Scapes 10–40 cm., usually many from each rosette, erect to decumbent, with 0–2 small lvs, the branches slightly enlarged beneath the heads and with a few scale-like bracts. Heads 12–15 mm. long, opening widely only in full sun; involucral bracts narrow, with whitish margins and dark tips, about equalling the bright yellow florets. Achenes reddish-brown, muricate. Fl. 6–10. Grassy fields, derelict arable land, heaths, fixed dunes, etc., on sandy soils; locally frequent northwards to Moray and Inverness; Derry only in Ireland.

38. LEONTODON L.

Usually perennial scapigerous herbs with lvs confined to a basal rosette; lvs and scapes commonly with forked hairs. Involucral bracts in several rows. Receptacular scales 0. Achenes little flattened, strongly ribbed, narrowed above and sometimes beaked; pappus usually of 2 rows of hairs, the inner feathery, the outer simple or 0, but the outermost achenes sometimes with a small cup of scarious scales instead of hairs.

1 Lvs glabrous or with simple hairs; scape usually branched and
 bearing 2 or more heads; pappus of a single row of feathery hairs.
                1. autumnalis
 Lvs usually with forked hairs; scape with a single terminal head;
 pappus of central achenes with an inner row of feathery and an
 outer of shorter simple hairs.   2
2 Scape usually densely hairy above; involucre more than 10 mm.
 long; outer florets orange or reddish beneath; achenes all with
 feathery hairs.   **2. hispidus**
 Scape sparsely hairy, especially below; involucre less than 10 mm.
 long; outer florets grey-violet beneath; outermost achenes with
 pappus of scales.   **3. taraxacoides**

1. L. autumnalis L.   Autumnal Hawkbit.

Perennial, with a branched stock and several rosettes of lvs oblanceolate in outline, distantly toothed to deeply pinnatifid, glabrous or with simple hairs. Scapes 5–60 cm., ascending, usually branched, glabrous or sparsely hairy, somewhat enlarged and hollow above and bearing numerous scale-like bracts just beneath the heads. Heads 12–35 mm. diam. Involucral bracts narrow, dark green, glabrous to woolly. Fls golden-yellow, outer usually red-streaked beneath. Achenes with no beak and all with a pappus of a single row of feathery hairs. Fl. 6–10. Meadows, pastures, waysides, screes, etc., to 3200 ft. Abundant throughout the British Is.

2. L. hispidus L.   Rough Hawkbit.

Perennial, with a branched stock and several rosettes of lvs oblanceolate in outline, distantly toothed to runcinate-pinnatifid, usually rough with forked hairs. Scapes 10–60 cm., unbranched, usually densely hairy at least above and with 0–2 small bracts beneath the single head. Head 25–40 mm. diam. Involucral bracts narrow, dark green, ± hairy. Fls golden-yellow, usually orange or reddish beneath. Achenes narrowing from below the middle but not beaked, longitudinally ribbed and transversely ridged; pappus of an inner row of feathery and an outer of shorter simple hairs. Fl. 6–9. Meadows, pastures, grassy slopes, etc., especially on calcareous soils, throughout Great Britain and in Ireland northwards to Mayo and Meath.

3. L. taraxacoides (Vill.) Mérat (*L. leysseri* G. Beck)   Hairy Hawkbit

Perennial, rarely biennial, with short erect stock and a basal rosette of narrowly oblanceolate lvs, distantly toothed to runcinate-pinnatifid, glabrous or with ciliate margins and scattered forked hairs. Scapes 2·5–30 cm., ascending, slender, bractless, usually sparsely hairy below. Head 12–20 mm. diam., solitary. Involucral bracts narrow, glabrous or with bristly midrib and ciliate margins. Fls pale yellow, the outermost grey-violet beneath. Central achenes chestnut, straight, with a pappus of feathery hairs; outer pale brown, curved, with a pappus of scarious scales. Fl. 6–9. Dry grassland, especially on basic soils and sand-dunes. Throughout the British Is northwards to Aberdeen and Argyll.

39. PICRIS L.

Stiffly hairy herbs with flowering stems bearing spirally arranged entire or sinuate-toothed lvs. Involucral bracts in many rows, the innermost longest. Florets yellow. Achenes curved, ribbed; pappus of 2 rows of hairs, the inner always feathery, deciduous.

Outer involucral bracts 3–5, large, ovate-cordate, resembling an
  epicalyx; achenes long-beaked.   **1. echioides**
Outer bracts small, narrow, ±spreading; achenes not or very shortly
  beaked.   **2. hieracioides**

1. P. echioides L.   Bristly Ox-Tongue.

Annual or biennial, with stout furrowed erect stem, 30–90 cm., covered with short rigid swollen-based hairs trifid and minutely hooked at the tip. Lower lvs oblanceolate, narrowed into a stalk-like base; middle and upper lvs lanceolate to oblong, sessile, somewhat clasping at the base; all coarsely toothed or sinuate, bristly-ciliate and very rough with scattered bristles on white swollen bases. Heads 2–2·5 cm. diam. in an irregular corymb. Outer bracts 3–5, broadly cordate-acuminate, rough like the lvs, not quite equalling the narrow awned inner bracts. Florets yellow, the outermost purplish beneath. Achenes 2·5–3·5 mm., transversely wrinkled, with a beak about as long as the achene; pappus of white feathery hairs. Fl. 6–10. Doubtfully native. Roadsides, hedgebanks, field-margins and waste places, especially on stiff and calcareous soils. Locally common in lowland England and Wales and in extreme S.E. Scotland; S. and E. Ireland to Clare and Antrim.

2. P. hieracioides L.   Hawkweed Ox-Tongue.

Biennial to perennial, with stout furrowed erect stem, 15–90 cm., rough with short forked and hooked bristles especially below and with spreading corymbose branches above. Lower lvs 10–20 cm., oblanceolate, narrowed into a stalk-like base; middle and upper lvs lanceolate, usually broadened and somewhat clasping at the base; all bristly at least on the veins beneath and with sinuate-toothed, waved and bristly-ciliate margins. Heads 2–3·5 cm. diam., solitary at the somewhat thickened ends of the main stem and branches. Outer bracts short, narrow, usually spreading or recurved, blackish-haired; inner lanceolate, with bristles and white hairs down a central strip. Florets bright yellow. Achenes 3–5 mm., finely wrinkled; beak very short; pappus of cream-coloured feathery hairs. Fl. 7–9. Grassland, especially on calcareous slopes, waysides, etc. Locally common in lowland England and Wales and reaching S. Scotland; introduced in Ireland.

40. TRAGOPOGON L.

Herbs with abundant latex and linear or linear-lanceolate entire long-pointed sheathing lvs resembling those of leeks. Heads large, yellow or purple. Involucre of 1 row of lanceolate-acuminate bracts united at their base. Achenes 5–10-ribbed with a long beak ending upwards in a hairy ring; pappus of 1 row of hairs simple below and densely feathery at the tips except for 5 which are longer than the rest and simple throughout.

Fls yellow; stem and branches slightly enlarged beneath heads.
                                                     1. pratensis
Fls purple; stem and branches conspicuously enlarged beneath heads.
   Introduced as a vegetable for its tap-roots. Salsify.   **T. porrifolius* L.

1. T. pratensis L.   Goat's-Beard, Jack-go-to-bed-at-noon.

Annual to perennial with long brownish tap-root and erect stem, 30–70 cm., ± glabrous, somewhat glaucous. Basal lvs linear-lanceolate, somewhat sheathing at the broadened base, conspicuously white-veined; stem-lvs more abruptly narrowed from a half-clasping base. Heads terminal on main stem and branches which are slightly enlarged beneath them. Florets yellow, usually falling short of the spreading lanceolate-acuminate involucral bracts. Achenes usually 10–12 mm., with a beak equally long; pappus very large, with the feathery hairs interwoven. Fl. 6–7, the heads usually closing round noon. Meadows, pastures, dunes, roadsides, waste places, etc. Locally common in lowland Great Britain and Ireland and extending to Sutherland and Caithness.

41. LACTUCA L.

Herbs with abundant white latex and panicles of smallish heads. Involucre cylindrical, of many rows of bracts. Fls yellow or bluish. Achenes strongly compressed and abruptly contracted into a beak; pappus of 2 equal rows of soft white simple hairs.

1   Margins of lvs and lf-lobes prickly-toothed.   *2*
       Margins of lvs or lf-lobes entire or nearly so, not prickly-toothed.   *3*

2   Stem-lvs usually held vertically (often in one plane), the lower often
       pinnatifid with narrow distant lateral lobes; bracts with spreading
       auricles; ripe achenes olive-grey, shortly bristly at the apex.
                                                                    1. serriola

       Stem-lvs not held vertically, obovate-oblong or pinnatifid with
       broad lobes; bracts with appressed rounded auricles; ripe achenes
       blackish, ± glabrous at the apex. Annual to biennial, 60–200 cm.,
       with pale yellowish heads, 10 mm. diam., in a long panicle. Local
       and chiefly coastal in Great Britain northwards to Perth. Wild
       Lettuce.   *L. virosa* L.

3   Stem-lvs simple, ovate to orbicular; infl. a dense corymbose panicle;
       florets pale yellow, often violet-streaked. Introduced as a salad
       plant and often escaping. Garden Lettuce.   **L. sativa* L.*
       Upper stem-lvs linear-lanceolate with sagittate base; basal lvs un-
       divided to pinnatifid with narrow distant lobes; infl. a narrow
       spike-like panicle with short erect branches; florets pale yellow,
       often reddish beneath. A very local plant, chiefly near the sea in
       S. and E. England. Least Lettuce.   *L. saligna* L.

1. L. serriola L.   Prickly Lettuce.

Annual to biennial, with tap-root and stiffly erect lfy stem, 30–150 cm.,
glabrous or somewhat prickly below. Lower lvs narrowly obovate-oblong in outline, undivided, or ± deeply pinnate-lobed with acuminate
terminal lobe and a few pairs of distant acute lateral lobes curving back
at their tips so that each is shaped like a curved bill-hook; upper leaves
less lobed; all rigid, ± glaucous, glabrous, but spinous-ciliate on the
veins beneath, all but the lowermost sagittate or hastate at base and
clasping the stem. Stem-lvs of plants in full sun held vertically. Infl. a
long pyramidal panicle; bracts with spreading auricles. Heads 11–
13 mm. diam. Involucral bracts narrow, glabrous, glaucous, often
violet-tipped. Fls pale yellow, often violet-tinged. Achenes elliptical,
olive-grey when ripe, ribbed, shortly bristly at the tip; beak white,
equalling the achene. Fl. 7–9. Locally abundant in waste places and on
walls and dunes in S. and E. England and Wales.

42. Mycelis Cass.

Like *Lactuca* but the involucre of only 2 rows of bracts, the outer row short and resembling an epicalyx; and the pappus of 2 unequal rows of simple hairs, the inner longer.

1. M. muralis (L.) Dumort. Wall Lettuce.

A perennial herb with a short erect stock and an erect stem, 25–100 cm., glabrous. Lower lvs lyrate-pinnatifid with long winged stalks, the lobes rhombic or hastate with the terminal lobe much larger than the laterals; middle and upper lvs sessile and ± clasping; all thin, glabrous, reddish, their lobes triangular-toothed. Heads in a large open panicle. Involucre 7–10 mm., narrowly cylindrical, the inner bracts linear, the outer very small and spreading. Fls usually 5, yellow. Achene 3–4 mm., including the short pale beak. Fl. 7–9. Walls and rocks and sometimes in woods, particularly beechwoods on chalk; usually on base-rich substrata. Scattered throughout the British Isles except the Highlands of Scotland and the neighbouring islands.

43. Sonchus L.

Annual to perennial herbs like *Lactuca* but with larger heads and unbeaked achenes. Involucral bracts in several rows. Achenes flattened; pappus of 2 equal rows of simple white hairs.

1 Annual to biennial; achenes with 3 longitudinal ribs on each face. 2
   Perennial; achenes with 5 longitudinal ribs on each face. 3

2 Stem-lvs with rounded auricles; achenes smooth. **3. asper**
   Stem-lvs with pointed auricles; achenes transversely wrinkled.
   2. oleraceus

3 Plant with bud-bearing roots; stem-lvs with rounded auricles; glands
     on involucre usually yellow; achenes brown. **1. arvensis**
   Plant with ± erect stock; stem-lvs with pointed auricles; glands on
     involucre usually blackish-green; achenes yellow. A tall herb
     (90–300 cm.) with pale yellow heads in a dense corymbose panicle.
     Rare and local in marshes and fens in S.E. England. Marsh Sow-
     Thistle. *S. palustris* L.

1. S. arvensis L. Field Milk-Thistle.

Perennial, with far-creeping roots bearing erect stems, 60–150 cm., furrowed, hollow, glabrous or at first cottony below, glandular-hairy above. Basal lvs narrowed into a winged stalk, runcinate-pinnatifid with short triangular-oblong spinous-ciliate and spine-toothed lobes;

stem-lvs similar but less divided and sessile, the cordate-clasping base having rounded appressed auricles. Heads 4–5 cm. diam. in a loose corymb; involucre and infl.-branches densely covered with yellowish glandular hairs. Involucre bell-shaped, its bracts oblong-lanceolate. Fls golden-yellow. Achenes dark brown, narrowly elliptical, strongly 5-ribbed on each face. Fl. 7–10. Stream-sides, drift-lines on salt and brackish marshes, banks, arable land, etc. Common throughout the British Is.

2. S. oleraceus L.   Milk- or Sow-Thistle.

Annual or overwintering herb with a long pale tap-root and stout erect glabrous stems, 20–150 cm., angular, hollow. Lvs very variable; basal lvs usually ovate, stalked; lower stem-lvs runcinate-pinnatifid with the terminal lobe usually wider than the uppermost pair of laterals and with a short winged stalk and pointed spreading auricles; upper stem-lvs smaller with more broadly winged stalk; all glabrous at least when mature, dull, never spinous. Infl. corymbose. Heads 2–2·5 cm. diam. Involucral bracts usually glabrous, the outer broadly lanceolate, shorter and more acute than the inner. Fls yellow. Achenes oblanceolate, first yellow then brown, transversely wrinkled. Fl. 6–8. Cultivated soil, waysides, waste places, etc.; common throughout the British Is.

3. S. asper (L.) Hill   Spiny Milk- or Sow-Thistle.

Closely resembling *S. oleraceus* but differing in the following ways: lvs dark glossy green above with rounded appressed auricles and usually crisped and spinous-ciliate margins; when pinnate-lobed the terminal lobe is narrower than the uppermost pair of laterals; fls bright yellow; achenes obovate, usually brown, not transversely wrinkled. Fl. 6–8. Cultivated soil, waste places, etc.; common throughout the British Is.

44. HIERACIUM L.   Hawkweeds.

Perennial herbs, sometimes stoloniferous, with vertical to horizontal stocks and stout fibrous roots. Lvs spirally arranged on the flowering stems or some or all in a basal rosette. Infl. a cymose and often corymbose panicle or a few-fld forking cyme or the heads solitary, terminal. Involucral bracts in few to several rows, the outermost shortest. Stem, lvs, infl. and involucre glabrous or clothed with simple hairs, toothed hairs, glandular hairs, sessile glands or soft stellate down in varying proportions. Receptacular scales 0. Florets usually yellow (brick-red or brownish-red in two introduced and often naturalized stoloniferous many-headed species). Achenes 1·5–5 mm., cylindrical, 10-ribbed, truncate above, neither appreciably narrowed upwards nor beaked; pappus-hairs in 1–2 rows, simple, rigid, brittle, usually pale brownish.

The genus *Hieracium* is the most difficult taxonomically of those represented in the British flora. This arises largely from the peculiar mode of reproduction, seeds being produced apomictically, that is, independently of fertilization, so that each hawkweed and its offspring become more or less completely closed units and there is nothing comparable with the ordinary outbreeding species. The following key and descriptions include only the commonest of the lowland species and species-aggregates. It must therefore be used with caution, since a hawkweed collected even in lowland England may not be referable to any of the described forms. Students who seek more detailed information should consult the larger *Flora* or H. W. Pugsley's *Prodromus of the British Hieracia*.

1 Plants with above-ground stolons and well-defined basal lf-rosettes;
      flowering stem with lvs 0 or bract-like and a solitary terminal pale
      yellow head; achenes 1·5–2 mm.; pappus-hairs equal, in 1 row.
                                                                1. pilosella
      Stolons 0; rosette-lvs present or 0; heads more than 1 except in very
      small plants; achenes 2·5–5 mm.; pappus-hairs of various lengths,
      in 2 rows.   *2*

2 No rosette formed, or rosette-lvs withered at time of flowering,
      rarely a few persisting; stem-lvs 8–many, rarely fewer in small
      plants.   *3*
      Rosette-lvs present at flowering time; stem-lvs 0–8(–10).   *6*

3 Lvs all similar in shape, lanceolate to linear-lanceolate, narrowed to
      a sessile base, their margins revolute and rough; heads in an
      umbellate panicle; all but the innermost involucral bracts with
      recurved tips; styles yellow.   **2. umbellatum**
      Lower lvs stalked; middle and upper lvs short-stalked or sessile,
      lanceolate to broadly ovate, their margins not revolute; invo-
      lucral bracts very rarely with recurved tips; styles usually dark.
                                                                          4

4 No rosette-lvs; stem-lvs numerous, crowded at least below; upper
      stem-lvs short with broad rounded bases.   *5*
      Rosette-lvs withering early or a few persisting to flowering time;
      stem-lvs usually not very numerous, rather widely spaced in the
      upper half or confined to the lower half; all lvs narrowing to the
      base, usually toothed in the middle.   **5. trichocaulon**

5 Involucral bracts with long whitish hairs and many glandular hairs.
                                                          3. perpropinquum
      Involucral bracts with no or few long white hairs and with no
      glandular hairs.   **4. vagum**

6 Stem-lvs 0–1(–3); rosette-lvs several to many, ± broad, often with
      concave-sided spreading or reflexed basal teeth, the base of the
      blade cordate, truncate or rounded and thus clearly demarcated
      from the shaggy stalk.   **1. exotericum**

Stem-lvs 2–5(–10); rosette-lvs not numerous, their blade broadly to
　　narrowly lanceolate, narrowing gradually into the stalk, with
　　outwardly or forwardly directed teeth.   7

7　Involucral bracts with numerous whitish hairs and usually some
　　glandular hairs also.   **7. vulgatum**
　Involucral bracts densely glandular with few or no simple hairs.
                                                           6. lachenalii

1. H. pilosella L.   Mouse-ear Hawkweed.

Rhizome long, slender. Stolons numerous, long, above-ground, bearing small distant lvs and terminating in overwintering rosettes. Scape
5–30 cm., floccose especially above and with both glandular and simple
hairs. Rosette-lvs linear to obovate-oblong, narrowed below into a long
stalk-like base, entire, downy to tomentose beneath, with sparse long
stiff white hairs on both sides, on the margins and on the stalk; stem-lvs
0; rarely 1–3, and then small, hairy and bract-like. Involucral bracts
floccose and with both glandular and simple hairs. Florets pale yellow,
the outer reddish beneath. Styles yellow. Achenes 2 mm. Grassy
pastures and heaths, banks, rocks, walls, etc.; locally common throughout the British Is.

2. H. umbellatum L.

Stem commonly 30–80 cm., erect, wiry, often somewhat woody below,
usually hairy, not glandular; stolons 0. Rosette-lvs 0. Stem-lvs
numerous (up to 50), crowded at least below, diminishing in size upwards, all similar, linear to linear-lanceolate, with rough revolute
margins, subentire or with 2–3 distant teeth on each side, narrowed to a
sessile base, floccose and hairy beneath, becoming glabrous above.
Heads in an umbellate panicle, their scaly and floccose stalks thickened
just beneath the head. Involucral bracts usually blackish-green, narrow,
blunt, almost glabrous, all but the innermost with recurved tips. Styles
usually yellow. Achenes 3–4 mm. Roadsides, banks, open woods and
copses, heaths, etc., chiefly lowland; throughout most of the British Is.

3. H. perpropinquum (Zahn) Druce

Stems commonly 50–100 cm., with dense, spreading, whitish, bulbous-
based hairs at least below; stolons 0. Rosette-lvs 0. Stem-lvs 20–40(–70),
crowded at least below, decreasing rapidly in size upwards; the lower
lanceolate-acuminate, finely toothed at least near the middle, narrowing
into a long stalk-like base; middle lvs elliptic-lanceolate, acuminate, with
rounded sessile to half-clasping base; upper lvs shorter, ovate-acuminate
with sessile base; all dark or dull green and subglabrous above, paler

and hairy beneath. Heads several in a corymbose panicle with floccose and hairy branches. Involucral bracts dark green, narrow, blunt, with numerous long whitish hairs and many fine glandular hairs; tips of bracts not recurved. Styles dark. Achenes 3–4 mm. Fl. 8–10. Woods, especially along margins and in clearings, copses, hedgerows, railway-banks, rocks, etc., chiefly at low levels; common in England and Wales and reaching Ross; Ireland.

4. H. vagum Jordan

Like *H. perpropinquum* but the lvs firmer and all but the lower held sub-erect, the infl. more compact and the blackish involucral bracts almost glabrous and lacking glandular hairs. Fl. 8–10. Woods, banks, hedge-rows, etc., chiefly in C. and N. England and Wales but reaching Moray.

5. H. trichocaulon agg. (incl. *H. eboracense* Pugsley, *H. trichocaulon* (Dahlst.) Roffey and *H. calcaricola* (Hanb.) Roffey)

Stem commonly 40–100 cm., often reddish, hairy especially at the base and in lf-axils; stolons 0. Rosette-lvs few, withering before flowering, or 0. Stem-lvs 6–30, often crowded below; lower lvs ovate- to elliptic-lanceolate, narrowed into a winged stalk; middle and upper lvs narrower, sessile; all acuminate and usually with 3–5 teeth in the basal ⅔ of each side, deepish green, paler beneath, hairy at least beneath and with margins usually thickened and ciliate or rough. Infl. a corymbose panicle with floccose but scarcely hairy or glandular branches. Involu-cral bracts narrow, bluntish, olive-green, with many whitish dark-based hairs mixed with glandular hairs. Styles yellow to dark. Achenes 3·5–4 mm. Fl. 8–10. Woods, copses, hedge-banks, walls, rocks, etc., throughout most of the British Is.

6. H. lachenalii agg. (incl. *H. lachenalii* C. C. Gmelin, *H. diaphanoides* Lindeb., *H. anglorum* (A. Ley) Pugsley, etc.)

Stem 30–100 cm., hairy especially below, often downy and somewhat glandular above; stolons 0. Rosette-lvs usually few, oblong to ovate-lanceolate, cuneate-based, toothed. Stem-lvs 2–7(–10), the lowest ovate-lanceolate to lanceolate, acuminate, often sharply toothed towards the cuneate base, stalked; the upper smaller, narrower and ± sessile; all with ciliate margins and often with hairs on both surfaces. Infl. a corymbose panicle with floccose and glandular branches. Involucral bracts dark green, narrow, ± acute, densely glandular, downy or not and with few or no dark simple hairs. Styles yellow to dark. Achenes 2·5–3·5 mm. Fl. 8–10. Woods, shaded rocks and walls; throughout England and Wales and reaching C. Scotland; Ireland.

7. H. vulgatum Fries

Stem 20–80 cm., usually rather slender, hairy especially below and floccose especially above; stolons 0. Rosette-lvs narrowly oblong to lanceolate, narrowing to the base, commonly with sharp ascending teeth in the lower half; stem-lvs 2–4(–5), the lowest stalked, the upper narrower, sessile, ± toothed towards the narrow base; all with scattered hairs on both sides and ciliate margins. Infl. a corymbose or subumbellate panicle with densely floccose branches. Involucral bracts grey-green, narrow, subacute, floccose, with numerous whitish dark-based hairs and usually a few glandular hairs amongst them. Styles dark. Achenes 2·5–3 mm. Fl. 7–10. Woods, banks, walls and rocks. Common in N. Wales, C. and N. England, and Scotland; Ireland.

8. H. exotericum agg. (incl. *H. exotericum* Jordan, *H. grandidens* Dahlst., *H. sublepistoides* (Zahn) Druce, etc.)

Stem 30–80 cm., hairy and somewhat floccose, with glandular hairs above; stolons 0. Rosette-lvs thin, yellowish-green, ovate to oblong, blunt to acute, often with large ± reflexed teeth close to the cordate to rounded base; stem-lf usually 1, ovate-lanceolate, acuminate, sharply toothed; all ± softly hairy above and beneath, ciliate, with shaggy stalks. Infl. corymbose, its branches floccose and glandular. Involucral bracts densely glandular, sparsely floccose but with very few or no simple hairs. Styles yellow to dark. Achenes 3 mm. Fl. 6–8. Woods, shaded rocks, stream-sides, banks, roadsides, etc. Common in S. and C. England and Wales and extending northwards to Stirling; Ireland.

H. pellucidum Laest., like *H. exotericum* but with the rosette-lvs dark green, shining and almost glabrous above, often purplish beneath, stem-lvs 0–1, involucral bracts blackish-green, styles dark and pappus conspicuously white, is widespread especially in N. England.

45. CREPIS L.

Usually herbs with spirally arranged often runcinate lvs, the basal ones commonly in a rosette, and lfy flowering stems. Heads usually in panicles or corymbs, rarely solitary; involucral bracts many, equal, in 1 row apart from a few small basal bracts; receptacular scales 0. Florets all ligulate, yellow. Achenes cylindrical, ribbed, narrowed above, with or without a beak; pappus of simple hairs in many rows, usually white.

1 Perennial herbs of streamsides, moist woods and flushes in Wales, N. England, Scotland and Ireland.   2

Annual, biennial or rarely short-lived perennial herbs of arable
fields, waste places, dry grassland, heaths, walls, etc.   3

2 Basal lvs acute, short-stalked; stem-lvs distinctly toothed, the
upper with long-pointed auricles; pappus brownish, brittle.
                                                        4. paludosa
Basal lvs blunt, long-stalked; stem-lvs ±entire, the upper with
rounded auricles; pappus pure white, soft. A rare plant of
mountainous districts in Wales, N. England and Scotland. Soft
Hawk's-beard.   *C. mollis* (Jacq.) Ascherson

3 Involucral bracts with distinct and paler scarious margins and
shortly pubescent inner surface; lowermost bracts lanceolate,
spreading horizontally.   4
Involucral bracts with indistinct scarious margins and ±glabrous
inner surface; lowermost bracts linear, appressed or nearly so.   5

4 Heads 1·5–2·5 cm. long; style-arms greenish-brown; ripe achenes
3·5 mm., light brown, tapering gradually into a slender beak.
                                        1. vesicaria ssp. taraxacifolia
Heads 2·5–3·5 cm. long; style-arms yellow; ripe achenes 4–8 mm.
orange-brown, tapering but not beaked.   **2. biennis**

5 Heads few, long-stalked, drooping in bud; involucre pubescent;
marginal achenes short-beaked and clasped by the inner bracts,
central long-beaked. Lvs dandelion-like, foetid, densely hairy. A
rare plant of waysides and waste places in S.E. England and
East Anglia. Stinking Hawk's-beard.   *C. foetida* L.
Heads many, erect in bud; involucre sometimes glandular or
bristly but not pubescent; achenes not beaked and not clasped by
the inner bracts.   **3. capillaris**

*1. C. vesicaria L. ssp. **taraxacifolia** (Thuill.) Thell.
                                        Beaked Hawk's-beard.
Usually biennial, with erect stems 15–80 cm., hispid and purplish below.
Basal lvs stalked, oblanceolate in outline, lyrate-pinnatifid with the
lobes very variable in length and width, sometimes merely toothed;
stem-lvs short-stalked or sessile, pinnate-lobed or not, the middle ones
clasping the stem; all finely pubescent on both sides. Heads 1·5–2·5 cm.
diam., erect in bud, in terminal corymbs. Involucral bracts downy and
often glandular, the outer narrower and spreading. Achenes all gradually narrowed into a beak about as long as the achene; pappus white,
soft, exceeding the involucral bracts. Fl. 5–7. Introduced. Waysides,
walls, railway banks, waste places, etc., especially on calcareous soils.
Locally common in England and Wales northwards to Yorks and Lancs,
and in C. Ireland; spreading rapidly.

2. C. biennis L.   Rough Hawk's-beard.

Biennial, with an erect stout grooved stem, 30–120 cm., ± roughly hairy, often purplish below. Basal lvs 15–30 cm., irregularly pinnate-lobed, stalked; stem-lvs smaller and sessile, half-clasping but not or scarcely sagittate at base; all rough with scattered hairs. Heads 2–3·5 cm. diam., at first crowded, then in corymbs. Inner involucral bracts downy within; outer narrow, spreading. Achenes narrowed above but not beaked, 13-ribbed; pappus equalling or slightly exceeding the involucral bracts, white, soft. Fl. 6–7. Probably native. Pastures, waysides, clover and lucerne fields, waste places. Locally frequent in lowland Great Britain northwards to Aberdeen; introduced in Ireland.

3. C. capillaris (L.) Wallr.   Smooth Hawk's-beard.

Usually annual with erect or ascending stems, 20–90 cm., glabrous or ± hairy below. Basal and lower lvs 5–25 cm., very variable, usually lanceolate or oblanceolate in outline, narrowed into a stalk-like base, pinnate-lobed with distant often toothed lobes, or merely toothed; middle and upper stem-lvs lanceolate, acute, sessile and clasping with a sagittate base; all subglabrous. Heads 1–1·3(–2·5) cm. diam., erect in bud, in lax corymbs. Involucre contracted above in fr., usually downy and often glandular-bristly; inner bracts glabrous within, outer one-third as long, appressed. Achenes 1·5–2·5 mm., 10-ribbed; pappus about equalling the involucral bracts, snow-white, soft. Fl. 6–9. Grassland, heaths, walls, waste places, etc. Common throughout the British Is.

4. C. paludosa (L.) Moench   Marsh Hawk's-beard.

Perennial, with an oblique blackish stock and an erect grooved glabrous stem, 30–90 cm., often reddish below. Basal and lower stem-lvs broadly elliptical, acute, narrowed into a short winged stalk; middle stem-lvs acuminate with broad sessile clasping base; upper clasping with long-pointed auricles; all thin, toothed, glabrous. Heads 1·5–2·5 cm. diam., in a terminal corymb. Involucral bracts blackish-green, woolly and with black glandular hairs, the outer about half as long as the inner and appressed. Achenes 4–5 mm., hardly narrowed at the ends, 10-ribbed; pappus stiff, brittle, brownish (and so like *Hieracium*, but differing in plant almost completely glabrous). Fl. 7–9. Stream-sides, wet woods, flushes, wet meadows, etc. Locally common in N. Britain from Glamorgan and Worcester to Caithness and throughout Ireland.

46. Taraxacum Weber   Dandelions.

Perennial herbs with tap-roots, upon which adventitious shoots arise readily, and spirally arranged entire to pinnate-lobed commonly runcinate lvs confined to a basal rosette. Scape unbranched, with a solitary terminal head. Inner involucral bracts erect, equal, in 2 rows; outer shorter, often spreading or reflexed. Receptacular scales 0. Fls yellow. Achenes fusiform-cylindrical, ribbed, the ribs usually muricate above; beak long; pappus of many rows of simple rough white hairs.

Most British forms are apomictic (fruits ripening independently of fertilization) and present great difficulties to the taxonomist. The following key and notes deal only with the four main aggregates of micro-species, but forms may be found that seem intermediate.

1   Inner bracts with an appendage on the outer side just beneath the tip, so that they appear bifid; achene abruptly narrowed into a cylindrical cusp between its tip and the beak.

                                            3. laevigatum

    Inner bracts not appendaged beneath the tip; cusp of achene cylindrical or conical.   2

2   Outer bracts mostly narrowly lanceolate or linear, usually more than 3 times as long as broad and usually spreading or reflexed; achenes 3·5–4 mm., including the short conical cusp; beak 2·5–4 times as long as the achene.   **1. officinale**

    Outer bracts mostly broadly or narrowly ovate, usually less than 3 times as long as broad and commonly appressed; achene 4–6 mm. including the cusp; beak 1·5–2·5 times as long as the achene.   3

3   Outer bracts broadly ovate, appressed; achene contracted abruptly into a long slender cusp; lvs narrow in outline.   **2. palustre**

    Outer bracts narrowly ovate, appressed or spreading; achenes narrowed gradually into a short conical cusp; lvs broad in outline, commonly not deeply lobed, ± roughly hairy above, midrib reddish. Chiefly in moist places in mountainous areas of the north and west.   *T. spectabile* Dahlst., *sensu lato*

1. T. officinale Weber, *sensu lato*   Common Dandelion.

This aggregate includes the common forms of pastures, meadows, lawns, waysides, waste places, etc., througout the British Is. The plants are usually robust and the lvs are usually runcinate-pinnatifid, though very variable in the amount and depth of lobing. The outer involucral bracts are mostly narrow and spreading or reflexed.

2. T. palustre (Lyons) DC., *sensu lato*

Narrow-leaved Marsh Dandelion.

Includes forms in which the outer bracts are broadly ovate and firmly appressed and the inner bracts are not appendaged at the tip. Lvs narrowly oblanceolate in outline, variously cut or lobed but often almost entire. Locally abundant in marshes and fens and by streams throughout the British Is.

3. T. laevigatum (Willd.) DC., *sensu lato*   Lesser Dandelion.

Forms differing from the foregoing in having a short appendage on the outer side of the inner involucral bracts, just beneath the tip, so that the tip of the bract appears bifid; outer bracts narrowly ovate, appressed to somewhat recurved. Achene commonly purplish-red. Often small plants with narrow deeply cut lvs and small pale yellow heads, locally abundant in dry places on sandy or calcareous soils, heaths, walls, etc., probably throughout lowland Britain.

MONOCOTYLEDONES

121. ALISMATACEAE

Herbs living in water or wet places. Fls actinomorphic, usually hermaphrodite. Perianth in 2 whorls, differing from each other, the outer
persistent. Stamens free. Ovary superior; carpels usually free. Fr. a
head or whorl of achenes, rarely dehiscing at base.

1 Lvs rounded at apex; carpels 10 or fewer, c. 13 mm., with a long
  spreading beak. Rare.   *Damasonium alisma* Miller
  Lvs acute; carpels numerous, not as above.   *2*

2 Fls unisexual; lvs often sagittate.   3. SAGITTARIA
  Fls hermaphrodite; lvs never sagittate.   *3*

3 Fls in more than 2 whorls; infl. at least twice branched; carpels in
  one whorl.   2. ALISMA
  Fls solitary or in 1 or 2 whorls; infl. simple or once branched;
  carpels crowded in a head.   *4*

4 Stem and lvs not floating; lvs narrowly lanceolate, acute; fls c. 1 cm.
  diam.   1. BALDELLIA
  Stem and lvs floating; lvs ovate, blunt; fls 2·5–3 cm. diam. Very
  local.   *Luronium natans* (L.) Rafin.

1. BALDELLIA Parl.

Lvs in a basal rosette. Fls long-stalked, umbellate or (rarely) in 2 simple
whorls, sometimes solitary. Stamens 6. Carpels numerous, oblong-
ovoid when ripe.

1. B. ranunculoides (L.) Parl.   Lesser Water-Plantain.

Usually 5–20 cm. Lvs long-stalked, narrow, blade 2–4 cm. Fls pale
purplish; stalks up to c. 10 cm., unequal. Achenes in ± spherical heads.
Fl. 5–8. In damp places, widely distributed and locally common.

2. ALISMA L.

Lvs in a basal rosette. Infl. much-branched. Stamens 6. Carpels
numerous, in one whorl, strongly flattened when ripe.

1 Style spirally coiled; fr. broadest near apex; lvs usually very narrow.
  Very local.   *A. gramineum* Lejeune
  Style ± straight; fr. ovate; lvs lanceolate to ovate.   *2*

2   Lvs ovate, rounded to subcordate at base; style arising below the
    middle of the fr.; anthers c. twice as long as broad.
                                              1. plantago-aquatica
    Lvs lanceolate, gradually narrowed at base; style arising above the
    middle of the fr.; anthers about as long as broad. Local; mainly
    S. and E. England.   *A. lanceolatum* With.

1. A. plantago-aquatica L.   Water Plantain.
20–100 cm., glabrous. Lvs long-stalked, blades 8–20 cm.; first lvs with
small narrow blades, later sometimes floating in deep water. Infl.
branches ± straight, usually ascending. Fls c. 1 cm. diam., usually pale
lilac. Outer per. segs oblong, inner rounded. Carpels c. 20 in a ± flat
head. Fl. 6–8. In shallow water and on mud beside slow-flowing rivers,
etc., widely distributed except in N. Scotland.

3. SAGITTARIA L.

Lvs in a basal rosette. Infl. of several simple or rarely slightly branched
whorls. Fls unisexual. Stamens 9 or more. Carpels numerous, spirally
arranged, strongly flattened.

Lvs typically sagittate; inner per. segs with a dark violet patch at base.
                                              1. sagittifolia
Lvs never sagittate; inner per. segs entirely white or slightly yellowish
    at base. Naturalized near Exeter.   **S. rigida* Pursh

1. S. sagittifolia L.   Arrow-head.
30–90 cm., perennating by means of turions borne at the end of long
slender runners in the mud. Turions bright blue with yellow spots. Submerged lvs grass-like, translucent; floating lvs lanceolate to ovate;
aerial lvs long-stalked, blade 5–20 cm., sagittate, with lobes about as
long as main part. Fls monoecious, 3–5 in a whorl, c. 2 cm. diam.;
female in lower part of infl., stalks c. 5 mm.; male above, stalks c.
20 mm. Mature head of carpels hemispherical. Fl. 7–8. In shallow water,
England and Wales, rare in the north.

122. BUTOMACEAE

Herbs, usually with latex, living in water or wet places. Fls actinomorphic, solitary or in umbels. Perianth in 2 whorls. Stamens usually
8–9. Ovary superior; carpels usually free.

1. Butomus L.

Lvs narrow, erect. Fls hermaphrodite, in umbels. Per. segs persistent. Stamens usually 8(-9). Carpels 6-9, cohering at base.

1. B. umbellatus L.   Flowering Rush.

Up to 150 cm., rhizomatous. Lvs about as long as stems, in a basal rosette, triquetrous, twisted, sheathing at base. Umbel with an involucre of acuminate bracts. Fls 2·5–3 cm. diam.; stalks up to c. 10 cm., unequal. Per. segs pink with darker veins. Fl. 7–9. In shallow water, rather local; doubtfully native in Scotland.

123. HYDROCHARITACEAE

Wholly or partially submerged aquatic herbs. Fls actinomorphic, usually dioecious, subtended by a solitary bifid or two opposite bracts forming a spathe. Male fls usually numerous, female solitary. Per. segs usually in 2 series, 3 in each series. Ovary inferior.

1 Lvs stalked, orbicular-reniform, floating.   **1. Hydrocharis**
　 Lvs sessile, submerged or partly so.   2
2 Lvs much more than 1 cm., all arising from the base.   3
　 Lvs c. 1 cm., spaced out along the stem.   **2. Elodea**
3 Lvs rigid, spiny along margin, tapering from base. Very local. Water
　 Soldier.   *Stratiotes aloides* L.
　 Lvs flaccid, minutely toothed at tip, ribbon-shaped. Naturalized in
　 a few canals.   **Vallisneria spiralis* L.

1. Hydrocharis L.

Fls white, male 2–3 in a spathe, female solitary, both stalked. Sepals greenish, smaller and narrower than petals. Stamens 12, 3–6 outer usually sterile. Female fls with 6 staminodes. Fr. fleshy.

1. H. morsus-ranae L.   Frog-bit.

Floating stoloniferous herb with lvs and roots in bunches at nodes. Lf-blades c. 3 cm. diam.; stipules large, translucent. Fls c. 2 cm. diam.; petals obovate, crumpled, with a yellow spot near base. Fl. 7–8. Ponds and ditches mainly in calcareous districts, locally common in England and Wales.

2. ELODEA Michx

Submerged herbs with sessile whorled lvs. Fls solitary in tubular spathes arising in axils of lvs. Petals narrower than sepals. Stamens 3–9; filaments short or 0.

Dark green; lvs 3 (rarely 4) in a whorl; usually 3–4 mm. wide, acute or bluntish.   **1. canadensis**
Pale green; lvs 3–5 (often 4) in a whorl, usually c. 2 mm. wide, acuminate. Rare. *E. nuttallii* (Planchon) St John (*Hydrilla verticillata* auct.)

*1. E. canadensis Michx.   Canadian Pondweed.

Up to 3 m., usually less, very brittle. Lvs c. 10 × 3–4 mm., oblonglanceolate, translucent, finely toothed. Fls 5 mm. diam., floating, greenish-purple; axis up to 30 cm., thread-like. Fl. 5–10. Naturalized in canals, rivers etc., throughout most of the British Is.

124. SCHEUCHZERIACEAE (p. xxxiii)

125. JUNCAGINACEAE

Marsh or aquatic herbs. Lvs narrow, mostly basal, sheathing at base. Fls small, in spikes or racemes, hermaphrodite or unisexual. Per. segs in 2 series, green or reddish. Stamens 6 or 4, filaments very short. Ovary superior.

1. TRIGLOCHIN L.

Rhizomatous herbs. Lvs erect, ½-cylindrical. Infl. a raceme. Fls 3-merous. Per. segs deciduous. Fr. dehiscing by the carpels separating from the central axis.

Lvs very narrow; fr. 8–10 × 1 mm., appressed to scape.   **1. palustris**
Lvs up to c. 3 mm. wide; fr. 3–4 × 2 mm., spreading.   **2. maritima**

1. T. palustris L.   Marsh Arrow-grass.
15–50 cm., slender. Fl.-stalks 3–4 mm., elongating (like the raceme) after fl. Per. segs purple-edged. Carpels 3 sterile and 3 fertile, remaining attached at the top after dehiscence. Fl. 6–8. Local but widespread in marshes, usually among tall grass.

2. T. maritima L.   Sea Arrow-grass.
15–50 cm., rather stout and rigid. Racemes scarcely elongating after fl. Fl.-stalks 1–2 mm., elongating. All 6 carpels fertile, separating completely at dehiscence. Fl. 7–9. Salt-marsh turf and grassy places on rocky shores, all round the coasts.

126. ZOSTERACEAE

Perennial submerged marine herbs. Stems flattened. Lvs grass-like, sheathing at base. Fls unisexual, borne on one side of the flattened axis, ± enclosed in a lf-sheath. Perianth 0 or very small. Male fls consisting of 1 anther. Female fls consisting of an ovary with 2 stigmas.

1. ZOSTERA L.

Rhizome creeping and rooting at nodes. Lvs alternate, distichous. Fl. shoots annual, erect.

1	Fl. stems branched; sheaths entire; seeds ribbed.	2
	Fl. stems unbranched; sheaths split; seeds smooth.	**3. noltii**
2	Lvs usually 5–10 mm. wide; stigma twice as long as style; seed 3–3·5 mm.	**1. marina**
	Lvs usually 1–2 mm. wide; stigma as long as style; seed 2·5–3 mm.	**2. angustifolia**

1. Z. marina L.   Eel-grass, Grass-wrack.

Lvs dark green, grass-like; those of non-flowering shoots 20–50 (–100) cm., (2–)5–10 mm. wide, rounded at apex; those of flowering shoots shorter and narrower, sometimes emarginate. Fl. stems up to 60 cm., much-branched. Infl. (4–)9–12(–14) cm., membranous margins of sheath 0·5–1·0 mm. wide. Seed pale brown or bluish-grey. Fl. 6–9. On fine gravel, sand or mud from low water spring tides down to 4 m., local.

2. Z. angustifolia (Hornem.) Reichenb. (*Z. hornemanniana* Tutin)

Slender. Lvs of non-flowering shoots 15–30 cm. × 1–2 mm. in summer, 5–12 cm. × 1 mm. in winter, rounded at apex when young, later emarginate. Lvs of flowering shoots 4–15 cm. × 2–3 mm. Fl. shoots 10–30 cm., branched. Infl. 8–11 cm., membranous margins of sheath 1·5–2 mm. wide. Seed pale brown. Fl. 6–11. On mud-flats, usually in estuaries from half-tide mark to low-tide mark, rarely below; local.

3. Z. noltii Hornem. (*Z. nana* auct.)

Slender. Lvs of sterile shoots (4–)6–12(–20) cm. × 1 mm. Fl. stems 0·5–2 cm., unbranched or rarely branched from very near the base, each bearing one infl. 3–6 cm. enclosed in the inflated lf-sheath. Seed 2 mm., dark brown. Fl. 6–10. In similar habitats to the preceding, local.

127. POTAMOGETONACEAE

Aquatic herbs, chiefly of fresh water, with alternate or opposite distichous lvs. Fls in axillary or terminal bractless spikes, inconspicuous, hermaphrodite, regular. Per. segs 4; stamens 4, sessile on the claws of the per. segs; carpels (1–3–)4, free, superior, each with 1 ovule; stigma ± sessile. Fr. of 4 or fewer brownish drupes or achenes; seed non-endospermic.

Lvs opposite or in whorls of 3; stipules 0 except in the flowering region.
                                                                 2. GROENLANDIA
Lvs alternate (or subopposite only in the flowering region); stipules
    present in the axils of at least all young lvs.   1. POTAMOGETON

1. POTAMOGETON L.   Pondweed.

Chiefly perennial and rhizomatous, overwintering both by the rhizomes and by specialized winter-buds (*turions*) borne directly on the rhizome, on rhizome-stolons or on the lfy stems. In some spp. perennation is chiefly or only by turions. Lvs usually alternate (but subopposite in the flowering region of many spp.), all submerged, thin and translucent, or some lvs floating and then usually ± leathery and opaque. In most spp. the lf has in its axil a delicate membranous sheathing scale which may be free throughout (*stipule*) or adnate to the lf-base in its lower part (*stipular sheath*) and free above (*ligule*); in either case the basal part may be open, with rolled overlapping margins, or closed and tubular. Spikes ovoid to cylindrical, dense, lax or interrupted, submerged or not. Perianth of 4 free rounded short-clawed segments.

The following key includes only three of the numerous hybrids reported from the British Is. These are the most frequently occurring, but other sterile intermediate forms may be found.

1   Floating lvs present.   2
    Floating lvs 0.   10

2   Submerged lvs all linear, channelled, usually with no expanded
       blade; floating lvs with the margins running for a short distance
       down the stalk, making a flexible joint.   **1. natans**
    Submerged lvs with a ± expanded translucent blade; floating lvs   3
       not jointed.

3   Stems compressed; submerged lvs all linear, their midribs bordered
       on each side by a well-marked band of air-tissue; floating lvs
       narrowly oblong or oblong-lanceolate, usually rounded at the
       tip. A rare plant of lakes in the Outer Hebrides; introduced in
       S.W. Yorks.   *P. epihydrus* Rafin.
    Stems not compressed; submerged lvs not all linear.   4

4 All lvs thin, translucent, distinctly and finely net-veined; floating
   and most of submerged lvs ± broadly elliptical, all short-stalked.
                                                                 3. coloratus
   Floating lvs ± leathery, opaque, not translucent.   5

5 Submerged lvs mostly 10–16 × 3·5–5 cm., elliptical-lanceolate,
   transparent, very conspicuously but delicately net-veined; floating lvs of similar shape; all lvs long-stalked. Known only in slow-
   flowing base-rich rivers of S.W. England.   *P. nodosus* Poiret
   Submerged lvs translucent but not conspicuously net-veined unless
   held against the light; floating lvs usually broader than submerged lvs.   6

6 All lvs distinctly stalked; usually in very shallow water and then
   most or all lvs of the floating type, leathery and opaque, broadly
   elliptical to elliptical-ovate.   **2. polygonifolius**
   Some or all submerged lvs sessile.   7

7 At least the lower submerged lvs rounded and half-clasping at the
   base; a sterile hybrid.   **7. × nitens**
   Submerged lvs narrowed to the base, sessile or short-stalked.   8

8 Submerged lvs blunt, with quite entire margins, commonly 6–12 ×
   1–2 cm., often reddish; stipules 2–6 cm., broad, blunt, robust;
   infl.-stalks not thickened upwards.   **8. alpinus**
   Submerged lvs acute, cuspidate or mucronate, with microscopically toothed margins; infl.-stalks thickened upwards.   9

9 Submerged lvs commonly 2·5–10 × 0·5–1 cm., acute or acuminate;
   plant much branched at the base.   **6. gramineus**
   Submerged lvs commonly 8–12 × 1·5–2·5 cm., cuspidate or mucronate; plant not much branched at the base.   **5. × zizii**

10 Grass-leaved: lvs narrowly linear or thread-like, not exceeding
   6 mm. wide, parallel-sided.   19
   Lvs usually exceeding 6 mm. wide, or, if narrower, linear-
   lanceolate, not parallel-sided.   11

11 At least the lower lvs sessile and ± clasping at the base.   12
   Lvs not clasping.   15

12 Lvs usually narrowly to broadly ovate, all cordate and ± completely clasping; stipules small (up to 1 cm.), fugacious.
                                                          10. perfoliatus
   Lvs oblong or lanceolate, at least the lower half-clasping at the
   base; stipules usually more than 1 cm., persistent.   13

13 Lvs tapering gradually from the rounded half-clasping base to a
   blunt and distinctly hooded tip, their margins quite entire;
   stipules blunt, not keeled; fertile.   **9. praelongus**
   Lvs not hooded at the tip, their margins minutely toothed, at least
   when young; sterile hybrids.   14

14   Lvs usually not exceeding 8 cm. long, oblong-lanceolate, narrowed
     to a cuspidate tip, usually rounded and ±cordate at the base;
     stipules 1–2 cm., acute not keeled.   **7. ×nitens**
    Lvs very variable, 3–20 × 1·5–4 cm., oblong, ±rounded at the tip;
     stipules 2·5–3 cm., blunt, slightly 2-keeled. *P. lucens × perfolia-*
     tus. Found with the parents in scattered localities over much of
     the British Is.   *P. × salicifolius* Wolfg.

15   Stem compressed; lvs commonly 4–9 × 1–1·5 cm., lanceolate, with
     margins distinctly toothed and often strongly waved; beak
     equalling the rest of the fr.   **17. crispus**
    Stem ±cylindrical; lf-margin not distinctly toothed to the naked
     eye; beak shorter than the rest of the fr.   16

16   Lvs blunt, often reddish, their margins quite entire; infl.-stalks not
     or hardly thickened upwards.   **8. alpinus**
    Lvs not blunt, their margins microscopically toothed; infl.-stalks
     distinctly thickened upwards.   17

17   Plant much branched at the base; lvs commonly 2·5–10 × 0·3–1 cm.,
     acute or acuminate; stipules acute, not keeled.   **6. gramineus**
    Plant not much branched at the base; lvs commonly 8–10 ×
     1·5–5 cm.; stipules blunt, 2-keeled.   18

18   Lvs commonly 8–12 × 1·5–2·5 cm., cuspidate or rounded and
     mucronate at the tip.   **5. ×zizii**
    Lvs commonly 12–20 × 3·5–5 cm., acuminate or cuspidate.
                              4. lucens

19   Lvs with 2 large air-filled longitudinal canals, one on each side of
     the midrib, occupying the greater part of their volume; stipules
     adnate below to the lf-base, forming a stipular sheath with a
     free ligule above.   20
    Lvs not as above; stipules free from the lf-base throughout their
     length.   21

20   Stipular sheath open, with overlapping whitish rolled margins; fr.
     3–5 × 2–4 mm. with a short lateral beak.   **11. pectinatus**
    Stipular sheath forming a closed tube below when young; fr. 2–3 ×
     2 mm. with an extremely short almost central beak; lvs not more
     than 1 mm. wide. Local in lakes and streams in Anglesey, Scot-
     land, Hebrides, Orkney, Shetland and N. Ireland.
                             P. filiformis Pers.

21   Lvs with 3(–5) principal and many faint intermediate longitudinal
     veins; stems strongly compressed or even winged.   22
    Lvs with only 3–5 longitudinal veins or apparently 1-veined
     especially when very narrow, but never with many faint inter-
     mediate longitudinal veins; stems not or slightly compressed.   23

22 Lvs 10–20 cm. × 2–4 mm., usually rounded and cuspidate at the
tip, sometimes acuminate; principal lateral veins usually distinctly joining the midrib below the tip of the lf; stipules blunt;
fr. smooth, with a straight beak.   **16. compressus**

     Lvs 5–13 cm. × 2–3(–4) mm., finely acuminate; principal lateral
veins usually not distinctly joining the midrib near the tip;
stipules acute; fr. toothed near the base of the ventral margin
and tubercled along the dorsal margin; beak recurved. Rare in
calcareous ponds, streams, ditches, etc., in S. and E. England.
P. acutifolius Link

23 Lvs mostly 5-veined, the laterals closer to the margin and to each
other than to the midrib; stipular sheath a closed tube below.
11. friesii

     Lvs mostly 3-veined or apparently 1-veined, rarely 5-veined and
then with equal spacing.   24

24 Lvs 2–3 mm. wide, 3(–5)-veined, the laterals joining the midrib at a
wide angle close below the blunt mucronate tip; stipular sheath
open, with rolled overlapping margins.   **13. obtusifolius**

     Lvs mostly less than 2 mm. wide.   25

25 Stipular sheath a closed tube below; nodal glands (p. 566) present
or 0.   26

     Stipular sheath open, with rolled overlapping margins; nodal
glands present.   27

26 Stipules tubular for at least ⅔ of their length; lf-tip gradually acute
with a narrow blunt terminal cusp; nodal glands 0.   **12. pusillus**

     Stipules tubular only towards the base; lf narrowed to a sharply
acuminate tip; nodal glands present. Shetland and Outer
Hebrides only.   *P. rutilus* Wolfg.

27 Lvs usually exceeding 1 mm. wide, subacute to rounded and
mucronate at the tip, with air-filled spaces bordering the midrib;
nodal glands conspicuous; fr. smooth, c. 4 per fl. **14. berchtoldii**

     Lvs rarely more than 1 mm. wide, narrowing to the base and
tapering to a long fine point; usually without air-filled spaces
bordering the midrib; fr. often toothed below and tubercled on
the dorsal margin, usually 1 per fl.   **15. trichoides**

1. P. natans L.

Rhizomes extensively creeping. Lfy stems to 100 cm. or more, ± cylindrical, not or little branched. Submerged lvs linear (1–3 mm. wide),
channelled, rarely with a small blade. Floating lvs stalked, with the
leathery blade elliptical to ovate-lanceolate, rounded to cordate at the
base (which is inrolled at first), with 2 wings running for a short distance
down the stalk so that it appears jointed just below the blade; transverse veins indistinctly visible against the light. Stipules 5–12(–18) cm.,

free, persistent, at length fibrous. Fr.-spike 3–8 cm., cylindrical, dense; its stalk stout, not thickened upwards. Fr. 4–5 mm.; beak short, straight. Fl. 5–9. Lakes, ponds, rivers and ditches, especially on a highly organic substratum and usually in water less than 1 m. deep. Common throughout the British Is.

2. P. polygonifolius Pourret

Rhizomes extensively creeping. Lfy stems commonly to 20 cm., slender, cylindrical, branched. Lvs all with blade and stalk. Submerged lvs with blade commonly 8–20 × 1–3 cm., ± narrowly lanceolate-elliptical (but very variable), membranous, translucent. Floating lvs with blade commonly 2–6 × 1–4 cm. and stalk one-half to twice as long, broadly elliptical to lanceolate, not jointed below the blade, not translucent; transverse veins plainly visible against the light. Stipules 2–4 cm., free, persistent, blunt. Fr.-spike 1–4 cm., cylindrical, dense; its stalk slender, not thickened upwards. Fr. c. 2 mm., reddish, hardly beaked. Fl. 5–10. Ponds, bog-pools, ditches and small streams with acid and usually shallow water. Common throughout the British Is.

3. P. coloratus Hornem.

Resembling *P. polygonifolius* in habit but all lvs with stalk usually shorter than blade, ± blunt, thin, translucent, finely and distinctly net-veined. Submerged lvs 6–10(–18) cm., linear-lanceolate to narrowly elliptical, often subsessile. Floating lvs with blade 2–7(–10) × 1·5–5 cm., ovate-elliptical. Stipules 2–4 cm., free, blunt, persistent. Fr.-spike 2·5–4 cm., cylindrical, dense, its stalk slender, not thickened upwards. Fr. 1·7–2 mm., green; beak very short, curved. Fl. 6–7. Shallow ponds and pools in fen peat, especially in calcareous water. Local throughout the British Is.

4. P. lucens L.

Rhizome extensively creeping. Lfy stems to 2 m. or more, stout and tough. Submerged lvs commonly 10–20 × 2·5–6 cm., oblong-lanceolate, the thin translucent shining blade running down the short stalk and appearing subsessile, acuminate or cuspidate, margins minutely toothed. Floating lvs 0. Stipules 3–8 cm., free, blunt, prominently 2-keeled. Fr.-spike 5–6 cm., cylindrical, stout, dense, its stalk stout, thickened upwards. Fr. c. 3·5 mm., olive-green; beak short. Fl. 6–9. Lakes, ponds, canals and slow streams on base-rich inorganic substrata; commonly chalk-encrusted. Locally common throughout much of the British Is.

5. P. × zizii Roth (*P. gramineus × lucens*)

Rhizome creeping. Lfy stems commonly 0·5–2 m., stout, branching below. Submerged lvs sessile or very short-stalked, their blades 3·5–15 ×

0·7–2·5 cm., oblanceolate to oblong-lanceolate, narrowed to the base and cuspidate or rounded and mucronate at the tip, their margins finely and irregularly toothed and waved. Floating lvs often produced, their short stalks never longer than the leathery, oblong-lanceolate, usually cuspidate blade with finely toothed margins. Stipules 1·5–4 cm., free, broad, blunt, 2-keeled. Fr.-spike 3–6 cm., cylindrical, ripening a variable number of achenes, sometimes 0; its stalk stout, thickened upwards. Fr. c. 3 mm.; beak very short. Fl. 6–9. Local in ponds, lakes, streams, etc., throughout much of the British Is.

6. P. gramineus L.

Rhizome creeping. Lfy stems to 1 m. or more, slender, flexuous, with very many short, non-flowering branches near the base. Submerged lvs commonly 2·5–8 × 0·3–1·2 cm., usually linear-lanceolate or elliptical-oblong, narrowed into a sessile base, acuminate or cuspidate, with minutely toothed margins. Floating lvs few or 0, long-stalked; blade 2·5–7 × 1–2·5 cm., usually broadly elliptical-oblong, ± rounded at the base, leathery, with transverse veins visible against the light. Stipules 2–5 cm., broadly lanceolate, acute, not keeled. Fr.-spike 2·5–5 cm., cylindrical, dense, its stalk stout, thickened upwards. Fr. 2·5–3 mm., green; beak short, straightish. Fl. 6–9. Lakes, ponds, streams, canals, etc., chiefly in acid water. Locally common throughout much of the British Is.

7. P. × nitens Weber (*P. gramineus × perfoliatus*)

Rhizome creeping. Lfy stems to 120 cm., slender, sparingly branched. Submerged lvs commonly 2–8 × 0·5–1·5 cm., oblong-lanceolate, rounded, ± cordate and usually half-clasping at the sessile base, acute or cuspidate, their margins waved and (at least when young) minutely toothed. Floating lvs few or 0, elliptical-oblong, leathery. Stipules 1–2 cm., free, lanceolate, acute, not keeled, persistent. Spike sterile, its stalk usually widest at or above the middle. Fl. 6–8. Lakes, ponds, streams, etc.; rather local throughout the British Is.

8. P. alpinus Balb.

Rhizome creeping. Lfy stems 15–200 cm., slender, little branched. Submerged lvs usually 6–15 × 1–2 cm., often reddish, narrowly oblong-elliptical, narrowed to each end, subsessile or short-stalked, always blunt, quite entire. Floating lvs 3–8 × 0·8–2 cm., broadest above the middle, short-stalked, blunt, entire, with the transverse veins easily visible against the light; sometimes 0. Stipules 2–6 cm., free, ovate, blunt, not keeled, robust. Fr.-spike 2–4 cm., dense, its stalk slender, not thickened upwards. Fr. 3 mm., becoming pale reddish; beak longish, curved. Fl. 6–9. Lakes, ditches, streams, etc., especially in acid water and on organic substrata; throughout the British Is, but rare in S. and E. England.

9. P. praelongus Wulfen

Rhizome creeping. Lfy stems commonly to 2 m., stout, somewhat branched above. Submerged lvs 6–18 × 2–4·5 cm., green, not shining, strap-shaped to oblong, rounded at the sessile half-clasping base and narrowing gradually to the blunt and distinctly hooded tip, their margins slightly waved and quite entire. Floating lvs 0. Stipules 0·5–6 cm., free, blunt, thin, not keeled. Fr.-spike 3–7 cm., cylindrical, dense, its stalk 15–40 cm., stoutish, not thickened upwards. Fr. 4–6 mm., green; beak short, straight. Fl. 5–8. Lakes, rivers, canals and ditches, on mainly inorganic substrata; rather local and absent from much of S. England.

10. P. perfoliatus L.

Rhizome extensively creeping. Lfy stems commonly to 2 m., usually stout, branching above. Submerged lvs commonly 2–6 × 1·3–4 cm., all sessile and ± completely clasping at the wide cordate base, commonly ovate (but very variable in shape), blunt or rarely mucronate, microscopically and irregularly toothed. Floating lvs 0. Stipules up to 1 cm., free, blunt, not keeled, very delicate and soon disappearing from all but the uppermost axils. Fr.-spike 1–3 cm., stout, dense, its stalk 3·5–13 cm., stout, not thickened upwards. Fr. 3·5–4 mm., olive-green; beak very short, curved. Fl. 6–9. Lakes, ponds, streams, canals, etc., especially on moderately organic substrata. Common throughout the British Is.

11. P. friesii Rupr.

Rhizome 0; stem 20–100 cm., usually from a turion, strongly compressed, producing numerous very short lfy branches or clusters of lvs which later develop turions. Submerged lvs commonly 4–6·5 cm. × 2–3 mm., linear, sessile, subacute or abruptly mucronate; longitudinal veins usually 5, the distance between the midrib and the nearest lateral veins being almost twice that between the laterals and between the further lateral and the margin; there are transverse veins but no faint intermediate longitudinal veins. Floating lvs 0. Stipules 0·7–1·5 cm., tubular below at first but soon splitting, whitish, fibrous-persistent. Fr.-spike 0·7–1·5 cm., with 3–4 remote whorls of fr., its stalk flattened, somewhat thickened upwards. Fr. 2–3 mm., olive; beak short, erect or recurved. Fl. 6–8. Lakes, ponds, canals, etc., especially on a muddy substratum; common in suitable lowland waters throughout Great Britain, rare in Ireland.

12. P. pusillus L., sec. Dandy & Taylor (*P. panormitanus* Biv.)

Rhizome usually 0; stem 20–100 cm., slightly compressed, very slender, usually with many long branches from near the base. Submerged lvs commonly 1–4 cm. × 0·3–1 mm., narrowly linear, sessile, tapering gradually to a long but blunt tip; longitudinal veins 3 (rarely 5), the laterals joining the midrib at a narrow angle and at a distance of 2–3 lf-widths below the tip; midrib (except of uppermost lvs) usually not bordered by pale bands of air-tissue. Floating lvs 0. Stipules 6–17 mm., tubular to above half-way, splitting later, persistent. Spike 6–12 mm., interrupted, of c. 2–8 fls in 2–4 whorls, its stalk threadlike, not thickened upwards. Fr. 2–2·5 mm., pale olive, smooth; beak almost central, straightish. Fl. 6–9. Turions chiefly axillary, narrowly spindle-shaped. Lakes, ponds, canals, streams, etc., especially in highly calcareous or brackish waters; not uncommon throughout the British Is.

13. P. obtusifolius Mert. & Koch

Rhizome 0; stem 20–100 cm., compressed (c. 2:1), with many short lateral branches. Submerged lvs 3–9 cm. × 2–4 mm., linear, narrowed to the sessile base, rounded and shortly apiculate at the tip, dark green, very translucent; longitudinal veins 3(–5), the laterals joining the midrib at a wide angle (sometimes greater than a right angle) and usually close below the tip; no faint intermediate longitudinal veins; midrib bordered by pale bands of air-tissue especially towards the base. Floating lvs 0. Stipules 1·3–2 cm., open, broad, blunt. Spike 6–13 mm., dense and continuous, its stalk slender, straight, not thickened upwards. Fr. 3–4 mm., brownish-olive, 3-keeled when dry; beak short, straight. Fl. 6–9. Turions terminal, narrowly fan-shaped. Lakes, ponds, streams, canals, ditches, etc.; local throughout much of the British Is.

14. P. berchtoldii Fieber (*P. pusillus* auct. mult., non L., sec. Dandy & Taylor)

Rhizome usually 0; stem 10–100 cm., very slender, very little compressed (less than 2:1). Submerged lvs 2–5·5 cm. × 0·5–2 mm., linear, sessile, usually shortly mucronate at the tip, dark green; longitudinal veins always 3, the laterals close to the margins and meeting the midrib almost at right angles and at ½–1 lf-width below the tip; no faint intermediate longitudinal veins; midrib bordered at least near the base by pale bands of air-tissue. Floating lvs 0. Stipules 3–10 mm. or more, open, with rolled margins, blunt, persistent only in the uppermost axils. Fr.-spike 2–8 mm., subglobose, continuous or slightly interrupted, its stalk commonly 0·5–3 cm., threadlike, hardly thickened upwards. Fr.

2–2·5 mm., dark olive, tubercled below and bluntly keeled when dry; beak short. Fl. 6–9. Turions terminal, spindle-shaped. Lakes, ponds, canals, streams, ditches etc.; common throughout the British Is.

15. P. trichoides Cham. & Schlecht.

Rhizome threadlike or 0. Stems 20–75 cm., threadlike, ± cylindrical, repeatedly wide-branching. Submerged lvs commonly 2–4 cm. × 0·5–1 mm., narrowly linear, sessile or half-clasping at the base and tapering to a long fine point, spreading, rather rigid, deep dull green; longitudinal veins 3, the laterals very faint; midrib usually not bordered by pale bands. Floating lvs 0. Stipules c. 10 mm. or more, narrow, acute, with rolled overlapping margins. Spikes 1–1·5 cm., 3–6-fld, ± ovoid, interrupted, usually ripening only 1 fr. per fl.; stalk 5–10 cm., threadlike, not thickened upwards, often curved above. Fr. 2·5 mm., often toothed near the base and tubercled down the obscurely keeled back; beak short, straight. Fl. 6–9. Ponds, canals, ditches, etc.; chiefly in S. and E. England.

16. P. compressus L.

Rhizome present or 0. Stem 50–200 cm. × 3–6 mm., strongly flattened and ± winged. Submerged lvs 10–20 cm. × 2–4 mm., linear, sessile, rounded and cuspidate or sometimes acuminate at the tip; main longitudinal veins usually 5, with many faint intermediate longitudinal veins. Floating lvs 0. Stipules 2·5–3·5 cm., very blunt, 2-keeled, persistent, open, with rolled overlapping margins. Spike 1–3 cm., cylindrical, continuous, its stalk compressed, stout, not thickened upwards. Fr. 3–4·5 mm., bluntly 2-keeled, usually not toothed; beak almost central, very short, stout. Fl. 6–9. Turions with the inner lvs much longer than the outer. Lakes, slow streams, canals, ponds, ditches, etc., chiefly in C. and E. England.

17. P. crispus L.

Rhizome creeping or 0. Stem 30–120 cm., slender, compressed and ± 4-angled with the broader sides furrowed, repeatedly forked above. Submerged lvs commonly 3–9 cm. × 8–15 mm., lanceolate to linear-lanceolate, sessile, usually rounded and blunt at the tip, their margins serrate and often strongly waved when mature, often reddish, shining; longitudinal veins 3–5, with no faint intermediates. Floating lvs 0. Stipules 1–2 cm., soon becoming torn and decayed. Fr.-spike 1–2 cm., oblong-ovoid, rather lax, its stalk slender, compressed, narrowed upwards and often curved. Fr. 2–4 mm. (excl. beak), dark olive; beak equalling the rest of the fr., tapering, sickle-shaped. Fl. 5–10. Turions 1–5 cm., with thick horny spiny-bordered spreading lvs. Lakes, ponds, streams, canals, etc.; common throughout the British Is.

18. P. pectinatus L.

Rhizome extensively creeping. Lfy stem 40–200 cm., very slender, ± cylindrical, usually much branched. Submerged lvs 5–20 cm. × 0·25–2(–5) mm., dark green, narrowly linear (stem-lvs always wider than branch-lvs, and lower lvs than upper), with 2 wide longitudinal air-filled canals, one on each side of the midrib, occupying the greater part of the interior. Floating lvs 0. Stipules 2–5 cm., adnate below to the lf-base but open and rolled, with whitish margins; ligule long, blunt, soon shed. Spike 2–5 cm. of 4–8 usually 2-fld whorls, ± interrupted, especially below, its stalk threadlike, not thickened upwards. Fr. 3–5 mm., olive tinged with orange, 1–3-keeled; beak short. Fl. 5–9. Usually perennates by tuberous lateral rhizome-buds. Ponds, rivers, canals, ditches, etc.; abundant in base-rich or brackish lowland waters but absent from mountain districts of Wales and Scotland and from much of Ireland.

2. GROENLANDIA Gay

Aquatic herbs like *Potamogeton* but with lvs opposite or whorled, and stipules 0 except in the flowering region.

1. G. densa (L.) Fourr. (*Potamogeton densus* L.)

Rhizome creeping. Lfy stem 10–30 cm., ± cylindrical, forking above. Submerged lvs opposite or in whorls of 3, commonly 1·5–2·5 × 0·5–1·5 cm., ovate-triangular to lanceolate, commonly folded lengthwise and recurved, their margins minutely toothed especially near the tip; longitudinal veins 3–5. Floating lvs 0. Stipules 0, except in flowering region. Spike usually of only 4 fls, ovoid in fr., its stalk slender, shorter than the lvs, at first erect then strongly recurved. Fr. 3 mm., olive, strongly compressed and sharply keeled; beak short, central, recurved. Fl. 5–9. Clear swift streams, ditches, ponds; locally abundant in lowland Great Britain and S. Ireland.

128. RUPPIACEAE

Submerged herbs of salt or brackish pools, rarely in fresh water. Lvs very narrow, serrulate at apex. Infl. a short terminal raceme, appearing subumbellate. Fls hermaphrodite, small. Perianth 0. Stamens 2, filaments short and broad. Ovary superior, carpels 4 or more, becoming long-stalked in fr.

1. RUPPIA L.

The only genus.

Infl.-stalk several times as long as stalk of the fr., often spirally coiled;
  lvs acute. Rather rare.

R. spiralis Dumort.

Infl.-stalk about as long as stalk of the fr., not spirally coiled; lvs.
  acute.   **1. maritima**

1. R. maritima L.

30 cm. or more, slender, much-branched. Lvs c. 1 mm. wide, dark
green; sheaths dilated, brownish. Infl. stalk 0·5–5 cm. Ventral side of fr.
convex, ending in a long beak, dorsal side strongly gibbous at base. Fl.
7–9. Brackish ditches and salt-marsh pools, local but widely distributed.

129. ZANNICHELLIACEAE

Submerged perennial aquatic herbs of fresh, brackish or sea water, with
slender creeping rhizomes. Lvs linear with sheathing bases and usually
with a ligule at the junction of blade and sheath. Fls very small, unisexual. Perianth of 3 small scales or 0; stamens hypogynous, 3–1;
carpels 1–9, free, each with 1 ovule. Pollination hydrophilous. Fr. of
1–9 sessile or stalked achenes; seeds non-endospermic.

1. ZANNICHELLIA L.

Monoecious herbs of fresh or brackish water. Lvs mostly opposite,
linear, entire, with axillary stipules. Fls axillary in a transparent deciduous cup-shaped spathe, 1 male and 2–6 female fls often in the same
spathe. Male fl. of 1 stamen. Female fl. of 1 carpel with a flattened
stigma. Perianth 0. Fr. a stalked ± curved achene.

1. Z. palustris L.   Horned Pondweed.

Submerged, with very slender much-branched lfy shoots commonly to
50 cm., sometimes creeping and rooting over much of their length. Lvs
1·5–5(–10) cm. × 0·4–2 mm., tapering to a fine point; stipule clasping
the stem, tubular below, scarious, soon falling. Style stout below, tapering upwards; stigma peltate with a waved and crenate margin. Achenes
in groups of 2–6, subsessile or stalked; dorsal margin ± toothed. Fl.
5–8. Very variable. Rivers, streams, ditches and pools of fresh or brackish water; locally common throughout the British Is.

130. NAJADACEAE

1. Najas L.

The only genus. Three spp. occur, all very local; they may be distinguished as follows:

1 Lvs with many large spinous teeth. A few of the Norfolk Broads.
N. marina L.
  Lvs minutely toothed or nearly entire. 2

2 Lvs very minutely toothed or nearly entire, 2–3 together. A few
  lakes in the Lake District, Scotland and W. Ireland.
N. flexilis (Willd.) Rostk. & Schmidt
  Lvs minutely spinous-toothed, mostly tufted on short lateral shoots.
  Reddish Canal, Lancs.; probably extinct. **N. graminea* Delile

131. ERIOCAULACEAE (p. xxiv)

132. LILIACEAE

Herbs, rarely shrubby, sometimes climbing. Lvs various arranged.
Infl. usually racemose (an umbel in *Allium*). Fls usually hermaphrodite
and actinomorphic, 3-merous, rarely 2-, 4- or 5-merous. Perianth usually
petaloid, in two similar whorls. Stamens in two whorls, free or inserted
on the perianth. Ovary superior, usually 3-celled. Ovules usually
numerous. Fr. a capsule or berry. Seeds endospermic.

1 Lvs normally developed; herbs with the aerial stems unbranched
  or branched only in infl. 2
  Lvs scale-like, the assimilating organs being lf-like or needle-like
  and borne in the axils of the scales; much-branched herbs or
  shrubs. 18

2 Fls racemose or solitary, rarely subumbellate and then bracts lf-
  like; plant not smelling of onion. 3
  Fls umbellate; infl. enclosed before flowering in a scarious spathe
  of 1 or more bracts; plant smelling of onion or garlic.
10. ALLIUM

3 Lvs alike on both surfaces, vertical, iris-like. Wet places. 4
  Lvs ± horizontal, the two surfaces differing. Drier ground. 5

4 Styles 3, free; fls whitish; filaments glabrous. A rare plant of
  mountains in N. England and Scotland. Scottish Asphodel.
Tofieldia pusilla (Michx) Pers.
  Style 1, simple; fls golden-yellow; filaments woolly.
1. NARTHECIUM

5 Fl. stem lfy or with lf-like bracts.   6
   Lvs all basal or lvs absent at flowering or stem with a few scale-
      like lvs.   12

6 Per. segs 4, white; fls small, in an erect terminal raceme; lvs ovate,
      deeply cordate at the base. A very rare woodland plant. May
      Lily.   *Maianthemum bifolium* (L.) F. W. Schmidt
   Per. segs 6; lvs not cordate.   7

7 Fls in the axils of the lvs, greenish-white, almost cylindrical.
                                                          3. POLYGONATUM
   Infl. terminal, or fls terminal and solitary.   8

8 Infl. a raceme; fls large, with recurved per. segs.   5. LILIUM
   Infl. not a raceme; per. segs not recurved.   9

9 Fls solitary (rarely 2).   10
   Fls subumbellate, yellow.   6. GAGEA

10 Fls yellow. Naturalized in meadows and orchards. Wild Tulip.
                                                          Tulipa sylvestris L.
   Fls not yellow.   11

11 Fls drooping, bell-shaped, their per. segs with a check pattern of
      reddish-purple on a paler ground, rarely white; lvs linear.
      A very local plant of damp meadows in S. and C. England.
      Fritillary; Snakeshead.   *Fritillaria meleagris* L.
   Fls erect, white with purplish veins; lvs filiform. A very rare alpine
      of rock-ledges in the Snowdon range.
                                          Lloydia serotina (L.) Reichenb.

12 Fls solitary, from the ground; plant lfless at flowering.
                                                          11. COLCHICUM
   Fls not solitary; lvs present at flowering.   13

13 Infl. a panicle; per. segs white inside, purplish outside. A very rare
      linear-leaved plant of heathland in Co. Kerry and under pines
      near Bournemouth.   *Simethis planifolia* (L.) Gren. & Godron
   Infl. a raceme.   14

14 Lvs ovate-lanceolate, stalked; rhizomatous.   2. CONVALLARIA
   Lvs linear, sessile; bulbous.   15

15 Perianth 3–5 mm., dark blue, contracted at the mouth; upper fls
      paler, sterile. Dry grassland in E. Anglia and Oxon. Grape Hya-
      cinth. *Muscari atlanticum* Boiss. & Reuter (*M. racemosum* auct.)
   Perianth not contracted at the mouth; fls all fertile.   16

16 Fls blue or purple, rarely pure white.   17
   Fls white, marked with green outside.   7. ORNITHOGALUM

17 Bracts 0 or 1 to each fl.; per. segs quite free, 8 mm. or less.
                                                          8. SCILLA
   Bracts 2 to each fl.; per. segs united at base, more than 10 mm.
                                                          9. ENDYMION

18 Photosynthetic organs (cladodes) broad, bearing the fls in the
    middle of one surface. **4. RUSCUS**

Photosynthetic organs needle-like, in clusters; fls 1–2, axillary,
    bell-shaped, whitish; fr. a red berry. Asparagus.
                                                Asparagus officinalis L.

(Ssp. *prostratus* (Dumort.) E. F. Warburg, with prostrate stems, is
a very rare plant of maritime cliffs and sands; *ssp. *officinalis*,
with erect stems, is the introduced vegetable, often escaping.)

1. NARTHECIUM Hudson

Rhizomatous herbs. Lvs in 2 ranks, held vertically. Fls in racemes.
Per. segs 6, free or nearly so, persistent in fr. Stamens 6, with linear
anthers and woolly filaments. Style 1, simple. Fr. a loculicidal capsule.

1. N. ossifragum (L.) Hudson   Bog Asphodel.

Glabrous herb with creeping rhizome. Basal lvs 5–30 cm. × 2–5 mm.,
rigid, often curved; stem-lvs 4 cm. or less, few, sheathing. Fl.-stem
5–40 cm. Infl. 2–10 cm.; bracts lanceolate. Per. segs 6–8 mm., yellow,
spreading in fl., erect in fr. Anthers orange. Whole infl. becoming deep
orange after fl. Capsule c. 12 mm., narrow, 6-grooved, mucronate. Fl.
7–9. Bogs, wet heaths and moors and wet acid places on mountains.
Common throughout most of the British Is except for some of the
eastern and midland counties.

2. CONVALLARIA L.

Rhizomatous herbs. Fls drooping, in a one-sided axillary raceme.
Perianth shortly bell-shaped, the 6 segs joined to about the middle and
their free lobes recurved. Stamens inserted on the perianth-base, included. Style simple; stigma capitate. Fr. a berry.

1. C. majalis L.   Lily-of-the-Valley.

Rhizome far-creeping, much branched, producing several scale-lvs and
2(–3) long-stalked ovate-lanceolate or elliptic lvs 8–20 × 3–5 cm., from
each upwardly-directed apex; stalk of lower lf ensheathing that of upper
lf and resembling an aerial stem. Scape lfless, from axil of a scale-lf;
bracts scarious, ovate-lanceolate. Fls 6–12, white, sweet-scented. Berry
globose, red. Fl. 5–6. Dry woods, chiefly on limestone. Widespread but
local in England, Wales and Scotland northwards to Dunbarton and
E. Inverness.

3. POLYGONATUM Miller

Rhizomatous herbs. Lvs numerous, all cauline. Fls axillary, solitary or in few-fld racemes. Perianth greenish-white, tubular, the 6 segs joined to near the base. Stamens 6, inserted on the perianth, included. Style slender; stigma 3-lobed. Fr. a berry.

1 Lvs in whorls of 3–6, linear-lanceolate; fr. red; fls 1–4. Very rare in
   mountain woods in N. England and Scotland. Whorled Solomon's Seal.   *P. verticillatum* (L.) All.
   Lvs not whorled; fr. blue-black.   *2*

2 Stem angled; fls 1–2; perianth 18–22 mm., not contracted in the
   middle; filaments glabrous. Very local in limestone woods, rarely
   on sand-dunes, in N. and W. England and Wales; Inner
   Hebrides. Angular Solomon's Seal.   *P. odoratum* (Miller) Druce
   Stem terete; fls 2–5; perianth 9–15 mm., contracted in the middle;
   filaments pubescent.   **1. multiflorum**

1. P. multiflorum (L.) All.   Solomon's Seal.

Stem 30–80 cm., terete, arching. Lvs 5–12 cm., alternating in 2 rows, ovate to elliptic-oblong, acute, sessile. Fr. c. 8 mm. Fl. 5–6. Local in woods over most of England and Wales; Jersey; naturalized in Scotland. (The plant commonly cultivated is *P.* × *hybridum* Brügger.)

4. RUSCUS L.

Rhizomatous shrubs. Lvs reduced to small scarious scales with axillary flattened lf-like shoots (cladodes). Fls on surface of cladodes, dioecious. Inner per. segs smaller than outer. Stamens 3, with filaments joined in a cup (anthers 0 in female fl.). Style short; stigma capitate. Fr. a berry.

1. R. aculeatus L.   Butcher's Broom.

Stems 25–80 cm., stiff, much-branched, green, grooved, glabrous. Scale-lvs less than 5 mm., thin, brownish. Cladodes 1–4 cm., ovate, entire, thick, rigid, spine-tipped, dark green, twisted at base. Fls 1–2 on upper surface of cladode in the axil of a scale-like bract. Perianth c. 3 mm., greenish. Stamen-cup violet. Fr. c. 1 cm., red, globose. Fl. 1–4. Dry woods and among rocks. Widespread but local in S. England north-wards to N. Wales, Leicester and Norfolk.

5. LILIUM L.

Bulbous herbs, the bulb-scales numerous, imbricate, non-sheathing. Lvs numerous, all cauline. Fls large. Per. segs 6, all alike. Stamens 6;

anthers versatile. Style long; stigma ±3-lobed. Fr. a loculicidal capsule with numerous flat seeds.

Lvs in distant whorls of 5–10; fls 3–10 in a terminal raceme, dull purple, the per. segs strongly recurved. Possibly native in woods in S. England but widely naturalized. Martagon Lily.   *L. martagon* L.

Lvs dense, spirally arranged; fls 1–8 in a terminal raceme, yellow with black dots, the per. segs strongly recurved. Introduced and locally naturalized. Yellow Martagon Lily.   **L. pyrenaicum* L.

6. GAGEA Salisb.

Bulbous herbs. Lf basal only or the stem bearing 1 lf in addition to lf-like bracts. Infl. often subumbellate. Fls erect. Per. segs 6, yellow, spreading. Stamens 6. Fr. a loculicidal capsule with roundish seeds.

1. G. lutea (L.) Ker-Gawler   Yellow Star-of-Bethlehem.

Flowering bulb c. 1 cm. diam., yellowish, usually with numerous small daughter bulbs at the base. Basal lf solitary, commonly 15–45 cm. × 7–12 mm., suddenly narrowed to a hooded acuminate tip, strongly 3–5(–7) ribbed. Stem 8–25 cm. Infl. subumbellate, the 1–5 fls having unequal stalks 1·5–5 cm. Perianth 10–15 mm., the narrowly oblong segs with a broad green band on the outside. Fl. 3–5. Damp woods and pastures on base-rich soils. Very local; commonest in C. and N. England and extending northwards to Stirling.

7. ORNITHOGALUM L.

Bulbous herbs with lvs all basal. Infl. racemose. Per. segs 6, free, obscurely veined, persistent, usually whitish marked with green. Stamens 6, with flattened filaments. Fr. a loculicidal capsule with roundish seeds.

1 Infl. corymbose with the lower fl.-stalks much longer than the upper; lvs green with a white stripe down the midrib; scape 10–30 cm. Probably native in grassy places in E. England, naturalized elsewhere and widespread but local. Star-of-Bethlehem.
    O. umbellatum L.
   Infl. with all fl.-stalks about equally long; scape 25–100 cm.   2

2 Fls large, 2–3 cm., 2–12; drooping; filaments with 2 teeth at apex; lvs with white central stripe; scape 25–60 cm. Introduced in grassy places, local. Drooping Star-of-Bethlehem.   **O. nutans* L.
   Fls 6–10 mm., more than 20, ±erect; filaments without teeth; lvs without white stripe; scape 50–100 cm. Very locally abundant in woods and scrub. Bath Asparagus.   *O. pyrenaicum* L.

8. SCILLA L.

Bulbous herbs with lvs all basal. Infl. racemose. Per. segs 6, free, spreading, with a prominent midrib, usually bluish. Stamens 6, inserted at base of perianth, filaments slender or widened at base. Fr. a loculicidal capsule with roundish or angled black seeds.

Bracts conspicuous; fl. April–May. **1. verna**
Bracts 0; fl. July–Sept.; lvs produced after the purple fls. Very local in dry short grassland, usually near the sea, in S. England, S. Wales and Channel Is. Autumnal Squill. *S. autumnalis* L.

1. S. verna Hudson
Spring Squill.

Bulb 1·5–3 cm. Lvs 3–20 × 2–4 mm., 3–6, linear, produced before the fls. Scape 5–15 cm. Infl. 2–12-fld. Bracts bluish, longer than the fl.-stalks. Per. segs 5–8 mm., violet-blue. Anthers violet-blue; filaments lanceolate. Capsule c. 4 mm. Fl. 4–5. Dry grassy places near the coast, locally from Cornwall and Devon to Shetland and from Northumberland to Moray; E. Ireland.

9. ENDYMION Dumort.

Like *Scilla* but with 2 bracts to each fl. and a bell-shaped perianth whose segs are united at the base. Stamens inserted near the middle of the perianth.

Infl. drooping at tip; per. segs ± parallel below; anthers cream.
1. non-scriptus
Infl. ± erect; per. segs divergent from the base; anthers blue. Introduced and naturalized locally. Garden Bluebell.
*E. hispanicus (Miller) Chouard

1. E. non-scriptus (L.) Garcke
Bluebell.

Bulb 2–3 cm., ovoid, renewed annually, its scales tubular. Lvs 20–45 cm. × 7–15 mm., linear, glabrous. Scape 20–50 cm. Infl. 4–16-fld, 1-sided, the fls erect in bud, nodding when open. Fl.-stalks c. 5 mm., later elongating and erecting. Bracts paired, bluish. Per. segs 1·5–2 cm., violet-blue, rarely pink or white, ± parallel so that the lower part of the fl. is cylindrical. Anthers cream. Fr. c. 15 mm., ovoid. Fl. 4–6. Common in woods, hedge-banks, etc., rarely in pastures; throughout the British Is, except Orkney and Shetland, and often dominant in coppiced woods on light soils.

10. ALLIUM L.

Usually bulbous herbs smelling of onion or garlic. Lvs all basal, often with long sheaths. Infl. an umbel, enclosed before fl. in a spathe which later splits. Per. segs 6, all alike, free or nearly so. Fr. a loculicidal capsule with black seeds.

1	Lvs linear or cylindric, sessile.	*2*
	Lvs ± elliptic, stalked.	**3. ursinum**
2	Scape terete; perianth 12 mm. or less, or infl. with bulbils only.	*3*
	Scape 3-angled; perianth 10–18 mm., white.	*13*
3	Infl. of bulbils only, without fls.	*4*
	Infl. with fls.	*5*
4	Spathe scarious, shorter than infl., usually in 1 piece, often falling.	
		1. vineale var. *compactum*
	Spathe in 2 parts with long lf-like points.	**2. oleraceum**
5	Inner filaments divided at apex into 3 long points, the middle one bearing the anther, outer entire.	*6*
	Filaments all entire or the inner with 2 small teeth at base.	*10*
6	Lvs flat, solid.	*7*
	Lvs cylindric or semicylindric, hollow.	*9*
7	Spathe in 1 piece, falling before fl.; plant robust with lvs 12–35 mm. broad. (Western.)	*8*
	Spathe in 2 pieces, present at fl.; plant rather slender with lvs 7–15 mm. broad; umbels with few reddish-purple fls and purple bulbils. Grassland and scrub on dry soils from N. Lincoln and Cheshire to Perth, very local; probably introduced in Ireland. Sand Leek. *A. scorodoprasum* L.	
8	Umbel subglobose, many-fld, without or with a few small bulbils; scape 60–200 cm.; perianth pale purple or whitish. Rocky and waste places near the coast in S.W. England and S. Wales and Guernsey, very rare. Wild Leek. *A. ampeloprasum* L.	
	Umbel irregular, few-fld, with numerous large bulbils, otherwise like *A. ampeloprasum*. S.W. England and W. Ireland, in rocky or sandy places near the coast; rare and endemic.	
		A. babingtonii Borrer
9	Inner stamens with lateral points shorter than anther; bulbils 0; perianth reddish-purple; scape 30–80 cm. Limestone rocks near Bristol and in Jersey. Round-headed Leek.	
		A. sphaerocephalon L.
	Inner stamens with lateral points longer than anther; bulbils nearly always present; perianth pink or greenish-white.	**1. vineale**

10   Spathe shorter than fls; per. segs 7–12 mm.   *11*
       Spathe with long lf-like points, much longer than fls; per. segs
          5–7 mm.   *12*

11   Lvs cylindric, hollow, 1–3 mm. broad; scape 15–40 cm.; per. segs
          7–12 mm., spreading, pale purple or pink. A tufted plant of
          rocky pastures, usually on limestone, superficially resembling
          Armeria maritima; very local in W. England and Wales from
          Cornwall to Westmorland; Northumberland, Berwick, Co.
          Mayo. Chives.   *A. schoenoprasum* L.
       Lvs flat, 4–12 mm. broad; scape 30–80 cm.; per. segs 10–12 mm.,
          pink, becoming scarious. Naturalized in a few places.
                                                              *A. roseum L.

12   Anthers included.   **2. oleraceum**
       Anthers exserted; lvs flat; per. segs bright pink. Naturalized in
          several places.   **A. carinatum* L.

13   Lvs 2–5; infl. without bulbils, 3–15-fld; per. segs 12–18 mm., white
          with green line. Thoroughly naturalized in hedge-banks and
          waste places in Channel Is, S.W. England, S. Wales and S.W.
          Ireland. Triquetrous Garlic.   **A. triquetrum* L.
       Lf 1; infl. with numerous bulbils, 1–4-fld; per. segs 10 mm., white.
          Naturalized in a few places.   **A. paradoxum* (Bieb.) G. Don

1. A. vineale L.   Crow Garlic.

Bulb with offsets. Lvs 20–60 cm. × c. 2 mm., subcylindric, somewhat
grooved, hollow, their bases sheathing the lower part of the terete
scape 30–80 cm. Spathe usually in 1 piece and soon falling, with a beak
about as long as itself, not or scarcely exceeding the fls. Umbels rather
lax, usually with bulbils only, often fls and bulbils, rarely fls only; fl.-
stalks several times as long as fls. Perianth c. 5 mm., bell-shaped, pink
or greenish-white. Stamens exserted; outer filaments slender, entire;
inner broad, with 3 points, the central anther-bearing, about half as
long as the 2 lateral. Stigma entire. Seeds angular. Fl. 6–7. Fields,
roadsides, grassy banks. Rather common in England and Wales and
often a serious weed; local in Scotland and Ireland; Channel Is.

2. A. oleraceum L.   Field Garlic.

Bulb ovoid, usually with offsets. Lvs 15–30 cm. × 2–3 mm., semi-terete,
grooved, usually hollow, at least below, (less commonly flat and to
4 mm. broad). Scape 25–80 cm., cylindric, sheathed by lf-bases in the
lower half. Spathe in 2 parts with long lf-like points much longer than
fls. Umbel lax, with fls and bulbils, rarely fls 0; fl.-stalks unequal, much
longer than fls. Perianth 5–7 mm., bell-shaped, pinkish, greenish or
brownish. Stamens included in perianth; filaments all slender, entire.

Fr. rarely or never produced. Fl. 7–8. Dry grassy places northwards to Wigtown and Moray, with a distinct eastern tendency; in Ireland very local in eastern coastal counties.

3. A. ursinum L. Ramsons.

Bulb narrow, solitary, consisting of a single lf-base. Lvs 10–25 × 4–7 cm., 2(–3), elliptic or ovate-elliptic, acute, bright green, with stalk 5–20 cm., twisted through 180°. Scape 10–45 cm., 3-angled or hemicylindrical and 2-angled, sheathed by lf-stalks at base. Spathe scarious, in 2 parts, each ovate, acuminate, shorter than fls. Infl. 6–20-fld, flat-topped, without bulbils; fls shorter than their stalks. Per. segs 8–10 mm., spreading, white, lanceolate, acute. Stamens shorter than perianth; filaments all slender and entire. Seeds globose. Fl. 4–6. Damp woods and shady places, especially on basic soils. Rather common throughout the British Is, except Orkney, Shetland and Channel Is.

11. COLCHICUM L.

Plant with corm. Lvs all basal. Fls 1–3, from the ground. Per. segs all alike, united below into a long tube. Styles 3, slender, free from the base. Fr. a septicidal capsule.

1. C. autumnale L.

Meadow Saffron, Naked Ladies, Autumn Crocus.

Glabrous. Corm 3–5 cm., with brown outer scales. Plant lfless at fl.; lvs produced in spring as the fr. ripens. Perianth pale purple, lobes oblong, 3–4·5 cm.; tube 5–20 cm. Scape elongating in fr. and appearing above the ground, sheathed by lf-bases. Lvs 12–30 × 1·5–4 cm., oblong-lanceolate, bright glossy green. Fr. 3–5 cm., obovoid; seeds numerous. Fl. 8–10. Damp meadows and woods on basic and neutral soils, local but sometimes in quantity, northwards to Cumberland and Durham; S.E. Ireland; sometimes naturalized in Scotland.

133. TRILLIACEAE

Rhizomatous herbs with erect simple stems. Lvs opposite or in a single whorl near top of stem, net-veined. Fls solitary or umbellate, terminal; hermaphrodite, regular. Per. segs free, unequal. Stamens equal in number and opposite the per. segs. Ovary superior, 1-celled with parietal, or 3–6-celled with axile, placentae; ovules numerous. Fr. a berry or fleshy loculicidal capsule; seeds endospermic.

1. PARIS L.

Rhizome creeping, monopodial. Lvs 4 or more in a single whorl. Fls solitary, 4(–6)-merous; outer per. segs sepaloid, inner petaloid, narrow. Styles free. Fr. a fleshy loculicidal capsule.

1. P. quadrifolia L.   Herb Paris.

Glabrous. Stems 15–40 cm. Lvs 6–12 cm., obovate, shortly acuminate, subsessile, 3–5-veined with a network between. Sepals 2·5–3·5 cm., green, lanceolate. Petals almost as long, very narrow. Ovary 4–5-celled. Fr. berry-like, black, globose, finally dehiscent. Fl. 5–8. Damp woods on calcareous soils; local. Throughout much of Great Britain but with an eastern tendency.

134. JUNCACEAE

Herbs, usually perennial, frequently tufted or with creeping rhizomes. Lvs long and narrow; terete, channelled or grass-like, with sheathing base; sometimes reduced to scales. Fls few to many, usually in numerous crowded monochasial cymes, sometimes in heads; actinomorphic, hermaphrodite, protogynous, wind-pollinated. Per. segs 6, in 2 whorls, usually greenish or brownish. Stamens free, in 2 whorls of 3, or with the inner whorl missing; pollen in tetrads. Ovary 1- or 3-celled; stigmas 3, brush-like. Fr. a loculicidal capsule with 3 to many often appendaged seeds; endosperm starchy.

Lvs glabrous, rarely flat and grasslike; seeds many.   1. JUNCUS
Lvs sparsely hairy, at least when young; flat or channelled and grasslike;
  seeds 3.   2. LUZULA

1. JUNCUS L.

Glabrous perennial herbs or dwarf annuals; erect, ± tufted, often rhizomatous. Stem-lvs with sheathing bases; blade channelled, flattened or ± terete, sometimes 0. Infl. a cluster of terminal or apparently lateral cymes, sometimes condensed into a head. Capsule many-seeded.

1 Flowering stems bearing only brown sheaths below the infl., which
      is apparently lateral and exceeded by a bract continuing the stem.   2
    Lvs on flowering stems (when present) with green blade; infl.
      obviously terminal or, if apparently lateral, then lvs with prickly
      points.   7

2 Stem very slender, 1 mm. or less in diam.; infl. at middle of the
      apparent stem or lower. A rare plant of stony lake shores in the
      Lake District and S. Scotland. Slender Rush.   *J. filiformis* L.
    Stem stouter; infl. above the middle of the apparent stem.   3

3   Far-creeping, growing in straight lines; infl. usually 5–20-fld.
    A local plant, chiefly of dune-slacks on the E. and N. coasts of
    Scotland and in the Hebrides. Baltic Rush.   *J. balticus* Willd.
    Densely tufted; infl. usually many-fld.   **4**

4   Pith interrupted; stem glaucous, with 12–16 prominent ridges.
                                                                        6. inflexus
    Pith continuous, at least in the lower part of the stem; stem not
    glaucous, usually with more than 18 ridges or striae.   **5**

5   Stem with 18–45 ridges; capsule with no fertile seeds.
                                                        J. inflexus × *effusus*
    Stem usually with more than 40 ridges or striae; capsule fertile.   **6**

6   Stem when fresh strongly ridged, especially just below the infl.;
    infl. dark reddish; capsule shortly mucronate.   **8. subuliflorus**
    Stem when fresh striate, scarcely ridged; infl. usually pale in
    colour; capsule not mucronate.   **7. effusus**

7   Annuals, rarely over 30 cm., readily uprooted (lowland).   **8**
    Perennials, usually over 30 cm., difficult to uproot (except for some
    small alpine plants).   **10**

8   Infl. a much-branched lfy panicle, occupying the greater part of the
    plant, rarely in terminal heads; seeds usually roundly ovoid.
                                                                        5. bufonius
    Fls in 1 or a few terminal heads; seeds twice as long as broad.   **9**

9   Per. segs less than 4 mm., with fine often recurved points; lf-sheath
    without auricles. A dwarf tufted annual, 1–5 cm., with very
    slender basal lvs, 0·5–4 cm., and a single terminal head. A very
    local plant of damp heaths in Cornwall, Anglesey and the
    Hebrides. Capitate Rush.   *J. capitatus* Weigel
    Per. segs 5 mm. or more, acute; lf-sheath with pointed auricles.
    A dwarf annual, 2–8 cm., with lvs mostly basal and fls in 1 or
    few heads. Lizard (Cornwall) and Raasay (Hebrides). Dwarf
    Rush.   *J. mutabilis* Lam. (*J. pygmaeus* L. C. M. Richard)

10 Lvs and lowest bracts ending in stiff prickly points; infl. many fld,
    apparently lateral.   **11**
    Lvs and lowest bracts not prickly pointed; infl. terminal.   **12**

11 Fls reddish-brown; capsule much exceeding the perianth.
                                                                        10. acutus
    Fls straw-coloured; capsule not or barely exceeding the perianth.
                                                                        9. maritimus

12 Fls solitary and terminal or in head-like clusters of 6 or fewer;
    seeds with appendages (alpine plants).   **13**
    Fls neither solitary nor in heads, usually in panicles; seeds without
    appendages (lowland or alpine).   **16**

13   Fl.-stems densely tufted; fls 1–3 together between axils of 2–3 long
       slender bracts. Slender, grasslike, often in circular patches.
       Tops of high mountains in Scotland and the Scottish islands.
       Three-leaved Rush.   *J. trifidus* L.
       Fl.-stems solitary or in small not dense tufts; longest bract not or
       only shortly exceeding the fls.   **14**

14   Stoloniferous; outer per. segs acute. Fls very large, dark chestnut-
       brown. Marshes by springs and streams on high Scottish
       mountains. Chestnut Rush.   *J. castaneus* Sm.
       Not stoloniferous; outer per. segs obtuse or bluntly pointed.   **15**

15   Fls 2–3, almost at one level, usually exceeding the lowest bract; lvs
       3–10 cm., all basal, in section showing 2 tubes; capsule shortly
       exceeding perianth. Local in bogs, stream-sides and wet rock-
       ledges on high mountains northwards from N. Wales, Lake
       District and Teesdale. Three-flowered Rush.   *J. triglumis* L.
       Fls (1–)2, one just below the other, usually exceeded by the lowest
       bract; lvs 3–6 cm., all basal, in section showing 1 tube; capsule
       much exceeding perianth. Very local in wet stony places and
       rock-ledges on Scottish mountains and in Inner Hebrides. Two-
       flowered Rush.   *J. biglumis* L.

16   Lvs solid, not septate; channelled or horizontally flattened, not
       laterally flattened nor very slender.   **17**
       Lvs hollow with or without internal septa; subterete, laterally
       flattened or very slender.   **21**

17   Lvs all basal, sharply reflexed above the sheathing base; lowest
       bract usually less than ½ as long as the infl.   **1. squarrosus**
       1–2 lvs on the stem; lowest bract at least ½ as long as infl.   **18**

18   Fls greenish or straw-coloured; per. segs very acute.   **19**
       Fls dark brown; per. segs blunt.   **20**

19   Lf-sheaths prolonged above into long scarious auricles.   **2. tenuis**
       Lf-sheaths with short brownish auricles, not scarious, broader than
       long. Introduced and naturalized locally in mid-Perth and
       Rhum. Dudley's Rush.   **J. dudleyi* Wiegand

20   Usually in salt-marshes; style as long as capsule; perianth nearly
       equalling capsule.   **4. gerardi**
       Usually not on saline soils; style shorter than capsule; perianth ½–⅔
       capsule.   **3. compressus**

21   Lvs with internal septa and usually distinctly jointed (but the
       septa indistinct in the very slender lvs of *J. bulbosus*); infl. a
       ±spreading panicle of many small stalked clusters of heads
       of fls.   **22**
       Lvs subterete with no persistent internal septa and therefore not
       jointed; infl. a long narrow interrupted panicle of separate fls.
       Salt-marsh plant with creeping rhizome and erect aerial shoots

to 1 m. or more, glaucous, furrowed. Known only from one locality on the N. Somerset coast and probably a recent arrival.
                                                              J. subulatus Forskål

22 Lvs very slender, in section (examine with lens) showing 2 tubes; plant less than 25 cm.   **14. bulbosus**
    Lvs not very slender, in section showing either 1 or more than 2 tubes; plant usually more than 25 cm.   23

23 Lvs with longitudinal as well as transverse septa; perianth straw-coloured to light reddish-brown.   **1. subnodulosus**
    Lvs without longitudinal septa; perianth dark brown or blackish, never straw-coloured.   24

24 Per. segs blunt, the outer mucronate; capsule obtuse, mucronate.   25
    At least the outer per. segs acute; capsule acute or acuminate.   26

25 Some fls stalked; capsule barely equalling perianth. A very rare plant of 2 localities in Scotland.   *J. nodulosus* Wahlenb.
    All fls subsessile; infl. with few branches arising from 2 points about 4 cm. apart; capsule slightly exceeding perianth. Rare in gravelly stream-beds and marshy places on mountains in Teesdale and Scotland. Alpine Rush.   *J. alpinoarticulatus* Chaix

26 Lvs strongly flattened laterally; capsule fertile, abruptly acuminate.
                                                              13. articulatus
    Lvs subterete, slightly flattened; capsule acute, or if acuminate then without fertile seeds.   27

27 Fls usually less than 7 in a head; capsule sterile, either much shorter than perianth or longer and abruptly acuminate.
                                                13. acutiflorus × articulatus
    Fls 6–12 in a head; capsule many-seeded, gradually tapered to a very acute point.   **12. acutiflorus**

1. J. squarrosus L.   Heath Rush.

Tough wiry perennial forming dense low tufts. Lvs 8–15 cm., usually all basal from a very short upright stock, awl-shaped, deeply channelled, rigid, sharply reflexed above the sheathing base. Fl.-stems 15–50 cm., erect, very stiff. Lowest bract less than half as long as the lax infl. Per. segs 4–7 mm., dark chestnut-brown, lanceolate, slightly exceeding the obovoid mucronate capsule. Fl. 6–7. Moors, bogs and moist heaths, confined to acid soils. Abundant in suitable habitats throughout the British Is.

2. J. tenuis Willd.

Perennial. Lvs 10–25 cm., usually all basal from a very short usually erect stock, narrowly linear, channelled, with a broad sheathing base prolonged above into blunt scarious auricles several times as long as wide. Fl.-stems 15–35 cm., erect. Fls in a terminal panicle much exceeded by at least one of the very narrow lf-like bracts. Per. segs 3–4 mm., greenish, becoming straw-coloured, narrowly lanceolate, very acute. Capsule ovoid, mucronate, shorter than the perianth. Fl. 6–9. Introduced and naturalized (or native in S.W. Ireland) on roadsides, waste ground, field paths, etc. northwards to Inverness and the Outer Hebrides and locally abundant but rare or absent in C. and E. England; E. and S.W. Ireland,

3. J. compressus Jacq.   Round-fruited Rush.

Tufted rhizomatous perennial rarely forming extensive patches. Lvs narrowly linear, horizontally flattened. Fl.-stems 10–30 cm., curved, not stiffly erect, flattened throughout, bearing 1–2 lvs. Infl. a terminal panicle usually shorter than its lowest bract. Per. segs 1·5–2 mm., ovate, very blunt, light brown. Anthers slightly shorter than their filaments. Style shorter than the ovary. Capsule very glossy, subglobose, c. 1½ times as long as the perianth, very shortly mucronate. Fl. 6–7. Marshes, alluvial meadows and grassy places, chiefly on non-acid soils: Great Britain northwards to Ross.

4. J. gerardii Loisel.   Mud Rush.

Like *J. compressus* but usually taller and forming larger patches; fl.-stems straight, stiffly erect, flattened below but 3-angled above; infl. usually exceeding its lowest bract; per. segs dark brown to blackish; anthers 3 times as long as their filaments; style as long as ovary or slightly longer; capsule acuminate, not or slightly longer than perianth. Fl. 6–7. Abundant in the upper parts of salt-marshes all round the coasts of the British Isles; rarely inland.

5. J. bufonius L.   Toad Rush.

Slender annual, 3–25 cm., erect or ± prostrate, the filiform stems often much branched below and forking above. Lvs 1–5 cm., very slender, deeply channelled, usually all basal except for 1 on the stem. Infl. a much-branched panicle with lfy bracts, occupying most of the plant. Fls on upper side of branches, mostly sessile and solitary. Per. segs 2·5–6 mm., narrow, fine-pointed, pale green with hyaline border, usually longer than the oblong blunt capsule. Fl. 5–9. Paths, roadsides, arable land, mud by ponds, etc.; abundant throughout the British Is.

6. J. inflexus L.   Hard Rush.

Perennial, in large dense grey-green tufts. Rhizomes matted, with short internodes. Stems 25–60 cm. × 1–1·5 mm., slender, stiffly erect, with

12–18 prominent ridges, dull, glaucous, pith interrupted. Lf-sheaths dark brown or blackish, glossy. Infl. apparently lateral, lax. Per. segs 2·5–4 mm., lanceolate with fine points, unequal. Capsule dark chestnut-brown, glossy, ovoid-acuminate, mucronate, about equalling perianth. Fl. 6–8. Chiefly in damp pastures on heavy base-rich soils. Abundant in most of England and Wales, more local in Scotland and Ireland.

7. J. effusus L.   Soft Rush.

Densely tufted, stiffly erect perennial. Stems 30–150 cm. × 1·5–3 mm., bright to yellowish green, glossy and smooth when fresh, with 40–90 striae; pith continuous. Lf-sheaths reddish to dark brown, not glossy. Infl. apparently lateral, many-fld, lax or condensed into a single rounded head. Per. segs 2–2·5 mm., lanceolate, fine-pointed. Capsule yellowish to chestnut-brown, broadly ovoid, retuse, not mucronate. Fl. 6–8. Very abundant and locally dominant in wet pastures, bogs, damp woods, etc., throughout the British Is.

8. J. Subuliflorus Drej. (*J. conglomeratus* auct.)   Conglomerate Rush.

Like *J. effusus* but usually less robust. Stems bright to greyish green, not glossy, with numerous ridges which are especially prominent just below the infl. Sheathing base of bract widely expanded. Infl. usually condensed into a dark reddish rounded head. Capsule with the remains of the style on a small mounded elevation in the hollowed top. Fl. 5–7. In similar habitats to *J. effusus* but more narrowly restricted to acid but not extremely base-poor soils; locally abundant, probably throughout the British Is.

9. J. maritimus Lam.   Sea Rush.

Erect, densely tufted, very tough perennial. Stem 30–100 cm., light green, smooth when fresh, lfy only near base; pith continuous. Lowest stem-lvs with brown glossy sheaths only; upper with terete green blade, sharp-pointed. Infl. apparently lateral, many-fld, usually shorter than the sharp-pointed bract. Per. segs 3–4·5 mm., straw-coloured, lanceolate, outer acute, inner blunt. Capsule ovoid-trigonous, mucronate, about equalling perianth. Seeds with large appendage. Fl. 7–8. On salt-marshes above tide-mark, often extensively dominant. Abundant on coasts of British Is northwards to Inverness.

10. J. acutus L.   Sharp Rush.

Tall very robust perennial forming dense prickly tussocks, somewhat like *J. maritimus* but with stouter and taller stems, 25–150 cm., bearing a few very sharply pointed lvs, and with a more compact infl. Per. segs 2·5–4 mm., reddish-brown, ovate-lanceolate, with broad scarious margin, becoming almost woody. Capsule broadly ovoid, at least twice as long as perianth. Local on sandy shores and in dune-slacks, sometimes on salt-marshes; coasts of England and Wales from Caernarvon to Norfolk and N.E. Yorks; S.E. coast of Ireland from Cork to Wicklow.

11. J. subnodulosus Schrank   Blunt-flowered Rush.

Tall erect rather soft perennial far-creeping in extensive patches, not tufted. Stem and lvs similar, 50–120 cm., bright green, terete, smooth, hollow, with both longitudinal and transverse septa. Sterile stems with 1 lf, flowering stems with 1–2 lvs, each with 35–60 transverse septa. Infl. repeatedly compound, of many heads of 3–12 fls, the secondary branches widely diverging. Fls pale straw-coloured at first, then darkening. Per. segs c. 2 mm., incurved, blunt. Capsule light brown, broadly ovoid, acuminate and shortly beaked, slightly longer than perianth. Fl. 7–9. Fens, marshes and dune-slacks with basic ground-water, often on calcareous peat. Locally abundant northwards to S. Hebrides and S. Scotland; local in N. Ireland.

12. J. acutiflorus Ehrh. ex Hoffm.   Sharp-flowered Rush.

Tall stiffly erect perennial, far-creeping, not densely tufted. Stem 30–100 cm., with 2–4 deep green, straight, subterete lvs, each with 18–25 conspicuous transverse septa. Infl. richly branched, of many small short-stalked heads; secondary branches diverging at an acute angle. Fls chestnut-brown, evenly tapered to an acute point, longer than perianth. Seeds c. 12 per capsule. Fl. 7–9. Wet meadows, moorland and swampy woods; abundant throughout the British Is, especially on acid substrata.

13. J. articulatus L.   Jointed Rush.

Ascending, decumbent or prostrate perennial, very variable in size and habit, often subcaespitose. Stem to 80 cm., terete, bearing 2–7 laterally flattened usually curved deep green lvs with 18–25 transverse septa. Infl. repeatedly compound, with few to many stalked heads each of 4–8 fls; branches diverging at an acute angle. Fls dark chestnut-brown to blackish. Per. segs lanceolate, acute, the inner with broad colourless margins. Capsule long-ovoid, beaked, usually blackish and shining. Seeds c. 40 per capsule. Fl. 6–9. Wet ground, especially when grazed or mown, dune-slacks, and margins of damp tracks, etc.; abundant throughout the British Is.

The hybrid *J. acutiflorus × articulatus*, intermediate between the parents but sterile, is often very abundant.

14. J. bulbosus L.   Bulbous Rush.
Small grass-like perennial, very variable in size and habit, erect and densely tufted, procumbent and rooting at the nodes, or floating and much branched. Stems slender, slightly swollen at base. Lvs 3–10 cm., slender, with many very indistinct septa. Infl. simple or branched, often proliferating, with or without fls. Per. segs 2 mm., pale, acute, mucronate. Stamens usually 3; anthers about equalling filaments. Capsule 2·5–3 mm., yellowish-brown, oblong, blunt, not sharply 3-angled. Fl. 6–9. Moist heaths, bogs, cart-ruts, woodland rides, etc., on acid soils; abundant throughout the British Is.

More robust forms with usually 6 stamens and sharply 3-angled capsules are sometimes separated as *J. kochii* Schultz.

2. LUZULA DC.

Tufted grasslike perennials, sometimes stoloniferous. Lvs mostly basal, sheathing at base, without auricles; blades flat or channelled, fringed with long colourless hairs. Capsule with 1 seed in each cell; seeds shining, often appendaged.

1	Fls borne singly in the infl., rarely in pairs.	2
	Fls in heads or clusters of 3 or more.	3
2	Infl.-branches drooping to one side; capsule acuminate, not suddenly contracted above the middle.	**2. forsteri**
	Infl.-branches spreading; capsule truncate, suddenly contracted above the middle.	**1. pilosa**
3	Perianth white or whitish, sometimes tinged with red.	4
	Perianth chestnut or yellowish-brown.	5
4	Fls dirty white, sometimes tinged with red; per. segs 2·5–3 mm., about equalling capsule. Naturalized in several places in Great Britain in woods and by streams chiefly on acid soils. White Woodrush.	*L. luzuloides* (Lam.) Dandy & Wilmott
	Fls snowy white; per. segs 4·5–5·5 mm., twice as long as capsule. Introduced and occasionally naturalized. Snow-white Woodrush.	*L. nivea* (L.) DC.
5	Robust; fl.-stem often more than 40 cm.; lvs 6–12 mm. broad.	**3. sylvatica**
	Fl.-stem rarely to 40 cm.; lvs less than 6 mm. broad.	6

6   Infl. drooping, spike-like, dense. Tufted perennial, 2–20 cm., of
    mountains northwards from N. Wales and Lake District. Spiked
    Woodrush.   *L. spicata* (L.) DC.
    Infl. of stalked clusters or, if compact, then not drooping.   7

7   Dwarf alpine, 3–8 cm., with channelled lvs; fls drooping in a sub-
    umbellate panicle of many clusters of 2–5 fls. Local on high
    Scottish mountains, chiefly above 3500 ft. Arcuate Woodrush.
                                                        L. arcuata Sw.
    Lvs not channelled; fls in each cluster more numerous.   8

8   Stoloniferous; anthers 2–6 times as long as filaments.   **4. campestris**
    Not stoloniferous; anthers not longer than filaments.   9

9   Fls 2·5–3 mm., chestnut-brown; capsule almost spherical.
                                                          5. multiflora
    Fls 2 mm., pale yellowish-brown; capsule obovoid. Very local in
    Hunts and introduced in Surrey. Pale Woodrush.
                                                     L. pallescens Swartz

1. L. pilosa (L.) Willd.   Hairy Woodrush.

Tufted perennial with short erect stock and slender stolons. Basal lvs
3–4 mm. broad, sparsely hairy, with a small truncate swelling at the
tip. Fl.-stem 15–30 cm. Infl. a lax cyme, with unequal slender spreading branches reflexed in fr. Fls usually single, dark chestnut-brown.
Per. segs 3–4 mm., ovate-lanceolate, not longer than the fr. Capsule
ovoid, very broad below, suddenly contracted above the middle to a
truncate conical top. Seeds with a hooked appendage. Fl. 4–6. Woods,
hedge-banks, etc., throughout the British Is.

2. L. forsteri (Sm.) DC.   Forster's Woodrush.

Like *L. pilosa* but usually smaller, with basal lvs only 1·5–3 mm.
broad, infl.-branches drooping to one side and remaining erect in fr.,
fls reddish chestnut-brown, per. segs lanceolate, usually longer than the
ovoid capsule which is not suddenly contracted nor truncate, and seeds
with a straight appendage. Fl. 4–6. Woods and hedge-banks in S.
England and S. Wales, local. Hybrids with *L. pilosa* not uncommon.

3. L. sylvatica (Hudson) Gaudin   Greater Woodrush.

Robust perennial forming bright green mats or tussocks, with short
obliquely erect stock and numerous stolons. Basal lvs 10–30 cm. ×
6–12 mm. or more, gradually tapering to a very acute point, sparsely
hairy. Fl.-stems 30–80 cm. Fls chestnut-brown, in clusters of 3–4 and
forming a lax terminal cyme. Per. segs 3–3·5 mm., lanceolate, about
equalling the finely beaked ovoid capsule. Fl. 5–6. Woods on acid soil

and peat, open moorlands and by rocky streams; throughout the British Is. but most abundant in the west and north.

4. L. campestris (L.) DC. Field Woodrush.

Somewhat loosely tufted perennial with short stock and shortly creeping stolons. Lvs usually 2–4 mm. broad with a small truncate swelling at the tip, bright green, thinly covered with long colourless hairs. Fl.-stem rarely more than 15 cm., bearing a loose panicle of 1 sessile and 3–6 stalked spherical clusters of 3–12 fls; infl.-branches ± curved, reflexed in fr. Fls 3–4 mm., chestnut-brown. Per. segs lanceolate, longer than capsule. Anthers 2–6 times as long as filaments. Capsule 2·5–3 mm., obovoid. Seeds nearly spherical, with a large white basal appendage. Fls 3–6. Very common in grassy places throughout the British Is.

5. L. multiflora (Retz.) Lej. Many-headed Woodrush.

Densely tufted perennial with few or no stolons, taller than *L. campestris*. Lvs to 6 mm. broad, sparsely hairy. Fl.-stems 20–40 cm., erect, wiry. Infl. somewhat umbellate, of up to 10 ovate or elongate 8–16-fld clusters on straight slender erect branches (or sometimes subsessile in a rounded or lobed head). Fls 2·5–3 mm., chestnut-brown. Per. segs broadly lanceolate, longer than capsule. Anthers slightly shorter than filaments. Capsule almost spherical. Seeds oblong, with a large white basal appendage. Fl. 5–6. Heaths and moorlands, woods, chiefly on acid and peaty soils; abundant throughout the British Is.

135. AMARYLLIDACEAE

Bulbous herbs. Lvs all basal. Fls solitary or umbellate, enclosed before flowering in a usually scarious spathe, hermaphrodite, usually regular, 3-merous. Perianth petaloid, in 2 whorls, sometimes with a corona. Stamens in 2 whorls, opposite the per. segs, inserted on them or free. Ovary inferior, 3-celled, with axile placentation. Ovules usually numerous. Style simple; stigma capitate or 3-lobed. Seeds endospermic.

1   Perianth without a corona.   2
    Perianth with a trumpet-like or ring-like corona inside the 6 segs.
                                         3. NARCISSUS

2   Inner per. segs much smaller than the outer.   2. GALANTHUS
    Per. segs all alike and equal.   1. LEUCOJUM

1. LEUCOJUM L.

Fls solitary to several, nodding. Perianth bell-shaped, segs all alike, white, tipped with green; no corona; tube 0 or very short.

Scape 15–20 cm.; fls 1(–2); per. segs 20–25 mm. Very rare in damp scrub and hedge-banks in Dorset and S. Somerset. Spring Snowflake.
                                                                L. vernum L.

Scape 30–60 cm.; fls (2–)3–7; per. segs 14–18 mm. Local in wet meadows and willow thickets in S. England and Ireland. Loddon Lily.   *L. aestivum* L.

2. GALANTHUS L.

Fls solitary, nodding. Outer per. segs somewhat spreading, separated; inner whorl bell-shaped, much shorter; no corona; tube 0 or very short.

1. G. nivalis L.   Snowdrop.

Lvs 10–25 cm. × c. 4 mm., linear, glaucous. Scape 15–25 cm. Spathe 2-fid, green, with broad scarious margins. Outer per. segs 14–17 mm., pure white; inner about half as long, deeply emarginate, white with a green spot at the incision. Fl. 1–3. Probably native. Local in damp woods and by streams northwards to Dunbarton and Moray; commonly planted and usually only naturalized.

3. NARCISSUS L.

Fls solitary or several, usually horizontal. Per. segs all alike, usually spreading: corona trumpet-like or ring-like, between the perianth and stamens; tube distinct.

1 Fl. solitary; corona about as long as the perianth.   2
   Fls usually 2; the yellow corona less than ½ as long as the cream or
      yellowish-white perianth. Naturalized locally in grassy places all
      over England and Wales, in S. Ireland, in Midlothian and in
      Jersey. Almost certainly a hybrid between *N. maialis* Curtis and
      N. tazetta L.   Primrose Peerless.   **N. × biflorus* Curtis

2 Corona darker yellow than the perianth; fl.-stalks 3–10 mm.,
      strongly deflexed.   **1. pseudonarcissus**
   Corona and perianth both deep yellow, the margin of the corona
      spreading at the mouth; fl.-stalks 10–15 mm., slightly curved.
      Possibly native but may be of early garden origin. Pastures near
      Tenby, Pembroke.   Tenby Daffodil.   *N. obvallaris* Salisb.

1. N. pseudonarcissus L. Lent Lily, Wild Daffodil.

Lvs 12–35 cm. × 5–12 mm., linear, erect, glaucous. Scape 20–35 cm., ± compressed, 2-edged. Fls solitary, drooping to nearly horizontal. Fl.-stalks 3–10 mm., strongly deflexed. Perianth 35–60 mm., pale yellow; tube 15–22 mm.; corona as long as or slightly shorter than per. segs, deep yellow, scarcely expanded or spreading at the mouth, irregularly lobed. Fl. 3–4. Damp woods and grassland throughout England and Wales and in Jersey, but rather local; naturalized in Scotland and Ireland.

136. IRIDACEAE

Perennial herbs with rhizomes, corms or bulbs. Lvs sessile, sheathing, often distichous and overlapping at base. Fls 3-merous, hermaphrodite, usually actinomorphic with 1 or 2 bracts at their base forming a spathe. Per. segs in 2 series, withering but persistent after fl., usually forming a longer or shorter tube. Stamens 3. Ovary 3-celled, inferior; style 3-lobed, branches entire or divided, sometimes petaloid.

1   Outer per. segs distinctly longer than inner, if nearly equal then
     outer recurved, inner incurved.   2
  All per. segs nearly or quite equal, never curving in opposite
     directions.   3
2   Rhizomatous; lvs flattened; ovary 3-locular.   **1.** I<small>RIS</small>
  Tuberous; lvs 4-angled; ovary 1-locular. Naturalized locally in
     S.W. England.   **Hermodactylus tuberosus* (L.) Miller
3   Fls almost sessile on the corm; corolla-tube long and slender; mid-
     rib of lf white, at least beneath.   **2.** C<small>ROCUS</small>
  Fls stalked; corolla-tube short; midrib green on both surfaces.   4
4   Fl. stems lfless; fls terminal.   5
  Fl. stems bearing lvs; fls lateral in a secund spike-like infl.   6
5   Fl. stems flattened, winged; lvs flattened. W. Ireland. Locally
     naturalized in England.   *Sisyrinchium bermudiana* L.
  Fl. stems terete; lvs almost bristle-like. S. Devon and Channel Is;
     very local.   *Romulea columnae* Sebastiani & Mauri
6   Fls deep orange; tube straight or nearly so.   **3.** C<small>ROCOSMIA</small>
  Fls crimson-purple; tube distinctly curved. Very local in S. England.
                 Gladiolus illyricus Koch

1. I<small>RIS</small> L.

Fls actinomorphic, large and showy. Outer per. segs ('falls') usually deflexed and larger than the often erect inner per. segs ('standards');

tube short. Stamens inserted at base of outer per. segs. Style-branches ('crest') broad, petaloid, bifid at tip.

1 Outer per. segs bearded, nearly equalling inner; tube longer than
      ovary. ± Naturalized.   *I. germanica* L. and garden hybrids
   Outer per. segs not bearded, larger than inner; tube shorter than
      ovary.   2
2 Growing in dry places; lvs dark green, polished, evergreen, at least
      some basal 1·5 cm. or more wide; seeds orange-red.
                                                                    1. foetidissima
   Growing in wet places; lvs glaucous or c. 1 cm. wide, mostly dying
      in autumn; seeds brown.   3
3 Style-branches yellow.   **2. pseudacorus**
   Style-branches violet or pale pinkish-purple.   4
4 Style-branches violet; claw of outer per. segs twice as long as limb.
      Lincoln and Dorset; very local.   *I. spuria* L.
   Style-branches pale pinkish-purple; claw of outer per. segs about as
      long as limb. Naturalized in a few places.   *I. versicolor* L.

1. I. foetidissima L.   Gladdon, Stinking Iris.

30–80 cm., tufted, with a strong unpleasant smell when bruised. Fl.-stem angled on one side. Spathes with a narrow scarious margin, 2–3-fld. Fl.-stalks 4–5 times as long as ovary. Fls c. 8 cm. diam., purplish-livid, rarely yellow; tube very short; style-branches yellowish. Fl. 5–7. Hedge-banks, sea cliffs and in open woods, chiefly on calcareous soils, locally common.

2. I. pseudacorus L.   Yellow Flag.

40–150 cm., glaucous; rhizome often 3–4 cm. diam. Fl.-stem terete but somewhat flattened. Spathes broadly scarious towards the top, 2–3-fld. Fl.-stalks about as long as ovary. Fls 8–10 cm. diam., pale yellow to almost orange; tube short; style-branches yellow. Fl. 5–7. Marshes, swampy woods and shallow water, common and widely distributed in suitable habitats.

2. CROCUS L.

Two spp. are locally naturalized. They may be distinguished as follows:

Fls appearing in spring, with or shortly after the lvs.
                                                                  C. purpureus Weston
Fls appearing in autumn, long before the lvs.   *C. nudiflorus* Sm.

3. Crocosmia Planchon

Corm covered by a reticulate fibrous tunic. Infl. long, spike-like, secund. Spathes membranous, emarginate. Fls weakly zygomorphic, showy. Tube usually shorter than lobes, widening gradually upwards.

*1. C. × crocosmiflora (Lemoine) N.E. Br. Montbretia.

30–90 cm., freely stoloniferous. Corms nearly 2 cm. diam., often several together in a row. Lvs 5–20 mm. wide. Bracts small, oblong, reddish. Fls 2·5–5 cm., deep orange suffused with red, funnel-shaped. Fl. 7–8. Raised at Nancy, France, by crossing *C. pottsii* (female) with *C. aurea* (male); flowered for the first time in 1880. Commonly naturalized, particularly in the west, in hedge-banks, woods, by lakes, etc.

137. DIOSCOREACEAE

Usually slender herbaceous or woody climbers with tuberous rhizomes or stocks or thick woody stem-tubers. Lvs usually spirally arranged, often cordate, entire, lobed or palmately divided, with palmate main veins and a network of smaller veins. Fls small, in spikes, racemes or panicles, unisexual (usually dioecious), actinomorphic. Perianth bell-shaped of 3 + 3 equal segments united below into a short tube; stamens 3 + 3 or 3, borne on the base of the perianth-tube; ovary inferior, 3-celled with 2 ovules in each cell; styles 3, or 1 with 3 stigmatic lobes. Fr. a 3-valved capsule or berry; seeds often flattened or winged, endospermic.

1. Tamus L.

Perennial dioecious herbs with large stem-tubers, slender annual twining stems and entire cordate lvs. Fls small, in axillary racemes or spikes. Stamens 6, rudimentary in female fl.; style 1 with three 2-lobed stigmas. Fr. a berry, incompletely 3-celled, with few globose seeds.

1. T. communis L. Black Bryony.

Tuber up to 20 (rarely to 60) cm. diam., irregularly ovoid, blackish, up to 20 cm. below soil-surface. Stems 2–4 m., slender, angled, unbranched, glabrous, twining to the left. Lvs 3–10 × 2·5–10 cm., broadly ovate, deeply cordate at base, finely acuminate, dark shining green, with 3–9 curving main veins; stalk long with 2 stipule-like emergences at base. Fls yellowish-green in axillary racemes, the male stalked, long, erect or spreading, the female almost sessile, recurved, shorter and fewer-fld.;

male fls 5 mm., female 4 mm. diam. Perianth of 6 narrow somewhat recurved lobes. Stigmas recurved. Berry c. 12 mm. diam., pale red, glabrous, with 1–6 pale yellow seeds. Fl. 5–7. Wood-margins, scrub, hedgerows, etc., in moist well-drained fertile soils. Through England and Wales northwards to Cumberland and Northumberland; in Ireland only in Sligo and Leitrim, probably introduced.

138. ORCHIDACEAE

Perennial herbs with rhizomes, vertical stocks or tuberous roots; all British spp. terrestrial, always mycorrhizic and sometimes saprophytic. Stems often swollen at the base (*pseudobulbs*). Lvs entire, often with a sheathing base and sometimes with dark spots or blotches; in saprophytic spp. reduced to scales or membranous sheaths and lacking chlorophyll. Infl. a spike or raceme. Fls zygomorphic, epigynous and usually hermaphrodite, often very striking in form and colour. Perianth of 6 segments in 2 whorls, usually all petaloid though often green; the median (posterior) segment of the inner whorl (*labellum*) is commonly larger and different in shape from the rest, and is usually on the lower side of the fl. and directed downwards owing to the twisting of the ovary or its stalk through 180°; the base of the labellum is often spurred. The anthers and stigmas are borne on the *column* (figure 4, p. 572; stamens 2 (in *Cypripedium*) or 1, with ± sessile 2-celled anthers seated behind or upon the summit of the column, the contents of a cell usually forming 1–4 granular or waxy masses (*pollinia*), each often narrowed at its apical or basal end into a stalk-like *caudicle*; ovary inferior, 1-celled, with numerous minute ovules on 3 parietal placentae; style 0; stigmas 3 fertile (*Cypripedium*) or 2 fertile and the other sterile and forming a beak-like process, the *rostellum*, between the anther and the fertile stigmas; in some genera the rostellum forms 1 or 2 viscid bodies (*viscidia*) to which the pollinia are attached, and it may be represented only by viscidia or be quite lacking. Fr. a capsule opening by 3 or 6 longitudinal slits; seeds very numerous and very minute.

1   Plant lacking green lvs (saprophytic).   2
     Plant with green lvs (or with green bract-like scales on a green
        stem).   4

2   Labellum directed upwards, pink; spur fairly long, directed up-
        wards: stem swollen above the base. An extremely rare orchid of
        woods in S. and W. England. Spurred Coral-root.
                                        Epipogium aphyllum Swartz
     Labellum directed downwards; spur 0 or very short and adnate to
        the ovary; stem not swollen.   3

3 Stem covered with numerous brownish scales; fls brown; per. segs
     connivent into an open hood.   5. NEOTTIA
     Stem with 2–4 long sheathing scales; fls yellowish-green with a
     whitish labellum; outer lateral per. segs curved downwards close
     to the labellum. Rare in damp peaty woods and moist dune-
     slacks from Northumberland and Westmorland northwards.
     Coral-root.   *Corallorhiza trifida* Chatel.

4 Fl. spurred.   5
     Fl. not spurred.   15

5 Labellum with a long narrow ribbon-like mid-lobe, 3–5 cm. ×
     2 mm., coiled in bud like a watch-spring. A tall plant with long
     spikes of many greenish-grey fls often streaked and spotted with
     purple and with a goaty smell. Rare but widely scattered on the
     chalk and limestone of S. and E. England. Lizard Orchid.
                   Himantoglossum hircinum (L.) Sprengel
     Labellum not as above.   6

6 Fls greenish-white.   7
     Fls green, white, pink, red, etc., but not greenish-white.   8

7 Fls large, strongly fragrant, with spreading per. segs, entire strap-
     shaped labellum and long slender spur, 15–30 mm.
                            8. PLATANTHERA
     Fls in a narrow cylindrical spike, very small, ± campanulate, faintly
     vanilla-scented, with connivent per. segs, a 3-lobed labellum
     and a short blunt conical spur. Very rare in S. England but
     locally frequent in hill pastures northwards from N. Wales and
     Yorks. Small White Orchid.   *Leucorchis albida* (L.) Schur

8 Spur very short (c. 2 mm.); per. segs connivent into a hood.   9
     Spur exceeding 5 mm.; per. segs connivent or spreading.   11

9 Fls green, often red-tinged; labellum narrowly oblong with 2(–3)
     parallel distal lobes.   6. COELOGLOSSUM
     Fls not green; labellum with 3 or 5 ± spreading lobes.   10

10 Lvs unspotted; labellum 5-lobed; per. segs dark purplish-brown at
     first but becoming paler, so that the tip of the spike appears burnt
     or scorched; labellum white with dark reddish spots.
                       10. *Orchis ustulata*
     Lvs sometimes with rows of small dots; labellum 3-lobed, the
     central lobe longer and forked; fls small, whitish or pink, in a
     dense spike 2·5–8 cm. long. Only in W. Ireland and I. of Man.
     Dense-flowered Orchid.   *Neotinea intacta* (Link) Reichenb. f.

11 Spur c. 12 mm., filiform; labellum with 3 subequal lobes; 2 fertile
     stigmas borne on lateral lobes of the column.   12
     Spur less than 12 mm., or if as long then not filiform; 2 fertile
     stigmas ± confluent on the front of the column.   13

12 Spike markedly conical; fls not sweet-scented; labellum with vertical plates decurrent on its base from the lateral lobes of the column; viscidium 1, strap-shaped. 13. ANACAMPTIS

Spike ±cylindrical; fls strongly and sweetly scented; no vertical plates on labellum: viscidia 2, naked. 7. GYMNADENIA

13 All per. segs except the labellum connivent into a helmet over the column. 10. ORCHIS

At least the outer lateral per. segs spreading or upturned. 14

14 Emerging fl.-spike enclosed by thin spathe-like lvs; bracts membranous; spur ±ascending. 10. ORCHIS

Emerging spike not enclosed in spathe-like lvs; bracts lfy; spur descending. 12. DACTYLORHIZA

15 Labellum with an insect-like ±velvety mid-lobe; per. segs spreading. 9. OPHRYS

Labellum concave, slipper-shaped, yellow; fls very large, solitary; per. segs spreading, reddish, 6–9 cm. Extremely rare in limestone woods in N. England. Lady's Slipper Orchid.
Cypripedium calceolus L.

Labellum neither insect-like and velvety nor concave and slipper-shaped; per. segs much less than 6 cm. long. 16

16 Labellum on the upper side of the fl. and directed upwards in some or all of the small yellowish-green fls. 17

Labellum on the lower side of the fl. 18

17 Lvs 0·5–1 cm., obovate, rounded and concave, the margin fringed with tiny green bulbils; labellum always on the upper side of the fl.; stem 3–12 cm., with basal pseudo-bulb; usually growing in *Sphagnum*. Throughout the British Is, but very local and easily overlooked. Bog Orchid. *Hammarbya paludosa* (L.) O. Kuntze

Lvs 2·5–8 cm., oblong-elliptical, in a subopposite pair; stem 6–20 cm., with basal pseudo-bulb; fls few in a lax raceme, yellow-green, with the labellum variously directed. In wet fens and dune-slacks in E. Anglia, N. Devon and S. Wales. Fen Orchid.
Liparis loeselii (L.) L. C. M. Richard

18 Fls small, whitish, in 1 or more spirally-twisted rows; infl. narrowly cylindrical. 19

Fls not obviously in spirally-twisted rows. 20

19 Stoloniferous; lvs ovate, stalked, conspicuously net-veined; labellum free from the column, with a saccate base and entire narrow spout-like distal part. Locally in pine-woods, rarely under birch or on moist dunes, from Cumberland and Northumberland northwards; very local in E. Anglia. Creeping Ladies' Tresses. *Goodyera repens* (L.) R.Br.

Not stoloniferous; lvs not conspicuously net-veined; labellum adnate to and embracing the column at its base, its distal part entire, recurved. 3. SPIRANTHES

20 Labellum consisting of a concave basal part (hypochile) and a
    tongue-like distal part (epichile)—see figure 4A, p. 572.   *21*
    Labellum not as above; fls greenish.   *22*

21 Fls erect, ±sessile; hypochile embracing the column; ovary
    twisted.   **1. CEPHALANTHERA**
    Fls spreading or hanging, stalked; hypochile not embracing the
    column; ovary straight.   **2. EPIPACTIS**

22 Labellum shaped like a man, with slender lobes resembling arms
    and legs.   **11. ACERAS**
    Labellum not shaped like a man.   *23*

23 Lvs 2, opposite or nearly so, borne some way up the stem; labellum
    forked distally, sometimes with a small tooth in the sinus.
        4. LISTERA
    Lvs 2, basal, with 0–2 smaller stem-lvs; labellum 3-lobed, the
    central lobe longer than the laterals; spike slender, dense, of
    small greenish-yellow fls whose per. segs are incurved and conni-
    vent. Local in chalk and limestone pastures chiefly in S.E.
    England. Musk Orchid.   *Herminium monorchis* (L.) R.Br.

1. CEPHALANTHERA L. C. M. Richard

Shortly rhizomatous herbs with lfy stems and lax spikes with a few large
suberect white or pink fls which never open widely. Labellum with a
constriction between the suberect concave *hypochile*, which clasps the
base of the column, and the forwardly-directed *epichile* which is longitudinally crested and has a recurved tip; basal bosses 0; spur 0. Column
long, erect; rostellum 0. Pollinia 2, clavate, each ± completely divided
into longitudinal halves; caudicles 0; pollen-grains single, in powdery
masses.

1 Fls bright rose-pink; ovary pubescent. S. England; very rare. Red
    Helleborine.   *C. rubra* (L.) L. C. M. Richard
    Fls white; ovary ± glabrous.   *2*

2 Lvs ovate or ovate-lanceolate; bracts exceeding the ovary; outer per.
    segs blunt.   **1. damasonium**
    Lvs lanceolate; bracts shorter than the ovary; outer per. segs acute.
    Local and rare but widespread. Long-leaved Helleborine.
        C. longifolia (L.) Fritsch

1. C. damasonium (Miller) Druce   White Helleborine.

Stem 15–50 cm., glabrous, with 2–3 brown sheathing basal scales. Lvs
narrowing up the stem from ovate-oblong to lanceolate and grading
into the lanceolate bracts which much exceed all but the uppermost fls.
Fls 3–12, creamy-white, scentless, closed and ± tubular except for a

brief period. Labellum shorter than the per. segs, with an orange-yellow blotch in the hollow of the hypochile and 3–5 interrupted orange-yellow crested ridges running along the cordate crenate epichile. Ovary glabrous. Fl. 5–6. Woods and shady places on calcareous soils; commonly under beech. Local in England northwards to E. Yorks and Cumberland.

2. EPIPACTIS Zinn

Rhizomatous herbs whose lfy stems have basal sheathing scales. Largest lvs near the middle of the stem with a few shorter broad lvs below and narrower lvs above grading upwards into the bracts. Fls rather inconspicuous, in 1-sided racemes. Labellum usually of two parts: a basal cup (*hypochile*) and a terminal triangular to cordate downwardly-directed lobe (*epichile*); spur 0. Column short. Rostellum placed centrally above the prominent oblong stigma, large and globular; sometimes evanescent or 0. Pollinia 2, tapering towards their apices and each divided into longitudinal halves; caudicles 0; pollen-grains spherical, in friable masses of tetrads. Ovary straight.

1 A fen plant with long creeping rhizomes; fls brownish outside; hypochile with 2 lateral lobes, connected with the epichile by a narrow hinge.   **1. palustris**
    Not fen plants; rhizome short; fls greenish or purplish outside; hypochile without lateral lobes, connected broadly and rigidly with the epichile.   **2**

2 Fls entirely dull reddish-purple; infl.-axis with dense whitish pubescence; epichile broader than long, with 3 large strongly rugose basal bosses; ovary pubescent. A local limestone plant. Dark-red Helleborine.   *E. atrorubens* (Hoffm.) Schultes
    Fls not entirely reddish; infl.-axis glabrous to pubescent but not densely whitish-pubescent; basal bosses of labellum smooth or nearly so.   **3**

3 Lvs spirally arranged; rostellum prominent and persistent unless removed by insects with the pollinia.   **4**
    Lvs in 2 opposite ranks; rostellum 0 in mature fls.   **5**

4 Stems 1–3, rarely more; lvs broadly ovate to ovate-lanceolate, usually dark green; per. segs green to deep purplish; hypochile dark maroon or dark green inside.   **2. helleborine**
    Stems commonly 6–10 or more; lvs ovate-lanceolate to lanceolate, grey-green, often violet-tinged; per. segs pale whitish-green; hypochile ± purplish inside.   **3. purpurata**

5 Infl.-axis obviously pubescent; fls spreading; hypochile purplish inside.   **6**

Infl.-axis glabrous or sparsely pubescent; fls hanging ± vertically downwards; hypochile greenish-white inside. Variable and very local. Green Helleborine.   *E. phyllanthes* G. E. Sm.

6 Fls opening widely. Local in woods in S. England. Narrow-lipped Helleborine.   *E. leptochila* (Godf.) Godf.
   Fls rarely opening widely.   7

7 Fls remaining partially closed and therefore bell-shaped; epichile with a recurved tip. A rare plant of sand-dunes in Lancs and Anglesey. Dune Helleborine.
                              E. dunensis (T. & T. A. Steph.) Godf.
   Fls usually remaining closed; epichile flat. A rare beechwood plant in Gloucestershire. Closed Helleborine.
                              E. cleistogama C. Thomas
   (Probably a form of *E. leptochila* (Godf.) Godf.)

1. E. palustris (L.) Crantz   Marsh Helleborine.

Rhizome long-creeping. Stem 15–45(–60) cm., pubescent above, often purplish below. Lvs 4–8, the lower 5–15 cm., ovate to oblong-lanceolate, acute, often purple beneath; uppermost narrow, acuminate, grading into the bracts; all ± erect and folded, with 3–5 prominent veins beneath. Fls 7–14, drooping in bud and in fr., ± horizontal when open; lowest bracts about equalling the fls, the rest falling short. Outer per. segs brownish-purple and hairy outside, paler inside; inner per. segs shorter and narrower, whitish with pink veins at the base, glabrous. Hypochile with an erect triangular lobe at each side, white with rose veins and many bright yellow raised spots down the middle; epichile joined by a narrow hinge, white with red veins, broadly ovate with waved and frilled upturned sides and with a basal furrowed boss. Ovary narrowly pear-shaped, very shortly pubescent. Fl. 6–8. Fens and dune-slacks. Locally frequent northwards to Perth; Ireland.

2. E. helleborine (L.) Crantz   Broad Helleborine.

Rhizome short. Stems 1–3 or more, 25–80 cm., pubescent above, often purplish below. Lvs spirally arranged, the largest 5–17 cm., usually broadly ovate-elliptical to almost orbicular, acute or shortly acuminate, somewhat crowded near the middle of the stem; upper lvs lanceolate-acuminate; all dull dark green with about 5 prominent veins beneath. Fls 15–50, drooping; lowest bracts equalling the fls, the rest falling short. Per. segs green to dull purple. Hypochile green outside, dark maroon or green inside; epichile broader than long, rose or greenish-white, with a recurved tip and 2–3 basal bosses, smooth or nearly so. Rostellum large, whitish, persistent. Ovary glabrous. Fl. 7–10. Woods, wood-margins, hedge-banks, etc. Locally frequent throughout most of Great Britain and Ireland.

3. E. purpurata Sm. (*E. sessilifolia* Peterm.)   Violet Helleborine.

Rhizome vertical. Stems 1–20 or more, 20–70 cm., violet-tinged, pubescent above. Lvs spirally arranged, the largest 6–10 cm., ovate-lanceolate to lanceolate, acute or shortly acuminate; all grey-green and often violet-tinged, especially beneath; main veins 3–7. Fls numerous, rather crowded, slightly fragrant; bracts narrow, acuminate, often violet-tinged, equalling or exceeding the fls. Per. segs pale, whitish-green, sometimes rose-tinged. Hypochile green or purple outside, ± purplish inside; epichile at least as long as broad, whitish, with a recurved tip and 2–3 confluent and slightly violet-tinged basal bosses. Rostellum whitish, persistent. Ovary rough with short hairs. Fl. 8–9. Woods, especially of beech, on calcareous soils. Locally throughout England northwards to Yorks and Cumberland.

3. SPIRANTHES L. C. M. Richard

Small herbs with erect stocks, 2–6 ± tuberous roots, lfy stems and small fls in spirally twisted spike-like racemes. Per. segs cohering to form the upper half of a 2-lipped trumpet-like tube round the column. Labellum frilled and ± recurved distally, furrowed below and embracing the base of the column to form the lower part of the perianth-tube; spur 0: column horizontal with the circular stigma on its underside. Rostellum projecting beyond the stigma and consisting of a narrow viscidium supported between 2 long narrow teeth which are left behind like the prongs of a fork when the viscidium is removed; pollinia 2, each of 2 plates of coherent pollen tetrads; caudicles 0.

1   Flowering stem bearing only bract-like sheathing appressed scales,
       the true lvs of the current season being in a lateral basal rosette.
                                                                    1. spiralis

   Flowering stems bearing true lvs.   *2*

2   Fls in a single spirally-twisted row. New Forest; very rare. Summer
       Lady's Tresses.   *S. aestivalis* (Poiret) L. C. M. Richard
       Fls in 3 spirally-twisted rows. Very local in Devon, Ireland,
       Inverness, Colonsay and Coll. Drooping Lady's Tresses.
                                                      S. romanzoffiana Cham.

1. S. spiralis (L.) Chevall.   Autumn Lady's Tresses.

Roots 2–3(–5), radish-shaped, pale brown, hairy. Stem 7–20 cm., glandular-hairy especially above, with several pale green appressed bract-like scales, and sometimes with the withered remains of last season's rosette-lvs still visible. Lvs (4–5) of the current season appearing with or after the fls in a lateral basal rosette; each c. 2·5 cm., ovate, acute, stiff, bluish-green, glabrous. Spike 3–12 cm., of 7–20 small (4–5 mm.) white day-scented fls in a single spirally-twisted row. Bracts sheathing the ovary and about equalling it. Per. segs narrow, blunt. Labellum broadened, rounded and recurved distally, pale green with a broad white irregularly crenate or fringed margin. Capsule 6 mm., obovoid. Fl. 8–9. Hilly pastures, downs, moist meadows and grassy coastal dunes, usually on a calcareous substratum. Throughout England and Wales northwards to Westmorland and N.E. Yorks, but common only in the west; Ireland.

4. LISTERA R.Br.

Herbs with short rhizomes and erect stems usually with a pair of ± opposite lvs and spike-like racemes of inconspicuous greenish or reddish fls. Labellum long and narrow, forking distally into 2 narrow segments and sometimes with 2 lateral lobes near the base; spur 0; column short, erect; rostellum broad, flat, blade-like, arching over the stigma. Pollinia 2, club-shaped, each divided into longitudinal halves, caudicles 0; pollen in loosely bound tetrads.

Plant 20–60 cm.; lvs 5–20 cm., broadly ovate-elliptical; spike long with
    many green fls.   **1. ovata**
Plant 6–20 cm.; lvs 1–2·5 cm., ovate-deltate; spike short; fls few,
    reddish-green.   **2. cordata**

1. L. ovata (L.) R.Br.   Twayblade.

Stem 20–60 cm., rather stout, pubescent above, with 2–3 basal sheaths. Lvs 5–20 cm., in a subopposite pair rather below the middle of the stem, broadly ovate-elliptical, sessile; upper part of stem with 1–2 tiny bractlike lvs. Racemes 7–25 cm.; bracts minute; fls numerous, short-stalked, greenish-yellow, the per. segs subconnivent. Labellum 10–15 mm., yellow-green, directed forwards at the cuneate base, then turning abruptly and almost vertically downwards, broadening slightly downwards and deeply forked almost to half-way into 2 narrow parallel segments, sometimes with an intervening tooth. Capsule almost globular. Fl. 6–7. Moist woods and pastures on base-rich soils, and sand-dunes. Common throughout the British Is, except Shetland.

2. L. cordata (L.) R.Br.   Lesser Twayblade.

Stem 6–20 cm., slender, angled and slightly pubescent above, with 1–2 close basal sheaths. Lvs 1–2·5 cm., in a subopposite pair rather below the middle of the stem, ovate-deltate with a rounded horny tip and basal corners and a broadly cuneate sessile base; shining above, somewhat translucent. Raceme 1·5–6 cm., lax, of 6–12 short-stalked fls; bracts minute. Outer per. segs green, the upper one hood-like, the 2 lateral narrower and forward-spreading like the inner segs, which are reddish inside. Labellum c. 4 mm., reddish, very narrow, with 2 linear-oblong basal lobes and with the central lobe forking about half-way into 2 widely divergent linear tapering segments. Capsule globular, 6-ribbed. Fl. 7–9. Mountain woods, especially of Scots pine, and peaty moors, especially under heather and often amongst sphagnum. Very rare in S. England, more frequent northwards from N. Wales and Derby and locally common in Scotland; Ireland.

5. NEOTTIA Guett.

Perennial saprophytic herbs with short creeping rhizomes concealed in a mass of short thick fleshy blunt roots ('bird's nest'), stems densely covered with brownish scales, and spike-like racemes of pale brownish fls. Per. segs connivent to an open hood. Labellum saccate at the base, with 2 distal lobes; spur 0. Column long, erect; rostellum broad, flat, blade-like, arching over the stigma. Pollinia 2, slender, almost cylindrical, each divided into longitudinal halves; pollen tetrads in very friable masses.

1. N. nidus-avis (L.) L. C. M. Richard   Bird's-nest Orchid.

Stem 20–45 cm., somewhat glandular above. Raceme 5–20 cm.; bracts lanceolate-acuminate, scarious, falling short of the ovary. Per. segs ovate-oblong. Labellum c. 12 mm., twice as long as the per. segs, with 2 small lateral teeth at the base and 2 oblong blunt distal lobes extending almost half-way and diverging. Capsule 12 mm., erect. Fl. 6–7. Shady woods, especially of beech and especially on humus-rich calcareous soils. Throughout Great Britain northwards to Banff and Inverness; Ireland.

6. COELOGLOSSUM Hartman

Small herbs with palmate root-tubers, ovate or oblong lvs, and small green and brown fls whose per. segs are connivent into a hood. Labellum narrowly oblong; spur short, saccate. Column short, erect; rostellum of 2 widely separated protuberances, one on each side of the upper edge of the reniform stigma. Pollinia club-shaped, converging above but narrowing downwards into widely separated caudicles each attached basally to one of the 2 oblong viscidia which are only partially enclosed in small pouches; pollen in packets of tetrads.

1. C. viride (L.) Hartman   Frog Orchid.

Tubers 2, ovoid, usually palmately lobed with 2–4 tapering segments. Stem 6–25(–35) cm., angled and often reddish above, with 1–2 brown sheathing basal scales. Lower lvs 1·5–5 cm., broadly oblong, blunt, the lowest sometimes almost orbicular; upper lvs smaller and narrower, acute; all unspotted. Spike 1·5–6(–10) cm.; lowest bracts about equalling the greenish inconspicuous slightly scented fls. Outer per. segs brownish or greenish-purple; inner green, linear, almost hidden by the outer. Labellum 3·5–6 mm., oblong, straight and parallel-sided, hanging almost vertically, 3-lobed near its tip, the outer lobes narrow and parallel, the central much shorter; colour variable, uniformly green or edged with chocolate-brown or brownish in the lower half. Spur 2 mm., ovoid-conical, translucent greenish-white. Fl. 6–8. Pastures and grassy hillsides especially on calcareous soils; sometimes on sand-dunes and rock-ledges; to 3300 ft. Throughout the British Is.

7. GYMNADENIA R.Br.

Herbs with palmately lobed root-tubers, lfy stems and dense spikes of small fragrant fls. Outer lateral per. segs spreading; the rest connivent into a hood. Labellum shortly 3-lobed, directed downwards; spur long and slender. Column short, erect, with 2 stigmas on oval lateral lobes; rostellum elongated. Pollinia 2, convergent and narrowed below into caudicles each attached basally to one of the long linear naked viscidia which lie close together; pollen in packets of tetrads.

1. G. conopsea (L.) R.Br.   Fragrant Orchid.

Root-tubers with 3–6 tapering blunt segments. Stem 15–40(–60) cm., glabrous, with 2–3 close brown basal sheaths. Lower lvs 3–5, 6–15 cm., ± narrowly oblong-lanceolate, keeled and folded, slightly hooded, blunt or subacute; upper lvs 2–3, smaller, bract-like, appressed to the stem; all unspotted, with minutely toothed margins. Spike ± cylindrical; bracts about equalling the small reddish-lilac very fragrant fls. Labellum c. 3·5 mm., its 3 lobes subequal and rounded. Spur c. 12 mm., very slender, somewhat curved below, almost twice as long as the ovary. Fl. 6–8. Base-rich grassland, especially on chalk or limestone, fens and marshes; locally common throughout the British Is.

8. Platanthera L. C. M. Richard

Herbs with entire tapering root-tubers, stems with 2(–3) broad un-spotted lower lvs and much smaller upper lvs grading into the bracts, and lax spikes of strongly scented whitish fls. Outer lateral per. segs spreading, the rest connivent into an ovate ± erect hood. Labellum narrowly strap-shaped, entire; spur usually long and slender. Column rather short; rostellum represented only by the 2 lateral naked viscidia which project beyond the sides of the single transversely elongated stigma. Pollinia 2, narrowed downwards into slender caudicles, each attached laterally, close to its base, to a viscidium; pollen in packets of tetrads.

Fls 18–23 mm. across; labellum 10–16 mm.; pollinia divergent down-
    wards; viscidia c. 4 mm. apart, circular.   **1. chlorantha**
Fls 11–18 mm. across; labellum 6–10 mm.; pollinia parallel; viscidia c.
    1 mm. apart, oval.   **2. bifolia**

1. P. chlorantha (Cust.) Reichenb.   Greater Butterfly Orchid.

Root-tubers 2, ovoid, narrowed below. Stem 20–40(–60) cm., glabrous, with 1–3 brown basal sheathing scales. Lower lvs usually 2, 5–15(–20) cm., elliptical or elliptical-oblanceolate, blunt; upper lvs 1–5. Spike 5–20 cm., lax, greenish; bracts ovate-lanceolate, acuminate, blunt, about equalling the ovary. Fls 18–23 mm. across, heavily fragrant, especially at night. Per. segs greenish-white; outer lateral segs ovate. Labellum 10–16 × 2–2·5 mm., slightly tapering downwards, rounded at the tip, green near the base, greenish-white distally. Spur 19–28 mm., very slender, curved downwards and forwards, sometimes almost into a semi-circle. Pollinia 3–4 mm. in overall length, sloping forwards and outwards and so diverging from c. 2 mm. apart above to c. 4 mm. between the large circular viscidia. Fl. 5–7. Woods and grassy slopes, especially on base-rich or calcareous soils. Throughout the British Is, except Orkney and Shetland, but more frequent in the south.

2. P. bifolia (L.) L. C. M. Richard   Lesser Butterfly Orchid.

Like *P. chlorantha* but smaller in all its parts. Stem 15–30(–45) cm. Lower lvs usually 2, 3–9(–15) cm., elliptical; upper 1–5. Spike 2·5–20 cm., lax, whitish. Fls 11–18 mm. across, night-scented. Outer lateral per. segs lanceolate. Spur 15–20 mm., slender, almost horizontal. Pollinia c. 2 mm. in overall length, vertical, parallel, their small oval viscidia c. 1 mm. apart. Fl. 5–7. Grassy and heathy places and open woods. Throughout most of the British Is, but commoner in the north and more tolerant of acid soils than *P. chlorantha*.

9. Ophrys L.

Herbs with entire root-tubers, lfy stems and lax-fld spikes. Per. segs spreading. Labellum large, entire or 3-lobed, often convex, velvety, usually dark-coloured and conspicuously marked; spur 0. Column long, erect; rostellum represented by 2 separated pouches above the large single depressed stigma. Pollinia 2, narrowed downwards into long caudicles which are attached basally to separate globose viscidia enclosed in the distinct rostellar pouches; pollen in firmly bound packets of tetrads.

1 Outer per. segs green; labellum reddish-brown with glabrous bluish
        markings.   2
    Outer per. segs rose-pink or whitish; labellum with glabrous yellow
        or pale brown markings.   3
2 Inner lateral per. segs ± filiform, reddish-brown, velvety; labellum
        longer than broad, 3-lobed, the long central lobe bifid at its tip.
                                                                                       2. insectifera
    Inner lateral per. segs strap-shaped, greenish, undulate; labellum
        about as broad as long, hardly lobed, the bluish markings often
        H- or horseshoe-shaped. A rare plant of the chalk in S. Eng-
        land. Early Spider Orchid.   *O. sphegodes* Miller
3 Labellum strongly convex, ending in 2 short blunt lobes and an
        intervening long narrow tooth turned up between and behind them
        so as to be invisible from above.   **1. apifera**
    Labellum ± flat, squarish or broader in front, ending in a variously
        shaped, often cordate, appendage which lies flat or turns up in
        front so as to be clearly visible from above. A rare plant of the
        chalk in Kent. Late Spider Orchid *O. fuciflora* (Crantz) Moench

1. O. apifera Hudson   Bee Orchid.

Stem 15–45(–60) cm., glabrous. Lvs 3–8 cm., elliptical-oblong, di-minishing rapidly up the stem and grading into the bracts; all un-spotted. Spike 3–12 cm., of 2–5(–10) rather distant fls. Outer per. segs ovate-oblong, rose-pink or whitish, greenish on the back; inner lanceolate, pinkish-green. Labellum 12–15 mm., resembling a bumble-bee, 3-lobed, the lateral lobes small, hairy, rounded, with a blunt apex directed forwards and upwards; central lobe much larger, ending in 2 truncate lateral segments and a long narrow central tooth which is curled up behind; the central lobe velvety, reddish-brown to dark brown and marked with glabrous yellow spots distally and greenish-yellow collars or horseshoes basally. Self-pollinated. Fl. 6–7. Pastures, field-borders, banks and copses on chalk or limestone, especially on recently disturbed soils; also on base-rich clays and calcareous dunes. Local throughout England and Wales and northwards to Lanark and Renfrew; Ireland.

2. O. insectifera L.   Fly Orchid.

Stem 15–60 cm., subglabrous. Lvs 4–12·5 cm., few, oblong to elliptical. Spike long, of 4–12 distant fls. Outer per. segs oblong, yellowish-green; inner filiform, purplish-brown, velvety. Labellum 12 mm., 3-lobed near the middle; lateral lobes narrowly triangular, spreading; middle lobe ± deeply bifid, longer than broad; the whole reddish-brown, downy, with a broad glabrous shining bluish transverse patch or bar just below the insertion of the lateral lobes. Fl. 5–7. Woods, copses, field-borders, spoil-slopes, banks and grassy hillsides on chalk or limestone; not uncommon in suitable habitats throughout lowland England; local in Wales and C. Ireland, rare in Scotland.

10. ORCHIS L.

Herbs with rounded unforked root-tubers and sheathing lvs, the lower lvs in a rosette. Fls in a spike which is usually enclosed by thin spathe-like lvs during emergence. Bracts membranous. All per. segs except the labellum connivent into a helmet over the column, or only the outer lateral segs spreading or turned upwards. Labellum usually 3-lobed, directed downwards, with descending or ascending spur. Column erect; stigma ± 2-lobed, roofing the entrance to the spur; rostellum over-hanging the stigma as a single pouch enclosing 2 separate ± globular viscidia. Anther joined to the top of the column; pollinia 2, each narrowed below to a caudicle and attached by a basal disk to one of the viscidia; pollen in packets of tetrads united by elastic threads. Capsule erect.

1   All per. segs except the labellum connivent to form a 'helmet' over
      the column; spur descending.   2
    Outer lateral per. segs erect or spreading; spur ± ascending.   6
2   Labellum shaped somewhat like a man, the central lobe much longer
      than the ± slender lateral lobes (arms) and forking distally into
      2 branches (legs) often with a short tooth in the sinus between
      them; outer per. segs usually ± coherent.   3
    Labellum not shaped like a man; outer per. segs green-veined, not
      coherent.   **1. morio**

3 Helmet of young fls dark reddish-brown (maroon); outer per. segs free to the base; labellum white, c. 6 mm.; spur about $\frac{1}{4}$ the length of the ovary; bracts long. A small plant of chalk downs and limestone pastures, very local throughout England. Burnt or Dwarf Orchid. *O. ustulata* L.

Helmet not dark reddish-brown; outer per. segs coherent below; labellum at least 10 mm.; spur about $\frac{1}{2}$ the length of the ovary; bracts minute. *4*

4 Helmet heavily blotched with dark reddish-purple, almost black in unopened fls; labellum usually tinged with pink or pale purple and with darker spots, its distal branches broadly rhombic. A tall plant of copses and open woods or rarely of grassland on the chalk of S. England; very local. Lady Orchid.

O. purpurea Hudson

Helmet pale, flushed, veined and spotted with pale purple or rose; distal branches of labellum linear or oblong, not broadly rhombic. *5*

5 Lobes of labellum all very narrow, reddish; helmet whitish, $\pm$ flushed or marked with rose. A very rare plant of the chalk, probably now only in Oxfordshire and Kent. Monkey Orchid.

O. simia Lam.

Distal branches of labellum oblong, shorter than and 3–4 times as wide as the basal lobes, purple or rose; helmet greyish, $\pm$ flushed or marked with pale purple. Very rare. Wood margins or open grassland on chalk, perhaps now only in Suffolk and Bucks. Soldier or Military Orchid. *O. militaris* L.

6 Bracts 1-veined or the lowest sometimes 3-veined; spur equalling or exceeding the ovary; labellum 3-lobed, the central lobe largest.

2. mascula

Bracts 3-veined; spur shorter than the ovary; labellum with the central lobe shorter or 0. A dark purple-fld plant frequent in wet meadows in the Channel Is. Jersey Orchid. *O. laxiflora* Lam.

1. O. morio L. Green-winged Orchid.

Root-tubers almost globular. Stem 10–40 cm. Lvs 3–9 cm., elliptical-oblong to lanceolate, unspotted. Spike 2·5–8 cm., rather lax. Per. segs connivent in a 'helmet', reddish-purple with conspicuous green veins, rarely flesh-coloured or white. Labellum broader than long, about equally 3-lobed, reddish-purple; the lateral lobes somewhat reflexed, broadly cuneate, the central paler with dark dots; all crenulate. Spur almost equalling the ovary, horizontal or ascending, almost straight, enlarged and blunt at the tip. Fl. 5–6. Meadows and pastures, especially on calcareous soils. Throughout England and Wales, locally abundant in the south but becoming rare in the north; Ireland.

2. O. mascula (L.) L.   Early Purple Orchid.

Root-tubers ovoid. Stem 15–60 cm., rather stout. Lvs 5–20 cm., broadly to narrowly oblanceolate-oblong, bluntish, commonly with rounded dark spots. Spike 4–15 cm., rather lax. Outer lateral per. segs at first spreading, later folded back; the rest connivent; all reddish-purple. Labellum 8–12 mm., about as broad as long, 3-lobed, reddish-purple, paler at the base and dotted with darker purple; lateral lobes somewhat reflexed; the central slightly larger, notched; all crenate. Spur at least equalling the ovary, stout, straight or upwardly curved, horizontal or ascending, blunt. Fl. 4–6. Woods, copses and open pastures, chiefly on base-rich soils. Common throughout the British Is.

11. ACERAS R.Br.

Herbs with entire root-tubers, lfy stems and long narrow spikes of greenish fls. All per. segs except the labellum connivent into a 'helmet' over the column. Labellum shaped like a man, with long slender lateral lobes (arms) near its base, and a longer narrow central lobe which is forked distally into 2 slender segments (legs) rather shorter than the lateral lobes; spur 0. Structure of column as in *Orchis*.

1. A. anthropophorum (L.) Aiton f.   Man Orchid.

Root-tubers ovoid. Stem 20–40(–60) cm. Lower lvs 6–12 cm., crowded, oblong to oblong-lanceolate, ± acute, glossy on both sides; upper lvs smaller, erect, grading into the bracts; all unspotted. Spike up to half the length of the stem, narrowly cylindrical, becoming lax; bracts shorter than the ovary. Fls greenish-yellow, often edged or ringed with reddish-brown. Labellum c. 12 mm., hanging almost vertically. Fl. 6–7. Local on field-borders, grassy slopes and disused quarries on chalk and limestone, rarely in scrub and open woods; from Hants to Kent and northwards to Northampton and Lincoln.

12. DACTYLORHIZA Nevski

Like *Orchis* but with palmately lobed or divided root-tubers, basal lvs not in a definite rosette at time of flowering, emerging spike not enclosed in spathe-like lvs, bracts always lfy, per. segs apart from labellum erect or spreading and not connivent into a 'helmet', spur always ± descending and usually less than 10 mm. long.

1   Stem solid; 1–6 or more small bract-like transitional lvs between normal lvs and bracts; lowest bracts usually not exceeding the fls; outer lateral per. segs spreading horizontally or drooping.   *2*

Stem usually hollow; transitional lvs 0–1; lowest bracts usually exceeding the fls; outer lateral per. segs ± erect.   *3*

2   Lower lvs broadly elliptical, blunt; upper narrower, subacute; all commonly heavily marked with ± transversely elongated dark blotches, but blotches sometimes pale or 0; labellum with 3 subequal lobes, the middle usually longer than the lateral.
                                         1. fuchsii

Lower lvs narrowly lanceolate, subacute, lightly marked with ± circular spots, but spots sometimes very faint or 0; labellum broad, its middle lobe much smaller and usually shorter than the rounded lateral lobes; spur slender.   **2. maculata**

3   A majority of plants with lvs spotted on both sides, the spots often elongated and parallel to the side of the lf; fls lilac-purple; labellum entire or obscurely 3-lobed, its sides often reflexed. Very local on calcareous peat in W. Ireland.
                            D. cruenta (O. F. Mueller) Soó

Lvs not spotted on both sides.   *4*

4   Lvs yellow-green, long, erect, keeled, tapering to a long narrowly hooded tip, never spotted; fls flesh-coloured, rose-red, pale-brick-red, ruby-red or, less commonly, bright reddish-purple; sides of labellum usually strongly reflexed.   **3. incarnata**

Lvs mid-green or somewhat bluish, usually spreading, not or broadly hooded at the tip, spotted or not; sides of labellum usually not strongly reflexed in old fls.   *5*

5   Lvs short, spreading, broadly hooded at the tip, unspotted or with very small circular spots especially near the tip; fls usually rich deep purple; labellum broadly rhombic, 5–7 mm. long by 7–9 mm. wide, entire or shortly 3-lobed.   **5. purpurella**

Labellum not rhombic, usually exceeding 7 mm. in length and 9 mm. in width.   *6*

6   Plants usually with 3–5 narrow lvs; longest lf averaging less than 1·5 cm. in width, rarely exceeding 2 cm.; lvs unspotted or with transversely elongated spots; fls reddish-purple; labellum 3-lobed, the central lobe bluntly triangular. Very local. Narrow-leaved Marsh Orchid.   *D. traunsteineri* (Sauter) Soó

Plants usually with 5–8 lvs, the longest averaging more than 2 cm. in width and rarely less than 1·5 cm. wide.   *7*

7   Lvs 10–20(–30) cm. long, usually unspotted, rarely ring-spotted; fls rich reddish-lilac; labellum with a short rounded middle lobe or almost entire, often concave, with darker dots and fine lines usually confined to a central elliptical area; England and Wales.
                                     4. praetermissa

Lvs 4–10(–17) cm. long, often heavily spotted or blotched, but populations usually have a proportion of plants with unspotted leaves; fls pale to dark purple; labellum usually distinctly 3-lobed, not concave, with irregular dark lines and dots, its lateral margins somewhat angular and often notched. Two groups of marsh orchids closely resembling the early-flowering continental *D. majalis* (Reichenb.) Soó: Irish Marsh Orchid, *D. Kerryensis* (Wilmott) P. F. Hunt & Summerh. broad-leaved and fl. 5–6, widespread in Ireland and also in the Hebrides and Sutherland; and Welsh Marsh Orchid, *D. majalis* ssp. *cambrensis* (R. H. Roberts) R. H. Roberts, with longer and narrower leaves and fl. 6–7, very local in N. Wales.

1–2. D. maculata agg.   Spotted Orchid.

Root-tubers palmately lobed or divided. Stem usually solid, 15–60 cm. Lvs commonly spotted. Several (1–6) erect and appressed transitional lvs between the normal lvs and bracts. Lower bracts usually not exceeding the fls. Infl. distinctly pointed above until all the fls have opened. Fls usually ranging from white to bright rose-pink, rarely darker; outer lateral per. segs spreading horizontally or drooping; labellum ± distinctly 3-lobed; spur ± cylindrical, straight. Fl. 6–8; usually the latest palmate-tubered orchids to flower. Two species in the British Is.

1. D. fuchsii (Druce) Soó   Common Spotted Orchid.

Lowest lvs usually broadly elliptical, elliptical-oblong or obovate-oblong, blunt; the rest oblong-lanceolate, becoming narrower up the stem, slightly hooded, subacute; all commonly marked with ± transversely elongated dark blotches, sometimes unmarked. Labellum distinctly and often deeply and narrowly divided into 3 subequal lobes, the lateral ± rhombic, the central triangular and usually somewhat longer; usually clearly marked with a symmetrical pattern of ± continuous reddish lines on a paler pink or whitish background. Spur cylindrical, rather slender. Base-rich fens, marshes, damp meadows, grassy slopes and woods. Throughout the British Is, but more abundant in the south. There are 3 local sspp. in W. Ireland and the north-west, distinguished especially by fl.-colour and labellum-shape.

2. D. maculata (L.) Soó ssp. **ericetorum** (Linton) P. F. Hunt & Summerh.   Heath Spotted Orchid.

Lvs all narrowly oblong-lanceolate to linear-lanceolate, folded upwards, not or slightly hooded, subacute, marked with dark, almost circular spots or unspotted. Labellum 3-lobed, the central lobe much smaller and usually shorter than the broadly rounded entire or notched lateral lobes; rather lightly marked with numerous small reddish dots and short lines on a pale pink or whitish background, giving the effect of a pattern in dots (rather than in lines as in *D. fuchsii*), but sometimes with a symmetrical line-pattern. Spur very slender. Moist acid peaty soils throughout the British Is.

3. D. incarnata (L.) Soó   Marsh Orchid.

Stem 15–50(–100) cm., with a wide hollow. Lvs 10–20 cm., yellow-green, unspotted, erect or erect-spreading, oblong-lanceolate, broadest near the base and tapering gradually to a markedly and narrowly hooded apex. Transitional lvs between true lvs and bracts usually 1 or 0. Spike dense, rounded above. Lowest bracts much exceeding the fls. Fls commonly salmon-pink or flesh-coloured, but pale brick-red to ruby-red in the dwarfed and small-flowered form of dune-slacks, ssp. **coccinea** (Pugsley) Soó, and sometimes reddish-purple, straw-coloured or white. Outer lateral per. segs almost erect. Labellum obscurely 3-lobed, its sides strongly reflexed soon after the fl. opens so that it appears very narrow from the front; marked with darker lines and dots. Spur conical, somewhat curved. Fl. 5–7. Wet meadows, marshes and fens throughout the British Is.

4. D. praetermissa (Druce) Soó   Fen Orchid.

Stem 15–60 cm., distinctly hollow. Lvs 10–20(–30) cm., mid-green, usually unspotted, ± spreading, oblong-lanceolate, broadest near the base and tapering to a flat or slightly hooded tip. Transitional lvs 1–0. Spike dense, with most bracts exceeding their fls. Fls a fairly rich reddish-lilac, redder and paler than in *D. purpurella* or the purple forms of *D. incarnata*. Outer lateral per. segs ± erect. Labellum shortly 3-lobed to almost entire, commonly concave until a late stage in flowering, marked with dots and fine lines down a central elliptical area. Spur conical, blunt, often very stout. Fl. 6–8. Locally abundant on base-rich peat in S. and C. England and Wales. Often hybridizes with *D. fuchsii* to give vigorous plants which commonly have ring-spotted lvs and deep-pink fls with a clearly marked labellum.

5. D. purpurella (T. & T. A. Steph.) Soó   Northern Fen Orchid.

Stem usually 10–25 cm., sometimes to 40 cm., with a narrow hollow. Lvs 5–10 cm., dark green, unspotted or with small circular spots especially in the upper half, rather stiffly spreading, broadly lanceolate or oblong-lanceolate, narrowing to the closely sheathing base and often with a broadly hooded tip; lower lvs often with very short internodes. Spike short, dense; bracts usually purplish. Fls rich deep purple. Outer lateral per. segs ± erect. Labellum c. 6 × 8 mm., broadly diamond-shaped, flat or slightly concave, entire or shortly 3-lobed, occasionally with the central lobe projecting like a tongue; rather indistinctly marked with irregular dark reddish lines and spots. Spur conical, stout. Fl. 6–7. Marshes, fens and damp pastures, especially on base-rich soils. A northern species, extending from Carmarthen and S.E. Yorks to the Hebrides and Shetland; Ireland. *D. purpurella* and *D. praetermissa* overlap only in a narrow belt across north-central England and Wales.

13. ANACAMPTIS L. C. M. Richard

Herb with entire root-tubers, lfy stems and conical spikes of small fls. Outer lateral per. segs spreading. Labellum deeply 3-lobed with 2 obliquely erect plates running down from the rounded lateral lobes of the column which bear the 2 stigmas; spur long and very slender; rostellum a centrally-placed pouch between the bases of the stigmas and partially closing the entrance to the spur. Pollinia 2 with caudicles attached by their bases to a single transversely elongated narrowly saddle-shaped viscidium enclosed in the rostellum-pouch; pollen in packets of tetrads.

1. A. pyramidalis (L.) L. C. M. Richard   Pyramidal Orchid.

Root-tubers ovoid. Stem 20–50(–75) cm. Lower lvs 8–15 cm., narrowly oblong-lanceolate, acute; upper lvs smaller, acuminate, grading into the bracts; all unspotted. Spike 2–5 cm., at first markedly conical, dense-fld, with a foxy smell; bracts slightly exceeding the ovary. Outer median and inner per. segs connivent; all per. segs deep rosy purple, becoming paler. Labellum c. 6 mm., pale rose, broadly cuneate, with 3 subequal oblong lobes. Spur filiform, acute, equalling or exceeding the ovary. Fl. 6–8. Locally frequent in chalk or limestone grassland or on calcareous dunes throughout Great Britain northwards to Fife; Hebrides; Ireland.

139. ARACEAE

Herbs, rarely woody climbers. Fls small, hermaphrodite or unisexual, usually crowded on a thickened axis (*spadix*) which is generally ± enclosed by a spathe. Perianth present but small in hermaphrodite fls, usually 0 in unisexual fls. Stamens 2, 4 or 8. Ovary superior.

1 Lvs not stalked, narrow, with wavy margins; spadix lateral, taper-
    ing upwards; spathe 0; smelling of tangerines when crushed.
    Naturalized locally in shallow water. Sweet Flag.
                                                  *Acorus calamus L.

   Lvs stalked; spadix terminal.   2

2   Lvs cordate, suborbicular; spathe flat, white inside; spadix covered
     with fls to the tip. ± Naturalized in a pond in Surrey.
                                                   *Calla palustris L.
    Lvs hastate or sagittate; spathe greenish, enclosing the lower part of
    the spadix; upper part of spadix without fls.   1. Arum

1. Arum L.

Stock tuberous. Lvs net-veined, stalks sheathing at base. Spadix terete.
Fls all unisexual: female below, the upper sterile; male above, the upper
sterile. Per. segs 0.

Lvs appearing in spring, midrib dark green; spathe twice as long as the
    dull purple (rarely yellow) spadix.   **1. maculatum**
Lvs well developed by December, midrib pale yellow-green; spathe
    3 times as long as orange-yellow spadix. S. coast of England and
    Channel Is; very local.   *A. italicum* Miller

1. A. maculatum L.   Lords-and-Ladies, Cuckoo-pint.

30–50 cm., glabrous. Tuber c. 2 cm., oblique, a fresh one produced each
year. Lvs long-stalked, blade 7–20 cm., triangular-hastate, often
blackish-spotted. Spathe 15–25 cm., erect, pale yellow-green, edged
and sometimes spotted with purple. Spadix 7–12 cm., upper part dull
purple, rarely yellow. Fr.-spike 3–5 cm.; fr. scarlet, fleshy, ripening
after lvs are dead. Fl. 4–5. Woods and shady hedge-banks, especially
on base-rich soils; common in the south, rare in the north.

140. LEMNACEAE

Small floating, aquatic herbs; roots simple or 0. Fls monoecious;
perianth 0. Male fls of 1–2 stamens. Female fls of a solitary ovary.

Thallus ± flattened; roots present.   1. Lemna
Thallus subglobose, not more than 1 mm.; roots 0. Rare.
                                    Wolffia arrhiza (L.) Wimmer

1. Lemna L.

Frequently forming a green carpet on the surface of stagnant water. Infl.
minute.

1   Plants floating on the surface; thallus not stalked.   2
    Plants floating submerged; old thalli distinctly stalked.   **2. trisulca**
2   Several roots to each thallus.   **1. polyrrhiza**
    One root to each thallus.   3

3 Thallus nearly flat on both sides.   **3. minor**
   Thallus convex, typically much swollen beneath.   **4. gibba**

1. L. polyrrhiza L.   Great Duckweed.

Thallus 5–8(–10) mm. diam., flat and shiny, often purplish beneath, ovate or almost orbicular. Roots up to 3 cm. Fl. 7. In still water, local and mainly in the south and east.

2. L. trisulca L.   Ivy Duckweed.

Thallus submerged, translucent, tapering at base into a stalk when mature, several thalli joined together by their stalks. Two young thalli arise on opposite sides of, at right angles to, and in the same plane as each old one. Thallus (5–)7–12(–15) mm., ± acute and finely toothed at apex, narrowing abruptly to the stalk. Fertile thalli floating, smaller than sterile. Fl. 5–7. In still water, widely distributed except for N. Scotland.

3. L. minor L.   Duckweed, Duck's-meat.

Thallus 1·5–4·0 mm. diam., opaque, obovate or suborbicular, with a small projection at point of attachment to parent thallus. Roots up to 15 cm., usually less. Fl. 6–7. In still water, by far the commonest sp. and widely distributed.

4. L. gibba L.   Gibbous Duckweed.

Thallus 3–5 mm., ovate, usually asymmetrical and rounded at base, convex above, with fine reticulate markings, typically strongly swollen beneath, with large whitish cells. Roots up to 6 cm. In unfavourable conditions the thalli are flat and difficult to distinguish from the preceding. Fl. 6–7. In still waters, local; absent from S.W. England and much of Scotland.

141. SPARGANIACEAE

Rhizomatous aquatic herbs. Stems lfy; lvs long and narrow, sheathing at base. Fls unisexual, crowded in separate globose heads, female towards the base in each infl. Perianth of small scales. Ovary superior. Fr. dry, indehiscent.

1. SPARGANIUM L.

The only genus.

1 Stem branched, the branches bearing male and female heads.
                                                                  1. erectum
   Stem unbranched, but female heads sometimes stalked.   *2*

2   Lvs thin and flat, not keeled; male head 1, rarely 2.   **4. minimum**
      Lvs triangular or semicircular in section; male heads more than 1.   *3*

3   Floating lvs 0 or distinctly keeled at base; male part of infl. elongate;
        anthers 6–8 times as long as broad.   **2. emersum**
      Floating lvs always present, never keeled; male part of infl. short;
        anthers 3–4 times as long as broad.   **3. angustifolium**

1. S. erectum L. Bur-reed.

Stem (30–)60–100(–150) cm. Lvs usually all erect, rather longer than
stems, rarely some floating, triangular in section, 10–30 mm. wide, apex
broadly rounded or truncate; cross veins very obscure; longitudinal
veins pellucid, with no dark border. Male heads up to 17, deciduous.
Female heads all in the lf-axils. Fl. 6–8. Common and widely distributed at margins of rivers, canals, etc.

2. S. emersum Rehman Unbranched Bur-reed.

Stem 20–60(–100) cm., erect or floating. Lvs 3–12 mm. wide; erect ones
triangular in section, narrowed but blunt at apex; cross veins distinct;
longitudinal veins pellucid, with a dark green border; floating lvs usually
present, distinctly keeled; sheaths not inflated. Male heads 3–8; female
2–5, not all in lf-axils. Fl. 6–7. In shallow water; widely distributed.

3. S. angustifolium Michx. Floating Bur-reed.

A long slender perennial with floating lvs and stems. Basal lvs rather
thick and usually semi-cylindrical, rarely bluntly triangular in section,
but not keeled, c. 5 mm. wide, narrowed at apex; stem-lvs flat, sheaths
inflated. Male heads crowded. Fl. 8–9. In lakes and streams, mainly in
mountain districts; local and commoner in the north.

4. S. minimum Wallr. Small Bur-reed.

Stem 6–80 cm. Stems and lvs usually floating. Lvs 2–6 mm. wide,
narrowed at apex, midrib obscure; sheaths not inflated. Female heads
2–3(–4) on short stalks in the axils of lf-like bracts. Fl. 6–7. Lakes,
pools and ditches on peaty soils; local but widely distributed in suitable
places.

142. TYPHACEAE

Stout, rhizomatous herbs growing in wet places. Stems erect, simple.
Lvs long and narrow, slightly spirally twisted, distichous and sheathing
at base. Fls unisexual, small, very numerous, crowded in a dense terminal spike, male above, female below.

1. TYPHA L.   Reedmace, Bulrush.
The only genus.

Lvs (7–)10–18(–22) mm. wide; male and female parts of spike usually
   contiguous.   **1. latifolia**
Lvs c. 4 mm. wide; male and female parts of spike usually separated by
   a piece of stem.   **2. angustifolia**

1. T. latifolia L.   Cat's-tail, Great Reedmace.
1·5–2·5 m. Lvs rather glaucous, nearly flat, overtopping the infl. Fl.
6–7. In reed-swamp at edges of ponds, canals, etc., abundant in suitable
habitats.

2. T. angustifolia L.   Lesser Reedmace.
1–3 m. Lvs dark green, convex on back. Fl. 6–7. More local than *T.
latifolia*, but in similar habitats.

143. CYPERACEAE

Usually perennial herbs often growing in damp places. Stems usually
solid, often triangular in section. Lvs long and narrow, some or all
often reduced to sheaths. Fls hermaphrodite or unisexual, arising in the
axil of a bracteole (glume) and arranged in 1–many-fld spikelets. Spikelets solitary, terminal, or grouped in branched or spike-like infl.
Perianth of 1–many bristles or scales, or more often, 0. Stamens (1–)2–
3(–6); anthers basifixed. Ovary superior. Fr. indehiscent, globular or
trigonous in plants with 3 stigmas, biconvex in plants with 2 stigmas.

 1 Fls all unisexual; male and female in separate (usually dissimilar)
        spikes, or separate parts of the same spike; female fls enclosed by
        a flask-shaped or closely enfolding bracteole.   2
     Most or all fls hermaphrodite; female fls not surrounded by a
        bracteole.   3

 2   Female fls enclosed by perigynia which often end in a beak (many
        spp., several common).   12. CAREX
     Female fls closely enfolded by a bracteole, whose margins are not
        connate. Very local on base-rich mountains in N. England and
        Scotland.   *Kobresia simpliciuscula* (Wahlenb.) Mackenzie

 3 Stem hollow; margins and keel of lf very rough, readily cutting
        the skin.   11. CLADIUM
     Stem solid; margins and keel of lf not rough enough to cut the skin.   4

4   Per.-bristles long and silky, more than 6, conspicuous in fr.
                                                  1. ERIOPHORUM
     Bristles shorter than glumes, rarely longer and then only 6.   5

5   Infl. a solitary terminal spikelet; bract not forming an apparent
       prolongation of stem.   6
     Infl. of 2 or more spikelets, or (rarely) spikelet solitary but
       apparently lateral.   8

6   Water plant with elongate slender branched lfy stems.
                                      8. ELEOGITON
     Bog- or marsh-plants; tufted or with creeping rhizome; stems nearly
       or quite lfless.   7

7   Uppermost sheath on fl. stems with a short blade.
                                  2. TRICHOPHORUM
     Uppermost sheath on fl. stems without a blade.   3. ELEOCHARIS

8   Infl. with several flat or keeled lf-like bracts close together at base.   9
     Bracts not flat and lf-like or else solitary.   10

9   Spikelets terete, ovoid; glumes not distichous.   4. SCIRPUS
     Spikelets flattened, narrow and ± straight-sided; glumes distichous
       (rare).   CYPERUS (see p. 501)

10   Spikelets arranged in 2 rows in a flattened oblong head.
                                  5. BLYSMUS
     Spikelets spirally arranged or solitary; infl. often apparently
       lateral, branched, or if a terminal head then not flattened.   11

11   Infl. a compact blackish head, encircled at base by lowest bract;
       spikelets flattened, glumes distichous.   9. SCHOENUS
     Infl. reddish-brown or greenish; spikelets terete; glumes spirally
       arranged.   12

12   Stems lfy only at base or sometimes quite lfless; bract looking like a
       continuation of the stem and infl., apparently lateral.   13
     Stems lfy above base; bracts lf-like; infl. obviously terminal.
                              10. RHYNCHOSPORA

13   Plant slender, seldom more than 15 cm.; infl. of 1–3 spikelets.
                                  7. ISOLEPIS
     Plant stout, seldom less than 50 cm.; infl. of numerous spikelets.   14

14   Spikelets 5 mm. or more, few together, reddish-brown.
                           6. SCHOENOPLECTUS
     Spikelets 2–3 mm., crowded into dense, globular, greenish heads.
     Very local on sandy shores bordering the Bristol Channel.
                     Holoschoenus vulgaris Link

1. Eriophorum L.

Stems lfy, ± evergreen. Spikelets many-fld, solitary or umbellate. Fls hermaphrodite. Glumes spirally arranged, silver-slaty, thin. Per.-bristles elongating greatly after fl.

1   Spikelets solitary, erect; plant densely tufted.   **3. vaginatum**
    Spikelets several, ± nodding; plant with creeping rhizomes.   2
2   Lvs flat, uppermost without a ligule; bristle slightly yellowish,
    rough at tip (microscope).   **2. latifolium**
    Lvs deeply channelled or involute, uppermost with a short ligule;
    bristles quite white, smooth at tip (microscope).   3
3   Stem subterete at top; stalks of spikelets smooth.   **1. angustifolium**
    Stem sharply angled at top; stalks of spikelets rough with short
    forward-directed hairs. Very local, in S. and E. England; Ireland.
                                     E. gracile Roth

1. E. angustifolium Honck.   Common Cotton-grass.

20–60 cm., far-creeping. Lvs 3–6 mm. wide, channelled, narrowed into a long triquetrous point, reddish when old. Uppermost sheath ± inflated or funnel-shaped. Spikelets (1–)3–7. Glumes c. 7 mm., 1-nerved, lanceolate, acuminate, broadly hyaline. Bristles up to c. 4 cm. Fl. 5–6. Wet acid places, widely distributed but local in the south.

2. E. latifolium Hoppe   Broad-leaved Cotton-grass.

20–60 cm., rhizomes short. Stem sharply 3-angled (when fresh). Lvs 3–8 mm. wide, flat except for the short triquetrous point. Uppermost sheath close-fitting and cylindrical. Spikelets 2–12. Glumes 4–5 mm., 1-nerved, lanceolate, acuminate, with very narrow hyaline margins. Fl. 5–6. Wet places on base-rich soils, widely scattered but local.

3. E. vaginatum L.   Cotton-grass, Hare's-tail.

30–50 cm. Stems terete below, bluntly angled above. Lvs bristle-like, c. 1 mm. wide, angled. Upper sheaths strongly inflated but narrowed at mouth. Spikelet ovoid, rounded at base. Glumes c. 7 mm., 1-nerved, ovate-lanceolate, acuminate, silvery below, slaty-black above. Fl. 4–5. Damp peaty acid places, locally abundant.

2. Trichophorum Pers.

Lower sheaths lfless, only the uppermost with a short blade. Spikelet solitary, terminal, the lowest glume larger than the others, generally with a fl. in its axil. Fls hermaphrodite. Per.-bristles usually shorter than glumes.

Stems terete; bristles shorter than glumes.   **1. cespitosum**
Stems bluntly angled; bristles longer than glumes (probably extinct).
                                                    T. alpinum (L.) Pers.

1. T. cespitosum (L.) Hartman   Deer-grass.

5–35 cm., densely tufted. Stems slender, smooth, yellow-brown when
old. Lower sheaths light brown, shiny. Spikelet 3–6 mm., 3–6-fld.
Glumes subacute, 2 lower larger than rest. Two sspp. occur, of which
ssp. *germanicum* is by far the commoner. Fl. 5–6. Heaths, mountains,
moorlands, throughout most of Britain, often abundant.

3. ELEOCHARIS R.Br.

Stems terete or rarely 4-angled. At least the upper sheaths entirely
lfless. Spikelet solitary, terminal, the lower glume different from the
others and usually without a fl. in its axil. Fls hermaphrodite. Per.
bristles usually present, short.

1 Lowest glume at least ½ as long as spikelet.   2
   Lowest glume much less than ½ as long as spikelet.   4

2 Upper sheath conspicuous, brownish; rhizome brown.   3
   Upper sheath very delicate and inconspicuous; rhizomes thread-like,
      white, with a resting bud at tip. On estuarine mud; very local.
                     E. parvula (Roemer & Schultes) Bluff, Nees & Schau.

3 Stems very slender, 4-angled; glumes 2 mm., blunt (wet sandy
      places).   **1. acicularis**
   Stems stouter, terete; glumes 5 mm., acuminate (damp peaty
      places).   **2. quinqueflora**

4 Densely tufted; upper sheaths obliquely truncate; stigmas 3; nut
      triquetrous.   **3. multicaulis**
   Not densely tufted: upper sheath almost transversely truncate;
      stigmas 2; nut biconvex.   5

5 Lowest glume not more than ½ encircling base of spikelet.
                                                          4. palustris
   Lowest glume ±completely encircling base of spikelet.
                                                          5. uniglumis

1. E. acicularis (L.) Roemer & Schultes   Slender Spike-rush.
2–10(–20) cm. Spikelet 3–4 mm., 4–11-fld. Glumes c. 2 mm., ovate,
blunt, reddish-brown. Fl. 8–10. Widely scattered but rather local by
lakes and pools, sometimes submerged and then usually sterile.

2. E. quinqueflora (F. X. Hartmann) O. Schwarz (*E. pauciflora* (Lightf.)
Link)   Few-flowered Spike-rush.

5–30 cm., rather tufted but producing long slender runners. Upper
sheath obliquely truncate and blunt. Spikelet 5–7 mm., 3–7-fld, lowest
glume encircling its base; others c. 5 mm., ovate, acuminate, reddish-
brown with broad hyaline margins. Fl. 6–7. Damp peaty fairly base-
rich places; local.

3. E. multicaulis (Sm.) Sm.   Many-stemmed Spike-rush.

15–30 cm. Sheaths pale reddish or brownish, upper very obliquely
truncate and acute. Spikelet c. 10 mm., many-fld, often proliferating
vegetatively, lowest glume encircling its base; others c. 5 mm., ovate-
oblong, blunt, reddish-brown, ± hyaline, midrib green. Fl. 7–8. In acid
bogs, on wet sandy heaths, etc., locally common.

4. E. palustris (L.) Roemer & Schultes   Common Spike-rush.

10–60 cm., far-creeping, stems solitary or in many small tufts. Sheaths
yellowish-brown, upper nearly transversely truncate. Spikelet 5–20 mm.,
many-fld, lowest glume not more than ½ encircling its base; others ovate,
margins hyaline. 2 sspp. occur. Fl. 5–7. Marshes, ditches, etc., common and widely distributed.

5. E. uniglumis (Link) Schultes

Like **4** but stems usually shiny; lower sheaths reddish; lowest glume
± encircling base of spikelet; glumes usually darker brown with narrower
hyaline margins. Marshes, frequent near the coast, local inland.

4. Scirpus L.

Stems stout, lfy, 3-angled. Lvs with flat or keeled, well-developed blades.
Infl. terminal, mostly much-branched. Bracts several, lf-like.

Lvs keeled; infl. dense; bracts much longer than infl.; spikelets 10–
    20 mm., red-brown.   **1. maritimus**
Lvs flat; infl. lax; bracts about as long as infl.; spikelets 3–4 mm.,
    green or greenish-brown.   **2. sylvaticus**

1. S. maritimus L.   Sea Club-rush.

30–100 cm. Stems sharply angled, rough towards top. Lvs up to c.
10 mm. wide, margins rough. Infl. c. 5 cm., corymbose. Spikelets
rather few, ovoid. Glumes c. 7 mm., ovate, apex bifid and awned. Fl.
7–8. Shallow water at muddy margins of tidal rivers and ditches, etc.,
near the sea.

2. S. sylvaticus L. Wood Club-rush.

30–100 cm. Stems bluntly angled, smooth. Lvs up to 20 mm. wide, flat, margins rough. Infl. up to 15 cm., spreading. Spikelets very numerous, ovoid. Glumes 1·5 mm., ovate, apex entire, ± blunt. Fl. 6–7. Marshes, wet places in woods and by streams, local.

5. BLYSMUS Panzer

Stems subterete, lfy. Infl. a terminal spike of several distichous spikelets. Fls hermaphrodite, spirally arranged.

Lvs flat, keeled, rough; bracts (except lowest) shorter than spikelet.

1. compressus

Lvs involute, ± rush-like, smooth; bracts (except lowest) as long as
    spikelet. **2. rufus**

1. B. compressus (L.) Panzer ex Link Broad Blysmus.

10–35 cm., far-creeping. Lvs 1–2 mm. wide, tapering from base. Spikelets 5–7 mm., reddish-brown, 10–12 (rarely fewer). Glumes 3 mm., ovate, acute. Fl. 6–7. In marshy places, usually in rather open vegetation, locally abundant.

2. B. rufus (Hudson) Link Narrow Blysmus.

Lvs scarcely tapering. Spikelets dark brown, almost black, 5–8 in a spike. Glumes 5 mm., ovate, blunt. Fl. 6–7. Among short grass in salt-marshes, locally abundant.

6. SCHOENOPLECTUS (Reichenb.) Palla

Stems 3-angled or terete, usually nearly or quite lfless. Lower bract making what appears to be a continuation of the stem beyond the infl. Infl. apparently lateral, sessile or with short branches. Fls hermaphrodite, spirally arranged.

1 Stems sharply 3-angled. *2*
    Stems terete. *3*
2 Upper sheath usually with a short blade; glumes with blunt lateral
    lobes. Muddy banks of tidal rivers; very local.

S. triquetrus (L.) Palla

    Upper sheaths with blades up to 30 cm.; glumes with acute lateral
    lobes. Jersey. *S. americanus* (Pers.) Volkart
3 Stems not glaucous; glumes smooth (awn often rough); stigmas
    usually 3. **1. lacustris**
    Stems glaucous; glumes rough on back; stigmas 2.

2. tabernaemontani

1. S. lacustris (L.) Palla   Bulrush.

1–3 m., stout. Rhizome creeping, often bearing submerged lvs. Upper sheath often with a short blade. Lower bract usually shorter than mature infl. Spikelets 5–8 mm., ovoid, reddish-brown. Glumes 3–4 mm., broadly ovate, often fringed. Anthers bearded at tip. Fl. 6–7. Rivers, etc., usually where there is abundant silt, widely distributed.

2. S. tabernaemontani (C. C. Gmelin) Palla   Glaucous Bulrush.

Like 1 but 50–150 cm., rather slender; glumes densely beset with small dark brown papillae on back, specially near the mid-rib; anthers not or scarcely bearded. Fl. 6–7. Ditches, pools, etc., often in peaty places and specially near the sea, rather local.

7. ISOLEPIS R.Br.

Stems slender, terete. Lvs few, filiform, channelled. Infl. apparently lateral, a subterete green bract appearing as a prolongation of the stem. Spikelets small, 1–3 together. Fls hermaphrodite, spirally arranged.

Bracts distinctly longer than infl.; nut ribbed, shiny.   **1. setacea**
Bract shorter or not much longer than infl.; nut smooth, matt. Local;
  usually near the sea.   *I. cernua* (Vahl) Roemer & Schultes

1. I. setacea (L.) R.Br.   Bristle Scirpus.

3–15(–30) cm., tufted. Lvs 1–2, usually shorter than the stems from which they arise. Spikelets usually 2–3, c. 5 mm., ovoid. Glumes c. 1·2 mm., ovate, mucronate. Fl. 5–7. Damp bare sandy or gravelly places, sometimes in marshy meadows, rather local but widely distributed.

8. ELEOGITON Link

Similar to *Isolepis* but stem elongated, lfy, usually devoid of lfless sheaths, and spikelets always solitary, terminal and not overtopped by a stem-like bract.

1. E. fluitans (L.) Link   Floating Scirpus.

15–40 cm., slender, floating. Stems flattened, branched. Lvs c. 5 cm. × less than 1 mm. Spikelet 2–3 mm., narrow-ovoid, 3–5-fld. Peduncles c. 5 cm., terete below, 3-angled above. Glumes c. 2 mm., ovate, subacute. Fl. 6–9. Ditches, streams and ponds, widely distributed but local.

CYPERUS L.

Two rare spp. in Britain:

Perennial 50–100 cm.; lvs 4–7 mm. wide, about as long as stems; infl. a
compound umbel. Marshy places in S. and S.W. England. Galingale.

C. longus L.

Annual 5–20 cm.; lvs 1–3 mm. wide, shorter than stems; infl. subcapitate or a small dense umbel. Damp margins of ponds and ditches;
S. England.

C. fuscus L.

9. SCHOENUS L.

Perennials. Stems terete. Upper sheaths with at least a small blade.
Spikelets 1–4-fld, in ± flattened terminal heads surrounded by bracts.
Bract of lowest spikelet surrounding base of whole infl. Glumes disti-
chous, several lower without fls. Fls hermaphrodite.

Lvs at least ½ as long as stems; lf-like point of lowest bract 2–5 times as
long as infl.
1. nigricans

Lvs not more than ⅓ as long as stems; lf-like point of lowest bract
shorter to slightly longer than infl. Mid Perth; ? extinct.

S. ferrugineus L.

1. S. nigricans L. Bog-rush.

15–75 cm., densely tufted. Stems tough, wiry, with lvs only at base. Lvs
wiry, subterete. Lower sheaths dark red-brown or almost black, tough,
shiny. Infl. 1–1·5 cm., ovoid, blackish. Spikelets 5–8 mm., flattened, 5
or more in an infl. Fl. 5–6. Damp, usually peaty and base-rich places,
especially near the sea, locally abundant and widely distributed.

10. RHYNCHOSPORA Vahl

Perennial. Lvs well-developed. Infl. a compact terminal head or short
spike, sometimes with 1 or 2 lateral branches. Bracts lf-like, the lower
sheathing. Spikelets 1–2-fld. Fls hermaphrodite or the upper sometimes
unisexual. Several lower glumes without fls. Bristles 5–13.

Plant without a creeping rhizome; spikelets pale; bracts not or little
longer than terminal head; bristles 9–13.
1. alba

Plant with a creeping rhizome; spikelets dark red-brown; bracts 2–4
times as long as terminal head; bristles 5–6. Rare.

R. fusca (L.) Aiton f.

1. R. alba (L.) Vahl White Beak-sedge

10–50 cm., slender, ± tufted. Lvs narrow, channelled, margins rough.
Lower sheaths lfless, often with bulbils in their axils. Spikelets 4–5 mm.,

whitish, becoming pale reddish-brown, usually 2-fld. Terminal head usually broader than long. Glumes 4–5. Fl. 7–8. Wet, usually peaty places on acid soils, local but widely scattered.

11. CLADIUM Browne

Perennials. Lvs usually well-developed. Spikelets terete, few-fld. Fls usually hermaphrodite. Lower glumes sterile. Bristles 0.

1. C. mariscus (L.) Pohl   Saw Sedge, Sedge.

70–300 cm., stout and tough. Rhizome creeping. Stems hollow, ± terete. Lvs up to 2 cm. wide, evergreen, growing from base and dying away at top, grey-green, keeled, with sharp projecting teeth on margins and keel. Infl. much-branched, each branch terminated by a dense head of 3–10 spikelets. Spikelets 3–4 mm., 1–3-fld, reddish-brown, 2–3 lower glumes small and sterile, the remainder fertile. Fl. 7–8. Locally abundant in dense stands in reed-swamp and in fens on neutral or alkaline soils.

12. CAREX L.

Perennials. Lvs well-developed, usually narrow, keeled, or involute; lf-base usually sheathing and ligule present at junction of lf and sheath. Infl. various, from a much-branched panicle to a simple spike. Fls unisexual, in 1-fld spikelets, each in the axil of a glume; perianth 0. Male fls of 2–3 stamens. Female fls of 1 ovary surrounded by a globular, trigonous or flattened sac (*perigynium*), usually crowned by a longer or shorter beak from which the stigmas project. Nut trigonous and stigmas 3, or biconvex and stigmas 2, enclosed in the perigynium.

The male and female fls are variously arranged in the infl. Our spp. are, with one exception, monoecious. In the majority of spp. the terminal spike and sometimes some of the upper lateral spikes are male and the rest female. The female spikes sometimes have a few male fls at the top and the male spikes less frequently a few female fls at the base. The other spp. have male and female fls in the same spike, the male fls being either at the top or base of the spike.

In the following descriptions the fr. includes the nut and the perigynium surrounding it, and measurements of length include the beak.

1 Spikes more than 1, lateral ones sessile or stalked.   2
     Spike 1, terminal.   75

2 Spikes dissimilar in appearance, one or more terminal male (some-
        times ± concealed among the female), some or all the lateral

ones wholly or mainly female; fruiting spikes sometimes close together but never numerous, squarrose, not forming a roundish or ovoid ±lobed head.   3

Spikes all similar in appearance, usually all with male and female fls; fruiting spikes often numerous and either small and squarrose or forming a roundish or ovoid lobed head.   53

3   Fr. hairy, at least towards top.   4
     Fr. glabrous.   14

4   Fl. stems lfless with a few sheaths surrounding their bases; female spikes c. 2 mm. diam., overtopping the slender solitary male spike.   5
     Fl. stems with lvs in the lower part; female spikes 3–6 mm. diam., if narrower then concealed within sheathing bracts.   6

5   Lowest female spike rather distant from the others; fr. slightly longer than glume; basal sheaths purplish. Woods, etc., on chalk and limestone, very local.   *C. digitata* L.
     Female spikes clustered; fr. twice as long as glume; basal sheaths yellow-brown. On limestone, rarely in shady places; rare.
                                                                    C. ornithopoda Willd.

6   Lvs longer than stems; female spikes very slender, ±concealed in sheathing bracts. Small tufted plant in chalk and limestone grassland; locally abundant in the area Dorset–Hereford–Wilts–Hants.   *C. humilis* Leyss.
     Lvs usually shorter than stems; female spikes 3–6 mm. diam., not at all concealed by sheathing bracts.   7

7   Sheaths nearly always hairy, at least at top; male spikes 2–3; lvs ±flat; fr. 6–7 mm.   **18. hirta**
     Sheaths quite glabrous; male spike 1 (rarely 2 and then lvs very narrow and rolled); fr. 4 mm. or less.   8

8   Lvs distinctly glaucous beneath; fr. slightly hairy or merely papillose at top.   **17. flacca**
     Lvs not or scarcely glaucous; fr. distinctly hairy all over.   9

9   Female spikes not clustered round base of male, ±distant.   10
     Female spikes clustered round base of male, only the lowest some-times rather remote from others.   11

10   Growing in water; lvs very narrow, rolled; stems bluntly angled; glumes 4 mm., lanceolate.   **19. lasiocarpa**
     Growing in damp meadows; lvs 1·5–2 mm. wide, nearly flat; stems sharply angled; glumes 2 mm., broadly ovate. S. England; very local.   *C. filiformis* L.

11   Lower bract sheathing.   **21. caryophyllea**
     Lower bract not or scarcely sheathing.   12

12   Lowest bract green, resembling a small lf; glumes light brown, acuminate.   **20. pilulifera**

All bracts brown or bristle-like; glumes dark purple-brown, blunt but sometimes mucronate. 13

13 Stems smooth; lvs c. 3–4 mm. wide; glumes blunt, finely ciliate at top. Calcareous grassland; very local. *C. ericetorum* Poll.
Stems rough at top; lvs c. 2 mm. wide; glumes mucronate, not ciliate. Calcareous grassland; local. *C. montana* L.

14 Lower sheaths and lvs ± hairy beneath; lowest bract crimped at base. **15. pallescens**
Plant quite glabrous; lowest bract not crimped at base. 15

15 Lvs involute; stems subterete, curved (brackish marshes). **7. extensa**
Lvs not involute or stems sharply angled, straight. 16

16 Stigmas 3; nut trigonous; ripe fr. usually inflated or trigonous. 17
Stigmas 2; nut biconvex; ripe fr. usually flattened, frequently plano-convex. 48

17 Female spikes purplish-black, ovoid, ± nodding; male spikes similar but narrower. Base-rich mountains; very rare; Scotland.
C. atrofusca Schkuhr
Not as above. 18

18 Male spikes more than 1. 19
Male spike 1 (rarely with a second much smaller one close to its base). 25

19 Fr. broadest at or below middle, tapering into a distinct beak. 20
Fr. broadest above middle, rounded at top, minutely papillose; beak very short. **17. flacca**

20 Female spikes c. 3 times as long as broad; lvs 2–3 mm. wide. Higher Scottish mountains; local. 21
Female spikes mostly more than 3 times as long as broad, if less then lvs 4–6 mm. wide (usually lowland). 22

21 Ligule c. 1 mm., rounded or almost truncate; fr. smooth.
C. saxatilis L.
Ligule 3–4 mm., acute; fr. ribbed, often sterile.
C. stenolepis Less.

22 Margins of lvs inrolled. **10. rostrata**
Margins of lvs flat. 23

23 Male glumes obtuse to acute; fr. 4–6 mm.; male spikes 2–3. 24
Male glumes acuminate; fr. 8 mm.; male spikes commonly 5–6.
12. riparia

24 Male spikes 5–7 mm. diam.; fr. 4 mm., beak short, slightly notched. **13. acutiformis**
Male spikes 2–3 mm. diam.; fr. 5–6 mm., beak long, deeply notched. **11. vesicaria**

25 Female spikes not more than 6-fld; fr. 8 mm. On calcareous soils;
     very rare. S. England.   *C. depauperata* Curtis ex With.
     Fls more numerous; fr. not more than 6 mm., usually less.   26

26 Plant with creeping rhizomes, not forming dense tufts.   27
     Plant forming dense tufts.   33

27 Female spikes nodding, on slender stalks.   28
     Female spikes erect or lowest somewhat nodding; stalks stout.   30

28 Stem rough; lvs strongly channelled or rolled. Wet bogs; local.
                                          C. limosa L.
     Stem smooth; lvs flat, except for the keel.   29

29 Glumes lanceolate, acuminate or awned; fr. much broader than
     glumes. Bogs; rare and mostly in Scotland.
                                    C. paupercula Michx.
     Glumes obovate, almost truncate, apiculate; fr. narrower than
     glumes. Small mountain bogs in Scotland; rare.
                                 C. rariflora (Wahlenb.) Sm.

30 Beak of fr. notched; fr. symmetrical, ribbed.   31
     Beak of fr. ± truncate, fr. asymmetrical, not ribbed.   32

31 Glumes with broad scarious margins; fr. scarcely trigonous,
     yellow-green, ribs numerous, ± equally prominent.   **3. hostiana**
     Margins of glumes not scarious; fr. distinctly trigonous,
     reddish-brown or purplish with 2 prominent green lateral
     nerves.   **4. binervis**

32 Plant glaucous; beak entire and transversely truncate (common).
                                  16. panicea
     Plant green or yellow-green; beak obliquely truncate and slightly
     notched. Scottish mountains; local.   *C. vaginata* Tausch

33 Large plants, 50–150 cm.; lower lvs mostly 10–20 mm. wide;
     female spikes mostly 5–15 cm. × 5–10 mm. (in fr.).   34
     Not as above.   35

34 Female spikes 7–15 cm.; fr. c. 3 mm., erect.   **14. pendula**
     Female spikes 4–9 cm.; fr. c. 6 mm., deflexed.   **9. pseudocyperus**

35 Female spikes c. 2 mm. diam. (in fr.).   36
     Female spikes at least 4 mm. diam. (in fr.).   37

36 Female spikes (25–)40–80 mm., distant. Damp woods; rare.
                                  C. strigosa Hudson
     Female spikes up to 10 mm., clustered. Base-rich soil on
     mountains; very local.   *C. capillaris* L.

37 Female spikes ovoid, scarcely longer than broad.   38
     Female spikes cylindrical, at least twice as long as broad.   42

38 Male spike distinctly stalked.   39
     Male spike sessile, rarely 0.   41

39   Fr. (3·5)4–4·5(–5) mm. curved, deflexed; stems straight, erect.

5. lepidocarpa

Fr. (1–)3–3·5(–4) mm. not curved; stems curved, spreading or else plant small and fr. up to 2·5 mm.   *40*

40   Stems curved; fr. (2·5–)3–3·5(–4) mm.; beak 1 mm. (common).

6. demissa

Stems straight; fr. (1–)1·5–2·5 mm.; beak 0·25 mm. Scotland; very local.   *C. scandinavica* E. W. Davies

41   Plant usually 50–70 cm.; fr. c. 6 mm., curved. Rare.

C. flava L.

Plant usually 5–15 cm.; fr. c. 2–3 mm., straight. Local.

C. serotina Mérat

42   Stems little longer than lvs; lower sheaths reddish; spikes almost black. Scottish mountains; local.   *C. saxatilis* L.

Stems considerably longer than lvs; lower sheaths brown; spikes rarely blackish.   *43*

43   Stalk of lowest female spike at least twice as long as spike (in fr.); all spikes ±pendulous, 3–4 mm. diam.   **8. sylvatica**

Stalk of lowest female spike shorter to little longer than spike (in fr.); at least upper spikes erect, c. 5 mm. diam.   *44*

44   Lvs 5–10 mm. wide; glumes acuminate; fr. 5–6 mm., beak more than 1 mm.   **1. laevigata**

Lvs 2–5 mm. wide; glumes acute or mucronate; fr. 4–5 mm., beak not more than 1 mm.   *45*

45   Fr. with 2 or more strong ribs conspicuous when fresh.   *46*

Fr. smooth and shiny. By the sea; local.   *C. punctata* Gaudin

46   Fr. with 2 strong, green, submarginal nerves.   **4. binervis**

Fr. with several evenly-spaced nerves.   *47*

47   Densely tufted; glumes mucronate.   **2. distans**

Shortly creeping; glumes acute.   **3. hostiana**

48   Glumes (at least in the lower part of spike) with a long excurrent midrib. Tidal rivers in N. Scotland; very local.   *C. recta* Boott

Glumes acute or acuminate, but midrib not excurrent. Not estuarine.   *49*

49   Plant usually c. 10 cm., not tufted; rhizome creeping; stem stiff, stout, very sharply angled; lvs recurved, usually 4–5 mm. wide. Stony places on mountains from 2000 ft. upwards, local.

C. bigelowii Torrey ex Schweinf.

Not with the above combination of characters.   *50*

50   Plant forming dense tufts; leaf-sheaths with margins connected by filaments; lowest bract much shorter than infl.   **22. elata**

Plant rarely tufted; lf-sheaths not filamentous; lowest bracts ± equalling or longer than infl.   *51*

51 Plant usually 20–40 cm.; if taller, stem not more than 2 mm. diam.;
  stem-lvs 1·3–3·0 mm. wide.   **24. nigra**
  Plant usually 50–100 cm.; stem 3 mm. or more in diam.; stem-lvs
  3·0–6·0 mm. wide.   *52*

52 Stems not cracking in two when bent, sharply angled, very rough
  below infl.   **23. acuta**
  Stem cracking in two when bent, bluntly angled, smooth below infl.
  Margins of lakes and streams; local.   *C. aquatilis* Wahlenb.

53 Spikes few, either stalked or distant from one another; glumes
  usually blackish and fr. pale green (mountains, rare).   *54*
  Spikes usually numerous, either small and squarrose or forming a
  roundish or ovoid ±lobed head; glumes brown or greenish. *56*

54 Spikes clustered at top of stem, the lowest sometimes ± remote. *55*
  Spikes ± evenly spaced out, not or scarcely overlapping. Two bogs
  in Scotland.   *C. buxbaumii* Wahlenb.

55 Spikes ± stalked, lowest nodding; lowest bract longer than infl.
  Mountains; local.   *C. atrata* L.
  Spikes sessile or nearly so; lowest bract shorter than infl. Scottish
  mountains; rare.   *C. norvegica* Retz.

56 At least some spikes with male fls (usually easily recognized by the
  white filaments or narrower glumes) at top.   *57*
  At least the upper spikes with male fls at base.   *70*

57 Rhizome far-creeping, stems not tufted.   *58*
  Rhizome short, stems in dense tufts.   *62*

58 Spikes forming a small scarcely lobed head; bracts glume-like
  without a bristle-point.   *59*
  Spikes forming a distinctly lobed head; lower bracts lf-like or
  glume-like with a distinct bristle-point.   *60*

59 Lvs curved, as long as or longer than the curved stems. Sandy
  shores in the north; local.   *C. maritima* Gunnerus
  Lvs straight, much shorter than the straight stems. Bogs; W.
  Sutherland.   *C. chordorrhiza* L. fil.

60 Terminal spike entirely male.   **28. arenaria**
  Terminal spike female or mixed.   *61*

61 Stems rather stout, not wiry; infl. usually more than 2 cm.; beak
  narrowly winged in upper half (marshes).   **27. disticha**
  Stems slender, wiry; infl. up to 2 cm.; beak not winged. By
  ditches, etc. near the sea; local.   *C. divisa* Hudson

62 Lowest spike several times its own length below the next. *63*
  Lowest spike overlapping the next or at most twice its own length
  from it.   *64*

63 Lower bracts often longer than spikes; ripe fr. ± erect, not
  spreading or deflexed.   **29. divulsa**

Lower bracts usually shorter than spikes; ripe fr. spreading or deflexed. Dry calcareous grassland; local.

C. polyphylla Kar. & Kir.

64	Lvs mostly more than 4 mm. wide.	*65*
	Lvs rarely as much as 4 mm. wide.	*67*
65	Plant forming large tussocks; lvs ½-cylindrical. **25. paniculata**	
	Plant not forming conspicuous tussocks; lvs keeled.	*66*

66 Ligule 10–15 mm., longer than broad; at least some bracts bristle-like, conspicuous; fr. not readily dropping at maturity.

26. otrubae

Ligule 2–5 mm., broader than long; bracts short and inconspicuous; fr. readily dropping at maturity. Damp places; very local. *C. vulpina* L.

67	Plant growing in dry places.	*68*
	Plant growing in shallow water or very wet ground.	*69*

68 Glumes and bracts often tinged with purplish-red; fr. 4·2–5·3 mm., gradually narrowed at both ends, greenish when ripe.

30. spicata

Glumes and bracts never tinged with purplish-red; fr. 2·9–3·9 mm., rounded at base, brown when ripe. Local; chiefly on calcareous soils. *C. muricata* L.

69 Plant densely tufted; lvs yellow-green; beak of fr. not split down one side. Fens; very local.   *C. appropinquata* Schumacher

Plant loosely tufted; lvs grey-green; beak of fr. split down one side, the margins overlapping. Fens; very local.   *C. diandra* Schrank

70	At least the lower bracts lf-like. **32. remota**	
	Bracts never lf-like.	*71*

71 Spikes small, with up to c. 10 fruits; fr. spreading in a star.

31. echinata

	Spikes larger, with more than 10 fruits; fr. usually erect.	*72*

72 Spikes more than 4, ±widely spaced, forming an interrupted oblong infl.   *73*

Spikes 3–4, crowded together, forming a dense ±lobed ovoid infl.   *74*

73 Spikes pale greenish-white; fr. 2–3 mm.; beak rough.   **33. curta**

Spikes brown; fr. 3·5–4 mm.; beak smooth. Marshes and damp woods; rare. *C. elongata* L.

74 Plant 20–90 cm.; fr. 4–5 mm., almost winged; beak rough.

34. ovalis

Plant up to 20 cm.; fr. 3 mm.; beak smooth. Higher Scottish mountains, rare.; *C. lachenalii* Schkuhr

75 Spike unisexual; plant usually dioecious.   **36. dioica**

Spike male at top, the lower half or more female.   *76*

76 Glumes persistent; fr. not deflexed when ripe. Ledges of base-rich
    rocks; very rare.   *C. rupestris* All.
    Glumes soon falling; fr. deflexed when ripe.   77
77 Stigmas 2; fr. distinctly flattened, lanceolate, dark brown.
                                        35. pulicaris
    Stigmas 3; fr. not flattened, very narrow, yellowish.   78
78 Spike with 4–12 fruits; stout bristle from base of nut protruding
    along with stigmas from top of fr. Base-rich mountains in
    Scotland; very rare.   *C. microglochin* Wahlenb.
    Spike with 2–4 fruits; persistent stigmas protruding from top of fr.
    Sphagnum bogs; very local.   *C. pauciflora* Lightf.

1. C. laevigata Sm.   Smooth Sedge.

30–100 cm., stout. Sheaths brown, not fibrous. Female spikes 20–
50 × 5–10 mm., 2–4, distant, lower pendulous. Bracts lf-like, all longer
than their spikes, but not longer than infl., long-sheathing. Glumes
3–3·5 mm., ovate, acuminate. Fr. 5–6 mm., ovoid, green; beak bifid,
smooth or nearly so. Fl. 6. Marshes and damp woods, usually on base-
rich acid soils, local.

2. C. distans L.   Distant Sedge.

15–45 cm. Sheaths brown, scarcely fibrous. Female spikes 2–3, 10–20 ×
4–5 mm., distant, erect. Bracts lf-like shorter than infl., long-sheathing.
Glumes 3 mm., ovate, mucronate, brown or greenish. Fr. 4 mm.,
elliptic and bluntly angled; beak bifid, rough. Fl. 5–6. Marshes, etc.
chiefly near the coast, rather local.

3. C. hostiana DC.   Tawny Sedge.

15–50 cm., ± tufted. Sheaths blackish, rather fibrous in decay. Female
spikes 8–20 × 6–9 mm., (1–)2–3, very distant, erect. Lower bracts long
and lf-like, upper small, all sheathing. Glumes 3–4 mm., broadly ovate,
acute, brown with a broad hyaline margin. Fr. 5 mm., ovoid, rather
inflated; beak slender, notched, margins rough. Fl. 6. Fens, damp
pastures, etc., common in the north and west.

4. C. binervis Sm.   Ribbed Sedge.

30–60 cm., ± tufted. Sheaths brown, not fibrous. Female spikes 15–
35 × 5–7 mm., 2–3, very distant, ± erect. Lower bracts long and lf-like,
upper small, all sheathing. Glumes c. 3 mm., ovate, mucronate, usually
dark purple-brown. Fr. 4–5 mm., ovate, subtrigonous with broad
sharp lateral angles; beak broad, bifid, slightly rough. Fl. 6. Heaths,
moors, etc., on acid soils, common in suitable habitats.

5. C. lepidocarpa Tausch   Yellow Sedge.

15–50 cm., tufted, rather slender. Male spike stalked. Female spikes 8–12 × 5–7 mm., 1–3, somewhat distant, lowest sometimes stalked. Bracts lf-like. Glumes 2 mm., ovate, acute, brown or yellowish. Fr. 3·5–4·5 mm., gradually narrowed into the beak, all but the upper curved and deflexed, yellowish; beak 1·5 mm. Fl. 5–6. Locally abundant in fens and on damp base-rich soils, ssp. *lepidocarpa* in the lowlands, ssp. *scotica* E. W. Davies on mountains.

6. C. demissa Hornem.   Yellow Sedge.

Like **5** but 10–20 cm., female spikes 7–10 × 6–8 mm., 2–4, upper ± distant, lowest often very remote; glumes c. 3 mm., acute or subacute; fr. 3–3·5 mm., abruptly narrowed into the beak, lowest sometimes ± deflexed, others straight; beak 1 mm. Fl. 6. Damp places, usually on acid or neutral peaty or clayey soils, common in many districts.

7. C. extensa Good.   Long-bracted Sedge.

20–40 cm., rather rigid. Sheaths persistent, blackish. Male spike sessile. Female spikes 8–15 × 4–6 mm., 2–4, mostly crowded at top of stem, Bracts lf-like, much longer than spikes, narrow, rigid, spreading or deflexed. Glumes c. 2 mm., broadly ovate or suborbicular; mucronate. Fr. ovoid-trigonous, weakly veined, greenish or light brown; beak 0·5 mm., notched, smooth. Fl. 6–7. Grassy salt-marshes and damp coastal cliffs, locally common.

8. C. sylvatica Hudson   Wood Sedge.

15–60 cm., slender. Lvs 3–6 mm. wide. Sheaths brown, fibrous. Female spikes 20–50 × 3–4 mm., 3–4, distant, ± nodding, rather lax-fld; stalks often long. Lower bracts lf-like, sheathing, upper bristle-like. Glumes c. 3 mm., ovate, acute. Fr. 4–5 mm., trigonous, green; beak 1–1·5 mm., notched, smooth. Fl. 5–7. On clayey soils in woods, etc., common.

9. C. pseudocyperus L.   Cyperus Sedge.

40–90 cm., stout. Stems with sharp, very rough angles. Lvs 5–12 mm. wide. Sheaths dark brown, not fibrous. Female spikes 30–50 × 10 mm., 3–5, upper clustered, lowest ± distant, nodding. Bracts lf-like, lowest shortly sheathing, most longer than infl. Glumes 3·5–4·5 mm., narrowly obovate, long-acuminate. Fr. 5–6 mm., ± asymmetrical, broader than glumes, green or yellowish, many-ribbed, deflexed; beak c. 2 mm., deeply notched. Fl. 5–6. By slow-flowing rivers, ditches, etc., local and mainly in the south.

10. C. rostrata Stokes   Beaked Sedge, Bottle Sedge.

30–60 cm., rather glaucous. Stems bluntly angled. Lvs 3–7 mm. wide. Sheaths soon decaying. Female spikes 30–80 × 10 mm., 2–4, distant, sub-erect; stalks short. Bracts lf-like, lower shortly sheathing. Glumes 5 mm., oblong-lanceolate, acute. Fr. 5–6 mm., ovoid, rather inflated, yellow-green, broader than glumes; beak 1·5 mm., notched, smooth. Fl. 6–7. Wet peaty places and in water; widespread but rather local.

11. C. vesicaria L.   Bladder Sedge.

30–60 cm., dark green. Stems sharply angled. Lvs 4–5 mm. wide. Sheaths often reddish, becoming fibrous. Female spikes 15–35 × 7–10 mm., 2–3, distant, lower ± nodding; stalks slender, up to 4 cm. Bracts lf-like, sheathing or not. Glumes c. 3 mm., lanceolate, acuminate. Fr. 4–5 mm., ovoid, inflated, yellowish or brownish, longer and broader than glumes; beak c. 1 mm., notched, smooth. Fl. 6. Wet woods and swampy places; widespread but rather local.

12. C. riparia Curtis   Great Pond-sedge.

100–160 cm., stout, glaucous. Stems sharply angled. Lvs 6–15 mm. wide. Male glumes acuminate. Female spikes (30–)60–90 × 10–15 mm., 1–5, distant, lower nodding on rather long stalks. Bracts lf-like, lowest shortly sheathing, overtopping the stem. Glumes c. 7 mm., oblong-lanceolate to oblong-ovate with a long point, brown. Fr. c. 8 mm., ovate, strongly convex on back, brownish; beak c. 1·5 mm., notched, smooth. Fl. 5–6. By slow-flowing rivers, ponds, etc., common in the south, rarer in the north.

13. C. acutiformis Ehrh.   Lesser Pond-sedge.

60–150 cm., shortly creeping, glaucous. Lvs 7–10 mm. wide. Male glumes ± blunt. Female spikes 20–40 × 7–8 mm., 3–4, remote, all erect. Bracts lf-like, not or shortly sheathing, lowest overtopping the stem. Glumes 4–5 mm., oblong-lanceolate with a long point, brown. Fr. c. 4 mm., elliptic, flattened, pale green; beak c. 0·2 mm., shallowly notched. Fl. 6–7. In similar situations to the preceding but commoner than it is in the north; fens and fen-woods.

14. C. pendula Hudson   Pendulous Sedge.

60–150 cm., stout. Lvs 15–20 mm. wide. Female spikes 70–160 × 5–7 mm., 4–5, distant, pendulous, tapering towards the base; stalks within the sheaths or 0. Lower bracts lf-like, sheathing, upper small. Glumes 2–2·5 mm., ovate, acute or acuminate, red-brown. Fr. c. 3 mm., elliptic- or ovoid-trigonous, greenish; beak c. 0·2 mm., slightly notched. Fl. 5–6. Damp woods and shady stream banks, usually on clay soils, locally abundant in the south, rare in the north.

15. C. pallescens L.   Pale Sedge.

20–50 cm. Stems sharply angled, shortly hairy. Lvs up to 5 mm. wide. Male spike 1, ± concealed by female spikes. Female spikes 2–3, mostly clustered, 5–20 × 5–6 mm.; stalks slender, lowest up to c. 2 cm. Lowest bract lf-like, crimped at base, upper small. Glumes 3–3·5 mm., ovate, acuminate, greenish. Fr. c. 4 mm., ovoid-oblong, convex on both sides, bright green; beak very short. Fl. 5–6. Open woods, rough grassland, etc., widely distributed and not uncommon.

16. C. panicea L.   Carnation-Grass.

10–40(–60) cm. Lvs 2–5 mm. wide. Sheaths dull brown, not fibrous. Female spikes 10–15(–25) × 4–6 mm., 1–2(–3), distant, erect, rather few-fld (up to c. 20); stalks stiff, often completely within the sheaths. Lower bract lf-like, closely sheathing. Glumes c. 2 mm., broadly ovate, acute, brown. Fr. 3–4 mm., ovoid, olive-green, asymmetrical, brownish or purplish; beak short, entire. Fl. 5–6. Common and widely distributed in wet grassy places, fens and flushes.

17. C. flacca Schreber   Carnation-grass.

10–40 cm. Lvs 2–4 mm. wide. Sheaths reddish, not fibrous. Female spikes 15–40 × 4–6 mm., 2(–3), ± distant, erect or nodding, dense-fld; stalks slender, variable in length. Bracts lf-like, ± sheathing. Glumes 1·5–2 mm., oblong-ovate, abruptly narrowed to the acute or mucronate apex, blackish. Fr. 2–2·5 mm., elliptic to obovoid, asymmetrical, papillose, yellow-green to almost black; beak very short, entire. Fl. 5–6. Dry calcareous grassland, damp clayey woods, marshes and fens, common.

18. C. hirta L.   Hammer Sedge.

(15–)30–60 cm. Lvs 2–4 mm. wide. Lower sheaths mostly with a short blade, reddish, nearly glabrous. Female spikes 10–30 × 5–7 mm., 2–3(–5), distant, erect, lowest often near base of stem; stalks mostly within the sheaths. Bracts similar to lvs, all sheathing. Glumes 6–8 mm., ovate or oblong, awned, greenish. Fr. 6–7 mm., ovoid, ribbed, greenish; beak c. 2 mm., notched. Fl. 5–6. Rough grassy places, damp meadows, etc., common.

19. C. lasiocarpa Ehrh.   Slender Sedge.

45–120 cm. Stems slender, rigid. Lvs c. 1 mm. wide, grey-green. Sheaths numerous, dark purplish-brown. Female spikes 10–30 × 4–6 mm., 1–3, ± distant, erect; stalks short or 0. Bracts lf-like, very shortly sheathing. Glumes c. 4 mm., lanceolate, acuminate, chestnut-brown. Fr. c. 4 mm., ovoid, subtrigonous, densely grey-hairy; beak c. 0·5 mm., deeply notched. Fl. 6–7. In reed-swamp and wet peaty places, locally common.

20. C. pilulifera L.   Pill-headed Sedge.

10–30 cm. Stems sharply angled, ± curved. Lvs 2 mm. wide. Sheaths fibrous when old. Female spikes 5–6(–8) × 4–6 mm., 2–4, clustered, erect, few-fld; stalks 0. Lowest bract narrowly lf-like, green, not sheathing. Glumes 3–3·5 mm., broadly ovate, acuminate, brownish. Fr. c. 2·5 mm., almost globose but narrowed into a stout stalk, grey-green; beak 0·5–0·75 mm., slightly notched. Fl. 5–6. On acid sandy or peaty soils, locally common.

21. C. caryophyllea Latour.   Spring Sedge.

5–15(–30) cm. Lvs c. 2 mm. wide. Sheaths blackish, fibrous. Male spike long and stout. Female spikes 5–12 × 3–4 mm., 1–3, ± clustered, erect; stalks usually 0. Bracts bristle-like, shortly sheathing. Glumes c. 2·5 mm., broadly ovate, blunt and mucronate or ± acute, brown. Fr. c. 2·5 mm., obovate or elliptic, trigonous, narrowed below, olive-green; beak very short, slightly notched. Fl. 4–5. In dry grassland, locally common; variable.

22. C. elata All.   Tufted Sedge.

60–90 cm. Stems sharply angled. Lvs (3–)4–6 mm. wide. Lowest sheaths reddish. Female spikes 15–40 × 5–7 mm., usually 2, rather distant, erect; glumes and fr. arranged in conspicuous longitudinal rows; stalks usually 0. Bracts often bristle-like. Glumes c. 3 mm., ovate, subacute, nerve ending at or below apex, purple-brown. Fr. c. 3 mm., broadly ovate or suborbicular, flattened; beak short, entire. Fl. 5–6. Fen-ditches and in wet places by rivers and lakes, locally common.

23. C. acuta L.   Tufted Sedge.

60–100 cm. Like 22 but lowest sheaths with short blades; female spikes nodding in fl.; bracts often lf-like; glumes longer than fr., nerve usually shortly excurrent; fr. c. 2·5 mm. Fl. 5–6. In water and in wet grassy places, common in many districts.

24. C. nigra (L.) Reichard   Common Sedge.

7–70 cm. Lvs 2–3 mm. wide. Sheaths soon decaying to a blackish fibrous mass. Female spikes 10–20 × 4–5 mm., (1–)2–3(–4), ± clustered; stalks usually very short or 0. Lowest bract lf-like. Glumes 3–3·5 mm., lanceolate to obovate-oblong, obtuse to acute or acuminate, nerve ceasing below tip, black, rarely brown. Fr. 2·5–3 mm., broader than the glumes, ovate to suborbicular, green or purplish; beak very short, entire. Fl. 5–7. In damp grassy places, in bogs and beside water, common and widely distributed; very variable.

25. C. paniculata L.   Panicled Sedge.

60–150 cm., building big tussocks up to c. 1 m. diam. Stems spreading. Lvs 3–7 mm. wide, dark green. Sheaths dark brown, not fibrous. Infl. usually branched, spikes numerous, few-fld. Bracts small. Glumes c. 3 mm., triangular-ovate, sometimes mucronate, brownish. Fr. c. 3 mm., brownish, ovoid-trigonous, strongly corky and with many faint veins at base; beak c. 0·5 mm., broad, split to base. Fl. 5–6. Wet, often shady, places on peaty base-rich soils, widely distributed.

26. C. otrubae Podp.   False Fox-sedge.

Up to 1 m. Stems stout, sharply angled. Lvs 4–10 mm. wide, bright green becoming grey-green when dry. Sheaths brownish, soon decaying. Infl. branches sessile; spikes numerous, yellowish-green or light brown. At least some bracts long and conspicuous, though narrow. Glumes 4–5 mm., ovate, acuminate, brownish with green midrib. Fr. 5–6 mm., ovate, greenish becoming dark brown; beak notched. Fl. 6–7. Usually in damp grassy places, particularly on clay soils, common.

27. C. disticha Hudson   Brown Sedge.

20–80 cm. Stems rather stout, sharply angled. Lvs 2–4 mm. wide. Spikes numerous, sessile, forming a ± dense infl. up to c. 5 cm. Bracts small or sometimes the lower narrow, green and longer than infl. Glumes c. 4 mm., ovate, acute, brownish. Fr. 4–5 mm., ovate, plano-convex, distinctly many-ribbed, narrowly winged and toothed in upper half; beak c. 1 mm., notched. Fl. 6–7. Damp grassy places, fens, etc., locally common and widely scattered.

28. C. arenaria L.   Sand Sedge.

10–40 cm. Stems often curved, sharply angled. Lvs 1·5–3·5 mm. wide. Spikes 5–12, crowded, sessile, forming a dense infl. up to c. 4 cm. Bracts small, with bristle-points. Glumes 5–6 mm., ovate, acute or acuminate, brownish. Fr. 4–5 mm., ovate, plano-convex, distinctly many-ribbed, broadly winged and toothed in upper half; beak c. 1 mm., notched. Fl. 6–7. Sandy places near the sea, rarely inland, locally abundant.

29. C. divulsa Stokes   Grey Sedge.

30–60 cm. Stems slender, spreading, sharply angled. Lvs c. 2 mm. wide Sheaths becoming fibrous. Infl. 5–10 cm., often with 1 or 2 short branches at base, upper spikes crowded, lower distant. Bracts bristle-like. Glumes 3–3·5 mm., ovate, acute or acuminate, brownish. Fr. 4–5 mm., ovate, plano-convex, rather suddenly narrowed at base, taper-ing into the beak, brownish and ± erect when ripe; beak c. 1 mm., notched. Fl. 6–7. Rough pastures, hedge-banks, etc., locally common in England and Wales.

30. C. spicata Hudson (*C. contigua* Hoppe)   Spiked Sedge.

20–60 cm. Stems rather stout, ± erect, sharply angled. Lvs 3–4 mm. wide. Sheaths reddish, becoming brown and fibrous. Infl. 2–4 cm., simple; spikes sessile, all, or all but lowest, close together. Bracts bristle-like, often tinged with purplish-red at base. Glumes 5–6 mm., ovate, acute, mucronate or acuminate, brownish often tinged with purplish-red and with a broad green midrib. Fr. 4·2–5·3 mm., broadly ovate, plano-convex, distinctly narrowed and almost stalked at base, tapering above; beak short, notched. Fl. 6. Widely distributed and common on all but the driest base-rich soils.

31. C. echinata Murray   Star Sedge.

10–40 cm. Stems rather slender, spreading, bluntly angled. Lvs c. 2 mm. wide. Sheaths soon decaying. Spikes 3–4, somewhat distant, c. 5 mm., sessile, fr. spreading in a star. Bracts small. Glumes c. 2 mm., broadly ovate, acute, brownish. Fr. 4 mm., yellow-brown, ovate, plano-convex; beak broad, notched. Fl. 5–6. Damp acid meadows, bogs, etc.; wide-spread and common in the north and west.

32. C. remota L.   Remote Sedge.

30–60 cm. Stems bluntly angled, spreading. Lvs 2 mm. wide. Sheaths straw-coloured, not fibrous. Spikes sessile, usually 4–7, distant, lower 7–10 × 3–4 mm. Lower bracts lf-like, longer than infl. Glumes c. 2·5 mm., lanceolate to ovate, acute. Fr. c. 3 mm., ovate, plano-convex, greenish; beak short, broad, notched. Fl. 6. Damp shady places, wide-spread but rarer in the north.

33. C. curta Good.   White Sedge.

25–50 cm. Stems sharply angled. Lvs 2–3 mm. wide, pale green. Sheaths brown, soon decaying. Infl. 3–4 cm., of 4–8 sessile, ± distant spikes 5–8 mm. Bracts small. Glumes 2 mm., obovate to suborbicular, cuspidate, greenish. Fr. 2–3 mm., ovate, plano-convex, pale yellow-green with distinct yellow ribs; beak 0·5 mm., emarginate. Fl. 7–8. Bogs, acid fens, etc., locally common, especially in the north.

34. C. ovalis Good.   Oval Sedge.

20–90 cm. Stems bluntly angled, ± curved, lfy only near the base. Lvs 2–3 mm. wide. Sheaths brown, fibrous. Spikes c. 10 × 5 mm., (1–)3–9, sessile, crowded. Lower bracts often bristle-like. Glumes 3–4 mm., lanceolate, acute, brownish. Fr. 4–5 mm., erect, elliptic-ovate, plano-convex, almost winged, light brown, distinctly nerved; beak c. 1 mm., notched. Fl. 6. Rough grassy places on acid soils, common but local in the south and east; often on heathy tracks.

35. C. pulicaris L.   Flea-sedge.

10–30 cm., shortly creeping. Stems terete, slender, rigid. Lvs c. 1 mm. wide, dark green. Sheaths soon decaying. Spike 10–25 mm., solitary, male at top, with 3–10 female fls below. Bracts 0. Fr. 4–6 mm., lanceolate, flattened, dark brown, shiny; stalk short and stout; beak almost 0. Fl. 5–6. Damp calcareous grassland, fens, base-rich flushes, etc.; common except in the south and east.

36. C. dioica L.   Dioecious Sedge.

10–15 cm., shortly creeping. Stems terete, rigid. Lvs 0·5–1 mm. wide, dark green. Sheaths dark brown, not fibrous. Male spike 10–15 × 2–3 mm.; female spike 5–7 mm. wide, 20–30-fld. Bracts 0. Glumes 2·5–3 mm., persistent, ovate, acute, brown. Fr. c. 3·5 mm., spreading horizontally or slightly deflexed when ripe, flattened-ovoid, tapering to the broad notched beak, greenish-brown with numerous dark brown nerves. Fl. 5. Fens, base-rich flushes, etc.; rather local, infrequent in the south.

144. GRAMINEAE

Annual or perennial, mostly herbs. Stems solid at the nodes and usually hollow in the internodes, cylindrical or sometimes flattened. Lvs consisting of sheath, ligule and blade; sheath encircling the stem; *ligule* a small flap of thin tissue at junction of sheath and blade, sometimes replaced by a ring of hairs, rarely 0; blade usually long and narrow, sometimes with a thickened projection at each side of the base (*auricles*). Fls usually hermaphrodite, enclosed between 2 bracts, the lower called the *lemma* and the upper (sometimes 0) the *palea*, the whole forming the floret. Florets 1 to many, distichous, ± overlapping and sessile on a short slender axis (*rhachilla*), usually with 2 bracts at the base (*glumes*), the whole forming the *spikelet* (Fig. 5, p. 573). Spikelets pedicelled in panicles or racemes, or sessile in spikes; panicles sometimes spike-like with short branches, not readily apparent, bearing crowded spikelets.

1	Ligule a ring of hairs.	2
	Ligule membranous or 0.	6
2	Infl. a panicle with distinctly pedicelled spikelets.	3
	Infl. of 2–several spikes; spikelets sessile or nearly so.	5
3	Plant large, reed-like; sheaths loose; florets with silky hairs at base.	
		1. PHRAGMITES
	Not as above.	4
4	Stem-base swollen; glumes much shorter than spikelet.	
		2. MOLINIA
	Stem-base not swollen; glumes nearly as long as spikelet.	
		3. SIEGLINGIA
5	Stout, tufted; spikelets at least 10 mm.	44. SPARTINA
	Slender, creeping; spikelets c. 2 mm. Rare; in sandy places in S. England; Channel Is. Bermuda-grass.	
		Cynodon dactylon (L.) Pers.
6	Spikelets obviously of 2 or more florets.	7
	Spikelets of 1 floret, occasionally with vestiges of other florets.	47
7	Infl. a simple spike with the sessile or subsessile spikelets inserted singly or in pairs on the rhachis.	8
	Infl. a panicle, or spikelets clustered.	15
8	Glume 1 in lateral spikelets, 2 in terminal.	9
	Glumes always 2.	10
9	Glume 1 in all the lateral spikelets.	7. LOLIUM
	Glumes 2 in some of the lateral spikelets, though the one next the axis often small.	× FESTULOLIUM (see p. 526)
10	Tall perennials; plant tufted or with a far-creeping rhizome.	11
	Small annuals; plant neither tufted nor rhizomatous.	14
11	Spikelets edgeways on to rhachis or else terete.	12
	Spikelets broadside on to the rhachis, strongly flattened laterally.	13
12	Lemmas awned.	21 BRACHYPODIUM
	Lemmas awnless.	× FESTULOLIUM (see p. 526)

13 Spikelets 20–30 mm., in pairs; largest lvs c. 10 mm. wide.
23. ELYMUS
Spikelets (excluding awns) c. 10 mm., solitary; largest lvs c. 5 mm. wide.
22. AGROPYRON

14 Glumes subequal; plant stiff and wiry. 10. CATAPODIUM
Glumes unequal; plant not wiry. 8. VULPIA (and *Nardurus*, p. 528)

15 Glumes largely hyaline, greenish-golden; uppermost lf usually less than 5 mm.; margins of lemmas of 2 lower florets fringed with hairs; plant smelling strongly of coumarin when crushed. Rare; on wet banks in Scotland and Ireland. Holy-grass.
Hierochloë odorata (L.) Beauv.
Not as above. 16

16 Lemma awnless or with a straight awn from the tip, or from a notch at the tip. 17
Lemma of at least some florets with an awn sticking out from the back below the tip; awn usually geniculate. 40

17 Tufted perennials with narrow, dense, ±parallel-sided panicles; spikelets silvery; upper glume about equalling 1st floret.
26. KOELERIA
Not as above. 18

18 Spikelets crowded in dense one-sided masses towards the ends of the panicle branches; coarse tufted plant with strongly compressed vegetative shoots. 12. DACTYLIS
Not as above. 19

19 Infl. a ±spreading panicle; at least some branches easily seen. 20
Infl. spike-like; spikelets crowded in groups on short branches. 36

20 Spikelets not normally viviparous. 21
Spikelets viviparous (mountain districts). 35

21 Lemmas awned or at least with a bristle-like point. 22
Lemmas unawned. 26

22 Spikelets ovoid, subterete. 19. BROMUS
Spikelets lanceolate, widening upwards or ±parallel-sided, ±strongly compressed. 23

23 Lower glume 1-nerved. 24
Lower glume 3-nerved. 20. CERATOCHLOA

24 Spikelets up to 15 mm., lanceolate, somewhat compressed.
6. FESTUCA
Spikelets rarely less than 20 mm., parallel-sided or widening upwards, strongly compressed. 25

25 Annual; spikelets widening upwards. 18. ANISANTHA
Perennial; spikelets ±parallel-sided. 17. ZERNA

26 Nerves of the lemma ± parallel, prominent and extending nearly
        or quite to the apex (± aquatic).   27
      Nerves of the lemma converging towards the tip or else incon-
        spicuous or disappearing well below the apex.   28

27 Spikelets 5 mm. or more; lemma 7–9-nerved.   4. GLYCERIA
      Spikelets c. 2 mm.; lemma 3-nerved.   5. CATABROSA

28 Glumes nearly as long as spikelet; spikelets rather few; margins of
        sheaths connate.   15. MELICA
      Glumes much shorter than spikelet; spikelets usually numerous;
        margins of sheaths free.   29

29 Spikelets broadly conical or ovate, solitary on long, very slender
        pedicels; lemma cordate at base.   14. BRIZA
      Not as above.   30

30 Lemmas linear-lanceolate, acuminate. Rocky woods; very local.
                                                         6. *Festuca altissima*
      Lemmas broader, acute or obtuse.   31

31 Spikelets 10 mm. or more.   32
      Spikelets rarely more than 5 mm.   34

32 Spikelets subterete.   6. FESTUCA (part)
      Spikelets strongly flattened.   33

33 Lower glume 1-nerved. Rare alien.
                                             Zerna inermis (Leyss.) Lindman
      Lower glume 3-nerved. Rare alien, casual in waste places.
                                             20. *Ceratochloa unioloides*

34 Spikelets flattened; lemmas keeled; lvs flat when fresh (not salt-
        marshes).   11. POA (part)
      Spikelets subterete; lemmas rounded on back; lvs usually rolled up
        (salt- and brackish-marshes, rarely elsewhere).   9. PUCCINELLIA

35 Lvs setaceous.   6. *Festuca vivipara*
      Lvs flat.   11. POA (part)

36 Spikelets of 2 kinds; sterile lemmas stiff and distichous, fertile
        lemmas awned.   13. CYNOSURUS
      Spikelets all similar.   37

37 Lemmas ending in 5 small points; infl. bluish and shiny (local).
                                                                  16. SESLERIA
      Not as above.   38

38 Lemmas awned.   8. VULPIA (see also *Nardurus*, p. 528)
      Lemmas unawned.   39

39 Glumes subequal; wiry annuals.   10. CATAPODIUM
      Glumes very unequal; soft perennial.   × FESTULOLIUM (see p. 526)

40 Spikelets (excluding awns) 9 mm. or more.   41
      Spikelets (excluding awns) 5 mm. or less.   43

41 Upper glume 2 cm. or more; spikelets eventually pendulous;
       annual.   28. AVENA
    Upper glume not more than 1·5 cm.; spikelets erect; perennial.   42

42 Lvs long, soft, 4–6 mm. wide; lower floret usually male, awned;
       upper floret female or hermaphrodite, usually unawned.
                                                          30. ARRHENATHERUM
    Lvs short, often rather stiff, c. 2 mm. wide; 2–4 lower florets all
       hermaphrodite and awned.   29. HELICTOTRICHON

43 Spikelets shiny.   44
    Spikelets whitish or pinkish, not shiny.   46

44 Annual; 5–20(–30) cm.; lvs mostly dead at fl. or soon after; non-
       flowering shoots 0.   33. AIRA
    Perennial; normally taller; lvs green at fl. and fr.; non-flowering
       shoots present.   45

45 Tip of lemma ± truncate and torn; spikelets silvery or purplish.
                                                          32. DESCHAMPSIA
    Tip of lemma acute, ending in 2 bristle points; spikelets usually
       yellowish.   27. TRISETUM

46 Awn tapering to a sharp point; lvs flat, not glaucous; plant
       ± hairy.   31. HOLCUS
    Awn thickened towards the tip; lvs setaceous, very glaucous; plant
       glabrous. Rare; on coastal sand.
                                          Corynephorus canescens (L.) Beauv.

47 Spikelets inserted in 2 rows on the rhachis in groups of 2 or 3 at
       each node.   48
    Spikelets solitary or not inserted in 2 rows on the rhachis.   49

48 Central spikelet hermaphrodite, lateral male or sterile.
                                                               24. HORDEUM
    Central spikelet male or hermaphrodite, lateral hermaphrodite.
                                                            25. HORDELYMUS

49 Spikelets with 1–several conspicuous bristles from their pedicels
       (casual, rare).   SETARIA (p. 553)
    No bristles on pedicels.   50

50 Spikelets sunk in the rhachis of the solitary spike; small annuals;
       (clayey ground by the sea).   42. PARAPHOLIS
    Not as above.   51

51 Infl. ovoid, softly and densely hairy; annual. Channel Is. Hare's-
       tail.   Lagurus ovatus L.
    Not as above.   52

52 Infl. of 1 or more spikes, the spikelets sessile or nearly so and
       arranged in 1 or 2 rows.   53
    Infl. not as above.   56

53   Spike solitary.   54
       Spikes more than 1.   55

54   Lemma awnless; small annual. Channel Is, Anglesey; naturalized
       in Hants.; coastal.   *Mibora minima* (L.) Desv.
       Lemma awned; wiry perennial.   43. NARDUS

55   Spikes arranged in a raceme. Rare, casual. Cockspur.
                  **Echinochloa crus-galli* (L.) Beauv.
       Spikes digitate or nearly so. Rare casuals.   DIGITARIA (p. 553)

56   Spikelets clustered on short branches and forming a dense cylindri-
       cal or ovoid spike-like infl.   57
       Spikelets ± spread out along the easily apparent branches of the
       panicle.   66

57   Tall sand-binding grasses; spikelets more than 10 mm.   58
       Not as above; spikelets less than 10 mm.   59

58   Infl. whitish, obtuse.   34. AMMOPHILA
       Infl. purplish-brown, acute; E. Coast, W. Sutherland; local.
           × *Ammocalamagrostis baltica* (Schrader) P. Fourn.

59   Glumes very unequal, the lower much shorter than the spikelet;
       plant smelling of coumarin when crushed.   40. ANTHOXANTHUM
       Glumes subequal, both about as long as or longer than the lemma.   60

60   Glumes flattened, whitish with a green line, keel winged.
                  41. PHALARIS (part)
       Not as above.   61

61   Glumes awned.   62
       Glumes unawned.   63

62   Awns 5 mm. or more, conspicuous; glumes deciduous. Rare; S.
       England and Channel Is.   *Polypogon monspeliensis* (L.) Desf.
       Awns c. 1 mm.; glumes persistent. Rare; salt-marshes.
           × *Agropogon littoralis* (Sm.) C. E. Hubbard

63   Glumes narrow, acuminate, swollen and rounded at the base.
       Rare; usually near the sea.
           Gastridium ventricosum (Gouan) Schinz & Thell.
       Glumes not swollen and rounded at the base.   64

64   Infl. very dense and regular, cylindrical or ovoid; glumes ciliate or
       silky.   65
       Infl. rather lax and irregular; glumes rough on their surfaces,
       falling with the fr. Rare; naturalized in Channel Is, casual
       elsewhere.   **Polypogon semiverticillatus* (Forskål) Hyl.

65   Glumes tapering and curving inwards at the tip; awn of lemmas
       exserted or spikelets silky; palea 0.   38. ALOPECURUS
       Glumes abruptly contracted to the outward-curving tip, or, if as
       above, then lemma awnless and spikelets not silky; palea
       present.   37. PHLEUM

66 Ligule long; infl. purplish; tall reed-like grass in damp places.
                                        41. PHALARIS (part)
    Not as above.   67

67 Glumes 0; spikelets strongly flattened. Rare; in wet places in S.
    England. Cut-grass.   *Leersia oryzoides* (L.) Swartz
    Glumes 2.   68

68 Lower glume small, ±encircling the base of the spikelet; upper
    glume large, often awned and resembling the lemma. Casual.
    Cockspur.   *Echinochloa crus-galli* (L.) Beauv.
    Glumes ±equal.   69

69 Glumes rough on their surfaces, falling with the fr. Rare.
                Polypogon semiverticillatus (Forskål) Hyl.
    Glumes smooth, persistent.   70

70 Glumes ovate, obtuse or subacute; lemma shiny, becoming very
    hard in fr.   39. MILIUM
    Glumes lanceolate, acuminate; lemma tough but not hard in fr.   71

71 Lemma awned; awn c. 5 mm.; annuals. Rare.   APERA (p. 548)
    Lemma awned or not; awn not more than 2 mm.; perennials.   72

72 Spikelets with silky hairs at base of floret.   35. CALAMAGROSTIS
    Spikelets without silky hairs at base of floret.   36. AGROSTIS

1. PHRAGMITES Adanson

Large reeds with a ring of hairs at mouth of lf-sheath. Panicle large.
Spikelets acuminate, subterete. Glumes unequal, shorter than 1st floret,
3-nerved. Lemma 3-nerved. All florets except the lowest with long silky
hairs at base.

1. P. australis (Cav.) Steudel (*P. communis* Trin.)   Reed.

2–3 m. Rhizome stout, long. Lvs 10–20 mm. wide, tapering to long
slender points, deciduous. Sheaths loose, so that all the lvs point the
same way in the wind. Panicle 15–30 cm., nodding, soft, purple;
branches with scattered groups of long silky hairs. Spikelets 10–15 mm.;
florets (1–)3–6. Glumes and lemmas narrowly lanceolate, acuminate.
Silky hairs about as long as lemma. Fl. 8–9. In swamps, shallow water
and sometimes damp sand. In suitable habitats throughout the British
Is.

2. MOLINIA Schrank

Perennial herbs with a ring of hairs at mouth of lf-sheath. Panicle erect,
branches long, slender. Spikelets acuminate, subterete. Glumes subequal, shorter than 1st floret, 1–3-nerved. Lemma 3-nerved.

1. M. caerulea (L.) Moench   Purple Moor-grass.

30–130 cm., wiry and often forming large tussocks. Lvs with scattered long hairs, tapering from near base, deciduous. Base of culms swollen and bulbous. Panicle 3–30 cm., green or purplish; pedicels short, finely ciliate. Spikelets 6–9 mm.; florets 2–4. Glumes lanceolate, acute 1-nerved or upper 3-nerved. Lemma tapering, obtuse. Fl. 6–8. Fens, heaths, moors, etc., throughout the British Is.

3. SIEGLINGIA Bernh.

Perennial herb with a ring of hairs at mouth of lf-sheath. Panicle narrow, erect. Spikelets subterete, ovate. Glumes about equalling spikelet, subequal, 3-nerved. Lemma 7-nerved, apex with 3 short blunt points.

1. S. decumbens (L.) Bernh.   Heath grass.

10–40 cm., tufted. Lvs glaucous above, dark green beneath. Panicle 2–6 cm. Spikelets up to 10, 7–10 mm.; florets 4–5. Glumes lanceolate, obtuse. Lemma ovate, silky on margin in lower half and with a tuft of silky hairs at base. Fls usually cleistogamous. Fl. 7. In acid grassland throughout the British Is and locally on damp base-rich substrata.

4. GLYCERIA R.Br.

Perennial glabrous aquatic herbs. Margins of sheaths connate. Spikelets subterete; florets numerous. Glumes hyaline, unequal, 1-nerved, shorter than 1st floret. Lemma with 7–9 prominent parallel nerves, tip hyaline.

1	Apex of lemma distinctly toothed or lobed.	2
	Apex of lemma entire or nearly so.	3
2	Undehisced anthers 1 mm. or more; teeth of lemma blunt.	
		2. plicata
	Undehisced anthers 0·5 mm.; teeth of lemma acute.	**3. declinata**
3	Spikelets more than 10 mm.	**1. fluitans**
	Spikelets 10 mm. or less.	**4. maxima**

G. fluitans × plicata (*G. × pedicellata* Townsend) is ±intermediate between the parents but produces no seed and has bad pollen.

1. G. fluitans (L.) R.Br.   Flote-grass.

25–90 cm. Stems creeping or floating and rooting at nodes in lower part. Lvs rather suddenly contracted to the acute apex. Sheaths smooth; ligule up to 15 mm. Panicle 10–15 cm., simple or little branched.

Spikelets 15–30 mm., linear. Glumes oblong. Lemma 6–7 mm., oblong, entire, ± acute. Palea usually projecting beyond the lemma. Anthers 1·5–2 mm. Fl. 5–8. In stagnant or slow-flowing water throughout the British Is.

2. G. plicata Fries

Like *G. fluitans* but sheaths rough on ribs; lvs more gradually tapering; panicle usually more branched; lemma 3·5–4·5 mm., rounded, truncate or obscurely lobed at apex; palea shorter than lemma; and anthers 1 mm. Fl. 5–6. Habitat and distribution as for *G. fluitans*.

3. G. declinata Bréb.

Like *G. fluitans* but usually smaller; ribs of sheaths rough; lvs abruptly contracted, mucronate; panicle usually little branched; lemma 3·5–4·5 mm., with 3–5 distinct acute teeth at tip; anthers 0·5 mm. Fl. 6–9. Most commonly at the trampled margins of ponds, rather local.

4. G. maxima (Hartman) Holmberg　　　　　　　　　Reed-grass.

60–200 cm., stout, erect. Lvs up to 2 cm. wide, acute; margins thickened, serrate. Ligule c. 5 mm. Panicle 15–30 cm., much branched and spreading. Spikelets 5–8 mm., narrowly ovate. Lemma c. 3 mm., ovate, entire, obtuse, 9-nerved. Anthers 1 mm. Fl. 7–8. Reed-swamp, riverbanks, etc., scattered throughout the British Is, and abundant in most lowland areas.

5. Catabrosa Beauv.

Perennial glabrous aquatic herb. Margins of sheaths connate. Panicle lax, branches in half whorls of 3–5, successive whorls alternating. Spikelets subterete, florets usually 2. Glumes shorter than 1st floret, unequal. Lemma truncate with an erose hyaline tip and 3 prominent parallel nerves.

1. C. aquatica (L.) Beauv.　　　　　　　　　　Water whorl-grass.

5–70 cm. Stems creeping or floating and rooting at nodes in lower part. Lvs somewhat glaucous, short and broad, obtuse, up to 8 mm. wide. Ligule c. 5 mm., obtuse. Panicle up to 25 cm.; branches smooth. Spikelets 2–4 mm., florets rather distant. Glumes hyaline or purplish, the upper widening from the base to the broad truncate or notched apex. Fl. 6–8. In shallow streams and ditches, local but widely distributed.

6. FESTUCA L.

Perennial herbs. Spikelets subterete; florets 3 or more. Glumes soft, shorter than spikelet, acute; lower 1-, upper 1–3-nerved. Lemma tougher, 5–7-nerved, usually mucronate or awned from the tip.

1   Lvs of non-flowering shoots flat or sometimes folded, but not
    setaceous.   *2*
    Lvs of non-flowering shoots setaceous.   *6*

2   Spikelets downy. Dunes, very local.   *F. juncifolia* St Amans
    Spikelets glabrous.   *3*

3   Spikelets 6–7 mm.; ligule of uppermost lf c. 3 mm. Mountain
    woods; very local.   *F. altissima* All.
    Spikelets 8–20 mm.; ligule of uppermost lf very short.   *4*

4   Awn at least as long as lemma; spikelets green.   **3. gigantea**
    Awn 0 or shorter than lemma; some spikelets purplish.   *5*

5   Basal sheaths thin, brownish; panicle branches solitary or one of
    each pair bearing a solitary spikelet.   **1. pratensis**
    Basal sheaths tough, white; both panicle branches at each node
    usually with several spikelets.   **2. arundinacea**

6   Stem-lvs flat.   *7*
    Stem-lvs setaceous or convolute.   *8*

7   Not stoloniferous; spikelets pale green, shiny; top of ovary hairy.
    Locally naturalized.   **F. heterophylla* Lam.
    Stoloniferous; spikelets usually purplish; top of ovary glabrous.
            4. rubra

8   Sheaths closed nearly to top when young; spikelets usually more
    than 7 mm.   *9*
    Sheaths split more than half-way when young; spikelets usually
    less than 7 mm.   **5. ovina**

9   Lvs obtuse or subacute; panicle usually reddish or purplish; anthers
    2–3 mm.   **4. rubra**
    Lvs acute; panicle greenish; anthers 5 mm. Dunes; local.
          F. juncifolia St. Amans

1. F. pratensis Hudson   Meadow Fescue.

40–80 cm., glabrous. Basal sheaths dark brown, soon decaying. Lvs seldom more than 4 mm. wide, auricles glabrous, inconspicuous. Panicle 8–15 cm., slender, secund; branches 1–2 together, one of each pair usually with a single spikelet; rarely spike-like with solitary, sub-sessile, distichous spikelets. Spikelets 10–15(–20) mm., linear to lanceolate; florets 4–6(–10), rather distant. Fl. 6. Meadows, etc., throughout the British Is.

Hybridizes with *Lolium perenne* to produce × **Festulolium loliaceum** (Huds.) P. Fourn. Almost complete intergradation between the parents may on occasion be found.

2. F. arundinacea Schreber   Tall Fescue.

Usually larger and more tufted than *F. pratensis*. Basal sheaths tough, whitish, persistent. Lvs up to 10 mm. wide; auricles ciliate, prominent. Panicle (12–)20–30(–40) cm., usually spreading; branches in pairs, both with several spikelets. Spikelets 10–15 mm., lanceolate to ovate; florets (4–)5–6(–8), usually closely imbricate. Fl. 6–8. Meadows, etc., throughout the British Is.

3. F. gigantea (L.) Vill.   Tall Brome.

50–150 cm. Lvs up to 15 mm. wide, rough with forward-pointing projections; auricles prominent, reddish. Panicle 12–40 cm., very lax, nodding; branches long, bare below, 1–2 together. Spikelets 10–15 mm., lanceolate; florets 3–8. Lemma lanceolate, slightly rough; awn long, slender. Fl. 6–7. Woods and shady hedge-banks throughout most of the British Is.

4. F. rubra L.   Red Fescue.

10–70 cm. Usually stoloniferous. Lvs of non-flowering shoots setaceous or ± convolute, sometimes rather stiff; stem-lvs ± flat, 0·5–3 mm. wide, obtuse or subacute. Sheaths often shortly hairy, closed almost to top when young. Ligule very short, auricles small. Panicle 3–15 cm., erect. Spikelets 7–14 mm., often reddish or purplish; florets 4–8. Lemma lanceolate, obscurely nerved, glabrous or hairy; awn short. Fl. 5–7. Common, generally distributed and very variable.

5. F. ovina agg.   Sheep's Fescue.

10–50 cm., tufted. Roots brown. Lvs up to 1 mm. wide, all setaceous. Sheaths glabrous or shortly hairy, split more than half-way to base. Ligule very short; auricles rounded. Spikelets 3–7(–10) mm.; lemma awned or mucronate. Fl. 5–8. Common and generally distributed in short turf. In northern mountain districts **F. ovina** L., with normal florets, is largely replaced by **F. vivipara** (L.) Sm., with the florets always proliferating vegetatively. Two similar spp. with thicker (0·7–1 mm.) lvs are naturalized here and there.

7. LOLIUM L.

Glabrous annual or perennial herbs. Infl. a simple (rarely branched) spike with the spikelets edgeways on to the axis. Spikelets compressed, the lateral ones with the lower glume suppressed, the terminal one with 2 glumes; florets numerous. Glumes longer or shorter than the spikelet, 5–9-nerved, awnless. Lemma 5-nerved, awned or awnless.

Glume shorter than spikelet.   **1. perenne**
Glume as long as or longer than spikelet. Rare casual. Darnel.
                                 *L. temulentum L.

1. L. perenne L. (incl. *L. multiflorum* Lam.).   Rye-grass, Ray-grass.
25–80 cm., perennial or, in cultivated races, annual (or biennial). Stems tough. Ligule short, truncate. Infl. 8–20 cm. Spikelets 5–20 mm.; florets (3–)8–14. Glumes variable but always shorter than spikelet. Lemma lanceolate, acute or subacute, awned or not.

Ssp. **perenne**: perennial, lvs up to 3 mm. wide, folded when young; lemmas awnless. Fl. 5–8. Grassland, frequently sown, generally distributed.

Ssp. **multiflorum** (Lam.) Husnot: annual or biennial; lvs up to 10 mm. wide, curled when young; lemmas usually awned. Fl. 5–9. Introduced. Interfertile with ssp. *perenne* and commonly sown in leys where, however, hybrids between the 2 sspp. are most often used.

8. VULPIA C. C. Gmelin

Glabrous annuals with short convolute lvs and slender little-branched secund panicles. Spikelets shortly stalked; florets 4–6. Glumes unequal, not longer than 1st floret, lower usually small. Lemma tough, obscurely nerved; awn longer than lemma.

1   Upper sheaths strongly inflated; lower glume 0 or very small, lemma
     with 1 prominent nerve. Dunes, very local.
                          V. membranacea (L.) Dumort.
     Upper sheath not or slightly inflated; lower glume at least ⅙ of
     upper; lemma obscurely nerved.   2

2   Stem ridged; upper glume with prominent nerves.   **1. bromoides**
     Stem smooth or nearly so; upper glume obscurely nerved.   3

3   Panicle nodding; upper sheath not inflated; upper glume not more
     than 3 times as long as lower.   **2. myuros**
     Panicle erect; upper sheath somewhat inflated; upper glume 3–6
     times as long as lower. Rare.   *V. ambigua* (Le Gall) A. G. More

1. V. bromoides (L.) S. F. Gray   Barren Fescue.
10–25(–50) cm. Sheaths not inflated, upper usually some way below the
panicle at fl. Stems ribbed, ribs broad and shining. Panicle 2–10 cm.,
erect. Spikelets 6–10 mm. Glumes green with a narrow hyaline margin,
lower $\frac{1}{3}$–$\frac{1}{2}$ as long as upper; upper lanceolate, with 3 prominent nerves.
Lemma 4–6 mm. (excluding awn). Awn twice as long as lemma. Fl.
5–7. Locally common throughout the British Is in dry, rather open
habitats.

2. V. myuros (L.) C. C. Gmelin   Rat's-tail Fescue.
10–60 cm. Like *V. bromoides* but upper sheath usually reaching base
of panicle at fl.; stems obscurely ribbed, panicle nodding; upper
glume setaceous or subulate, obscurely nerved, lower not more than $\frac{1}{3}$ as
long as upper and awn 2–3 times as long as lemma. Fl. 5–7. Less common than the preceding sp.; in similar habitats, but mainly in the south.

Nardurus maritimus (L.) Murb., resembling *Vulpia* but differing in the
narrow, stiff, unilateral panicle, lemma 2·5–3 mm., and chasmogamous
florets, occurs locally among open vegetation on basic soils. It is
probably introduced.

9. PUCCINELLIA Parl.

Tufted or creeping perennials or annuals, mainly in salt-marshes.
Panicle compound; spikelets numerous, subterete. Glumes unequal,
shorter than 1st floret, 1–3-nerved. Lemma membranous, oblong or
ovate, obtuse, 5-nerved; apex hyaline.

1 Lemma 3–4 mm.   2
    Lemma less than 2·5 mm.   3
2 Stoloniferous; panicle with long, not specially stiff branches;
        anthers 2 mm.   **1. maritima**
    Tufted; panicle with short, very stiff branches; anthers 0·75–1 mm.
        Very local on muddy seashores.
                              P. rupestris (With.) Fernald & Weatherby
3 Panicle-branches 4–6 together, usually all long and bare below;
        midrib of lemma not reaching tip.   **2. distans**
    Panicle-branches 2–4 together, at least the short ones with spikelets
        to the base; midrib of lemma reaching tip. Very local in muddy
        places near the sea.   *P. fasciculata* (Torrey) Bicknell

1. P. maritima (Hudson) Parl.   Sea Poa.
Perennial 10–80 cm., stoloniferous. Lvs narrow, often convolute,
± obtuse. Ligule c. 1 mm., ovate, obtuse. Panicle 2–25 cm., branches
2–3 at each node, strict or little spreading. Spikelets 7–12 mm.; florets

usually 6–10. Lemma 3–4 mm., oblong, obtuse or apiculate, slightly silky towards base. Anthers 2 mm. Fl. 6–7. Salt-marshes and muddy estuaries, generally distributed in suitable habitats round the coast.

2. P. distans (Jacq.) Parl.   Reflexed Poa.

15–60 cm., tufted, stolons usually 0. Lvs narrow, flat, acute. Ligule c. 1 mm., ovate, obtuse. Panicle 4–15 cm.; branches 4–6 at each node, spreading or deflexed after fl., usually all long and devoid of spikelets below. Spikelets 4–6 mm.; florets 3–6. Lemma c. 2 mm., broadly ovate, obtuse or subacute, midrib not reaching tip. Anthers 0·5–0·75 mm. Salt-marshes, etc., occasionally on sandy ground inland. In suitable habitats round the coast.

10. CATAPODIUM Link

Glabrous annuals. Panicle rigid, secund. Spikelets somewhat compressed. Glumes subequal, 1–3-nerved. Lemma tough, with 3 prominent and 2 faint nerves.

Infl. usually branched and most of the spikelets distinctly stalked; spikelets linear-lanceolate.   **1. rigidum**

Infl. usually unbranched and most of the spikelets subsessile or sessile; spikelets ovate.   **2. marinum**

1. C. rigidum (L.) C. E. Hubbard   Hard Poa.

5–20 cm., rigid. Lvs narrow. Ligule 1–3 mm., ovate. Panicle 2–8 cm., narrow, extremely rigid; pedicels shorter than spikelets. Spikelets 2–4 mm., along one side of the branches; florets 3–6, somewhat spreading. Glumes acuminate. Lemma ovate, obtuse, obscurely nerved. Fl. 5–6. On dry rocks, walls, etc. Scattered throughout the British Is; locally common in the south.

2. C. marinum (L.) C. E. Hubbard   Darnel Poa.

1–13 cm., stout, stiff. Lvs narrow. Ligule ovate. Panicle 0·5–4·5 cm., pedicels very short or 0, rarely longer; rhachis flattened. Spikelets 4–8 mm., distichous, all directed to one side; florets 3–10, imbricate. Lower glume acute, upper obtuse. Lemma ovate, obtuse or mucronulate, nerves distinct. Fl. 6–7. On sand, shingle, rocks, etc., near the sea. Round the coast, local.

11. POA L.

Glabrous annuals or perennials with compound panicles. Spikelets compressed; florets (1–)2–5(–7). Glumes ± unequal, soft, keeled, lower

1–3-nerved, often with a tuft of long cottony hairs at base, tip usually hyaline, awnless.

1 Lvs at least 5 mm. wide; panicle 10 cm. or more; plant 50–100 cm.
    densely tufted. Locally naturalized.   **P. chaixii* Vill.
   Lvs normally up to c. 3 mm. wide; plant usually smaller.   **2**

2 Plant with long creeping rhizomes.   **3**
   Plant without rhizomes, tufted.   **4**

3 Stems and sheaths strongly compressed.   **3. compressa**
   Stems and sheaths ± terete.   **4. pratensis**

4 Stock stout, clothed with persistent ± fibrous sheaths; often
    viviparous.   **5**
   Stock slender, not as above; never viviparous.   **8**

5 Stem bulbous at base, tapering above; fl. 4–5; plant lfless June–
    Oct. Sandy places or dry shallow soils near the sea; very local.
                              P. bulbosa L.
   Stems cylindrical, not bulbous; fl. 7–8; plant green all summer (on
    higher mountains).   **6**

6 Some lvs at least 3 mm. wide, tapering rather abruptly at apex;
    lower glume ⅔ length of upper. Very local.   *P. alpina* L.
   Lvs usually not exceeding 2 mm. wide, tapering gradually at apex;
    glumes subequal.   **7**

7 Spikelets viviparous. Very rare. *P. × jemtlandica* (Almq.) K. Richter
   Spikelets not viviparous. Very rare.   *P. flexuosa* Sm.

8 Lower panicle-branches 1–2(–3) together.   **9**
   Lower panicle-branches (3–)4–6 together.   **12**

9 Panicle ovate to triangular; some lvs usually transversely wrinkled;
    plant soft, annual or short-lived perennial.   **11**
   Panicle narrowly lanceolate to almost linear; lvs never transversely
    wrinkled; plant usually rather stiff, always perennial.   **10**

10 Ligule of uppermost lf not more than 0·5 mm.; plant not usually
    glaucous; panicle generally nodding.   **2. nemoralis**
   Ligule of uppermost lf 1–3 mm.; plant very glaucous; panicle stiff
    and erect. Mountains, rare.   *P. glauca* Vahl

11 Lower panicle-branches spreading or reflexed after fl.; upper floret
    twice as long as pedicel; anthers c. 0·75 mm.   **1. annua**
   Lower panicle-branches not reflexed after fl.; upper floret about as
    long as pedicel; anthers c. 0·25 mm. Very local in sandy places
    near the sea; Cornwall, Scilly Is, Channel Is.   *P. infirma* Kunth

12 Ligule of uppermost lf not more than 0·5 mm.; panicle usually
    nodding.   **2. nemoralis**
   Ligule of uppermost lf 1 mm. or more; panicle erect.   **13**

13   Plant stiff and very glaucous; spikelets few, large.   *P. glauca* Vahl
     Plant not conspicuously stiff or glaucous; spikelets usually
       numerous, small.   *14*

14   Ligule acute; lemma distinctly 5-nerved.   **5. trivialis**
     Ligule truncate; lemma very obscurely nerved. Very local by rivers
       and ponds and in waste places.   *P. palustris* L.

1. P. annua L.   Annual Poa.

Annual or perennial, 5–30 cm., tufted or creeping. Lvs often transversely wrinkled; sheaths smooth, compressed; ligule 2–3 mm. Panicle
1–8 cm., ± triangular; branches 1–2(–4) together, spreading or deflexed
after fl. Spikelets 3–5 mm., lanceolate; florets 3–5, closely imbricate.
Lemma ovate, 5-nerved, glabrous or hairy on the nerves, margins and
tip broadly hyaline. Anthers 0·75 mm. Fl. 1–12. In ± open habitats
throughout the British Is.

2. P. nemoralis L.   Wood Poa.

Perennial, 10–90 cm., ± tufted, glaucous or green. Lvs broadest at base
and forming a right angle with sheath; ligule usually very short. Stems
terete; nodes dark brown or blackish. Panicle 2–10 cm., nodding, with
long branches and numerous spikelets in lowland forms or, in mountain
forms, erect with short branches and few spikelets. Spikelets 3–4 mm.
(–7 mm. in mountain forms), ovate; florets 1–5. Glumes and lemma
lanceolate to ovate, acute to acuminate, obscurely nerved. Variable.
Some mountain forms have been separated as *P. balfourii* Parn., but it is
doubtful if this can be maintained as a sp. Fl. 6–8. In shady places, on
walls and rock-ledges throughout most of the British Is.

3. P. compressa L.   Flattened Poa.

20–40 cm., stiff, rhizomatous, ± glaucous. Most lvs little wider than
their sheaths at base; ligule short. Stem flattened and 2-edged; nodes
pale to dark brown. Panicle 2–7 cm., erect, rather narrow and compact;
branches short, strict, 1–3 together. Spikelets 3–7 mm., ovate-oblong;
florets 2–7. Lemma oblong, obtuse, obscurely nerved. Fl. 6–8. On dry
banks and walls throughout most of the British Is.

4. P. pratensis L.   Smooth-stalked Meadow-grass.

10–80 cm., often tufted but sending out long rhizomes, glaucous or
green. Lvs flat or involute and ± setaceous; ligule usually 1 mm. or
less. Stems terete. Panicle 2–10(–25) cm., ovate or oblong; branches
2–5 together. Spikelets 4–6 mm., ovate; florets 3–5. Lemma lanceolate,

acute or acuminate, distinctly 5-nerved with long cottony hairs on keel and marginal nerves. Fl. 5–7. In meadows, by roads, on mountains etc. Common and generally distributed throughout the British Is. Very variable.

5. P. trivialis L.   Rough-stalked Meadow-grass.

20–60 cm., tufted, easily uprooted. Non-flowering shoots weak and usually decumbent, with numerous narrow lvs. Lvs flaccid; uppermost ligule up to 8 mm., acute. Stems rather weak. Panicle 5–10 cm., broadly ovate to oblong; branches 3–5 together. Spikelets 2–4 mm., ovate; florets 2–4. Lemma lanceolate, acuminate, distinctly 5-nerved, keel silky with cottony hairs at base, marginal nerves glabrous. Fl. 6. Common in meadows, disturbed ground, etc., throughout the British Is.

12. DACTYLIS L.

Perennials. Panicle compound, lower branches usually long. Spikelets compressed, crowded in dense masses at the ends of the branches. Glumes subequal, 3-nerved. Lemma 5-nerved, shortly awned.

Lvs usually stiff and glaucous; keel of lemma ciliate.   **1. glomerata**
Lvs usually soft and green; keel of lemma glabrous. Naturalized in a
    few places in S.E. England.   *D. polygama* Horvat.

1. D. glomerata L.   Cock's-foot.

Up to 100 cm., tufted, glabrous. Sterile shoots strongly flattened. Lvs rough, ± glaucous; ligule 2–10 mm. Stems terete. Panicle 3–15 cm., erect, lower branches usually long, horizontal or reflexed in fl., ascending in fr., upper very short. Spikelets 5–7 mm., secund, crowded at ends of branches. Lemma lanceolate, ciliate on keel; awn short, subterminal. Fl. 5–7. In grassland and by roads, etc. Generally distributed and common throughout the British Is.

13. CYNOSURUS L.

Annual or perennial, glabrous. Panicle dense, oblong or ovoid. Spikelets subsessile, dimorphic; upper spikelet on each branch fertile; lower spikelets sterile, with rigid, distichous lemmas. Fertile spikelets: glumes thin, subequal; lemmas terete, tough, awned.

Perennial; sheaths not inflated; panicle narrowly oblong.   **1. cristatus**
Annual; sheaths inflated; panicle ovoid, squarrose. Rare casual.
                                                           C. echinatus L.

1. C. cristatus L.   Crested Dog's-tail.

Wiry tufted perennial 15–75 cm. Lvs c. 2 mm. wide; sheaths not
inflated. Panicle dense and spike-like, narrowly oblong. Spikelets
3–4 mm., secund, compressed, sterile pectinate. Lemmas of fertile
spikelets terete, lanceolate, tough, obscurely nerved, shortly awned. Fl.
6–8. Common and generally distributed in grassland on acid and basic
soils.

14. BRIZA L.

Annual or perennial, glabrous. Panicle spreading, pedicels slender.
Spikelets ovoid or broadly triangular, compressed, often pendulous.
Glumes broad, obtuse, awnless, rounded on back. Lemma ovate,
obtuse, cordate at base and usually saccate, awnless.

1 Perennial; ligule less than 1 mm.   **1. media**
    Annual; ligule 2–5 mm.   *2*

2 Spikelets 2 mm., numerous. Very local in the south.   **B. minor* L.
    Spikelets 10 mm. or more, 3–12. Rare casual.   **B. maxima* L.

1. B. media L.   Quaking Grass, Doddering Dillies.

Tufted but shortly creeping perennial 20–50 cm. Lvs c. 2 mm. wide,
acute; ligule less than 1 mm., truncate. Panicle spreading; pedicels very
slender, up to twice as long as the spikelets and slightly thickened below
them. Spikelets 4–5 × 4–6 mm., obtuse, usually purplish. Glumes
unequal, shorter than lowest floret, margins white and shining. Lemma
tough and shining on the back, margins thinner, purplish then hyaline.
Fl. 6–7. Widespread in grassy places, particularly on dry calcareous or
damp peaty soils.

15. MELICA L.

Perennial, margins of sheaths connate. Infl. a panicle or raceme. Spikelets terete; florets 2–4, the upper sterile and club-shaped. Glumes
nearly as long as spikelet, thin, awnless, 3–5(–7)-nerved. Lemma tough,
obtuse, rounded on back, 7–9-nerved, awnless.

Panicle broadly spreading; lower branches long; glumes acute.
                                                          1. uniflora
Panicle narrow, secund; all branches short; glumes obtuse.   **2. nutans**

1. M. uniflora Retz.   Wood Melick.

30–70 cm. Lvs rough beneath; sheaths with a ligule-like projection on
the opposite side of the stem from the blade. Panicle 10–20 cm., broad,
spreading, lowest branches usually far below the others. Spikelets

4–5 mm., purplish-brown, erect; florets 2. Lower glume 3-nerved, acute, upper 5-nerved, mucronate. Lemma with 2 marginal nerves on each side close together. Fl. 5–6. In woods and shady hedge-banks, scattered throughout the British Is.

2. M. nutans L.   Mountain Melick.

20–40 cm. Lvs smooth beneath; sheath split for a short distance from the top. Panicle 5–15 cm., nearly simple, secund, drooping. Spikelets 6–7 mm., purplish-brown, drooping; florets 3–4. Both glumes obtuse and 5-nerved. Lemma with nerves equally spaced. Fl. 5–6. In woods on limestone and cracks of limestone pavement. In suitable places throughout Great Britain, rather local and absent from S.E. England.

16. SESLERIA Scop.

Perennials. Panicle ovoid and spike-like. Bracts sheathing the base of the lower panicle-branches. Spikelets somewhat compressed; florets few. Glumes subequal, keeled, longer than lemmas. Lemma keeled, 5-nerved, 3 central nerves close together, at least 3 of the nerves shortly excurrent.

1. S. albicans Schultes (*S. varia* Wettst., *S. caerulea* (L.) Ard. ssp. *calcarea* (Čelak.) Hegi)   Blue Sesleria.

15–40 cm., wiry, tufted. Lvs keeled, glaucous above, dark green beneath, abruptly narrowed at apex, mucronate; uppermost lf usually very short and far below infl. Panicle 1–2 cm., ovoid, blue-grey and glistening. Spikelets 5–8 mm. Glumes hyaline. Lemma hairy, slaty-blue and purplish towards the top, apex with 3–5 small points. Fl. 4–5. In limestone pastures in N. England and on base-rich mountains in Scotland.

17. ZERNA Panzer

Perennials. Spikelets subterete to compressed, narrower or only slightly broader at top than in middle; florets usually numerous. Glumes unequal, awnless, lower 1-, upper 3-nerved. Lemma usually 7-nerved. Awn shorter than lemma or frequently 0.

1   Lvs seldom more than 2 mm. wide, the upper broader than the
       lower; panicle with short erect branches.   **1. erecta**
      Lvs seldom less than 5 mm. wide, the upper narrower than the
       lower; panicle with long ± nodding branches.   *2*
2   Uppermost lf-sheath with long, spreading hairs; lower glumes
       glabrous.   **2. ramosa**
      Uppermost lf-sheath shortly pubescent; lower glumes pubescent.
       Local.   *Z. benekenii* (Lange) Lindman

1. Z. erecta (Hudson) Gray   Upright Brome.

60–100 cm. Lower lvs convolute, upper flat, broader than lower. Sheaths glabrous or with spreading hairs. Panicle 10–15 cm., erect, usually reddish or purplish, nearly simple. Spikelets 20–35 mm., narrowly oblong. Lemma 8–10 mm., glabrous, c. 3 times as long as awn. Fl. 6–7. On dry banks and downs, particularly on calcareous soils. Locally dominant in suitable habitats, chiefly in the southern half of England.

2. Z. ramosa (Hudson) Lindman   Hairy Brome.

60–190 cm. Lvs flat, lower broader than upper. Sheaths with long downward-directed hairs, upper sometimes glabrescent. Panicle 15–30 cm., often dark purplish, compound; branches nodding. Spikelets 20–30 mm., linear. Lemma 10–13 mm., c. twice as long as awn, the 3 prominent nerves pilose in their lower half. Fl. 7–8. In hedges and woods throughout most of the British Is on the more fertile soils.

18. ANISANTHA C. Koch

Like *Zerna* but annual; spikelets becoming distinctly broader towards the top; awn longer than lemma.

1   Longer panicle-branches with 4 or more spikelets. Naturalized in
     Norfolk; very local.   **A. tectorum* (L.) Nevski
    All panicle-branches with 1–2 spikelets.   2

2   Lemma less than 20 mm.; plant slender.   3
    Lemma more than 20 mm.; plant usually stout.   4

3   At least most panicle-branches as long as or longer than the spike-
     lets.   **1. sterilis**
    All or nearly all panicle-branches shorter than spikelets. Very local
     in S. and W. England and Wales.   *A. madritensis* (L.) Nevski

4   Panicle lax, spreading; glumes with hyaline margins; lemma 25–
     30 mm.; stamens 3 or 2. Local; in the south.
                             *A. diandra (Roth) Tutin
    Panicle dense, erect; glumes mostly hyaline except for nerves; lemma
     22–25 mm.; stamens always 2. Rare.   *A. rigida* (Roth) Hyl.

1. A. sterilis (L.) Nevski   Barren Brome.

10–100 cm. Lvs soft, lower soon withering. Stems glabrous. Panicle 50–15 cm., drooping, simple or slightly branched; branches usually much longer than the spikelets. Spikelets 20–25 mm., strongly compressed. Lemma 14–18 mm., linear-lanceolate, nerves equally spaced, margins hyaline; awn 1½ to twice as long as lemma. Fl. 5–7. Widely distributed in waste places, by roads and as a garden weed.

19. Bromus L.

Annuals or biennials. Spikelets somewhat compressed, ± ovoid, narrowing considerably towards the top; florets usually numerous. Glumes unequal, lower 3–5-nerved, upper 5–7-nerved. Lemma up to 13-nerved; awn shorter to little longer than lemma.

1 Caryopsis convolute and the lemma rolled round it, so that the whole is subterete; awns divaricate after fl.; spikelets disarticulating tardily.   **5. secalinus**

   Caryopsis thin, nearly flat, the lemma not completely enfolding it; awns rarely divaricate; spikelets disarticulating as soon as ripe.   2

2 Panicle short, erect; at least some pedicels shorter than their spikelets; nerves on back of lemma prominent.   3

   Panicle long, narrow, often nodding; most pedicels longer than their spikelets; nerves on back of lemma inconspicuous.   5

3 Spikelets distinctly stalked, not in groups of 3; palea entire.   4

   Many spikelets subsessile, often in groups of 3 at the end of a branch; palea split to base. Rare; chiefly in S. England.

                                                   B. interruptus (Hackel) Druce

4 Lemma 5·5–6·5 mm., strongly angled, with broad hyaline margins.
                                                                  2. lepidus

   Lemma 6·5–9 mm., rounded or weakly angled, margins narrowly hyaline.   **1. mollis**

5 Panicle erect, narrow; lemma c. 7 mm., rounded.   **3. racemosus**

   Panicle usually nodding, broader; lemma c. 9 mm., bluntly angled.
                                                              4. commutatus

1. B. mollis L.   Lop-grass.

5–80 cm., soft, ± hairy. Panicle 5–10 cm., erect, usually dense, rarely reduced to a single spikelet. Pedicels short, at least some much shorter than their spikelets. Spikelets 10–20 mm., glabrous or hairy. Lemma 6·5–9 mm., ovate, rounded or weakly angled; awn up to 10 mm. Fl. 5–7. In meadows, waste places, and on dunes, shingle banks and on cliffs throughout the British Is. **B. ferronii** Mabille, which is perhaps a distinct sp., occurs very locally near the sea in the south and west. It has a very dense panicle, the lemma 6·5–7·5 × 3·7–4 mm., weakly angled and hairy and the awn curving outwards in fr. The plant known as **B. thominii** Hard. seems to be the result of hybridization between *B. mollis* and *B. lepidus*.

2. B. lepidus Holmberg

15–70 cm., soft, ± hairy. Panicle 3–10 cm., erect, rather lax. Pedicels mostly much shorter than their spikelets. Spikelets 10–15 mm., nearly always glabrous. Lemma sharply angled about ⅔ of the way from the base, upper half with broad hyaline shining margins. Fl. 6–8. In waste places, less commonly in grassland. Commonly sown and widely distributed throughout the British Is.

3. B. racemosus L.   Smooth Brome.

20–90 cm., soft, ± hairy. Panicle 3–10 cm., erect, narrow, usually simple; pedicels up to c. 3 cm. Spikelets c. 15 mm., glabrous or nearly so; florets 1–1·5 mm. apart on the rhachilla. Lemma c. 7 mm., ovate, nerves obscure, margins rounded. Palea about as long as lemma. Fl. 6. In meadows, etc., sometimes in arable land. Scattered throughout the British Is; rather uncommon.

4. B. commutatus Schrader   Meadow Brome.

30–90 cm., soft, ± hairy. Panicle 7–20 cm., usually nodding and some-what spreading; pedicels up to 7 cm. Spikelets 18–28 mm., usually glabrous; florets 1·5–2 mm. apart on the rhachilla. Lemma c. 9 mm., broadly ovate, nerves obscure, margins broadly hyaline, bluntly angled. Palea usually shorter than lemma (6–8 mm.). Fl. 6. In similar places and with a similar distribution to *B. racemosus*.

*5. B. secalinus L.   Rye Brome.

30–60 cm. Panicle 5–20 cm., secund, nodding, lax. Spikelets 10–15 mm., glabrous or densely hairy; florets at first imbricate, spreading in fr. Lemma 7–9 mm., ovate, curled round the rolled caryopsis and so strongly convex in fr.; margins rounded; nerves obscure. Fl. 6–7. Introduced as a weed, now rare, usually in winter wheat.

20. CERATOCHLOA Beauv.

Two introduced spp. distinguished as follows:

Lower sheaths with scattered hairs; awns long. Locally naturalized;
    particularly by the Thames.   *C. carinata* (Hooker & Arnott) Tutin
Plant quite glabrous; awns 0 or up to 2 mm. Casual.
                                    C. unioloides (Willd.) Beauv.

21. BRACHYPODIUM Beauv.

Tufted perennials (Brit. spp.) with spike-like infl. Spikelets subsessile, distichous, terete, linear-lanceolate; florets numerous. Glumes unequal, shorter than 1st floret, prominently 5–7-nerved. Lemma acuminate or awned from tip, 7-nerved.

Lvs soft, yellow-green; awn equalling or longer than lemma.

1. sylvaticum

Lvs stiff, yellow-green to glaucous; awn much shorter than lemma.

2. pinnatum

1. B. sylvaticum (Hudson) Beauv.   Slender False-brome.

30–90 cm., hairy. Lvs flat, up to 13 mm. wide, soft, ± drooping, yellow-green; ligule c. 2 mm. Infl. 6–15 cm. Spikelets 12–25 mm., nearly straight. Lemma linear-lanceolate, acute; awn equalling or longer than lemma. Fl. 7. In woods and hedges and on railway banks. Throughout the British Is in calcareous or base-rich soils.

2. B. pinnatum (L.) Beauv.   Tor Grass, Heath False-brome.

30–60 cm., glabrescent. Lvs ± involute, usually less than 5 mm. wide, stiff, erect, ± glaucous; ligule very short. Infl. 6–17 cm. Spikelets 20–35 mm., usually curved away from the rhachis. Lemma linear-lanceolate, acuminate or, usually, shortly awned. Fl. 7. In calcareous grassland, locally dominant. Rather local and mainly in southern and central England.

22. AGROPYRON Gaertner

Tough perennials. Infl. a spike of many distichous spikelets. Spikelets compressed, solitary at the nodes of the rhachis and broadside on to it; florets 2–many. Glumes 3–9-nerved, shorter than spikelet. Lemma awned or awnless.

1	Plant densely tufted; glumes persistent when fr. is shed.	2
	Plant with far-creeping rhizomes; glumes falling with the fr.	3
2	Awn usually longer than lemma; stems finely hairy at or below some nodes.	**1. caninum**
	Awn much shorter than lemma; stems and nodes glabrous. Highlands of Scotland; very local.	*A. donianum* F. B. White
3	Ribs of lvs scarcely projecting, slender; lvs usually flat when dry, with long scattered hairs on upper surface.	**2. repens**
	Ribs of lvs prominent, broad and nearly concealing the upper surface of lf; lvs convolute when dry, hairs very short or 0.	4

4 Ribs of lvs with numerous short spreading hairs on upper surface.
4. junceiforme

Ribs of lvs glabrous and smooth, or rough with stiff points on upper
surface. 5

5 Ribs of lvs rough with stiff points on upper surface; most spikelets
overlapping by at least half their length. **3. pungens**

Ribs of lvs smooth; spikelets small, distant. S.E. coasts; local.
A. maritimum (Koch & Ziz) Jansen & Wachter

A. caninum × *donianum* occurs where the parents grow together.
A. junceiforme × *pungens*, *A. junceiforme* × *repens*, and *A. pungens* ×
repens occur locally in quantity, owing to vegetative spread. They may
be recognized by lvs intermediate between those of the parents and
sterile pollen.

1. A. caninum (L.) Beauv. Bearded Couch-grass.

30–100 cm., bright green, tufted. Some nodes or the stem below the
nodes finely hairy. Lvs flat, up to c. 13 mm. wide; ribs very slender.
Spike 10–20 cm., often nodding. Spikelets 10–17 mm., lanceolate;
florets 2–5. Lemma lanceolate; awn slender, flexuous. usually longer
than lemma. Palea emarginate. Fl. 7. In hedges and woods, scattered
throughout the British Is on the more fertile soils.

2. A. repens (L.) Beauv. Couch-grass, Scutch, Twitch.

30–100 cm., dull green or ± glaucous. Rhizomes abundant and far-
creeping. Stems and nodes glabrous. Lvs usually narrower than those
of *A. caninum*; ribs numerous, slender. Spike 5–15(–20) cm., usually
stiff and erect. Spikelets 10–15(–20) mm.; florets 3–6. Lemma lanceolate, obtuse and apiculate, acute or awned. Fl. 6–9. In fields and waste
places, common throughout the British Is.

3. A. pungens (Pers.) Roemer & Schultes Sea Couch-grass.

30–90 cm., usually glaucous. Rhizomes far-creeping. Stems and nodes
glabrous. Stems and non-flowering shoots erect. Lvs ± convolute,
rather stiff; ribs broad and prominent on the upper surface, glabrous
but with stiff points. Spike 5–12 cm., stout, resembling an ear of wheat;
rhachis not disarticulating at the nodes. Spikelets 10–18 mm., mostly
overlapping each other by at least half their length. Lemma lanceolate,
obtuse and apiculate or shortly awned. Fl. 7–9. Locally dominant on
dunes and drier parts of salt-marshes in England.

4. A. junceiforme (A. & D. Löve) A. & D. Löve

Sand Couch-grass.

25–50 cm., glaucous. Rhizomes far-creeping. Stems and nodes glabrous. Stems ± erect, non-flowering shoots decumbent. Lvs convolute when dry, stiff; ribs broad and prominent on the upper surface, densely clothed in short spreading hairs. Spike 5–15 cm., stout; rhachis fragile and readily disarticulating at the nodes when mature. Lemma lanceolate, obtuse and apiculate or subacute. Fl. 6–8. On young dunes. In suitable habitats round the coasts of the British Is.

23. ELYMUS L.

Perennial. Infl. a spike with 2–3 spikelets at each node. Spikelets broadside on to the rhachis; florets 3–4, upper usually sterile. Glumes 5-nerved, about equalling the spikelet and often placed side by side in front of it. Lemma tough, 5-nerved, rounded on back.

1. E. arenarius L.   Lyme-grass.

100–200 cm., stout, intensely glaucous, rhizomatous. Lvs broad, rigid, pungent, shortly and densely hairy on ribs above. Spike 15–30 cm., stout. Spikelets c. 20 mm., 2 at each node. Lemma lanceolate, hairy, acute or obtuse. Fl. 7–8. On dunes, often with *Ammophila*. Sandy coasts of the British Is, local.

24. HORDEUM L.

Annual or perennial. Infl. a compressed or subterete spike with alternate distichous groups of 3 spikelets at the nodes. Spikelets of 1 floret, subsessile and broadside on to the rhachis, central one of each group hermaphrodite, lateral male or sterile. Glumes equal, narrow, 1-nerved about as long as lemma. Lemma 5-nerved, awned.

1   Sheath of uppermost lf not inflated.   **1. secalinum**
      Sheath of uppermost lf inflated.   *2*

2   Glumes of central spikelet ciliate.   **2. murinum**
      Glumes of central spikelet scabrid. By the sea; very local.

H. marinum Hudson

1. H. secalinum Schreber   Meadow Barley.

Slender perennial 30–60 cm. Lvs flat, somewhat rough. Sheaths not inflated. Spike 2·5–5 cm., flattened. Spikelets c. 8 mm. Lemma lanceolate; awn 2–3 times as long as lemma. Lemma of sterile fls subulate, stalked, ending in a short point. Fl. 6–7. In meadows. Locally abundant in the south, rarer in the north.

2. H. murinum L.   Wall Barley.

Stout annual 20–60 cm. Lvs flat, hairy on both surfaces. Upper sheath inflated. Spike 4–10 cm., flattened. Spikelets 8–12 mm. Lemma ovate; awn 2–4 times as long as lemma. Lemma of sterile fls lanceolate. Fl. 6–7. In waste places and sandy ground near the sea. Throughout Great Britain but rare in the north.

25. HORDELYMUS (Jessen) Harz

Perennial. Like *Hordeum* but all spikelets hermaphrodite or middle one of a group male.

1. H. europaeus (L.) Harz   Wood Barley.

40–120 cm., tufted, stout. Lvs up to 10 mm. wide; upper sheath not inflated. Spike 5–10 cm., flattened. Spikelets 10–15 mm. Glumes subulate, ending in a bristle; lemma lanceolate; awn 2–3 times as long as lemma. Fl. 6–7. In woods and shady places, often on calcareous soils; locally abundant in England.

26. KOELERIA Pers.

Perennials. Panicle narrow, ± shiny. Spikelets laterally compressed; florets 2–3(–5). Glumes unequal, upper about equalling 1st floret, acute or shortly awned. Lemma tough, obscurely 5-nerved, subobtuse to shortly awned.

Stem-base covered with non-reticulate lf-sheaths; lvs with spreading
    hairs, particularly on margins and lower surface.   **1. cristata**
Stem-base covered with reticulate sheaths; lvs glabrous or nearly so.
    Only in N. Somerset.   *K. vallesiana* (Honck.) Bertol.

1. K. cristata (L.) Pers.   Crested Hair-grass.

10–40 cm., often glaucous. Basal sheaths entire or ± torn, not fibrous, with spreading hairs (like the lvs). Panicle 2–6·5 cm., narrowly oblong or ovoid, ± lobed. Spikelets 3–5 mm., usually purplish or green. Lemma acute to shortly aristate. Dune forms with whitish panicles have been referred to *K. albescens* DC. but do not seem to be specifically separable. Fl. 6–7. In turf in well-drained base-rich soils, locally common throughout the British Is.

27. TRISETUM Pers.

Perennials. Spikelets shiny, compressed; florets 2–4(–6). Glumes about as long as spikelet. Lemma 5-nerved, apex with 2 bristle points; awn dorsal, geniculate, arising above the middle of the lemma.

1. T. flavescens (L.) Beauv.   Yellow Oat.

20–50 cm., stoloniferous. Lower sheaths hairy, upper glabrous. Lvs flat. Panicle oblong or ovate, branches variable in length. Spikelets 3–6 mm., shiny, usually yellowish. Lemma lanceolate, margins broadly hyaline; awn long-exserted. Size of spikelet, glumes and lemma unusually variable. Fl. 5–6. In meadows, widely distributed.

28. AVENA L.

Stout annuals. Spikelets large, terete, eventually pendulous; florets 2–3. Glumes enclosing spikelet, lower 7-, upper 9-nerved. Lemma 7-nerved; awn geniculate, arising about the middle.

1	Lemma shortly bifid or entire and hyaline at apex.	*2*
	Lemma with 2 long bristles at apex. Rather rare; casual.	
		*A. strigosa Schreber
2	Florets separating readily at maturity; 2nd floret with callus-scar at base.	**1. fatua**
	Florets not readily separating; 2nd floret without scar at base.	
		2. ludoviciana

*1. A. fatua L.   Wild Oat.

30–90 cm. Lvs flat, broad, slightly rough. Panicle 15–20 cm., spreading. Spikelets 20–25 mm.; awns long and spreading. Glumes longer than florets, thin, acuminate. Lower part of florets usually covered with silky, often fulvous, hairs. Florets separating readily at maturity, 2nd with a callus-scar at base. Fl. 7–9. Naturalized in arable fields, locally common.

*2. A. ludoviciana Durieu   Wild Oat.

Like *A. fatua* but lvs very rough on both surfaces, spikelet 25–30 mm.; florets not readily separating, the 2nd without a scar at base. Fl. 6–7. A weed of arable land on heavy soils, locally abundant in S. England.

29. HELICTOTRICHON Besser

Perennials. Spikelets large, erect, subterete, shining; florets 2–3(–6). Glumes nearly as long as spikelets; lower 1–3-nerved, upper 5-nerved. Lemma 5-nerved, awned from the back; awn much longer than spikelet.

Sheaths glabrous; lvs stiff, glaucous.	**1. pratense**
At least the lower sheaths hairy; lvs soft, green.	**2. pubescens**

1. H. pratense (L.) Pilger   Meadow Oat.

30–60 cm., glabrous. Lvs ± channelled, glaucous above, dark green beneath, stiff; basal spreading. Sheath glabrous. Panicle 6–12 cm., strict, narrow, nearly simple, lower branches 1–2 together. Spikelets 12–20(–25) mm. Lemma lanceolate, with 2 bristles at tip; awn from about the middle. Fl. 6. In short turf, usually on chalk or limestone, rather local.

2. H. pubescens (Hudson) Pilger   Hairy Oat.

30–70 cm., ± hairy. Lvs flat, green, soft. At least the lower sheaths hairy. Panicle 6–14 cm., usually compound, lowest branches 5 together. Spikelets 10–15 mm. Lemma narrowly lanceolate with 4 points at tip; awn from about the middle. Fl. 6–7. In meadows and rough grassland, usually on chalk or limestone; locally abundant and widely distributed.

30. ARRHENATHERUM Beauv.

Similar to *Helictotrichon* but lower floret usually male, upper female or hermaphrodite, and usually awnless.

1. A. elatius (L.) Beauv. ex J. & C. Presl   Oat-grass.

60–120 cm. Lvs flat, long, flaccid. Panicle 10–20 cm., lax, nodding, rather narrow. Spikelets 7–10 mm. Glumes lanceolate, acuminate. Lemma lanceolate, acute, broadly hyaline. Variable. Fl. 6–7(–11). Common and generally distributed in rough grassy places and on scree and shingle.

31. HOLCUS L.

Perennials. Panicle compound. Spikelets compressed; florets 2. Glumes enclosing spikelet, strongly keeled, lower 1-, upper 3-nerved. Lemma tough, shiny, obscurely 5-nerved. Lower floret hermaphrodite, awnless, upper male, awned from just below the tip.

Nodes not bearded; awn hidden by the glumes.	**1. lanatus**
Nodes bearded; awn protruding.	**2. mollis**

1. H. lanatus L.   Yorkshire Fog.

20–60 cm., softly hairy. Sheaths hairy; nodes without a conspicuous tuft of hairs. Panicle 4–10 cm., ovate, often pink. Spikelets 3–4 mm., crowded. Glumes shortly hairy. Lemma with a few silky hairs at base, smooth and shiny. Awn short, hooked, not projecting beyond the glumes. Fl. 6–9. In waste places, fields and woods, often abundant, generally distributed.

2. H. mollis L.   Creeping Soft-grass.

20–60 cm., sparsely hairy, rather stiff. Sheaths usually glabrous; nodes with a tuft of downward-directed hairs. Panicle 4–10 cm., usually whitish. Spikelets 4–5 mm. Glumes nearly glabrous. Lemma conspicuously hairy at base. Awn projecting well beyond the glumes. Fl. 6–7. On acid soils, widely distributed but less common than *H. lanatus*.

32. DESCHAMPSIA Beauv.

Tufted perennials. Panicle compound. Spikelets somewhat compressed, shiny; florets 2. Glumes enclosing spikelet, 1-nerved, ± hyaline. Lemma obscurely 5-nerved, truncate and jagged at apex; awn dorsal.

1 Lvs flat; awns not or scarcely longer than glumes.   **1. caespitosa**
  Lvs setaceous; awn of lower floret distinctly longer than glumes.   2
2 Upper sheaths rough; ligule c. 1 mm., florets close together.
                                                                  2. flexuosa
  Upper sheaths smooth; ligule 4–5 mm., florets 1–1·5 mm. apart.
  Wet bogs; rare.   *D. setacea* (Hudson) Hackel

1. D. caespitosa (L.) Beauv.   Tufted Hair-grass.

50–200 cm., stout and densely tufted. Lvs flat, very rough above. Ligule 4–8 mm., obtuse. Panicle 15–50 cm., spreading, silvery or purplish. Lemma truncate and jagged at apex; awn about equalling lemma. Fl. 6–8. In meadows and woods, particularly on badly-drained clayey soils. Common and generally distributed. Mountain plants with viviparous spikelets and often with the lemma awned from near the apex have been distinguished as *D. alpina* (L.) Roemer & Schultes, but do not seem to be specifically distinguishable from *D. caespitosa*.

2. D. flexuosa (L.) Trin.   Wavy Hair-grass.

25–40 cm., rather slender, tufted. Lvs setaceous, rough on margins only. Ligule c. 1 mm., truncate. Panicle 5–10 cm., spreading, usually purplish. Lemma tapering to a slightly jagged apex; awn from near the base of the lemma and projecting beyond it. Fl. 6–7. On acid heaths, moors and in open woods, abundant and widely distributed.

33. AIRA L.

Slender, glabrous annuals with short narrow lvs. Panicle compound. Spikelets somewhat compressed; florets 2. Glumes as long as spikelet, 1-nerved. Lemma 3-nerved, apex with 2 bristle points; awn from below the middle of the lemma.

Sheaths smooth; panicle compact.   **1. praecox**
Sheaths rough; panicle spreading.   **2. caryophyllea**

1. A. praecox L.   Early Hair-grass.

5–12 cm. Lvs blunt, sheaths smooth. Panicle 0·5–2 cm., compact, oblong; branches little longer than spikelets. Spikelets c. 2 mm. Lemma lanceolate; awn arising ¼ distance from base to tip. Fl. 4–5. In open habitats, commonly on sand, throughout the British Is.

2. A. caryophyllea L.   Silvery Hair-grass.

10–30 cm. Lvs ± oblique at tip, sheaths rough. Panicle 2–8 cm., broadly ovate; branches much longer than spikelet. Spikelets c. 3 mm. Lemma ovate; awn arising ¼ distance from base to tip. Fl. 5. In open habitats, widely distributed but usually less common than *A. praecox*.

34. AMMOPHILA Host

Stout maritime perennials with convolute lvs. Panicle dense, ± cylindrical. Spikelets large, compressed; floret 1. Glumes longer than lemma, lower 1-, upper 3-nerved. Lemma 5–7-nerved, bifid at tip with a very short subterminal awn, hairy at base.

1. A. arenaria (L.) Link   Marram grass.

60–120 cm., extensively creeping, rooting at nodes and sand-binding. Lvs terete, rigid, pungent, under surface polished, upper glaucous. Ligule often more than 10 mm., acuminate. Panicle 7–12 cm., spike-like, blunt, whitish. Spikelets 12–14 mm. Glumes linear-lanceolate, acuminate. Lemma 3 times as long as the hairs at its base, half-lanceolate. Fl. 7–8. Abundant on dunes round the coasts of the British Is.

35. CALAMAGROSTIS Adanson

Perennials, often in damp places. Panicle compound. Spikelets subterete; floret 1. Glumes enclosing floret, tough, lower 1-, upper 3-nerved. Lemma thin, 3–5-nerved, bifid and awned from back or sinus, hairy at base.

1　Lvs hairy on upper surface. **2. canescens**
　　Lvs glabrous. 2
2　Lvs rough; hairs at base of floret twice as long as lemma.
　　　　　　　　　　　　　　　　　　　　　　1. epigejos
　　Lvs smooth; hairs at base of floret shorter or little longer than
　　　　lemma 3
3　Spikelets 3–4 mm.; glumes acute. Very local in bogs and marshes.
　　　　　　　　　　　　　　　　　C. stricta (Timm) Koeler
　　Spikelets 4·5–5 mm.; glumes acuminate. Very rare; bogs in
　　　　Caithness. *C. scotica* (Druce) Druce

1. C. epigejos (L.) Roth　　　　　　　　　　　　　　　Bushgrass.

60–200 cm., stout. Lvs flat, rough, tapering to a slender point. Ligule up to 12 mm., acute, torn. Stems rough just below panicle. Panicle 15–30 cm., ± spreading, purplish-brown. Spikelets 5–7 mm. Glumes 2–3 times as long as lemma, subulate. Hairs at base of floret about twice as long as lemma. Awn from near base of lemma. Fl. 7–8. In woods, rough grassland, etc., usually rather damp. Widely distributed in England, rarer in the north and west.

2. C. canescens (Weber) Roth　　　　　　　　　Purple Smallreed.

60–120 cm., rather slender. Lvs hairy above. Ligule 1–2 mm., obtuse, torn. Stems smooth. Panicle 5–15 cm., light brown or purplish. Spikelets 4–5 mm. Glumes similar to *C. epigejos*. Hairs at base of floret little longer than lemma. Awn very short, slender, arising from the sinus. Fl. 6–7. In damp shady places, local but fairly widespread in England, rare in Scotland.

36. AGROSTIS L.

Perennials. Spikelets small; panicle usually spreading, at least at fl.; floret 1. Glumes usually 1-nerved, enclosing the floret. Lemma ovate, truncate or obtuse, 3–5-nerved; awn dorsal or, more often, 0. Palea sometimes 0.

1　Ligules of stem-lvs acute; palea less than ¼ length of lemma. 2
　　Ligules of stem-lvs ± truncate; palea more than ½ length of lemma. 3

2 Basal lvs with 1 groove on upper surface; glumes rough.

1. setacea

Basal lvs with at least 4 grooves on upper surface; glumes smooth.

2. canina

3 Panicle narrow after fl.; rhizomes 0; stolons present.

5. stolonifera

Panicle spreading after fl.; rhizomes present; stolons 0.   4

4 Ligule of non-flowering shoots shorter than broad.   **3. tenuis**

Ligule of non-flowering shoots longer than broad.   **4. gigantea**

1. A. setacea Curtis   Bristle Agrostis.

10–60(–80) cm., tufted. Lvs setaceous, stiff and glaucous. Panicle 3–10 cm., narrow. Spikelets 3–4 mm. Glumes lanceolate, acute, rough. Lemma awned from near base. Palea minute. Fl. 6–7. Locally dominant on sandy or leached limestone soils in S. and S.W. England.

2. A. canina L.   Brown Bent-grass.

10–60(–80) cm., tufted and either stoloniferous or rhizomatous. Lvs flat, usually narrow. Panicle spreading at fl., rather narrow afterwards. Spikelets 1·5–4 mm. Glumes lanceolate, acute, smooth. Lemma awned from about the middle, or rarely awn 0. Palea minute. Fl. 6–7. In acid grassland; widely distributed.

3. A. tenuis Sibth.   Common Bent-grass.

(2–)20–50(–100) cm., ± tufted and rhizomatous. Lvs flat. Panicle 1–20 cm., spreading in fl. and fr. Spikelets 2–3·5 mm. Awn 0 or very short, rarely long, from near the top. Palea at least ½ as long as lemma. Fl. 6–8. Grassland, usually acid; widely distributed.

4. A. gigantea Roth   Common Bent-grass.

40–80(–120) cm., rhizomatous. Lvs flat, strongly furrowed above. Panicle up to 25 cm., spreading in fl. and fr. Spikelets 2–3 mm. Awn usually 0 or very short. Fl. 6–8. In grassy places and a common weed of arable land. Widely distributed in lowland districts.

5. A. stolonifera L.   Fiorin.

10–140 cm., stoloniferous. Lvs flat. Stems frequently decumbent and rooting at nodes. Panicle 1–30 cm., spreading in fl., contracted in fr. Spikelets 1·75–3 mm. Awn usually 0. Fl. 7–8. In grassy places, particularly damp ones. Common and generally distributed in lowland districts. Hybrids with *A. tenuis* occur frequently.

Apera Adanson

Two spp. occur in dry sandy fields, mainly in E. Anglia; they may be distinguished as follows:

Ligule acute, torn; panicle-branches spreading; spikelets usually purplish; anthers linear. Mainly as a weed in and near arable fields.

A. spica-venti (L.) Beauv.

Ligule truncate, toothed; panicle narrow; spikelets usually green; anthers ovate. Mainly on dry sandy banks, less frequently in arable fields.   *A. interrupta* (L.) Beauv.

37. Phleum L.

Annuals or perennials. Panicle cylindrical or ovoid, dense and spike-like. Spikelets strongly compressed; floret 1. Glumes 3-nerved, bristly on keel, often ending in short divergent points. Lemma thin. Palea equalling lemma.

1 Perennial; glumes truncate or obliquely truncate.   2
   Annual; glumes tapering gradually.   **3. arenarium**

2 Panicle ovoid or broadly cylindrical, usually dark purplish; awn c.
     3 mm. Mountains; rare.   *P. alpinum* L.
   Panicle cylindrical, rarely narrowly ovoid, but if so awn not exceed-
     ing 1 mm., usually green.   *3*

3 Lvs tapering gradually; glumes truncate; panicle-branches smooth.   *4*
   Lvs abruptly narrowed at tip; glumes obliquely truncate; panicle-
     branches ciliate. Dry grass heaths in S.E. England; very local.

P. phleoides (L.) Karsten

4 Plant up to c. 50 cm.; panicle (0·5)1–6(–8) cm. × 3–6 mm.; spikelets
     2–3 mm. (excluding awn).   **1. bertolonii**
   Plant usually exceeding 50 cm.; panicle 6–15(–30) cm. × 6–10 mm.;
     spikelets 3–4 mm. (excluding awn).   **2. pratense**

1. P. bertolonii DC. (*P. nodosum* auct.)   Cat's-tail.

Slender perennial 10–50 cm. Lvs 2–3 mm. wide. Panicle usually 1–6 cm. × 3–6 mm., cylindrical, blunt. Spikelets 2–3 mm. Glumes truncate with diverging awns 0·4–1 mm. Fl. 7. In pastures and short dry grass-land, locally common and widely distributed.

2. P. pratense L.   Timothy.

50–100 cm., stout. Like **1** but lvs 4–8 mm. wide; panicle usually 6–15 cm. × 6–10 mm.; spikelets 3–4 mm.; awns 1–2·5 mm. Fl. 7. In meadows and often sown for hay grass, common throughout most of the British Is.

3. P. arenarium L. Sand Cat's-tail.

Annual 3–15(–30) cm. Lvs short, smooth. Upper sheath inflated.
Panicle 0·5–3 cm., narrowly ovoid to cylindrical. Spikelets c. 3 mm.
Glumes tapering gradually and diverging in the upper part, shortly
awned. Lemma hairy. Fl. 5–6. On dunes and in sandy fields near the
sea; locally common round the coasts.

38. ALOPECURUS L.

Annual or perennial. Infl. cylindrical or ovoid, dense. Spikelets strongly
compressed, falling entire in fr.; floret 1. Glumes about as long as
floret, tough and often connate below the middle. Lemma usually
awned from the back. Palea 0.

1 Awn scarcely protruding beyond lemma or 0. 2
  Awn at least twice as long as lemma. 3

2 Panicle narrowly cylindrical, not conspicuously hairy. **4. aequalis**
  Panicle ovoid or broadly cylindrical, conspicuously hairy. Scottish
    mountains and Teesdale; very rare. *A. alpinus* Sm.

3 Annual; glumes connate for ⅓–½ their length. **1. myosuroides**
  Perennial; glumes free nearly to base. 4

4 Stem swollen and bulbous at base. Salt-marshes; local.
                                A. bulbosus Gouan
  Stem not bulbous at base. 5

5 Upright; not rooting at nodes; glumes and lemma acute.
                                2. pratensis
  Decumbent; rooting at lower nodes; glumes obtuse; lemma
    truncate. **3. geniculatus**

1. A. myosuroides Hudson Black Twitch.

Annual 20–70 cm. Panicle 4–12 cm., narrow, pointed. Spikelets 4–
7 mm. Glumes connate for ⅓–½ their length, shortly ciliate on nerves
and keel, oblong to lanceolate, acute. Awn from near base of lemma
and twice its length. Fl. 6–7. A weed in arable fields and waste places,
abundant in S.E. England, less frequent and often casual elsewhere.

2. A. pratensis L. Meadow Foxtail.

Perennial 30–90 cm. Panicle 3–6(–11) cm., silky, tapering somewhat
but obtuse. Spikelets 4–5 mm. Glumes free nearly to base, long-ciliate
on keel, shortly ciliate or nearly glabrous on nerves, lanceolate, acute or
acuminate. Awn from near base of lemma and twice its length. Fl. 4–6.
In meadows, pastures, etc., common and generally distributed.

3. A. geniculatus L.　　　　　　　　　　　　　　　　Marsh Foxtail.

Perennial 15–40 cm., creeping and rooting at the nodes then geniculate and ascending. Panicle 2–4 cm., blunt. Spikelets 2–3 mm. Glumes free nearly to base, with hairs on nerves, lanceolate, obtuse; tip of glumes and lemmas slaty-grey. Awn from near base of lemma and twice its length. Anthers ultimately yellow with violet tinge. Fl. 6–7. In wet meadows, at edges of ponds, etc., frequent and generally distributed.

4. A. aequalis Sobol.　　　　　　　　　　　　　　　Orange Foxtail.

Usual biennial or annual 10–25 cm., decumbent at base, often glaucous. Panicle 1·5–3 cm., narrow, obtuse. Spikelets 1–1·5 mm. Glumes free nearly to base, shortly silky and long-ciliate on keel, lanceolate, obtuse; grey-green sometimes with very narrow slaty band at tip. Awn from near middle of lemma and scarcely longer than it. Anthers at first white, later orange. Fl. 5–7. In damp places, less common than *A. geniculatus*.

39. MILIUM L.

Annual or perennial, glabrous. Spikelets subterete; floret 1. Glumes 3-nerved, longer than floret. Lemma 5-nerved, awnless, becoming extremely hard and shining in fr., margins folded round palea.

Perennial 50–120 cm.; sheaths smooth; panicle spreading.　**1. effusum**
Annual 2·5–15 cm.; sheaths rough; panicle narrow. Guernsey; rare.
　　　　　　　　　　　　　　　　　　M. scabrum L. C. M. Richard

1. M. effusum L.　　　　　　　　　　　　　　　　Wood Millet.

50–120 cm., tufted. Lvs flat, 5–10 mm. wide; sheaths smooth. Panicle 10–25 cm., with long spreading or deflexed branches up to 6 together. Spikelets 2–3 mm. Glumes ovate, acute, green. Fl. 6. In damp shady woods, rather local but widely scattered on all but the poorest soils.

40. ANTHOXANTHUM L.

Annual or perennial, smelling of coumarin (new-mown hay) when crushed. Panicle compact, ovoid or oblong. Spikelets lanceolate, acute, compressed; florets 3, 2 lower sterile and small, upper hermaphrodite. Glumes thin, very unequal. Sterile lemmas awned from back, fertile awnless.

Perennial; glumes hairy; awns usually nearly enclosed within spikelet.
　　　　　　　　　　　　　　　　　　　　　　　1. odoratum

Annual; glumes glabrous; awns always protruding well beyond spikelet. Mainly S. and E. England; very local.

A. puelii Lecoq & Lamotte

1. A. odoratum L. Sweet Vernal-grass.

20–50 cm., tufted. Lvs flat, slightly hairy, acuminate. Panicle (2–)4–
6(–7) cm., oblong, sometimes lobed at base. Spikelets 7–9 mm. Glumes
hairy, upper longer than floret and enfolding it, lower ½ its length. Fl.
4–6. Common and widely distributed in grassland and on heaths and
moors.

41. PHALARIS L.

Annual or perennial, glabrous. Panicle dense, lobed and with branches
spreading at fl., or ovoid or cylindrical. Spikelets strongly compressed;
florets 2 or 3, 1–2 lower sterile and terminal hermaphrodite, sometimes
all sterile or male. Glumes ± winged from the keel, subequal and longer
than florets. Lemma tough, 5-nerved, awnless.

1 Plant tall and reed-like, growing in damp places or in water; panicle
   distinctly lobed; branches spreading at fl. **1. arundinacea**
   Plant smaller, not reed-like, growing in dry places, usually on waste
   ground; panicle ovoid or cylindrical. 2

2 Uppermost sheath close below or enfolding base of panicle at fl.; at
   least some glumes long-acuminate. Rare casual. **P. paradoxa* L.
   Uppermost sheath well below base of panicle at fl.; glumes acute. 3

3 Spikelets nearly sessile; sterile lemmas equal. **2. canariensis**
   Pedicels at least ¼ as long as spikelet; sterile lemmas very unequal.
   Rare. Perhaps native in Channel Is, casual elsewhere.
   P. minor Retz.

1. P. arundinacea L. Reed-grass.

60–120 cm., stout, with long rhizomes. Lvs c. 10 mm. wide, acuminate.
Ligule up to 10 mm. Panicle 10–15 cm., lobed, purplish, branches
spreading at fl. Spikelets c. 5 mm. Glumes lanceolate, acuminate.
Lemma broadly lanceolate, acute. Fl. 6–7. The dead lvs persist through-
out the winter (compare *Phragmites*). In wet places throughout the
British Is.

*2. P. canariensis L. Canary Grass.

20–60 cm., annual, ± tufted. Lvs up to 12 mm. wide, acuminate. Ligule
up to c. 5 mm. Panicle 1·5–4 cm., dense, ovoid. Spikelets 5–8 mm.
Glumes half-ovate, strongly winged in upper half, straw-coloured with a
green band on keel. Lemma ovate, acute. Fl. 6–8. A casual on rubbish
tips and disturbed ground, widely distributed in the south.

42. PARAPHOLIS C. E. Hubbard

Slender glabrous maritime annuals. Spikelets placed broadside on to the rhachis and sunk in its concavities; floret 1. Glumes equal, placed side by side in front of the spikelet, tough, green. Lemma hyaline, 1-nerved.

Plant usually more than 15 cm.; uppermost sheath not inflated;
 anthers 2 mm. or more.   **1. strigosa**
Plant seldom exceeding 10 cm.; uppermost sheath ±inflated; anthers
 1 mm. or less. Rare; S. England. *P. incurva* (L.) C. E. Hubbard

1. P. strigosa (Dumort.) C. E. Hubbard   Sea Hard-grass.

15–40 cm., usually prostrate at base then ascending, freely branched. Uppermost sheaths not inflated, well below base of spike at fl. Spikelets 4–6 mm., rounded on back, flush with the rhachis except in fl. Glumes linear-lanceolate, ±asymmetrical, acuminate. Anthers 2–4 mm. Fl. 6–8. In salt-marsh turf and clayey places near the sea, rather local round the coasts.

43. NARDUS L.

Perennial, densely tufted. Infl. a unilateral spike. Spikelets sessile in the notches of the rhachis. Lower glumes very small, upper usually 0. Lemma tough, 3-nerved with a short terminal awn.

1. N. stricta L.   Mat-grass.

10–30 cm. Roots yellow, stout. Lvs setaceous, very hard, erect when young, later spreading at right angles to the sheaths, whitish and persistent when dead. Sheaths erect, persistent, whitish. Spike 2–10 cm. Spikelets 5–8 mm., narrow, acute. Lemma narrow, awl-shaped. Fl. 6–8. Abundant and widespread on poor sandy and peaty soils, chiefly on moors and mountains.

44. SPARTINA Schreber

Stout perennials with thick tough lvs. Ligule hairy. Infl. of a number of erect spikes arranged in a raceme. Spikelets large, distichous and alternate, pressed close to the flattened rhachis; floret 1. Glumes unequal, 3-nerved, upper longer than lemma. Lemma similar to upper glume.

1 Spikelets 14 mm. or less; anthers 5 mm.; plant usually purplish.   *2*
 Spikelets 15 mm. or more; anthers 10 mm. or 0; plant yellow-green.
                  1. × townsendii

2 Lvs shorter than infl.; spikes 2–3. Very local; in the south, mainly
 on the E. coast.   *S. maritima* (Curtis) Fernald
 Lvs longer than infl.; spikes 6–8. Introduced; 1 locality in South-
 ampton Water.   *S. alterniflora* Loisel.

1. S. × townsendii H. & J. Groves   Cord-grass.

50–130 cm., yellow-green. Lvs up to c. 8 mm. wide, very stiff, tapering
from base to a long slender point. Infl. 10–25 cm., of (2–)4–5 spikes.
Spikelets 15–20 mm.; rhachis extending beyond spikelets as a flexuous
bristle. Male-sterile and reproducing vegetatively. Fl. 6–8. Abundant
on tidal mud-flats, particularly on the south and east coasts.

 In addition to the widespread male-sterile hybrid, plants derived from
it, producing fertile pollen and reproducing by seed, also occur in many
parts of the country. These may be readily distinguished by their long
anthers.

DIGITARIA P. C. Fabr.

Two spp. occur rarely:

Lvs and sheaths all glabrous; upper glume as long as lemmas. Sandy
 fields in S. England; rare.   **D. ischaemum* (Schreber) Muhl.
At least the lower lvs and sheaths hairy; upper glume not more than
 half as long as lemmas. Rare casual near ports and in arable fields.
             *D. sanguinalis (L.) Scop.

SETARIA Beauv.

Three rare casuals, found chiefly near ports:

1 Fertile lemma conspicuously rugose.
        *S. lutescens (Weigel) C. E. Hubbard
 Fertile lemma almost smooth.   2

2 Teeth of bristles forward-pointing.   **S. viridis* (L.) Beauv.
 Teeth of bristles backward-pointing.   **S. verticillata* (L.) Beauv.

2 Lvs shorter than infl.; spikes 2–3. Very local; in the south, mainly
  on the E. coast.   S. maritima (Curtis) Fernald
  Lvs longer than infl.; spikes 6–8. Introduced. 1 locality in South-
  ampton Water.   S. alterniflora Loisel.

1. S. × townsendii H. & J. Groves.   Cord-grass.

50–130 cm., yellow-green. Lvs up to c. 8 mm. wide, very stiff, tapering
from base to a long slender point. Infl. 10–25 cm., of (2–)4–5 spikes.
Spikelets 15–20 mm.; rhachis extending beyond spikelets as a flexuous
bristle. Male-sterile and reproducing vegetatively. Fl. 6–8. Abundant
on tidal mud-flats, particularly on the south and east coasts.
  In addition to the widespread male-sterile hybrid, plants derived from
it, producing fertile pollen and reproducing by seed, also occur in many
parts of the country. These may be readily distinguished by their long
anthers.

DIGITARIA P.C. Fabr.

Two spp. occur rarely:

  Lvs and sheaths all glabrous; upper glume as long as lemmas. Sandy
    fields in S. England; rare.   *D. ischaemum (Schreber) Muhl.
  At least the lower lvs and sheaths hairy; upper glume not more than
    half as long as lemmas. Rare casual near ports and in arable fields.
                                             *D. sanguinalis (L.) Scop.

SETARIA Beauv.

Three rare casuals, found chiefly near ports:

1 Fertile lemma conspicuously rugose.
                                      *S. lutescens (Weigel) C. E. Hubbard
  Fertile lemma almost smooth.   2

2 Teeth of bristles forward-pointing.   *S. viridis (L.) Beauv.
  Teeth of bristles backward-pointing.   *S. verticillata (L.) Beauv.

GLOSSARY

With 3 figures (pp. 556–8) of explanatory illustrations, and 2 figures
(pp. 572–3) showing the floral structure of orchids and grasses

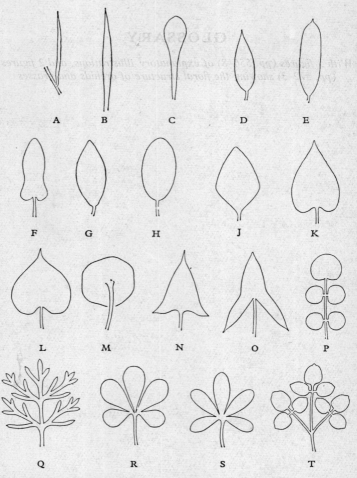

Fig. 1. Explanation of glossary—leaf shapes, etc. A, linear; B, ensiform; C, spathulate; D. lanceolate, apex acute; E, oblong, apex mucronate; F, panduriform; G. elliptic; H, oval; J, rhombic, base cuneate; K, ovate, base cordate; L, suborbicular, apex cuspidate; M, peltate; N, triangular, hastate, apex acuminate; O, sagittate; P, simply pinnate, segments orbicular; Q, bipinnatifid; R, pedate, leaflets obovate; S, palmate; T, biternate.

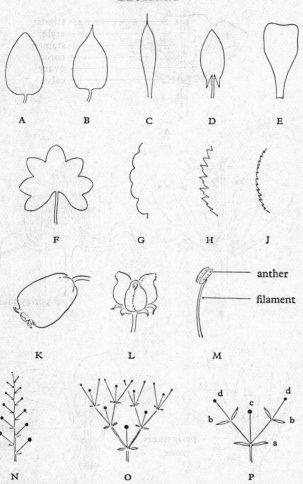

anther

filament

Fig. 2. Explanation of glossary. A, leaf with rounded base and obtuse apex; B, leaf with truncate base and cuspidate apex; C, aristate apex; D, sagittate leaf with subacute apex; E, leaf with retuse apex; F, lobed margin; G, crenate margin; H, dentate margin; J, serrate margin; K, urceolate corolla; L, carpels; M, stamen; N, racemose inflorescence; O, cymose inflorescence; P, bracts and bracteoles: *a*, bracts of flower *c*; *b*, bracteoles of flower *d*.

20-2

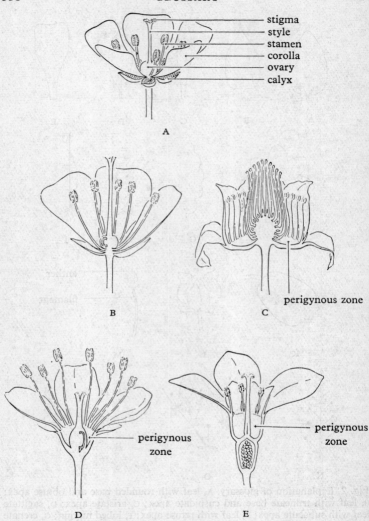

Fig. 3. Explanation of glossary. A, parts of the flower; B, hypogynous flower (ovary superior); C, D, perigynous flowers (ovary superior); E, epigynous flower (ovary inferior).

GLOSSARY

accrescent Becoming larger after flowering (usually applied to the calyx).

achene A small dry indehiscent single-seeded fr.

actinomorphic Radially symmetrical; having more than one plane of symmetry.

acuminate Fig. 1 N. **acumen**, the point of such a lf.

acute Fig. 1 D.

adnate Joined to another organ of a different kind.

alien Believed on good evidence to have been introduced by man and now ± naturalized.

allopolyploid A polyploid derived by hybridization between two different spp. followed by doubling of the chromosome number.

alternate Strictly, arranged in 2 rows but not opposite; commonly used (as often here) also to include spiral arrangement.

amphimixis Reproducing by seed produced by a normal sexual fusion; adj. *amphimictic*.

amplexicaul Clasping the stem.

anastomosing Joining up to form loops.

andromonoecious Having male and hermaphrodite fls on the same plant.

angustiseptate Of a fr. with the septum across the narrowest diameter.

annual Completing its life-cycle within 12 months from germination (cf. **biennial**).

annular Ring-shaped.

annulus Special thick-walled cells forming part of the opening mechanism of a fern sporangium, often forming a ring.

antesepalous Inserted opposite the insertion of the sepals.

anther The part of the stamen containing the pollen grains (Fig. 2 M).

antheridium The structure containing the male sexual cells.

apiculate With a small broad point at the apex.

apocarpous (ovary) Having the carpels free from one another.

apomixis Reproducing by seed not formed from a sexual fusion; adj. *apomictic*.

appressed Pressed close to another organ but not united with it.

archegonium The structure containing the female sexual cell in many land plants.

arcuate Curved so as to form about ¼ of a circle or more.

aristate Awned (Fig. 2 C).

ascending Sloping or curving upwards.

asperous Rough to the touch.

attenuate Gradually tapering.

auricles Small ear-like projections at base of lf (especially in grasses).

autotrophic Neither parasitic nor saprophytic.

awn A stiff bristle-like projection from the tip or back of the lemma in grasses, or from a fr. (usually the indurated style, e.g. *Erodium*), or, less frequently, the tip of a lf (Fig. 2c).

axile See **placentation**.

axillary Arising in the axil of a lf or bract.

basal (of lvs) Arising from the base of the stem or from a rhizome.

base-rich Soils containing a relatively large amount of free basic ions, e.g. calcium, magnesium, etc.

basifixed (of anthers) Joined by the base to the filament and not capable of independent movement.

berry A fleshy fr., usually several-seeded, without a stony layer surrounding the seeds.

biennial Completing its life-cycle within two years (but not within one year), not flowering in the first.

bifid Split deeply in two.

bog ('moss') A community on wet very acid peat.

bract Fig. 2p.

bracteole Fig. 2p.

bulb An underground organ consisting of a short stem bearing a number of swollen fleshy lf-bases or scale lvs, with or without a tunic, the whole enclosing the next year's bud.

bulbil A small bulb or tuber arising in the axil of a lf or in an infl. on the aerial part of a plant.

caducous Falling off at an early stage.

calcicole More frequently found upon or confined to soils containing free calcium carbonate.

calcifuge Not normally found on soils containing free calcium carbonate.

calyx The sepals as a whole (Fig. 3A).

campanulate Bell-shaped.

capillary Hair-like.

capitate Head-like.

capsule A dry dehiscent fr. composed of more than one carpel.

carpel One of the units of which the gynoecium is composed. In a septate ovary the number of divisions usually corresponds to the number of carpels (Fig. 2L).

cartilaginous Resembling cartilage in consistency.

caryopsis A fr. (achene) with ovary wall and seed-coat united (Gramineae).

casual An introduced plant which has not become established though it occurs in places where it is not cultivated.

cauline (of lvs) Borne on the aerial part of the stem, especially the upper part, but not subtending a fl. or infl.

chartaceous Of papery texture.

chlorophyll The green colouring matter of lvs, etc.

chromosomes Small deeply staining bodies found in all nuclei, which determine most or all of the inheritable characters of organisms. Two similar sets of these are normally present in all vegetative cells, the number (*diploid number*, 2*n*) usually being constant for a given sp. The sexual reproductive cells normally contain half this number (*haploid number*, *n*). Closely related spp. often have the same number, or a multiple of a common *basic number*.

ciliate   With regularly arranged hairs projecting from the margin.

cilium, cilia   A small whip-like structure by means of which some sexual reproductive cells swim.

circumscissile   Dehiscing transversely, the top of the capsule coming off as a lid.

cladode   A green ± lf-like lateral shoot.

clavate   Club-shaped.

cleistogamous   Fls which never open and are self-pollinated (opposite of chasmogamous).

commissure (in Umbelliferae)   The faces by which the two carpels are joined together.

compound   Of an infl., with the axis branched; of a lf, made up of several distinct lflets.

compressed   Flattened.

concolorous   Of approximately the same colour throughout.

connate   Organs of the same kind growing together and becoming joined, though distinct in origin.

connivent   Of two or more organs with their bases wide apart but their apices approaching one another.

contiguous   Touching each other at the edges.

contorted (of perianth lobes in bud)   With each lobe overlapping the next with the same edge and appearing twisted.

converging, convergent   Of two or more organs with their apices closer together than their bases.

convolute   Rolled together, coiled.

cordate   Fig. 1 K.

coriaceous   Of a leathery texture.

corm   A short, usually erect and tunicated, swollen underground stem of one year's duration, that of the next year arising at the top of the old one and close to it; cf. **tuber**.

corolla   The petals as a whole (Fig. 3 A).

corymb   A raceme with the pedicels becoming shorter towards the top, so that all the fls are at approximately the same level; adj. *corymbose*.

corymbose cyme   A flat-topped cyme, thus resembling a corymb in appearance though not in development.

cotyledon   The first lf or lvs of a plant, already present in the seed and usually differing in shape from the later lvs. Cotyledons may remain within the testa or may be raised above the ground and become green during germination.

crenate   Fig. 2 G; dimin. *crenulate*.

crisped   Curled.

cultivated ground (or **land**)   Includes arable land, gardens, allotments, etc.

cuneate   Fig. 1 J.

cuneiform Wedge-shaped with the thin end at the base.

cuspidate Fig. 1L, 2B.

cymose Of an infl., usually obconical in outline, whose growing points are each in turn terminated by a fl., so that the continued growth of the infl. depends on the production of new lateral growing points. A consequence of this mode of growth is that the oldest branches or fls are normally at the apex (Fig. 2o).

deciduous Losing its lvs in autumn; dropping off.

decumbent (of stems) Lying on the ground and tending to rise at the end.

decurrent Having the base prolonged down the axis, as in lvs where the blade is continued downwards as a wing on petiole or stem.

decussate (of lvs) Opposite, but successive pairs orientated at right angles to each other.

deflexed Bent sharply downwards.

dehiscent Opening to shed its seeds or spores.

deltate Shaped like the Greek letter Δ.

dentate Fig. 2H; dimin. *denticulate*.

dichasium A cyme in which the branches are opposite and approximately equal; adj. *dichasial*.

didymous Formed of two similar parts attached to each other by a small portion of their surface.

digitate See **palmate**.

dioecious Having the sexes on different plants.

diploid Having 2 basic sets of chromosomes.

disk The fleshy, sometimes nectar-secreting, portion of the receptacle, surrounding or surmounting the ovary.

distichous Arranged in two diametrically opposite rows.

divaricate Diverging at a wide angle.

diverging, divergent Of two or more organs with their apices wider apart than their bases.

dominant The chief constituents of a particular plant community, e.g. oaks in an oakwood or heather on a moor.

dorsifixed (of anthers) Attached by the back.

drupe A ± fleshy fr. with one or more seeds each surrounded by a stony layer, e.g. sloe, ivy.

ectotrophic See **mycorrhiza**.

eglandular Without glands.

ellipsoid Of a solid object elliptic in longitudinal section.

elliptic Fig. 1G.

emarginate Shallowly notched at the apex.

endemic Native only in one country or other small area. If used without qualification in this book means confined to the British Is.

endosperm The nutritive tissue in the seed of a flowering plant, formed after fertilization.

endotrophic See **mycorrhiza**.

ensiform Fig. 1B.

entire Not toothed or cut.

epicalyx A calyx-like structure outside but close to the true calyx.

epigeal Above ground; epigeal germination—when cotyledons are raised above the ground.

epigynous Of the other parts of a fl. with an inferior ovary; or, more loosely, of the whole fl.

epipetalous Inserted upon the corolla.

erose Appearing as if gnawed.

escape A plant growing outside a garden, but not well naturalized, derived from cultivated specimens either by vegetative spread or by seed.

exserted Protruding.

exstipulate Without stipules.

extrorse (of anthers) Opening towards the outside of the fl.

falcate Sickle-shaped.

fen A community on alkaline, neutral, or slightly acid wet peat.

fertile Producing seed capable of germination; or (of anthers) containing viable pollen.

filament The stalk of the anther, the two together forming the stamen (Fig. 2M).

filiform Thread-like.

fimbriate With the margin divided into a fringe.

flexuous Wavy (of a stem or other axis).

floccose Covered with a soft down of stellate hairs.

flush Wet ground, often on hillsides, where water flows but not in a definite channel.

follicle A dry dehiscent fr. formed of one carpel, dehiscing along one side.

fruit (fr.) The ripe seeds and structure surrounding them, whether fleshy or dry; strictly the ovary and seeds, but often used to include other associated parts such as the fleshy receptacle, as in rose and strawberry.

fugacious Withering or falling off very rapidly.

funicle The stalk of the ovule.

fusiform Spindle-shaped.

gamopetalous Having the petals joined into a tube, at least at the base.

geniculate Bent abruptly to make a 'knee'.

gibbous Of a solid object any part of which projects as a rounded swelling.

glabrescent Becoming glabrous.

glabrous Without hairs.

gland A small globular or oblong vesicle containing oil, resin or other liquid, sunk in, on the surface of, or protruding from any part of a plant. When furnished with a slender stalk usually known as a glandular hair.

glandular Furnished with glands.

glaucous Bluish.

glumaceous Resembling a glume.

glume The membranous or chaffy scales enclosing the spikelets of a grass or fl. of a sedge.

gynobasic A style which, because of the infolding of the ovary wall, appears to be inserted at the base of the ovary.

gynodioecious Having female and hermaphrodite fls on separate plants.

gynoecium The female part of the fl., made up of one or more ovaries with their styles and stigmas.

gynomonoecious Having female and hermaphrodite fls on the same plant.

hastate Fig. 1 N.

heath A lowland community dominated by heaths or ling, usually on sandy soils with a shallow layer of peat.

hemiparasite See under **parasite**.

herb Any vascular plant which is not woody.

herbaceous (of a plant organ) Soft, green, having the texture of lvs.

hermaphrodite Containing both stamens and ovary.

heterosporous Having spores of two distinct sizes.

heterostylous The length of the style in relation to the other parts of the fl. differing in the fls of different plants.

hexaploid Having 6 basic sets of chromosomes.

hilum The scar left on the seed by the stalk of the ovule.

hirsute Clothed with long, not very stiff, hairs.

hispid Coarsely and stiffly hairy, as many Boraginaceae.

homogamous The anthers and stigmas maturing simultaneously. See also p. 377.

homosporous Having all spores of approximately the same size.

hyaline Thin and translucent.

hybrid A plant originating by the fertilization of one sp. or ssp. by another.

hybrid swarm A series of plants originating by hybridization between two (or more) spp. and subsequently recrossing with the parents and between themselves, so that a continuous series of forms arises.

hypogeal Below ground; hypogeal germination—when the cotyledons remain below the ground.

hypogynous Of fls in which the stamens are inserted close to and beside or beneath the base of the ovary (Fig. 3 B).

imbricate, imbricating Of organs with their edges overlapping, when in bud, like the tiles on a roof.

impressed Sunk below the surface.

incurved Bent gradually inwards.

indehiscent Not opening to release its seeds or spores.

indumentum The hairy covering as a whole.

indurated Hardened and toughened.

indusium A piece of tissue ± covering or enclosing a sporangium or group of sporangia.

inferior (ovary) With perianth inserted round the top, the ovary being apparently sunk in and fused with the receptacle (Fig. 3 E).

inflexed Bent inwards.

inflorescence (infl.) Flowering branch, or portion of the stem above the last stem-lvs, including its branches, bracts and fls.

internode Part of the stem between two adjacent nodes.

interpetiolar Between the petioles.

interrupted Not continuous.

intrapetiolar Between the petiole and stem.

intravaginal Within the sheath.

introduced Not native; known to have been, or strongly suspected of having been brought into the British Is accidentally or intentionally by man within historic times.

introrse (of anthers) Opening towards the middle of the fl.

involucel See p. 374.

involucral Forming an involucre.

involucre Bracts forming a ±calyx-like structure round or just below the base of a usually condensed infl. (e.g. *Anthyllis*; Compositae); adj. *involucrate*.

involute With the margins rolled upwards.

isomerous The number of parts in two or more different floral whorls being the same, e.g. 5 petals, 5 stamens, and 5 carpels.

keel A sharp edge resembling the keel of a boat; the lower petal or petals when shaped like the keel of a boat (*Fumaria*, Papilionaceae).

lacerate Deeply and irregularly divided and appearing as if torn.

laciniate Deeply and irregularly divided into narrow segments.

lamina The blade of a lf or petal; a thin flat piece of tissue.

lanceolate Fig. 1 D.

latex Milky juice.

latiseptate Of a fr. with the septum across the widest diameter.

lax Loose; not dense.

lemma See p. 517 and Fig. 5 (p. 573).

lenticular Convex on both faces and ±circular in outline.

ligulate Strap-shaped.

ligule A small flap of tissue or a scale borne on the surface of a lf or per. seg. near its base; see also p. 377 (Compositae).

limb The flattened expanded part of a calyx or corolla the base of which is tubular.

linear Fig. 1 A.

lingulate Tongue-shaped.

lip A group of per. segs ±united and sharply divided from the remaining per. segs.

lobed (of lvs) Divided, but not into separate lflets (Fig. 2 F).

loculicidal Splitting down the middle of each cell of the ovary.

lyrate Shaped ±like a lyre.

marsh A community on wet or periodically wet but not peaty soils.

meadow A grassy field cut for hay.

measurements See Abbreviations (p. xxxvii).

megaspore A large spore giving rise to a prothallus bearing archegonia.

membranous Thin, dry and flexible, not green.

mericarp A 1-seeded portion split off from a syncarpous ovary at maturity.

-merous E.g. in 5-merous (=pentamerous): having the parts in fives.

microspore A small spore giving rise to a prothallus bearing antheridia.

monachasium A cyme in which the branches are spirally arranged or alternate or one is more strongly developed than the other; adj. *monochasial*.

monadelphous (of stamens) United into a single bundle by the fusion of the filaments.

monoecious Having unisexual fls, but both sexes on the same plant.

monopodial Of a stem in which growth is continued from year to year by the same apical growing point, cf. racemose.

moor Upland communities, often dominated by heather, on dry or damp but not wet peat.

mucronate Provided with a short narrow point (*mucro*), Fig. 1 E; dimin. *mucronulate*.

mull soil A fertile woodland soil with no raw humus layer.

muricate Rough with short firm projections.

mycorrhiza An association of roots with a fungus which may form a layer outside the root (ectotrophic) or within the outer tissues (endotrophic).

naked Devoid of hair or scales.

native Not known to have been introduced by human agency.

nerve A strand of strengthening and conducting tissue running through a lf or modified lf.

nodal glands Translucent glands, visible with a good hand-lens, one each side of the leaf-base in certain *Potamogeton* spp. Very conspicuous in *P. berchtoldii*, absent in *P. pusillus*.

node A point on the stem where one or more lvs arise.

nodule A small ± spherical swelling; adj. *nodular*.

nucellus The nutritive tissue in an ovule.

ob- (in combinations, e.g. obovate) Inverted; an obovate lf is broadest above the middle, an ovate one below the middle (Fig. 1 R).

obdiplostemonous The stamens in 2 whorls, the outer opposite the petals, the inner opposite the sepals.

oblong Fig. 1 E.

obtuse Blunt (Fig. 2 A).

opposite Of two organs arising at the same level on opposite sides of the stem.

orbicular Rounded, with length and breadth about the same (Fig. 1 P).

oval Fig. 1 H.

ovary That part of the gynoecium enclosing the ovules and consisting of one or more carpels (Fig. 3 A).

ovate Fig. 1 K.

ovoid Of a solid object which is ovate in longitudinal section; egg-shaped.

ovule A structure containing the egg and developing into the seed after fertilization.

palate See p. 317.

palea See pp. 517, 573.

palmate (of a lf) Consisting of more than 3 lflets arising from the same point (Fig. 1 s).

panduriform Fig. 1 F.

panicle Strictly a branched racemose infl., though often applied to any branched infl.

papillae Small elongated projections; adj. *papillose*.

parasite A plant which derives its food wholly or partially (hemiparasite) from other living plants to which it is attached.

partial infl. Any distinct portion of a branched infl.

pasture Grassy field grazed during summer.

pectinate Lobed, with the lobes resembling and arranged like the teeth of a comb.

pedate Fig. 1 R.

pedicel The stalk of a single fl.

peduncle The stalk of an infl. or partial infl.

peltate Of a flat organ with its stalk inserted on the under surface, not at the edge (Fig. 1 M).

pentaploid Having 5 basic sets of chromosomes.

perennating Surviving the winter after flowering.

perennial Living for more than 2 years and usually flowering each year.

perianth The floral lvs as a whole, including sepals and petals if both are present.

perianth segment (per. seg.) The separate lvs of which the perianth is made up, especially when petals and sepals are alike.

perigynous Of fls in which there is an annular region, flat or concave, between the insertion of the gynoecium and of the other floral parts (Fig. 3 C, D).

perigynous zone The annular region between the insertion of the gynoecium and of the other floral parts in perigynous or epigynous fls (Fig. 3 C–E).

perisperm The nutritive tissue in some seeds derived from the nucellus.

perispore A membrane surrounding a spore.

petal A member of the inner series of per. segs, if differing from the outer series, and especially if brightly coloured.

petaloid Brightly coloured and resembling petals.

petiole The stalk of a lf.

pilose Hairy with rather long soft hairs.

pinnate A lf composed of more than 3 lflets arranged in two rows along a common stalk or rhachis (Fig. 1 P); bipinnate (2-pinnate), a lf in which the primary divisions are themselves pinnate. Similarly, 3-pinnate, etc.

pinnatifid Pinnately cut, but not into separate portions, the lobes connected by lamina as well as midrib or stalk (Fig. 1 Q).

pinnatisect Like pinnatifid but with some of the lower divisions reaching very nearly or quite to the midrib.

placenta The part of the ovary to which the ovules are attached.

placentation The position of the placentae in the ovary. The chief types of placentation are: *apical*, at the apex of the ovary; *axile*, in the angles formed by the meeting of the septa in the middle of the ovary; *basal*, at the base of the ovary; *free-central*, on a column or projection arising from the base in the middle of the ovary, not connected with the wall by septa; *parietal*, on the wall of the ovary or an intrusion from it; *superficial*, when the ovules are scattered uniformly all over the inner surface of the wall of the ovary.

pollen The microspores of a flowering plant or Conifer.

pollinia Regularly shaped masses of pollen formed by a large number of pollen grains cohering.

polygamous Having male, female and hermaphrodite fls on the same or different plants.

polyploid A sp. having a chromosome number which is a multiple, greater than two, of the basic number of its group (see under **chromosomes**).

pome A fr. in which the seeds are surrounded by a tough but not woody or stony layer, derived from the inner part of the fr. wall, and the whole fused with the deeply cup-shaped fleshy receptacle (e.g. apple).

porrect Directed outwards and forwards.

premorse Ending abruptly and appearing as if bitten off at the lower end.

prickle A sharp relatively stout outgrowth from the outer layers. Prickles (unlike thorns) are usually irregularly arranged.

procumbent Lying loosely along the surface of the ground.

prostrate Lying rather closely along the surface of the ground.

protandrous Stamens maturing before the ovary.

prothallus A small plant formed by the germination of a spore and bearing antheridia or archegonia or both.

pruinose Having a whitish 'bloom'; appearing as if covered with hoar-frost.

pubescent Shortly and softly hairy; dimin. *puberulent*, *puberulous*.

punctate Dotted or shallowly pitted, often with glands.

punctiform Small and ± circular, resembling a dot.

pungent Sharply and stiffly pointed so as to prick.

raceme An unbranched racemose infl. in which the fls are borne on pedicels.

racemose Of an infl., usually conical in outline, whose growing points commonly continue to add to the infl. and in which there is usually no terminal fl. A consequence of this mode of growth is that the youngest and smallest branches or fls are normally nearest the apex (Fig. 2N).

raphe The united portions of the funicle and outer integument in an anatropous ovule.

ray The stalk of a partial umbel.

ray-floret See p. 377.

receptacle That flat, concave, or convex upper part of the stem from which the parts of the fl. arise; often to include the perigynous zone.

recurved Bent backwards in a curve.

regular Actinomorphic (q.v.).

reniform Kidney shaped.

replum See p. 50.

resilient Springing sharply back when bent out of position.

reticulate Marked with a network, usually of veins.

retuse   Obtuse or truncate and slightly indented (Fig. 2E).

revolute   Rolled downwards.

rhachilla   See pp. 517, 573.

rhachis   The axis of a pinnate lf or an infl.

rhizome   An underground stem lasting more than one growing season; adj. *rhizomatous*.

rhombic   Having ± the shape of a diamond in a pack of playing cards (Fig. 1J).

rotate (of a corolla)   With the petals or lobes spreading out at right angles to the axis, like a wheel.

rounded (of lf-base)   Fig. 2A.

rugose   Wrinkled; dimin. *rugulose*.

ruminate   Looking as though chewed.

runcinate   Pinnately lobed with the lobes directed backwards, towards the base of the lf.

runner   A special form of stolon consisting of an aerial branch rooting at the end and forming a new plant which eventually becomes detached from the parent.

saccate   Pouched.

sagittate   Fig. 1O, 2D.

salt-marsh   The series of communities growing on inter-tidal mud or sandy mud in sheltered places on coasts and in estuaries.

samara   A dry indehiscent fr. part of the wall of which forms a flattened wing.

saprophyte   A plant which derives its food wholly or partially (partial saprophyte) from dead organic matter.

scabrid   Rough to the touch; dimin. *scaberulous*.

scape   The flowering stem of a plant all the foliage lvs of which are basal; adj. *scapigerous*.

scarious   Thin, not green, rather stiff and dry.

schizocarp   A syncarpous ovary which splits up into separate 1-seeded portions (mericarps) when mature; adj. *schizocarpic*.

scrub (incl. thicket)   Any community dominated by shrubs.

secund   All directed towards one side.

seed   A reproductive unit formed from a fertilized ovule.

sensu lato   In the broad sense.

sepal   A member of the outer series of per. segs, especially when green and ± lf-like.

sepaloid   Resembling sepals.

septicidal   Dehiscing along the septa of the ovary.

septum   A partition; adj. *septate*.

serrate   Toothed like a saw (Fig. 2J); dimin. *serrulate*.

sessile   Without a stalk.

setaceous   Shaped like a bristle, but not necessarily rigid.

shrub   A woody plant branching abundantly from the base and not reaching a very large size.

simple Not compound.

sinuate Having a wavy outline.

sinus The depression between two lobes or teeth.

sorus A circumscribed group of sporangia.

spathulate Paddle-shaped (Fig. 1 c).

spermatozoid A male reproductive cell capable of moving by means of cilia.

spike A simple racemose infl. with sessile fls; adj. *spicate*.

spikelet See pp. 517, 573.

spine A stiff straight sharp-pointed structure.

sporangiophore A structure, not lf-like, bearing sporangia.

sporangium A structure containing spores; plur. *sporangia*.

spore A small asexual reproductive body, usually unicellular and always without tissue differentiation.

sporophyll A lf-like structure, or one regarded as homologous with a lf, bearing sporangia.

spur A hollow usually ±conical slender projection from the base of a per. seg. often a petal, or of a corolla; adj. *spurred*.

stamen One of the male reproductive organs of the plant (Fig. 2M, 3A).

staminode An infertile, often reduced, stamen.

standard See p. 148.

stellate Star-shaped.

sterile Not producing seed capable of germination; or (of stamens) viable pollen.

stigma The receptive surface of the gynaecium to which the pollen grains adhere (Fig. 3A).

stipel A structure similar to a stipule but at the base of the lflets of a compound lf.

stipitate Having a short stalk or stalk-like base.

stipule A scale-like or lf-like appendage usually at the base of the petiole, sometimes adnate to it.

stolon A creeping stem of short duration produced by a plant which has a central rosette or erect stem; when used without qualification is above ground; adj. *stoloniferous*.

stoma A pore in the epidermis which can be closed by changes in shape of the surrounding cells; plur. *stomata*.

stomium The part of the sporangium wall (in the ferns) which ruptures during dehiscence.

striate Marked with long narrow depressions or ridges.

strict Growing upwards at a small angle to the vertical.

strophiole A small hard appendage outside the testa of a seed.

style The part of the gynoecium connecting the ovary with the stigma (Fig. 3A).

stylopodium See p. 244.

sub- (in combinations, e.g. subcordate) Not quite, nearly, e.g. subacute (Fig. 2D), suborbicular (Fig. 1L).

subulate Awl-shaped, narrow, pointed and ±flattened.

sucker A shoot arising adventitiously from a root of a tree or shrub often at some distance from the main stem.

suffruticose Adjective from *suffrutex*, a dwarf shrub or undershrub.

superior (ovary) With perianth inserted round the base, the ovary being free (Fig. 3B–D).

suture The line of junction of two carpels.

sympodial Of a stem in which the growing point either terminates in an infl. or dies each year, growth being continued by a new lateral growing point, cf. **cymose**.

syncarpous (ovary) Having the carpels united to one another.

tendril A climbing organ often formed from the whole or a part of a stem, lf or petiole. Most frequently the terminal portion of a pinnate lf, as in many Leguminosae.

terete Not ridged, grooved or angled.

terminal Borne at the end of a stem and limiting its growth.

ternate lf A compound lf divided into 3 ±equal parts, which may themselves be similarly divided (2- or 3-ternate); Fig. 1T.

testa The skin or outer coat of a seed.

tetrad A group of 4 spores cohering in a tetrahedral shape or as a flat plate and originating from a single spore mother-cell.

tetraploid Having 4 basic sets of chromosomes.

thallus The plant body when not differentiated into stem, lf, etc.

thorn A woody sharp-pointed structure formed from a modified branch.

tomentum A dense covering of short cottony hairs; adj. *tomentose*.

tree A woody plant with normally a single main stem (trunk) bearing lateral branches and often attaining a considerable size.

triangular Having ±the shape of a triangle. Fig. 1N.

trifid Split into three.

trigonous Of a solid body triangular in section but obtusely angled.

triploid Having 3 basic sets of chromosomes.

triquetrous Of a solid body triangular in section and acutely angled.

trisect Cut into 3 almost separate parts.

truncate Fig. 2B.

tube The fused part of a corolla or calyx, or a hollow, cylindrical, empty prolongation of an anther.

tuber A swollen portion of a stem or root of one year's duration, those of successive years not arising directly from the old ones nor bearing any constant relation to them; cf. corm.

tubercle A spherical or ovoid swelling.

tuberculate With small blunt projections, warty.

tunic A dry, usually brown and ±papery covering round a bulb or corm; adj. *tunicated*.

turbinate Top-shaped.

turion A detached winter-bud by means of which many water plants perennate.

umbel An infl. in which the pedicels all arise from the top of the main stem.

Also used of compound umbels in which the peduncles also arise from the same point. An umbrella-shaped infl.

unarmed　Devoid of thorns, spines or prickles.

undulate　Wavy in a plane at right angles to the surface.

unilocular　Having a single cavity; similarly bilocular, etc., having 2, etc., cavities.

urceolate (corolla)　±globular to subcylindrical but strongly contracted at the mouth (Fig. 2ᴋ).

valvate　Of per. segs with their edges in contact but not overlapping in bud.

vein　See nerve.

versatile　With the filament attached near the middle of the anther so as to allow of movement.

villous　Shaggy.

viscid　Sticky.

viviparous　With the fls proliferating vegetatively and not forming seed.

waste place　Uncultivated ±open habitat much influenced by man. [Note: not used in Hooker's sense of almost any uncultivated place.]

whorl　More than two organs of the same kind arising at the same level; adj. *whorled*.

wing　The lateral petals in the fls of Leguminosae and Fumariaceae.

zygomorphic　Having only one plane of symmetry.

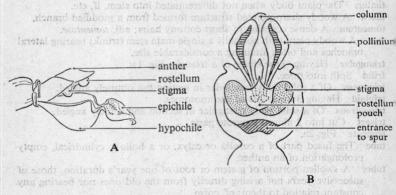

Fig. 4. A, Labellum and column of the orchid *Epipactis helleborine*; B, Column of *Orchis mascula*.

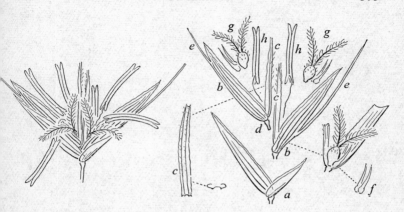

Fig. 5. A grass spikelet. *a*, glumes; *b*, lemma; *c*, palea; *d*, rhachilla; *e*, awn; *f*, lodicules; *g*, ovary and stigmas; *h*, stamen.

INDEX